CHROMATOGRAPHY OF ENVIRONMENTAL HAZARDS

Volume I

CHROMATOGRAPHY OF ENVIRONMENTAL HAZARDS

VOLUME I

CARCINOGENS, MUTAGENS AND TERATOGENS

LAWRENCE FISHBEIN

Chief, Analytical & Synthetic Chemistry Branch,
National Institute of Environmental Health Sciences,
Research Triangle Park, N.C., U.S.A.
Adjunct Professor of Entomology-Toxicology,
North Carolina State University,
Raleigh, N.C., U.S.A.

ELSEVIER PUBLISHING COMPANY

AMSTERDAM/LONDON/NEW YORK

1972

ELSEVIER PUBLISHING COMPANY
335 JAN VAN GALENSTRAAT
P. O. BOX 211, AMSTERDAM, THE NETHERLANDS

AMERICAN ELSEVIER PUBLISHING COMPANY, INC.
52 VANDERBILT AVENUE
NEW YORK, NEW YORK 10017

LIBRARY OF CONGRESS CARD NUMBER: 78–180000
ISBN 0-444-40948-3
WITH 239 ILLUSTRATIONS AND 210 TABLES

PRINTED IN THE GERMAN DEMOCRATIC REPUBLIC

CONTENTS

PREFACE

The incredible extent to which man has debased, and continues to debase, the environment in both a chemical and physical sense is becoming increasingly apparent. This has dictated an unparalleled need both to determine the parameters of this abuse as well as to marshall a multi-disciplined approach for its curtailment.

Chromatography of Environmental Hazards is a series of three volumes devoted to the elaboration of chromatographic procedures (primarily paper, thin-layer, and gas–liquid) that relate to the analyses of a broad spectrum of synthetic and naturally occurring toxicants that are of critical significance to the health of man and animals alike. The first volume centers on toxicants of carcinogenic, mutagenic, and teratogenic significance. The second and third volumes will focus on air, water, industrial pollutants, metals, pesticide residues, and drugs (narcotics, psychotropic agents, hallucinogens and miscellaneous drugs), respectively.

The main objective of this series is to provide the analytical chemist with both a practical text as well as a literature reference source of selected descriptive chromatographic procedures that stress the separation, detection, and determination of toxicants from environmental sources, biologic media, and mammalian and plant sources. However, relevant information has been included, where available, regarding the synthesis or occurrence, areas of utility, salient biological and physical properties, chemical stability, and degradation and metabolic fate of the toxicants that is of importance to investigators in diverse disciplines that include biochemistry, biology, genetics, molecular biology, toxicology and public health, all equally concerned, as well as others, with ameliorating or eliminating the effects of these environmental hazards.

I acknowledge with deep gratitude the efforts of Mrs. Peggy Sauls in the assembly, typing and proofing of the manuscript.

L. FISHBEIN

1

Chapter 1

INTRODUCTION

This initial volume deals with the treatment of the chromatographic as well as biological aspects of a number of chemical carcinogens, mutagens, and teratogens and related compounds of environmental significance, and for convenience is divided into areas of consideration that include alkylating agents, pesticides, drugs, food and feed additives and contaminants, and miscellaneous agents. It is well recognized that there are overlaps within the above categories as well as subject matter in succeeding volumes of this series (*e.g.* alkylating agents that are pesticides and drugs and air, water, and industrial pollutants, etc.). Because of their ubiquitous and unique ecological aspects, a number of carcinogenic and mutagenic agents (*e.g.* DDT, benzpyrene and organomercurials, etc.) are treated in Volume II of this series.

There are many parallels between the processes of carcinogenesis, mutagenesis, and teratogenesis. For example, chemical agents that produce malignancies may also disorganize normal behavior of embryonic cells to produce malformations. Although there are good biological reasons as well as circumstantial evidence to support the mutation hypothesis of chemical carcinogenesis, it must be stressed that the precise relationship between mutagenesis and carcinogenesis is not entirely clear.

The extent to which chromatography has been and is being utilized in the major aspects of separation, detection and determination of environmental toxicants can be gleaned from the enormous body of literature that exists in the area of pesticides as well as for DDT alone. While no attempt has been made to present an exhaustive review of the literature, the principal chromatographic techniques are presented with special focus on those procedures that pertain to the establishment of compound homogeneity, separation and identification from diverse environmental sources as well as from biologic media and degradation products, toxicological and forensic applications and finally those procedures that have utility in a general sense beyond the analysis of the toxicant *per se*. Hence, those chromatographic procedures based on chemical configurations or structural features of toxicants are of utility to the determination of related compounds not covered because of limitations of space.

The basic principles and the practical aspects of importance for analysis have been well described, *e.g.* for paper[1-3], thin-layer[4, 5], gas[6-10], and column chromatography[2]. The general literature is equally extensive with regard to chemical environmental hazards with focus on aspects of carcinogenesis[11-15], mutagenesis[16-24], teratogenesis[25, 26] and the induction of chromosomal aberrations and breakage[27-30], and the biochemistry[31] and detoxification of foreign compounds[32-34]. Salient reviews of chromatographic and biological aspects of diverse classes of chemical agents include carbamates[35], ureas[36], methylenedioxyphenyl derivatives[37], alkylating agents[38, 39], mold metabolites[40, 41], and triazines[42].

REFERENCES

1 I. SMITH, *Chromatographic and Electrophoretic Techniques*, Interscience, New York, 1960 Vol. I.
2 E. LEDERER AND M. LEDERER, *Chromatography*, 2nd edn., Elsevier, Amsterdam, 1957.
3 R. J. BLOCK, E. L. DURRUM AND G. ZWEIG, *Paper Chromatography and Electrophoresis*, Academic Press, New York, 1958.
4 E. STAHL, *Thin-layer Chromatography*, 2nd edn., Academic Press, New York, 1969.
5 K. RANDERRATH, *Thin-layer Chromatography*, Academic Press, New York, 1964.
6 J. TRANCHANT, *Practical Manual of Gas Chromatography*, Elsevier, Amsterdam, 1970.
7 L. S. ETTRE AND W. H. MCFADDEN (Eds.), *Ancillary Techniques of Gas Chromatography*, Wiley-Interscience, New York, 1970.
8 H. P. BURCHFIELD AND E. E. STORRS, *Biochemical Applications of Gas Chromatography*, Academic Press, New York, 1962.
9 C. P. STEWART AND A. STOLMAN, *Toxicology: Mechanisms and Analytical Methods*, Academic Press, New York, 1960, Vol. I.
10 L. R. GOLDBAUM, E. L. SCHLOEGEL AND A. M. DUMINGUEZ, in A. STOLMAN (Ed.), *Progress in Chemical Toxicology*, Academic Press, New York, 1963, Vol. I.
11 D. B. CLAYSON, *Chemical Carcinogenesis*, Little, Brown and Co., Boston, 1962.
12 W. C. HUEPER AND W. D. CONWAY, *Chemical Carcinogenesis and Cancers*, Charles C. Thomas, Springfield, 1964.
13 J. C. ARCOS, M. F. ARCOS AND G. WOLF, *Chemical Induction of Cancer*, Academic Press, New York, 1968.
14 P. BROOKES AND P. D. LAWLEY, *Brit. Med. Bull.*, 20 (1964) 91.
15 A. C. WALPOLE, *Ann. N. Y. Acad. Sci.*, 68 (1958) 70.
16 L. FISHBEIN, W. G. FLAMM AND H. L. FALK, *Chemical Mutagens*, Academic Press, New York, 1970.
17 R. VOGEL AND R. GUNTER, *Chemical Mutagenesis in Mammals and Man*, Springer-Verlag, New York, 1970.
18 A. HOLLANDER (Ed.), *Chemical Mutagens*, Plenum Press, New York, 1970.
19 C. AUERBACH, *Ann. N.Y. Acad. Sci.*, 68 (1958) 731.
20 A. LOVELESS, *Genetic and Allied Effects of Alkylating Agents*, Penn. State Univ. Press, University Park, 1966.
21 H. V. MALLING AND F. J. DE SERRES, *Ann. N.Y. Acad. Sci.*, 163 (1969) 288.
22 H. J. SANDERS, *Chem. Eng. News*, 47 (1969) 50.
23 H. J. SANDERS, *Chem. Eng. News*, 48 (1969) 54.
24 W. J. SCHULL, Mutations, *Second Macy Conf. Genetics, Univ. of Michigan, Ann Arbor, 1962*, Univ. of Michigan, Ann Arbor, Mich., 1963.
25 J. A. DI PAOLO, *Ann. N.Y. Acad. Sci.*, 163 (1969) 801.
26 H. KALTER, *Ann. N.Y. Acad. Sci.*, 151 (1968) 997.
27 K. S. KHERA AND D. J. CLEGG, *Can. Med. Assoc. J.*, 100 (1969) 167.
28 B. A. KIHLMAN, *Action of Chemicals on Dividing Cells*, Prentice-Hall, Englewood Cliffs, 1966.
29 M. W. SHAW, *Ann. Rev. Med.*, 21 (1970) 409.
30 A. K. SHARMA AND A. SHARMA, *Intern. Rev. Cytol.*, 10 (1960) 101.
31 D. V. PARKE, *The Biochemistry of Foreign Compounds*, Pergamon Press, New York, 1968.
32 R. T. WILLIAMS, *Detoxification Mechanisms*, 2nd edn., Chapman Hall, London, 1959.
33 D. SCHUGAR, *Biochemical Aspects of Antimetabolites and of Drug Hydroxylation*, Academic Press, New York, 1969, Vol. 16. Fifth FEBS Meeting, Prague, 1968.
34 W. H. FISHMAN (Ed.), *Metabolic Conjugation and Metabolic Hydrolysis*, Academic Press, New York, 1970.
35 L. FISHBEIN AND W. L. ZIELINSKI, JR., *Chromatog. Rev.*, 9 (1967) 37.
36 L. FISHBEIN, H. L. FALK AND P. KOTIN, *Chromatog. Rev.*, 10 (1968) 37.
37 L. FISHBEIN AND H. L. FALK, *Chromatog. Rev.*, 10 (1968) 175.

38   L. FISHBEIN AND H. L. FALK, *Chromatog. Rev.*, 11 (1969) 101.
39   L. FISHBEIN AND H. L. FALK, *Chromatog. Rev.*, 11 (1969) 365.
40   L. FISHBEIN AND H. L. FALK, *Chromatog. Rev.*, 12 (1969) 42.
41   W. A. PONS, JR. AND L. A. GOLDBLATT, in L. A. GOLDBLATT (Ed.), *Aflatoxin*, Academic Press, New York, 1969, p. 77.
42   L. FISHBEIN, *Chromatog. Rev.*, 12 (1970) 167.

Chapter 2

TABULAR SUMMARIES OF ENVIRONMENTAL HAZARDS

This chapter consists of tabular condensations of the environmental hazards (carcinogens, mutagens, teratogens and related compounds) arranged in terms of chemical or common name, structure, type(s) of hazard, and literature citations regarding their paper, thin-layer, gas–liquid and miscellaneous chromatographic analyses. The latter category includes column, ion-exchange and paper electrophoretic techniques.

The compounds are arranged in the order in which they will appear in the subsequent five chapters and are thus divided into the following tabular headings and subheadings.

(*1*) Alkylating agents (Table 1)
    (*a*) Nitrosamines
    (*b*) Aziridines
    (*c*) Epoxides and halohydrins
    (*d*) Aldehydes
    (*e*) Lactones
    (*f*) Phosphates
    (*g*) Pyrrolidizine alkaloids
(*2*) Pesticides (Table 2)
(*3*) Drugs (Table 3)
(*4*) Food and feed additives and contaminants (Table 4)
(*5*) Miscellaneous toxicants (Table 5)

TABLE 1

ALKYLATING AGENTS

(C = carcinogen, M = mutagen, T = teratogen)

Chemical or common name	Structure	Hazard			Literature references			
		C	M	T	Misc.	Paper	Thin-layer	Gas–liquid
(a) Nitrosamines								
Dimethyl nitrosamine	$(CH_3)_2N-NO$	×	×		27		1, 2, 4, 5, 12–16, 21, 26	4, 5, 9–12, 16–18, 21, 23
Alkyl and arylnitrosamines	$(R)_2-N-NO$ R = alkyl or aryl	×	×		27		1–8, 14–16, 19, 20, 23, 25	4, 5, 8–10, 12, 16, 17, 22–24
(b) Aziridines								
Ethylenimine (Aziridine)	CH_2-CH_2 / N–H (ring)	×	×		28		31	28–30
Tris(1-aziridinyl)-phosphine oxide (Tepa, APO, Aphoxide)	tris(aziridinyl)phosphine oxide structure	×			32, 43		31, 33–36	36, 37
Tris(2-methyl-1-aziridinyl)-phosphine oxide (Metepa, Mapo, Methaphoxide)	tris(2-methylaziridinyl)phosphine oxide structure	×				38, 39	31, 33, 34, 36	

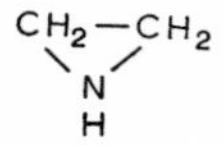

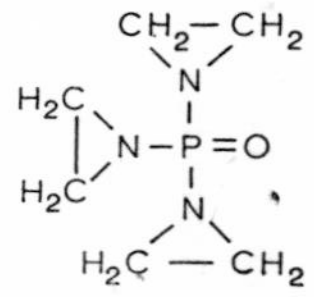

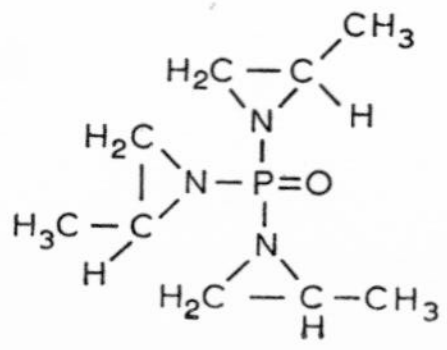

Compound	Structure					
Tris(1-aziridinyl)phosphine sulfide; triethylene thiophosphoramide (Thiotepa)	(CH₂–CH₂)N–P(=S) with two further N(CH₂–CH₂) aziridinyl groups	×	40, 41	43–46	36, 42	36
Triethylene melamine (2,4,6-tris-(1-aziridinyl)-s-triazine) (TEM)	s-triazine ring bearing three aziridinyl (CH₂–CH₂)N groups	×	47	47, 48		
(c) Epoxides and halohydrins						
Ethylene oxide	$CH_2\!-\!CH_2$ (epoxide, O)	×	56	53, 54		48–52, 54–57, 61–63, 97, 100, 101, 125
Propylene oxide	$H_2C\!-\!CHCH_3$ (epoxide, O)	× ×				52, 57, 60–62, 64, 97, 100
Ethylene chlorohydrin (2-chloroethanol)	$ClCH_2CH_2OH$	×	56	65		51, 56, 58, 59, 61–64, 66
Ethylene bromohydrin (2-bromoethanol)	$BrCH_2CH_2OH$	×				49, 67
2,3-Epoxy-1-propanol (glycidol)	$H_2C\!-\!CH\!-\!CH_2OH$ (epoxide, O)	×		68, 69		
1-Chloro-2,3-epoxy-propane (epichlorohydrin)	$H_2C\!-\!CHCH_2Cl$ (epoxide, O)	× ×	69, 70			71, 72

TABLE 1 (*cont.*)

Chemical or common name	Structure	Hazard C	Hazard M	Hazard T	Misc.	Paper	Thin-layer	Gas–liquid
Di(2,3-epoxypropyl) ether (diglycidyl ether)	$H_2C-CH-CH_2-O-CH_2-CH-CH_2$ (O)	×				69		
1,2:3,4-Diepoxybutane (butadiene diepoxide, DEB)	$H_2C-CH-CH-CH_2$ (O)	×	×					73
(d) Aldehydes								
Formaldehyde	HCHO	×	×		94, 95, 112, 114	84–87, 109, 110	88–93, 111	74–83, 98, 100, 103, 121
Acetaldehyde	CH_3CHO		×		94, 112–114	84–87, 109, 110	88, 91–93, 111, 115, 116	74, 83, 96–108, 118, 120, 121
Acrolein (acrylaldehyde, 2-propenal)	$CH_2=CHCHO$		×		114	84	88, 111	100, 101, 116–121
(e) Lactones								
β-Propiolactone (hydracrylic acid lactone, BPL)	H_2C-CH_2 / $O-C=O$	×	×			122, 123		
β-Butyrolactone	structure			×				125

γ-Butyrolactone
(γ-hydroxybutyric acid
lactone)

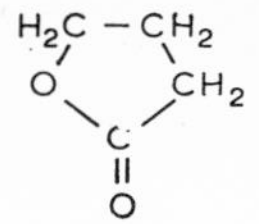

H₂C – CH₂ , CH₂, O, C=O   — X — 124 — — — 126–129

(f) Phosphates

Trimethylphosphate
(TMP)

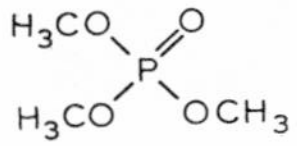

$H_3CO-P(=O)(OCH_3)-OCH_3$ — X — 130 — 130–135 — 135, 136, 202

(g) Pyrrolizidine alkaloids

Heliotrine
X X — 141, 142 — 137, 138 — 137–141 — 138

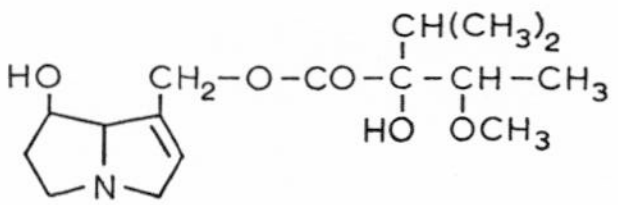

Lasiocarpine
X X — 142 — 138 — 138–140 — 138

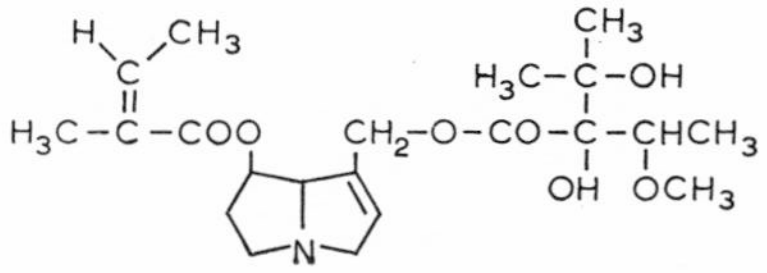

Monocrotaline
X X — 142 — 137, 138 — 137–140 — 138

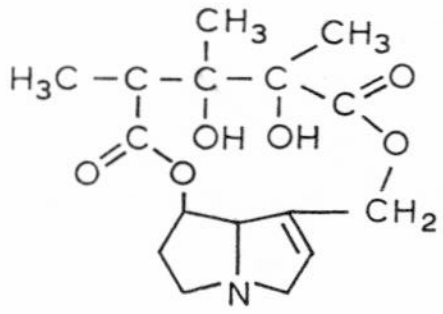

TABLE 2

PESTICIDES

(C = carcinogen, M = mutagen, T = teratogen)

Chemical or common name	Structure	Hazard			Literature references			
		C	M	T	Misc.	Paper	Thin-layer	Gas–liquid
Maleic hydrazide (1,2-dihydro-3,6-pyridazine-dione, MH)		×	×		146	143, 144, 148	145–147	149, 150
Captan (N-(trichloromethyl-thio)-4-cyclohexene-1,2-dicarboximide)		×	×				151, 155, 158–161	152–157, 161, 162
Hemel (2,4,6-tris-dimethyl-amino-1-triazine)		×			170, 171		36, 169, 170	169
Hempa (hexamethyl phosphoric triamide)		×			165	166	36, 163–168	36, 163–168
Aramite (2-(p-tert.-butyl-phenoxy)-1-methylethyl-2'-chloroethyl sulfite)		×	×		172	178	172, 173	172–177

Compound	Structure					
DDVP (2,2-dichlorovinyl dimethyl phosphate) (Dichlorovos, Vapona)	$(CH_3O)_2-P(=O)-OCH=CCl_2$	×	196		188–192, 196–200	156, 179–187, 196
Trichlorfon (dimethyl-2,2,2-trichloro-1-hydroxyethyl phosphonate) (Dipterex, Dylox)	$(CH_3O)_2-P(=O)-CH(OH)-CCl_3$	×	196		196–200	187, 193–196
2,4,5-T (2,4,5-trichlorophenoxy acetic acid)	Cl-substituted phenyl–OCH_2COOH	×	216		214, 215	156, 201–213
Dioxins						
2,3,7,8-Tetrachlorodibenzo-p-dioxin ("Dioxin") (TCDD)	tetrachlorodibenzo-p-dioxin ring (4 Cl)	×				222, 223
1,2,3,7,8,9-Hexachloro-dibenzo-p-dioxin (Chick edema factor)	hexachlorodibenzo-p-dioxin ring (6 Cl)		220, 221			217–223
Propanil (3,4-dichloropropion-anilide) (DCPA)	dichlorophenyl–$N(H)-C(=O)-CH_2CH_3$	×			224–227, 229	224–226, 228, 230
3-Amino-1H-1,2,4-triazole (aminotriazole, Amitrole)	triazole ring with NH_2	×	239, 241–243	231–238	240, 245	244
Chlorobenzilate (ethyl-4,4'-dichloro-benzilate)	bis(4-chlorophenyl)–$C(OH)-COOC_2H_5$	×			248, 249	246, 247

TABLE 2 (cont.)

Chemical or common name	Structure	Hazard				Literature references		
		C	M	T	Misc.	Paper	Thin-layer	Gas–liquid
Pentachloronitrobenzene (PCNB, Terrachlor)	(structure)	×	×				245, 251, 254, 259	250–258
Mirex (dodecachlorooctahydro-1,3,4-metheno-2H-cyclo-buta[cd]-pentalene)	(structure)	×					261	156, 176, 210, 217, 247, 260
β-Hydroxyethylhydrazine (Omaflora)	$H_2NNH-CH_2CH_2OH$			×				262

TABLE 3

DRUGS

(C = carcinogen, M = mutagen, T = teratogen)

Chemical or common name	Structure	Hazard				Literature references		
		C	M	T	Misc.	Paper	Thin-layer	Gas–liquid
Chloral hydrate	$Cl_3-C-CHOH$, OH	×			264			193, 263, 264
Cyclophosphamide (N,N-Bis-(2-chloroethyl)-N′,O-propylenephosphoric acid ester diamide) (Endoxan)	(structure)	×	×		266	265–267		
1,3-Bis(2-chloroethyl)-1-nitrosourea (BCNU)	$ClH_2CH_2C-N-C-N-CH_2CH_2Cl$, NO O H	×				268, 269	270, 271	272
Myleran (Busulfan) (1,4-Di(methane-sulfonyloxy)butane)	$CH_3SO_2O(CH_2)_4OSO_2CH_3$	×		×	276	273–279		
Ethyl methanesulfonate ("Half-Myleran") (EMS)	$CH_3CH_2OSO_2CH_3$	×	×			280		
Methyl methanesulfonate (MMS)	$CH_3OSO_2CH_3$		×		281, 283	281–283		
Antibiotics Mitomycin-C	(structure)		×		285	284–286	285	

TABLE 3 (cont.)

Chemical or common name	Structure	Hazard			Literature references			
		C	M	T	Misc.	Paper	Thin-layer	Gas–liquid
Streptozotocin		x	x		288	287		
Patulin (4-hydroxy-4H-furo [3,2-c]-pyran-2(6H)-one) (Clavicin)		x	x				289, 290	291
Griseofulvin (7-chloro-4,6,2′-tri-methoxy-6′-methyl-gris-2′-en-3:4′-dione)		x		x		292–294	295, 296	296
N-Hydroxyurea (hydroxyurea)			x	x		297–302	302	
Isonicotinyl hydrazide (isoniazid; INH)		x			305–307, 312–314	303–308	309, 314	310, 311
Thalidomide (α-phthalimidoglutarimide)			x	x		315–319, 321	320	322

Vinca alkaloids
Vincristine
Vinblastine

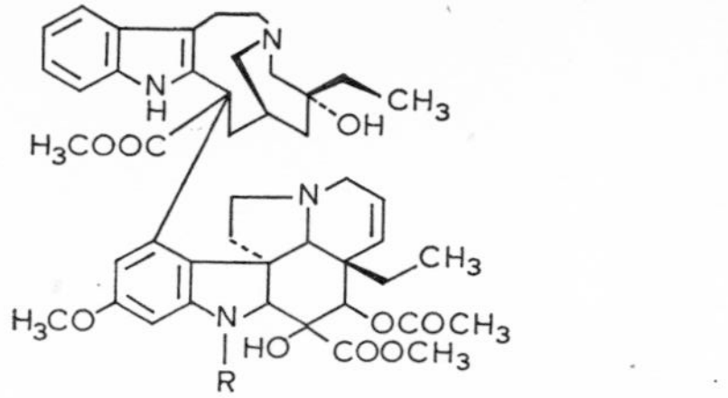

CH₃

OH

H₃COOC

N
H

N

CH₃

OCOCH₃
COOCH₃

HO
R

H₃CO

Vincristine, R=O=C–H   Vinblastine, R=CH₃

× ×   323   323–326

D-Lysergic acid diethylamide
(LSD; Delysid; Lysergide)

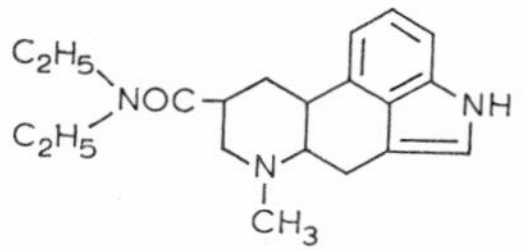

NH

C₂H₅–NOC
C₂H₅

N
CH₃

× ×   330, 334,   332, 333   327–331,   336, 337
335   336

References pp. 25–35

TABLE 4

FOOD AND FEED ADDITIVES AND CONTAMINANTS

(C = carcinogen, M = mutagen, T = teratogen)

Chemical or common name	Structure	Hazard				Literature references		
		C	M	T	Misc.	Paper	Thin-layer	Gas-liquid
Cyclamate (cyclohexanesulfamic acid, sodium salt, sodium cyclamate)	NHSO$_3$Na (cyclohexane)	×	×	×	350, 352, 353	338, 340, 348–351	338–341, 347	339, 340, 342–346
Cyclohexylamine	NH$_2$ (cyclohexane)		×	×			354, 355	339, 356–362
Dulcin (4-ethoxyphenylurea) (Sucrol, Valzin)	NHCONH$_2$ ... OC$_2$H$_5$	×			375	349, 363, 365	364, 366–375	376
Allyl isothiocyanate (volatile oil of mustard)	CH$_2$=CHCH$_2$NCS		×		377	384	377, 383	378–382
Safrole (4-allyl-1,2-methylenedioxybenzene)	CH$_2$CH=CH$_2$ (methylenedioxybenzene)	×					385–390, 394	391–393
EDTA (ethylenediaminetetraacetic acid, Versene)	HOOCCH$_2$—NCH$_2$CH$_2$N—CH$_2$COOH / HOOCCH$_2$— —CH$_2$COOH			×		395, 396, 399	398	397

Compound	Structure			References			
Caffeine (1,3,7-trimethylxanthine; methyltheobromine)		×	×	402, 411	402, 406, 407	400, 403–405, 408	401, 409, 410
Nitrofurazone (5-nitro-2-furaldehyde-semicarbazone)		×	×	413–415, 418, 419	415	412–414, 416	417
Cycasin and its aglycon methylazoxymethanol (MAM)		×	×	424	420–424	425	426, 427
Aflatoxins Aflatoxin-B₁		×	×	431, 432, 435, 452	454	428–450, 455–459, 462, 479	
Aflatoxin-B₂		×				428–443, 448, 457, 458	
Aflatoxin-G₁		×	×			428–440, 444, 449, 451, 457, 458, 479	

TABLE 4 (cont.)

Chemical or common name	Structure	Hazard				Literature references		
		C	M	T	Misc.	Paper	Thin-layer	Gas–liquid
Aflatoxin-G_2		×					428–430, 457, 458	
Aflatoxin-B_{2a}		×					428–430, 461	
Aflatoxin-G_{2a}		×					428–429, 462	
Aflatoxin-M_1		×				454	449, 453, 456–460	
Aflatoxin-M_2							458	

Ochratoxins

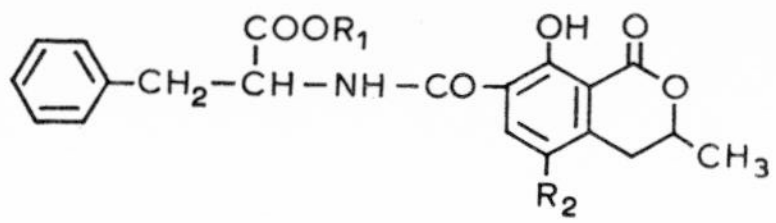

$$COOR_1$$
$$-CH_2-CH-NH-CO-$$

Ochratoxin A, $R_1 = H, R_2 = Cl$
Ochratoxin B, $R_1 = H, R_2 = H$
Ochratoxin C, $R_1 = C_2H_5, R_2 = Cl$

463   463–471, 478

X

Sterigmatocystin and
O-methylsterigmatocystin

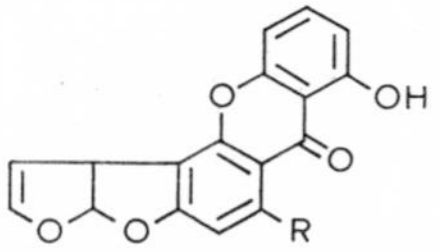

Sterigmatocystin, R = H
O-Methylsterigmatocystin, R = OCH_3

468   467, 468, 478

X

Aspertoxin

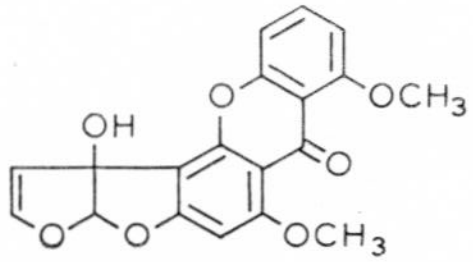

469, 470, 478

X

Zearalenone

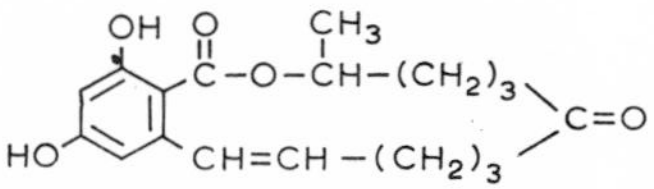

471, 472, 478,   471, 472

Fusarium toxins
Nivalenol

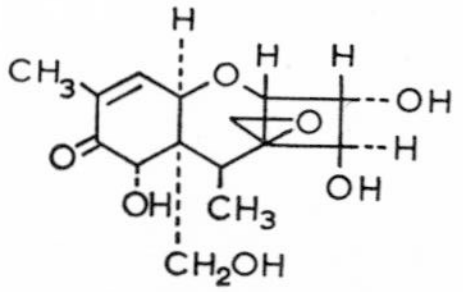

473, 474   473, 474   473, 474

TABLE 4 (cont.)

Chemical or common name	Structure	Hazard			Literature references			
		C	M	T	Misc.	Paper	Thin-layer	Gas–liquid
Fusarenon-X					475		475	
Penicillium toxins Cyclopiazonic acid					476		476, 478	
Islanditoxin				×			477	

TABLE 5

MISCELLANEOUS TOXICANTS

(C = carcinogen, M = mutagen, T = teratogen)

Chemical or common name	Structure	Hazard			Literature references			
		C	M	T	Misc.	Paper	Thin-layer	Gas–liquid
Hydrazine and its derivatives								
Hydrazine	H_2NNH_2	×	×			487		479–485
1,1-Dimethylhydrazine (UDMH)	$\begin{array}{c}H_3C\\ \\ H_3C\end{array}$N—$NNH_2$	×	×			487		480–483, 486
Naphthylamines and their metabolites								
2-Naphthylamine	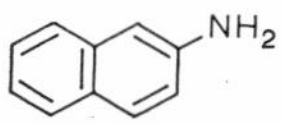 [structure: naphthalene with NH₂]	×			493	488, 493, 498	488–490, 493, 496, 497	492–495
1-Naphthylamine	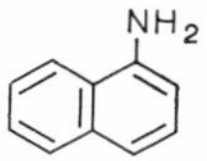 [structure: naphthalene with NH₂]				493	488, 493, 498	488, 489, 493, 496	492–495
N-Hydroxy-2-naphthylamine	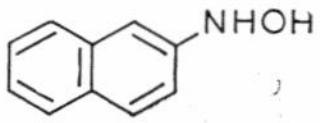 [structure: naphthalene with NHOH]	×				488, 498	488–491	491
N-Hydroxy-1-naphthylamine	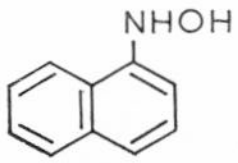 [structure: naphthalene with NHOH]		×			488, 498	488, 489, 491	491
Urethan (ethyl carbamate)	$H_2N-\underset{\underset{O}{\parallel}}{C}-OC_2H_5$	×	×	×		499, 507	499–503, 506	504, 505

TABLE 5 (cont.)

Chemical or common name	Structure	Hazard C M T	Hazard Misc.	Lit. ref. Paper	Lit. ref. Thin-layer	Lit. ref. Gas–liquid
N-Hydroxyurethan	HN–C–OC$_2$H$_5$, HO–O	× ×		499	499–503	504
Organic peroxides						
tert.-Butyl hydroperoxide	(CH$_3$)$_3$COOH	× ×		508–510, 512	515	518
Di-tert.-butyl peroxide	(CH$_3$)$_3$C–O–O–C(CH$_3$)$_3$	×		512	517	517
Cumene hydroperoxide	C(CH$_3$)$_2$OOH (phenyl)	×		508–512	514	516
Succinic acid peroxide	HOOC–CH$_2$CH$_2$–C(=O)–O–O–OH	×		511		
Disuccinyl peroxide	COCH$_2$CH$_2$COOH–O–O–O–COCH$_2$CH$_2$COOH	×				
Dihydroxydimethyl peroxide	HO–CH$_2$–O–O–CH$_2$OH	×		510		
Hydrogen peroxide	H$_2$O$_2$	×		508–511	513	519

REFERENCES

1 R. PREUSSMANN, D. DAIBER AND H. HENGY, *Nature*, 201 (1964) 502.
2 R. PREUSSMANN, G. NEURATH, G. WULF-LORENTZEN, D. DAIBER AND H. HENGY, *Z. Anal. Chem.*, 202 (1964) 187.
3 K. YASUDA AND K. NAKASHIMA, *Japan Analyst*, 17 (1968) 732.
4 N. P. SEN, D. C. SMITH, L. SCHWINGHAMER AND J. J. MARLEAU, *J. Assoc. Offic. Anal. Chemists*, 52 (1969) 47.
5 E. KRÖLLER, *Deut. Lebensm. Rundschau*, 10 (1967) 303.
6 L. HEDLER AND P. MARQUARDT, *Food Cosmet. Toxicol.*, 6 (1968) 341.
7 H. NEURATH, P. PIRMANN AND M. DÜNGER, *Chem. Ber.*, 97 (1964) 1631.
8 H. J. PETROWITZ, *Arzneimittel-Forsch.*, 18 (1968) 1486.
9 K. MÖHLER AND O. L. MAYRHOFER, *Z. Lebensm. Untersuch. Forsch.*, 135 (1968) 313.
10 O. L. MAYRHOFER AND K. MÖHLER, *Z. Lebensm. Untersuch. Forsch.*, 134 (1967) 246.
11 J. W. HOWARD, T. FAZIO AND J. O. WATTS, *J. Assoc. Offic. Anal. Chemists*, 53 (1970) 269.
12 N. P. SEN, D. C. SMITH, L. SCHWINGHAMER AND B. HOWSMAN, *Can. Inst. Food Technol. J.*, (1970) 66.
13 F. ENDER AND L. CEH, Alkylierend wirkende Verbindungen, *Second Conf. on Tobacco Research, Freiberg, 1967*, p. 83.
14 F. ENDER AND L. CEH, *Food Cosmet. Toxicol.*, 5 (1968) 569.
15 J. SAKSHAUG, E. SOGNEN, M. A. HANSEN AND N. KOPPANG, *Nature*, 206 (1965) 1261.
16 O. G. DEVIK, *Acta Chem. Scand.*, 21 (1967) 2302.
17 J. K. FOREMAN, J. F. PALFRAMAN AND E. A. WALKER, *Nature*, 225 (1970) 554.
18 N. P. SEN, *J. Chromatog.*, 51 (1970) 107.
19 M. J. H. KEYBETS, E. H. GROOT AND G. H. M. KELLER, *Food Cosmet. Toxicol.*, 8 (1970) 167.
20 N. P. SEN, D. C. SMITH AND L. SCHWINGHAMER, *Food Cosmet. Toxicol.*, 7 (1969) 301.
21 L. P. DU PLESSIS, N. R. NUNN AND W. A. ROACH, *Nature*, 222 (1969) 1198.
22 W. J. SERFONTEIN AND H. SMIT, *Nature*, 214 (1967) 169.
23 W. J. SERFONTEIN AND P. HURTER, *Cancer Res.*, 26 (1966) 575.
24 D. E. JOHNSON, J. D. MILLAR AND J. W. RHOADES, *Natl. Cancer Inst. Monograph*, 28 (1968) 181.
25 R. SCHOENTAL AND S. GIBBARD, *Nature*, 216 (1967) 612.
26 U. MOHR, J. ALTHOFF AND A. AUTHALER, *Cancer Res.*, 26 (1966) 2349.
27 G. EISENBRAND, K. SPACZYNSKI AND R. PREUSSMANN, *J. Chromatog.*, 51 (1970) 304.
28 G. F. WRIGHT AND V. K. ROWE, *Toxicol. Appl. Pharmacol.*, 11 (1967) 575.
29 V. S. TATARINSKII, N. N. MOSKVITIN, V. G. BEREZKIN AND V. N. ANDRONOV, *Sovrem. Metody Khim. Spektraln. Anal. Mater.*, (1967) 253; *Chem. Abstr.*, 68 (1968) 46024T.
30 A. D. LORENZO AND G. RUSSO, *J. Gas Chromatog.*, 6 (1968) 509.
31 M. BEROZA AND A. B. BORKOVEC, *J. Med. Chem.*, 7 (1964) 44.
32 S. C. CHANG, A. B. BORKOVEC AND C. W. WOODS, *J. Econ. Entomol.*, 59 (1966) 937.
33 C. W. COLLIER AND R. TARDIFF, *J. Econ. Entomol.*, 60 (1967) 28.
34 A. B. BORKOVEC, S. CHANG AND A. M. LIMBURG, *J. Econ. Entomol.*, 57 (1964) 815.
35 J. E. MAITLEN AND L. M. McDONOUGH, *J. Econ. Entomol.*, 60 (1967) 1391.
36 M. C. BOWMAN AND M. BEROZA, *J. Assoc. Offic. Anal. Chemists*, 49 (1966) 1046.
37 H. C. COX, T. R. YOUNG AND M. C. BOWMAN, *J. Econ. Entomol.*, 60 (1967) 1111.
38 F. W. PLAPP, JR., W. S. BIGLEY, G. A. CHAPMAN AND G. W. EDDY, *J. Econ. Entomol.*, 55 (1962) 607.
39 W. F. CHAMBERLAIN AND E. W. HAMILTON, *J. Econ. Entomol.*, 57 (1964) 800.
40 J. C. PARISH AND B. W. ARTHUR, *J. Econ. Entomol.*, 58 (1965) 976.
41 F. W. PLAPP, JR. AND J. E. CASIDA, *Anal. Chem.*, 30 (1958) 1622.
42 C. BENCK HUIJSEN, *Biochem. Pharmacol.*, 17 (1968) 55.
43 L. B. MELLETT AND L. A. WOODS, *Cancer Res.*, 20 (1960) 524.
44 L. B. MELLETT, P. E. HODGSON AND L. A. WOODS, *J. Lab. Clin. Med.*, 60 (1962) 818.
45 A. W. CRAIG, B. W. FOX AND H. JACKSON, *Biochem. Pharmacol.*, 3 (1959) 42.
46 R. R. PAINTER AND W. W. KILGORE, *J. Insect Physiol.*, 13 (1967) 1105.

47 E. I. GOLDENTHAL, M. V. NADKARNI AND P. K. SMITH, *J. Pharmacol. Exptl. Therap.*, 122 (1958) 431.
48 S. G. HEUSER AND K. A. SCUDAMORE, *Analyst*, 93 (1968) 252.
49 S. G. HEUSER AND K. A. SCUDAMORE, *Chem. Ind. (London)*, (1966) 1557.
50 B. BERCK, *J. Agr. Food Chem.*, 13 (1965) 375.
51 S. BEN-YEHOSHUA AND P. KRINSKY, *J. Gas Chromatog.*, 6 (1968) 350.
52 S. G. HEUSER AND K. A. SCUDAMORE, *J. Sci. Food Agr.*, 20 (1969) 566.
53 J. T. GORDON, W. W. THORNBURG AND L. N. WERUM, *J. Agr. Food Chem.*, 7 (1959) 196.
54 Y. OBI, Y. SHIMADA, K. TAKAHASHI, K. NISHIDA AND T. KISAKI, *Tobacco*, 166 (1968) 26.
55 M. MURAMATSU, Y. OBI, Y. SHIMADA, K. TAKAHASHI AND K. NISHIDA, *Japan Monopoly Cent. Res. Inst. Sci. Papers*, 110 (1968) 217.
56 D. J. BROWN, *J. Assoc. Offic. Anal. Chemists*, 53 (1970) 263.
57 N. ADLER, *J. Pharm. Sci.*, 54 (1965) 735.
58 A. HOLMGREN, N. DIDING AND G. SAMUELSSON, *Acta Pharm. Suecica*, 6 (1969) 33.
59 E. P. RAGELIS, B. S. FISHER AND B. A. KLIMECK, *J. Assoc. Offic. Agr. Chemists*, 49 (1966) 963.
60 R. MESTRES AND C. BARROIS, *Soc. Pharm. Montpellier*, 24 (1964) 47.
61 F. WESLEY, B. ROURKE AND O. DARBISHIRE, *J. Food Sci.*, 30 (1965) 1037.
62 J. F. SMITH, in R. P. W. SCOTT (Ed.), *Gas Chromatography, Proc. 3rd. Symp.*, Butterworths, London, 1960, p. 114.
63 E. P. RAGELIS, B. S. FISHER, B. A. KLIMECK AND C. JOHNSON, *J. Assoc. Offic. Anal. Chemists*, 51 (1968) 709.
64 P. MANCHON AND A. BUQUET, *Food Cosmet. Toxicol.*, 8 (1970) 9.
65 M. K. JOHNSON, *Biochem. Pharmacol.*, 16 (1967) 185.
66 J. S. PAGINGTON, *J. Chromatog.*, 36 (1968) 528.
67 S. G. HEUSER AND K. A. SCUDAMORE, *Chem. Ind. (London)*, (1969) 1054.
68 J. DYR AND J. MOSTEK, *Knasy Prumsyl*, 4 (1958) 121; *Chem. Abstr.*, 52 (1958) 19011.
69 W. SCHÄFER, W. NUCK AND H. JAHN, *J. Prakt. Chem.*, 20 (1969) 1.
70 W. SCHÄFER, W. NUCK AND H. JAHN, *J. Prakt. Chem.*, 11 (1960) 10.
71 H. R. COPPER AND J. G. ROBERTS, *J. Soc. Dyers Colourists*, 80 (1964) 428.
72 P. W. WEST, P. SEN, B. R. SANT, K. L. MALLIK AND J. G. S. GUPTA, *J. Chromatog.*, 6 (1961) 220.
73 B. L. VAN DUUREN, N. NELSON, L. ORRIS, E. D. PALMES AND F. L. SCHMITT, *J. Natl. Cancer Inst.*, 31 (1963) 41.
74 K. J. BOMBAUGH AND W. C. BULL, *Anal. Chem.*, 34 (1962) 1237.
75 K. JONES, *J. Gas Chromatog.*, 5 (1967) 432.
76 F. ONUSKA, J. JANKA, S. DURAS AND M. KROMAROVA, *J. Chromatog.*, 40 (1969) 209.
77 R. S. MANN AND K. W. HAHN, *Anal. Chem.*, 39 (1967) 1314.
78 S. SANDLER AND A. STROM, *Anal. Chem.*, 32 (1960) 1890.
79 H. CHERDON, L. HÖHR AND W. KERN, *Angew. Chem.*, 73 (1961) 215.
80 W. O. MCREYNOLDS, Pittsburgh Conf. Anal. Chem. Appl. Spectry. 1961, *Anal. Chem.*, 33 (1961) 78A.
81 T. IGUCHI AND T. TAKIUCHI, *Bunseki Kagaku*, 17 (1968) 1080.
82 M. P. STEVENS AND D. F. PERCIVAL, *Anal. Chem.*, 36 (1964) 1023.
83 S. HARRISON, *Analyst*, 92 (1967) 773.
84 D. F. MEIGH, *Nature*, 170 (1952) 579.
85 R. G. RICE, *Anal. Chem.*, 23 (1951) 194.
86 R. ELLIS, *Anal. Chem.*, 32 (1959) 1997.
87 A. M. GADDIS AND R. ELLIS, *Anal. Chem.*, 31 (1958) 870.
88 E. BLOEM, *J. Chromatog.*, 35 (1968) 108.
89 E. DENTI AND M. PLUBO, *J. Chromatog.*, 18 (1965) 325.
90 K. HARADA, S. SHIGETSUGA, Y. SHINODA AND K. YAMADA, *Nippon Suisan Gakkaishi*, 36 (1970) 188.
91 D. J. PIETRZYK AND E. P. CHAN, *Anal. Chem.*, 42 (1970) 41.
92 K. ONOE, *J. Chem. Soc. Japan.*, 73 (1952) 337.
93 D. P. SCHWARTZ, *Microchem. J.*, 72 (1963) 403.
94 J. RUSMUS, *J. Chromatog.*, 6 (1961) 187.
95 W. FREYTAG, *Fette Seifen Anstrichmittel*, 65 (1963) 603.

96 E. A. Demina, R. E. Podnebesnaya and Y. P. Grishutin, *Tr. Khim. Met. Inst. Akad. Nauk Kaz. SSR*, 2 (1968) 165; *Chem. Abstr.*, 70 (1969) 8677S.
97 R. N. Mokeeva and Y. A. Tsarafin, *Ind. Lab. (USSR) (Engl. Transl.)*, 31 (1965) 1306.
98 A. A. Kolidaev, *Sudebno-Med. Ekspertiza Min. Zdravookhr. SSSR*, 11 (1968) 24.
99 O. L. Lapitskaya and L. V. Ivanova, *Ferment. Spirt. Prom.*, 34 (1968) 24; *Chem. Abstr.*, 69 (1968) 49069T.
100 G. Kyryacos, H. R. Menapace and C. E. Boord, *Anal. Chem.*, 31 (1959) 222.
101 T. E. Bellar and J. E. Sigsby, Jr., *Environ. Sci. Technol.*, 4 (1970) 150.
102 C. F. Ellis, R. F. Kendall and B. H. Ekccleston, *Anal. Chem.*, 37 (1965) 511.
103 R. J. Soukup, R. J. Scarpellino and E. Danielczik, *Anal. Chem.*, 36 (1964) 2255.
104 L. A. Jones and R. J. Monroe, *Anal. Chem.*, 37 (1965) 935.
105 G. Freund and P. O'Hollaren, *J. Lipid Res.*, 6 (1965) 471.
106 Y. Fukui, *Japan. J. Legal Med.*, 23 (1969) 22.
107 M. K. Roach and P. J. Creaven, *Clin. Chim. Acta*, 21 (1968) 275.
108 R. N. Baker, A. L. Acenty and J. F. Zack, Jr., *J. Chromatog. Sci.*, 7 (1969) 312.
109 D. A. Buyske, L. H. Owen, P. Wilder, Jr. and M. E. Hobbs, *Anal. Chem.*, 28 (1956) 910.
110 M. Severin, *Bull. Inst. Agron. Sta. Rech. Gembloux*, 32 (1964) 122.
111 F. C. Hunt, *J. Chromatog.*, 40 (1969) 465.
112 D. P. Schwartz, O. W. Parks and M. Keeney, *Anal. Chem.*, 34 (1962) 669.
113 E. A. Corbin, D. P. Schwartz and M. Keeney, *J. Chromatog.*, 3 (1960) 322.
114 D. P. Schwartz, A. R. Johnson and O. W. Parks, *Microchem. J.*, 6 (1962) 37.
115 D. L. Mays, G. S. Born and J. E. Christian, *Bull. Environ. Contam. Toxicol.*, 3 (1968) 366.
116 L. A. Th. Verhaar and S. P. Lankhuijzen, *J. Chromatog. Sci.*, 8 (1970) 457.
117 M. P. Stevens, *Anal. Chem.*, 37 (1965) 167.
118 E. M. Bevilacqua, E. S. English and J. S. Gall, *Anal. Chem.*, 34 (1962) 861.
119 D. F. Gadbois, P. G. Scheurer and F. J. King, *Anal. Chem.*, 40 (1968) 1362.
120 J. W. Ralls, *Anal. Chem.*, 32 (1960) 332.
121 H. Kelker, *Angew. Chem.*, 71 (1959) 218.
122 C. E. Searle, *Brit. J. Cancer*, 15 (1961) 804.
123 N. H. Colburn and R. K. Boutwell, *Cancer Res.*, 28 (1968) 642.
124 P. G. Keeney, *J. Am. Oil Chemists' Soc.*, 34 (1957) 356.
125 B. L. van Duuren, L. Orris and N. Nelson, *J. Natl. Cancer Inst.*, 35 (1965) 707.
126 G. M. Gal'pern, G. A. Gudkova, E. Y. Shaposhnikova and E. B. Yabubskii, *Zavodsk. Lab.*, 32 (1966) 931; *Chem. Abstr.*, 65 (1966) 17049f.
127 R. Viani, F. Mueggler-Chavan, D. Reymond and R. H. Egli, *Helv. Chim. Acta*, 48 (1965) 1809.
128 D. Reymond, F. Mueggler-Chavan, R. Viani, L. Vauta and R. H. Egli, *Advan. Gas Chromatog., Proc. Symp. 3rd Intern., Houston, Texas, 1965*, Univ. Houston, Houston, Texas, 1966, p. 126.
129 V. M. Gianturco, A. S. Giamarino and P. Friedel, *Nature*, 210 (1966) 1358.
130 A. B. Jones, *Experientia*, 26 (1970) 492.
131 A. Lamotte and J. C. Merlin, *J. Chromatog.*, 45 (1969) 432.
132 A. Lamotte and J. C. Merlin, *J. Chromatog.*, 38 (1968) 296.
133 A. Lamotte, A. Francina and J. C. Merlin, *J. Chromatog.*, 44 (1969) 75.
134 R. Klement and A. Wild, *Z. Anal. Chem.*, 195 (1963) 180.
135 J. Askew, J. H. Ruzicka and B. B. Wheals, *J. Chromatog.*, 41 (1969) 180.
136 M. J. Shafik and H. F. Enos, *J. Agr. Food Chem.*, 17 (1969) 1186.
137 R. K. Sharma, G. S. Khajuria and C. K. Atal, *J. Chromatog.*, 19 (1965) 433.
138 A. H. Chalmers, C. C. J. Culvenor and L. W. Smith, *J. Chromatog.*, 20 (1965) 270.
139 A. R. Mattocks, *J. Chromatog.*, 27 (1967) 505.
140 A. R. Mattocks, *Anal. Chem.*, 39 (1967).
141 M. V. Jago, G. W. Lanigan, J. B. Bingley, D. W. T. Piercy, J. H. Whittem and D. A. Titchen, *J. Pathol.*, 98 (1969) 115.
142 J. L. Frahn, *Australian J. Chem.*, 22 (1969) 1655.
143 W. A. Andreae, *Can. J. Biochem. Physiol.*, 36 (1958) 71.
144 P. K. Biswas, O. Hall and B. D. Mayberry, *Physiol. Plantarum*, 20 (1967) 819.
145 D. L. Mays, G. S. Born, J. E. Christian and B. J. Liska, *J. Agr. Food Chem.*, 16 (1968) 356.

146 A. STOESSL, *Chem. Ind. (London)*, (1964) 580.
147 A. STOESSL, *Can. J. Chem.*, 43 (1965) 2430.
148 L. D. NOODEN, *Plant Physiol.*, 45 (1970) 46.
149 L. FISHBEIN AND W. L. ZIELINSKI, JR., *J. Chromatog.*, 18 (1965) 581.
150 W. L. ZIELINSKI, JR. AND L. FISHBEIN, *J. Chromatog.*, 20 (1965) 140.
151 D. V. RICHMOND AND E. SOMERS, *Ann. Appl. Biol.*, 62 (1968) 35.
152 W. W. KILGORE, W. WINTERLIN AND R. WHITE, *J. Agr. Food Chem.*, 15 (1967) 1035.
153 A. BEVENUE AND J. N. OGATH, *J. Chromatog.*, 36 (1968) 531.
154 I. H. POMERANTZ, L. J. MILLER AND G. KAVA, *J. Assoc. Offic. Anal. Chemists*, 53 (1970) 154.
155 I. H. POMERANTZ AND R. ROSS, *J. Assoc. Offic. Anal. Chemists*, 51 (1968) 1058.
156 J. A. BURKE AND W. HOLSWADE, *J. Assoc. Offic. Anal. Chemists*, 49 (1966) 374.
157 T. E. ARCHER AND J. B. CORBIN, *Food Technol.*, 23 (1969) 101.
158 T. E. ARCHER AND J. B. CORBIN, *Bull. Environ. Contam. Toxicol.*, 4 (1969) 55.
159 R. ENGST AND W. SCHNAAK, *Nahrung*, 11 (1967) 95.
160 L. FISHBEIN, J. FAWKES AND P. JONES, *J. Chromatog.*, 23 (1966) 476.
161 W. W. KILGORE AND E. R. WHITE, *J. Agr. Food Chemists*, 15 (1967) 1118.
162 E. SOMERS, D. V. RICHMOND AND J. A. PACKARD, *Nature*, 215 (1967) 214.
163 S. C. CHANG, P. H. TERRY, C. W. WOODS AND A. B. BORKOVEC, *J. Econ. Entomol.*, 60 (1967) 1623.
164 S. C. CHANG AND A. B. BORKOVEC, *J. Econ. Entomol.*, 62 (1969) 1417.
165 A. R. JONES, *Biochem. Pharmacol.*, 19 (1970) 603.
166 A. C. TERRANOVA AND C. H. SCHMIDT, *J. Econ. Entomol.*, 60 (1967) 1659.
167 S. AKOV, J. E. OLIVER AND A. B. BORKOVEC, *Life Sci.*, 7 (1968) 1207.
168 S. AKOV AND A. B. BORKOVEC, *Life Sci.*, 7 (1968) 1215.
169 S. C. CHANG, A. B. DE MILO, C. W. WOODS AND A. B. BORKOVEC, *J. Econ. Entomol.*, 61 (1968) 1357.
170 S. C. CHANG, C. W. WOOD AND A. B. BORKOVEC, *J. Econ. Entomol.*, 63 (1970) 1510.
171 G. T. BRYAN AND A. L. GORSKE, *J. Chromatog.*, 34 (1968) 67.
172 T. E. ARCHER, *Bull. Environ. Contam. Toxicol.*, 3 (1968) 71.
173 R. C. BLINN AND F. A. GUNTHER, *J. Assoc. Offic. Anal. Chemists*, 46 (1963) 204.
174 D. M. COULSON, L. A. CAVANAGH, J. E. DEVRIES AND B. WALTHER, *J. Agr. Food Chem.*, 8 (1960) 399.
175 J. BURKE AND L. JOHNSON, *J. Assoc. Offic. Anal. Chemists*, 45 (1962) 348.
176 M. C. BOWMAN AND M. BEROZA, *J. Assoc. Offic. Agr. Chemists*, 48 (1965) 943.
177 W. A. BOSIN, *Anal. Chem.*, 35 (1963) 833.
178 L. C. MITCHELL, *J. Assoc. Offic. Agr. Chemists*, 41 (1958) 781.
179 M. C. IVEY AND H. V. CLABORN, *J. Assoc. Offic. Anal. Chemists*, 52 (1969) 1248.
180 G. DRAEGER, *Pflanzenschutznachr. Bayer*, 21 (1968) 373.
181 J. W. MILES, *Chemical Memorandum No. 10, TDL 6-23-69*, U.S. Public Health Service, H.E.W., Savannah, Ga.
182 J. H. RUZICKA, J. THOMSON AND B. B. WHEALS, *J. Chromatog.*, 31 (1967) 37.
183 L. W. GETZIN AND I. ROSEFIELD, *J. Agr. Food Chem.*, 16 (1968) 598.
184 C. E. COOK, C. W. STANLEY AND J. E. BARNEY, II, *Anal. Chem.*, 36 (1964) 2354.
185 M. SALAME, *Ann. Biol. Clin.*, 26 (1968) 1011.
186 G. ZWEIG AND J. M. DEVINE, *Residue Rev.*, 26 (1969) 17.
187 J. ELLIS AND J. BATES, *J. Assoc. Offic. Agr. Chemists*, 48 (1965) 1115.
188 H. ACKERMANN, *J. Chromatog.*, 36 (1968) 309.
189 C. E. MENDOZA, D. L. GRANT, B. BRACELAND AND K. A. MCCULLY, *Analyst*, 94 (1969) 805.
190 G. F. ERNST AND F. SCHURING, *J. Chromatog.*, 49 (1970) 325.
191 M. RAMASAMY, *Analyst*, 94 (1969) 1078.
192 R. T. WANG AND S. S. CHOU, *J. Chromatog.*, 42 (1969) 416.
193 R. J. ANDERSON, C. A. ANDERSON AND T. J. OLSON, *J. Agr. Food Chem.*, 14 (1966) 508.
194 H. C. BARRY, J. G. HINDLEY AND L. Y. JOHNSON, *Pesticides Analytical Manual*, U.S. Food and Drug Administration, Washington, D.C., July, 1965, Vols. I and II.
195 A. R. A. EL-RAGAI AND L. GIUFFRIDA, *J. Assoc. Offic. Agr. Chemists*, 48 (1965) 374.
196 J. ASKEW, J. H. RUZICKA AND B. B. WHEALS, *Analyst*, 94 (1969) 275.
197 M. E. GETZ AND H. G. WHEELER, *J. Assoc. Offic. Anal. Chemists*, 51 (1968) 1101.

198 V.Y. KOLYAKOVA, *Tr. Vses. Nauchn.-Issled. Inst. Vet. Sanit. i Ektoparazitol.*, 29 (1967) 348.
199 H. ACKERMANN, R. ENGST AND G. FECHNER, *Z. Lebensm. Untersuch. Forsch.*, 137 (1968) 303.
200 H. ACKERMANN, *Nahrung*, 10 (1966) 273.
201 D. E. CLARK, *J. Agr. Food Chem.*, 17 (1969) 1168.
202 C. W. STANLEY, *J. Agr. Food Chem.*, 14 (1966) 321.
203 G. YIP, *J. Assoc. Offic. Agr. Chemists*, 47 (1964) 1116.
204 J. E. SCOGGINS AND C. H. FITZGERALD, *J. Agr. Food Chem.*, 17 (1969) 156.
205 J. M. DEVINE AND G. ZWEIG, *J. Assoc. Offic. Anal. Chemists*, 52 (1969) 187.
206 D. L. GUTNICK AND G. ZWEIG, *J. Chromatog.*, 13 (1964) 319.
207 T. P. GARBRECHT, *J. Assoc. Offic. Anal. Chemists*, 53 (1970) 70.
208 H. L. MORTON, F. S. DAVIS AND M. G. MERKLE, *Weed Sci.*, 16 (1968) 88.
209 C. H. FITZGERALD, C. L. BROWN AND E. G. BECK, *Plant Physiol.*, 42 (1967) 459.
210 J. F. THOMPSON, A. C. WALKER AND R. F. MOSEMAN, *J. Assoc. Offic. Anal. Chemists*, 52 (1969) 1263.
211 V. LEONI AND G. PUCCETI, *J. Chromatog.*, 43 (1969) 388.
212 L. E. ST. JOHN, JR., D. G. WAGNER AND D. J. LISK, *J. Dairy Sci.*, 47 (1964) 1267.
213 D. J. LISK, W. H. GUTENMANN, C. A. BACHE, R. G. WARNER AND D. G. WAGNER, *J. Dairy Sci.*, 46 (1963) 1435.
214 K. D. COURTNEY, *Pesticides Symposia of the 7th Inter-American Conference on Toxicology and Occupational Medicine, Miami Beach, Fla. (Aug. 1970)*, Halos, Miami Beach, Fla., 1970, p. 277.
215 K. ERNE, *Acta Vet. Scand.*, 7 (1966) 240.
216 G. B. CERESIA AND W. W. SANDERSON, *J. Water Pollution Control Federation*, 41 (1969) R34.
217 D. FIRESTONE, W. IBRAHIM AND W. HORWITZ, *J. Assoc. Offic. Agr. Chemists*, 46 (1963) 384.
218 J. C. WOOTON, N. R. ARTMAN AND J. C. ALEXANDER, *J. Assoc. Offic. Agr. Chemists*, 45 (1962) 739.
219 G. R. HIGGINBOTHAM, D. FIRESTONE, L. CHAVEZ AND A. D. CAMPBELL, *J. Assoc. Offic. Agr. Chemists*, 50 (1967) 874.
220 G. R. HIGGINBOTHAM, J. RESS AND D. FIRESTONE, *J. Assoc. Offic. Agr. Chemists*, 50 (1967) 884.
221 G. R. HIGGINBOTHAM, J. RESS AND D. FIRESTONE, *J. Assoc. Offic. Agr. Chemists*, 51 (1968) 940.
222 G. R. HIGGINBOTHAM, A. HUANG, D. FIRESTONE, J. VERRET, J. RESS AND A. D. CAMPBELL, *Nature*, 220 (1968) 702.
223 M. KONITA, S. UEDA AND M. NARISADA, *Yakugaku Zasshi*, 79 (1959) 186.
224 R. BARTHA AND D. PRAMER, *Science*, 156 (1967) 1617.
225 H. CHISAKA AND P. C. KEARNEY, *J. Agr. Food Chem.*, 18 (1970) 854.
226 R. BARTHA, *J. Agr. Food Chem.*, 16 (1968) 602.
227 R. Y. YIH, D. H. MCRAE AND H. F. WILSON, *Science*, 161 (1968) 376.
228 I. J. BELASCO AND H. C. PEASE, *J. Agr. Food Chem.*, 17 (1969) 1414.
229 J. R. PLIMMER, P. C. KEARNEY, H. CHISAKA, J. B. YOUNG AND U. I. KLINGEBIEL, *J. Agr. Food Chem.*, 18 (1970) 859.
230 R. BARTHA, H. A. B. LINKE AND D. PRAMER, *Science*, 161 (1968) 582.
231 S. C. FANG, M. GEORGE AND T. C. YU, *J. Agr. Food Chem.*, 12 (1964) 219.
232 S. C. FANG, S. KHANNA AND A. V. RAO, *J. Agr. Food Chem.*, 14 (1966) 262.
233 P. MASSINI, *Acta Botan. Neerl.*, 12 (1963) 64.
234 R. H. SHIMABUKURO AND A. J. LINCK, *Physiol. Plantarum*, 18 (1965) 532.
235 R. A. HERRETT AND W. P. BAGLEY, *J. Agr. Food Chem.*, 12 (1964) 17.
236 M. C. CARTER AND A. W. NAYLOR, *Botan. Gaz.*, 122 (1960) 138.
237 M. C. CARTER AND A. W. NAYLOR, *Physiol. Plantarum*, 14 (1961) 20.
238 F. M. ASHTON, *Weeds*, 11 (1963) 161.
239 R. A. HERRETT AND A. J. LINCK, *J. Agr. Food Chem.*, 9 (1961) 466.
240 J. L. HILTON, *J. Agr. Food Chem.*, 17 (1969) 182.
241 G. ZWEIG (Ed.), *Analytical Methods for Pesticides, Plant Growth Regulators and Food Additives*, Academic Press, New York, 1964, Vol. 4, p. 17.
242 R. W. STORHERR AND J. BURKE, *J. Assoc. Offic. Agr. Chemists*, 44 (1961) 196.
243 B. D. WILLS, *Analyst*, 91 (1966) 468.
244 D. M. COULSON, *J. Gas Chromatog.*, 4 (1966) 285.

245 T. SALO AND K. SALMINEN, *Z. Lebensm. Untersuch. Forsch.*, 129 (1966) 149.
246 H. J. HARRIS, *J. Agr. Food Chem.*, 3 (1955) 939.
247 E. S. WINDHAM, *J. Assoc. Offic. Anal. Chemists*, 52 (1969) 1237.
248 S. MIYAZAKI, G. M. BOUSH AND F. MATSUMURA, *Appl. Microbiol.*, 18 (1969) 972.
249 S. MIYAZAKI, G. M. BOUSH AND F. MATSUMURA, *J. Agr. Food Chem.*, 18 (1970) 87.
250 W. W. KILGORE AND E. R. WHITE, *J. Chromatog. Sci.*, 8 (1970) 166.
251 S. GORBACH AND U. WAGNER, *J. Agr. Food Chem.*, 15 (1967) 654.
252 T. P. METHRATTA, R. W. MONTAGNA AND W. P. GRIFFITH, *J. Agr. Food Chem.*, 15 (1967) 648.
253 L. E. ST. JOHN, JR., J. W. AMMERING, D. G. WAGNER, R. J. WAGNER AND D. J. LISK, *J. Dairy Sci.*, 48 (1965) 502.
254 E. J. KUCHAR, F. O. GEENTY, W. P. GRIFFITH AND R. J. THOMAS, *J. Agr. Food Chem.*, 17 (1969) 1237.
255 C. I. CHACKO, J. L. LOCKWOOD AND M. ZABIK, *Science*, 154 (1966) 893.
256 J. C. CASELEY, *Bull. Environ. Contam. Toxicol.*, 3 (1968) 180.
257 N. C. JAIN, C. R. FONTAN AND P. L. KIRK, *J. Pharm. Pharmacol.*, 17 (1965) 362.
258 N. C. JAIN AND P. L. KIRK, *Microchem. J.*, 12 (1967) 265.
259 N. V. FEHRINGER AND J. D. OGGER, *J. Chromatog.*, 25 (1966) 95.
260 J. F. THOMPSON, *J. Gas Chromatog.*, 6 (1968) 560.
261 K. C. WALKER AND M. BEROZA, *J. Assoc. Offic. Agr. Chemists*, 46 (1963) 250.
262 L. FISHBEIN AND W. L. ZIELINSKI, JR., *J. Chromatog.*, 28 (1967) 418.
263 E. R. GARRETT AND H. J. LAMBERT, *J. Pharm. Sci.*, 55 (1966) 812.
264 I. K. TSITOVICH AND E. A. KUZ'MENKO, *Zh. Analit. Khim.*, 22 (1967) 603.
265 E. H. GRAUL, H. HUNDESHAGEN AND H. WILLIAMS, *Proc. 3rd Intern. Congr. Chemotherapy, Stuttgart, 1963 (No. 2), 1964*, p. 1107.
266 H. M. RAUEN AND K. NORPOTH, *Arzneimittel-Forsch.*, 17 (1967) 599.
267 A. DEDE AND F. FARABOLLINI, *Bull. Soc. Ital. Biol. Sper.*, 43 (1967) 1489.
268 T. L. LOO, R. L. DIXON AND D. P. RALL, *J. Pharm. Sci.*, 55 (1966) 492.
269 T. L. LOO AND R. L. DIXON, *J. Pharm. Sci.*, 54 (1965) 809.
270 V. T. DE VITA, C. DENHAM, J. D. DAVIDSON AND V. T. OLIVERIO, *Clin. Pharmacol. Therap.*, 8 (1967) 566.
271 G. P. WHEELER, B. J. BOWDON AND T. C. HERREN, *Cancer Chemotherapy Rept.*, 42 (1964) 9.
272 J. A. MONTGOMERY, R. JAMES, G. S. MCCALEB AND T. P. JOHNSTON, *J. Med. Chem.*, 10 (1967) 688.
273 J. J. ROBERTS AND G. P. WARWICK, *Nature*, 179 (1957) 1181.
274 J. J. ROBERTS AND G. P. WARWICK, *Biochem. Pharmacol.*, 6 (1961) 217.
275 C. T. PENG, *J. Pharmacol. Exptl. Therap.*, 120 (1957) 229.
276 M. V. NADKARNI, E. G. TRAMS AND P. K. SMITH, *Cancer Res.*, 19 (1959) 713.
277 E. G. TRAMS, M. V. NADKARNI, V. DE QUATTRO, G. D. MAENGWYN-DAVIES AND P. K. SMITH, *Biochem. Pharmacol.*, 2 (1959) 7.
278 B. W. FOX, A. W. CRAIG AND H. JACKSON, *Biochem. Pharmacol.*, 5 (1960) 27.
279 B. W. FOX, A. W. CRAIG AND H. JACKSON, *Brit. J. Pharmacol.*, 14 (1959) 149.
280 J. J. ROBERTS AND G. P. WARWICK, *Biochem. Pharmacol.*, 1 (1958) 60.
281 E. A. BARNSLEY, *Biochem. J.*, 106 (1968) 18P.
282 D. J. PILLINGER, B. W. FOX AND A. W. CRAIG, in L. J. ROTH (Ed.), *Isotopes in Experimental Pharmacology*, Univ. of Chicago Press, Chicago, Ill., 1965, p. 415.
283 P. F. SWANN, *Nature*, 214 (1967) 918.
284 S. WAKAKI, H. MARUMO, K. TOMIOKA, G. SHIMIZU, E. KAO, H. KAMADA, S. KUDO AND Y. FUGIMOTO, *Antibiot. Chemotherapy*, 8 (1958) 228.
285 C. L. STEVENS, K. G. TAYLOR, M. E. MUNK, W. S. MARSHALL, K. NOLL, G. D. SHAH, L. G. SHAI and K. UZU, *J. Med. Chem.*, 8 (1964) 1.
286 H. S. SCHWARTZ AND F. S. PHILIPS, *J. Pharm. Exptl. Therap.*, 133 (1961) 335.
287 J. J. VAVRA, C. DE BOER, A. DIETZ, L. J. HANKA AND W. T. SOKOLSKI, *Antibiot. Ann.*, (1959–1960) 230.
288 R. R. HERR, T. E. EBLE, M. E. BERGY AND H. K. JAHNKE, *Antibiot. Ann.*, (1960) 236.
289 P. M. SCOTT AND E. SOMERS, *J. Agr. Food Chem.*, 16 (1968) 483.
290 A. E. POHLAND AND R. ALLEN, *J. Assoc. Offic. Anal. Chemists*, 53 (1970) 686.

291 A.E.POHLAND, K.SANDERS AND C.W.THORPE, *J. Assoc. Offic. Anal. Chemists*, 53 (1970) 692.
292 M.J.BARNES AND B.BOOTHROYD, *Biochem. J.*, 78 (1961) 41.
293 S.SYMCHOWICZ AND K.K.WONG, *Biochem. Pharmacol.*, 15 (1966) 1601.
294 S.SYMCHOWICZ, M.S.STAUB AND K.K.WONG, *Biochem. Pharmacol.*, 16 (1967) 2405.
295 L.J.FISCHER AND S.RIEGELMAN, *J. Chromatog.*, 21 (1966) 268.
296 R.J.COLE, J.W.KIRKSEY AND C.E.HOLADAY, *Appl. Microbiol.*, 19 (1970) 106.
297 R.H.ADAMSON, S.L.AGUE, S.M.HESS AND J.D.DAVIDSON, *J. Pharm. Exptl. Therap.*, 150 (1965) 322.
298 W.N.FISHBEIN, P.P.CARBONE, E.J.FRIEREICH, J.MISRA AND E.FREI, *Clin. Pharmacol. Therap.*, (1964) 574.
299 W.N.FISHBEIN AND P.O.CARBONE, *Science*, 142 (1963) 1069.
300 H.KOFOD, *Acta Chem. Scand.*, 9 (1955) 1575.
301 L.FISHBEIN AND M.A.CAVANAUGH, *J. Chromatog.*, 20 (1965) 283.
302 S.J.JACOBS AND H.S.ROSENKRANZ, *Cancer Res.*, 30 (1970) 1084.
303 R.C.R.BARRETO, *J. Chromatog.*, 7 (1962) 82.
304 R.C.R.BARRETO AND S.O.SABINO, *J. Chromatog.*, 9 (1962) 180.
305 R.C.R.BARRETO AND S.O.SABINO, *J. Chromatog.*, 11 (1963) 344.
306 R.C.R.BARRETO AND S.O.SABINO, *J. Chromatog.*, 13 (1964) 435.
307 A.LEWANDOWSKI AND H.SYBIRSKA, *Chem. Anal. (Warsaw)*, 13 (1968) 319.
308 T.A.LA RUE, *J. Chromatog.*, 32 (1968) 784.
309 A.ALESSANDRO, F.MARI AND S.SETTECASE, *Farmaco (Pavia)*, *Ed. Prat.*, 22 (1967) 437.
310 A.CALO, C.CARDINI AND V.QUERCIA, *J. Chromatog.*, 37 (1968) 194.
311 A.CALO, C.CARDINI AND V.QUERCIA, *Bull. Chim. Farm.*, 107 (1968) 296.
312 M.C.FAN AND W.G.WALD, *J. Assoc. Offic. Agr. Chemists*, 48 (1965) 1148.
313 S.INOUE, A.OGINO AND Y.ONO, *Yakuzaigaku*, 26 (1966) 302.
314 W.H.WU, T.F.CHIN AND J.L.LACH, *J. Pharm. Sci.*, 59 (1970) 1234.
315 S.FABRO, R.L.SMITH AND R.T.WILLIAMS, *Nature*, 215 (1967) 296.
316 R.T.WILLIAMS, *Lancet*, (1963-I) 723.
317 H.SCHUMACHER, R.L.SMITH, R.B.I.STAGG AND R.T.WILLIAMS, *Pharm. Acta Helv.*, 39 (1964) 394.
318 S.FABRO, R.L.SMITH AND R.T.WILLIAMS, *Biochem. J.*, 104 (1967) 565.
319 P.J.NICHOLLS, *J. Pharm. Pharmacol.*, 18 (1966) 46.
320 G.PISCHEK, E.KAISER AND H.KOCH, *Microchim. Acta*, (1970) 530.
321 M.FIEDLER AND W.HEINE, *Acta Biol. Med. Ger.*, 13 (1964) 1.
322 D.H.SANDBERG, S.A.BOCK AND D.A.TURNER, *Anal. Biochem.*, 8 (1964) 129.
323 C.T.BEER, M.L.WILSON AND J.A.BELL, *Can. J. Physiol. Pharmacol.*, 42 (1964) 368.
324 N.R.FARNSWORTH AND I.M.HILINSKI, *J. Chromatog.*, 18 (1965) 184.
325 I.M.JAKOVLJEVIC, L.D.SEAY AND R.W.SHAFFER, *J. Pharm. Sci.*, 53 (1964) 553.
326 N.J.CONE, R.MILLER AND N.NEUSS, *J. Pharm. Sci.*, 52 (1963) 688.
327 K.GENEST AND C.G.FARMILO, *J. Pharm. Pharmacol.*, 16 (1964) 250.
328 L.A.DAL CORTIVO, D.R.BROICH, A.DIHRBERG AND B.NEWMAN, *Anal. Chem.*, 38 (1959) 1966.
329 R.J.MARTIN AND T.G.ALEXANDER, *J. Assoc. Offic. Anal. Chemists*, 50 (1967) 1362.
330 R.J.MARTIN AND T.G.ALEXANDER, *J. Assoc. Offic. Anal. Chemists*, 51 (1968) 159.
331 D.L.ANDERSEN, *J. Chromatog.*, 41 (1969) 491.
332 A.S.CURRY AND H.POWELL, *Nature*, 173 (1954) 1143.
333 J.LOOK, *J. Assoc. Offic. Anal. Chemists*, 51 (1968) 1318.
334 A.STOLL AND A.HOFMAN, *US Patent 2,438,259*, March 23, 1948.
335 R.P.PIOCH, *US Patent 2,736,728*, Feb. 28, 1956.
336 C.RADECKA AND I.C.NIGAM, *J. Pharm. Sci.*, 55 (1966) 781.
337 M.A.KATZ, G.TADJER AND W.A.AUFRICHT, *J. Chromatog.*, 31 (1967) 545.
338 S.KOJIMA AND H.ICHIBAGASE, *Chem. Pharm. Bull. (Tokyo)*, 14 (1966) 971.
339 S.KOJIMA AND H.ICHIBAGASE, *Chem. Pharm. Bull. (Tokyo)*, 16 (1968) 1851.
340 S.KOJIMA AND H.ICHIBAGASE, *Chem. Pharm. Bull. (Tokyo)*, 17 (1969) 2620.
341 J.P.MILLER, L.E.M.CRAWFORD, R.C.SONDERS AND E.V.CARDINAL, *Biochem. Biophys. Res. Commun.*, 25 (1966) 153.
342 P.H.DERSE AND R.J.DAUN, *J. Assoc. Offic. Anal. Chemists*, 49 (1966) 1090.

343 R.C.SONDERS, R.G.WIEGAND AND J.C.NETNAL, *J. Assoc. Offic. Anal. Chemists*, 51 (1968) 136.
344 D.I.REES, *Analyst*, 90 (1965) 568.
345 M.L.RICHARDSON AND P.E.LUTON, *Analyst*, 91 (1966) 520.
346 S.KATO, T.KANEKO AND A.TANIMURA, *Shokuhin Eiseigaku Zasshi*, 11 (1970) 98.
347 D.K.DAS, T.V.MATTHEW AND S.N.MITRA, *J. Chromatog.*, 52 (1970) 354.
348 L.C.MITCHELL, *J. Assoc. Offic. Agr. Chemists*, 38 (1955) 943.
349 I.S.KO, I.S.CHUNG AND Y.H.PARK, *Rept. Natl. Chem. Lab. (Korea)*, 3 (1959) 72; *Chem. Abstr.*, 54 (1960) 10181L.
350 T.KOMODA AND R.TAKESHITA, *Shokuhin Eiseigaku Zasshi*, 3 (1961) 382; *Chem. Abstr.*, 60 (1964) 6130F.
351 D.K.DAS AND T.V.MATTHEW, *J. Inst. Chem. (Calcutta)*, 41 (1969) 192.
352 K.ASANO, M.TAIRA, H.NAKANISHI, E.SENDA, Y.SHIRAISHI AND R.TAKESHITA, *Nichidai Igaku Zasshi*, 22 (1963) 797; *Chem. Abstr.*, 61 (1964) 8813L.
353 H.YAMAGUCHI, *Nippon Kagaku Zasshi*, 82 (1961) 486; *Chem. Abstr.*, 56 (1962) 9927C.
354 T.H.ELLIOTT, N.Y.LEE-YOONG AND R.C.C.TAO, *Biochem. J.*, 109 (1968) 11P.
355 A.G.BLUMBERG AND A.M.HEATON, *J. Chromatog.*, 48 (1970) 565.
356 B.L.OSER, S.CARSON, E.E.VOGIN AND R.C.SONDERS, *Nature*, 220 (1968) 178.
357 R.E.WESTON AND B.B.WHEALS, *Analyst*, 95 (1970) 680.
358 S.W.GUNNER AND R.C.O'BRIEN, *J. Assoc. Offic. Anal. Chemists*, 52 (1969) 1200.
359 M.H.LITCHFIELD AND T.GREEN, *Analyst*, 95 (1970) 168.
360 J.HOWARD, T.FAZIO AND R.WHITE, *J. Assoc. Offic. Anal. Chemists*, 52 (1969) 492.
361 T.FAZIO AND J.W.HOWARD, *J. Assoc. Offic. Anal. Chemists*, 53 (1970) 701.
362 J.W.HOWARD, T.FAZIO, B.K.WILLIAMS AND R.H.WHITE, *J. Assoc. Offic. Anal. Chemists*, 52 (1969) 1197.
363 M.AGAGI, I.AOKI AND T.UEMATSU, *Chem. Pharm. Bull. (Tokyo)*, 14 (1966) 1.
364 M.AKAGI, I.AOKI, T.UEMATSU AND T.IYANAKI, *Chem. Pharm. Bull. (Tokyo)*, 14 (1966) 10.
365 T.SASAKI, Z.IIKURA AND T.YOKOTSUKA, *Chomi Kagaku*, 16 (1969) 6.
366 W.KAMP, *Pharm. Weekblad*, 101 (1966) 57.
367 S.C.LEE, *Hua Hsueh*, (3) (1966) 117; *Chem. Abstr.*, 67 (1967) 42602M.
368 E.LUDWIG AND U.FREIMUTH, *Nahrung*, 9 (1965) 569.
369 T.SALO AND U.SALMINEN, *Suomen Kemistilehti Sect. A*, 37 (1964) 161.
370 D.WALDI, in E.STAHL (Ed.), *Thin-Layer Chromatography*, Academic Press, New York, 1965, p.365.
371 T.KORBELAK AND J.N.BARTLETT, *J. Chromatog.*, 41 (1969) 124.
372 T.KORBELAK, *J. Assoc. Offic. Anal. Chemists*, 52 (1969) 489.
373 K.NAGASAWA, H.YOSHIDOME AND K.ANRYU, *J. Chromatog.*, 52 (1970) 173.
374 S.KOJIMA AND H.ICHIBAGASE, *Yakuzaigaku*, 26 (1966) 115.
375 R.TAKESHITA, Y.SAKAGAMI AND T.YAMASHITA, *Eisei Kagaku*, 15 (1969) 66.
376 R.NANIKAWA, S.KOTOKU AND T.YAMADA, *Japan. J. Legal Med.*, 21 (1967) 17.
377 S.KAWAKISHI AND M.NAMIKI, *Agr. Biol. Chem.*, 33 (1969) 452.
378 M.KOJIMA AND I.ICHIKAWA, *J. Ferment. Technol.*, 47 (1969) 262.
379 M.KOJIMA, Y.AKAHORI AND I.ICHIKAWA, *Nippon Nogeikagaku Kaishi*, 42 (1968) 18.
380 K.MODZELEWSKA AND F.MORDRET, *Tluszcze Jadalne*, 14 (1970) 127; *Chem. Abstr.*, 73 (1970) 119289Y.
381 D.L.ANDERSEN, *J. Assoc. Offic. Anal. Chemists*, 53 (1970) 1.
382 H.BINDER, *J. Chromatog.*, 41 (1969) 448.
383 A.KJAER AND A.JART, *Acta Chem. Scand.*, 11 (1957) 1423.
384 S.FISEL, F.MODREANU AND A.CARPOV, *Acad. Rep. Populare Romine, Studii Cercetari Stiint. Chim.*, 8 (1957) 277; *Chem. Abstr.*, 54 (1960) 18155.
385 L.FISHBEIN, J.FAWKES, H.L.FALK AND S.THOMPSON, *J. Chromatog.*, 29 (1967) 267.
386 E.O.OSWALD, L.FISHBEIN AND B.J.CORBETT, *J. Chromatog.*, 45 (1969) 437.
387 E.O.OSWALD, L.FISHBEIN, B.J.CORBETT AND M.P.WALKER, *Biochim. Biophys. Acta*, 230 (1971) 237.
388 S.KUWATSUKA AND J.E.CASIDA, *J. Agr. Food Chem.*, 13 (1965) 528.
389 E.G.ESSAC AND J.E.CASIDA, *J. Insect Physiol.*, 14 (1968) 913.
390 G.M.NANO AND A.MARTELLI, *J. Chromatog.*, 21 (1966) 349.

391   G. Riezebos, A. G. Peto and B. North, *Rec. Trav. Chim.*, 86 (1967) 31.
392   W. L. Zielinski, Jr. and L. Fishbein, *Anal. Chem.*, 38 (1966) 41.
393   Y. Saiki, A. Ueno, H. Sasaki, T. Morita, M. Suzuki, T. Saito and S. Fukushima, *Yakugaku Zasshi*, 88 (1968) 185.
394   S. W. Gunner and T. B. Hand, *J. Chromatog.*, 37 (1968) 357.
395   H. Foreman and T. T. Trujillo, *J. Lab. Clin. Med.*, 43 (1954) 566.
396   H. Foreman, M. Vier and M. Magee, *J. Biol. Chem.*, 203 (1953) 1045.
397   M. Mihara, R. Amano, T. Kondo, H. Tanabe, *Shokuhin Eiseigaku Zasshi*, 11 (1970) 88; *Chem. Abstr.*, 73 (1970) 97446T.
398   J. Vanderdeelen, *J. Chromatog.*, 39 (1969) 521.
399   J. Sykora and V. Eybl, *Collection Czech. Chem. Commun.*, 32 (1967) 352.
400   R. N. Warren, *J. Chromatog.*, 40 (1969) 468.
401   F. L. Grab and J. A. Reinstein, *J. Pharm. Sci.*, 57 (1968) 1703.
402   B. Weissmann, P. A. Bromberg and A. B. Gutman, *Proc. Soc. Exptl. Biol. Med.*, 87 (1954) 257.
403   U. M. Senanayake and R. O. B. Witesekera, *J. Chromatog.*, 32 (1968) 75.
404   V. M. Pechennikov and A. Z. Knizhnik, *Nauch. Tr. Aspir. Ordinatorov, 1-i Mosk. Med. Inst.*, (1967) 146; *Chem. Abstr.*, 70 (1969) 50490A.
405   W. N. French and A. Wehrli, *J. Pharm. Sci.*, 54 (1965) 1515.
406   M. Stuchlik, I. Csiba and L. Krasnek, *Cesk. Farm.*, 18 (1969) 91.
407   V. Vukcevic-Kovacevk and K. K. Anam, *Bull. Sci. Conseil Acad. Sci. Arts, RPF Yougoslavie, Sect. A*, 13 (1968) 77; *Chem. Abstr.*, 71 (1969) 42378A.
408   R. Bontemps, S. Leclercq and P. Teirlinck, *J. Pharm. Belg.*, 23 (1968) 512.
409   A. Monard, *J. Pharm. Belg.*, 23 (1968) 323.
410   J. M. Newton, *J. Assoc. Offic. Anal. Chemists*, 52 (1969) 653.
411   A. Pereira, Jr. and M. M. Pereira, *Garcia Orta*, 15 (1967) 41; *Chem. Abstr.*, 71 (1969) 122435G.
412   J. L. Antkowiak and A. L. Spatorico, *J. Chromatog.*, 29 (1967) 277.
413   D. W. Hammond and R. E. Weston, *Analyst*, 94 (1969) 921.
414   H. Knapstein, *Z. Anal. Chem.*, 217 (1966) 181.
415   T. Komoda and R. Takeshita, *Shokuhin Eiseigaku Zasshi*, 3 (4) (1962) 374.
416   T. Komoda and R. Takeshita, *Nippon Shokuhin Kogyo Gakkaishi*, 13 (1967) 201.
417   K. Nakamura, Y. Utsui and Y. Ninomiya, *Yakugaku Zasshi*, 86 (1966) 404.
418   H. F. Beckman, *J. Agr. Food Chem.*, 6 (1958) 130.
419   J. Brüggemann, K. Bronsch, H. Heigener and H. Knapstein, *Anal. Chem.*, 10 (1962) 108.
420   M. V. Riggs, *Chem. Ind. (London)*, (1956) 926.
421   D. K. Dastur and R. S. Palekar, *Nature*, 210 (1966) 841.
422   H. Matsumoto and F. M. Strong, *Arch. Biochem. Biophys.*, 101 (1963) 299.
423   K. Nishida, A. Kobayashi and T. Nagahara, *Kagoshima Daigaku Nogakubu Hokoku*, 15 (1956) 118.
424   A. Kobayashi and H. Matsumoto, *Arch. Biochem. Biophys.*, 110 (1965) 373.
425   M. Spatz and G. L. Laquer, *Proc. Soc. Exptl. Biol. Med.*, 127 (1968) 281.
426   U. Weiss, *Federation Proc.*, 23 (1964) 1357.
427   W. W. Wells, M. G. Yang, W. Bolzer and O. Mickelsen, *Anal. Biochem.*, 25 (1968) 325.
428   *Trop. Prod. Inst. Rept. No. 613*, May, 1965.
429   H. de Iongh, J. G. van Pelt, W. O. Ord and C. B. Barrett, *Vet. Record*, 76 (1964) 901.
430   W. T. Trager, L. Stoloff and A. D. Campbell, *J. Assoc. Offic. Agr. Chemists*, 47 (1964) 993.
431   S. Nesheim, *J. Assoc. Offic. Agr. Chemists*, 47 (1964) 586.
432   W. A. Pons, Jr., A. F. Cucullu, L. S. Lee, A. O. Franz and L. A. Goldblatt, *J. Assoc. Offic. Anal. Chemists*, 49 (1966) 554.
433   M. R. Heusinkveid, C. C. Shera and F. J. Baur, *J. Assoc. Offic. Agr. Chemists*, 43 (1965) 448.
434   R. H. Engebrecht, J. L. Ayres and R. O. Sinnbuber, *J. Assoc. Offic. Agr. Chemists*, 48 (1965) 815.
435   L. Stoloff, A. Graff and H. Rich, *J. Assoc. Offic. Anal. Chemists*, 49 (1966) 740.
436   R. M. Eppley, *J. Assoc. Offic. Anal. Chemists*, 49 (1966) 473.
437   W. A. Pons, Jr., A. F. Cucullu, A. O. Franz and L. A. Goldblatt, *J. Am. Oil Chemists' Soc.*, 45 (1968) 694.

438 W.A.Pons, Jr. and L.A.Goldblatt, *J. Am. Oil Chemists' Soc.*, 42 (1965) 471.
439 W.A.Pons, Jr., J.A.Robertson and L.A.Goldblatt, *J. Am. Oil Chemists' Soc.*, 43 (1966) 665.
440 J.L.Ayres and R.O.Sinnhuber, *J. Am. Oil Chemists' Soc.*, 43 (1966) 423.
441 R.B.A.Carnaghan, R.D.Hartley and J.O'Kelly, *Nature*, 200 (1963) 1101.
442 J.A.Robertson, W.A.Pons, Jr. and L.A.Goldblatt, *J. Agr. Food Chem.*, 15 (1967) 798.
443 J.A.Robertson and W.A.Pons, Jr., *J. Assoc. Offic. Anal. Chemists*, 51 (1968) 1190.
444 B.L. van Duuren, T.Chan and F.M.Irani, *Anal. Chem.*, 40 (1968) 2024.
445 P.J.Andrellos and G.R.Reid, *J. Assoc. Offic. Agr. Chemists*, 47 (1964) 801.
446 L.Stoloff, *J. Assoc. Offic. Anal. Chemists*, 50 (1967) 354.
447 A.E.Pohland, M.E.Cushmac and P.J.Andrellos, *J. Assoc. Offic. Anal. Chemists*, 51 (1968) 907.
448 E.V.Crisan, *Contrib. Boyce Thompson Inst.*, 24 (1968) 37.
449 P.J.Andrellos, A.C.Beckwith and R.M.Eppley, *J. Assoc. Offic. Anal. Chemists*, 50 (1967) 346.
450 H.T.Lee and K.H.Ling, *J. Formosan Med. Assoc.*, 66 (1967) 92.
451 T.W.Kwon and J.C.Ayres, *J. Chromatog.*, 31 (1967) 420.
452 S.Nesheim, *J. Assoc. Offic. Agr. Chemists*, 47 (1964) 1010.
453 H. de Iongh, R.O.Vles and J.G. van Pelt, *Nature*, 202 (1964) 466.
454 H.L.Falk, S.J.Thompson and P.Kotin, *Am. Assoc. Cancer Res. Proc.*, 6 (1965) Abst. No.18.
455 R.C.Shank and G.N.Wogan, *Federation Proc.*, 24 (1965) 627.
456 O.Bassir and F.Osiyemi, *Nature*, 215 (1967) 882.
457 R.Allcroft, H.Roger, G.Lewis, J.Nabney and P.E.Best, *Nature*, 209 (1966) 154.
458 C.W.Holzapfel, P.S.Steyn and I.F.H.Purchase, *Tetrahedron Letters*, 25 (1966) 2799.
459 M.S.Masri, D.E.Lundin, J.R.Page and V.C.Garcia, *Nature*, 215 (1967) 753.
460 T.C.Campbell, J.P.Caedo, Jr., J.Bulatao-Jayme, L.Salamat and R.W.Engel, *Nature*, 227 (1970) 404.
461 M.F.Dutton and J.G.Heathcote, *Chem. Ind. (London)*, (1968) 418.
462 D.A. van Dorp, A.S.M. van der Zijden, R.K.Beerthius, S.Sparreboom, W.O.Ord, K. de Jong and R.Keuning, *Rec. Trav. Chim.*, 82 (1963) 587.
463 K.J. van der Merwe and P.S.Steyn, *J. Chem. Soc.*, (1965) 7083.
464 P.S.Steyn and C.W.Holzapfel, *J. S. African Chem. Inst.*, 20 (1967) 186.
465 P.M.Scott and T.B.Hand, *J. Assoc. Offic. Anal. Chemists*, 50 (1967) 366.
466 W.Nel and I.F.H.Purchase, *J. S. African Chem. Inst.*, 21 (1968) 87.
467 L.J.Vorster and I.F.H.Purchase, *Analyst*, 93 (1968) 694.
468 H.J.Burkhardt and J.Furgacs, *Tetrahedron*, 24 (1968) 717.
469 J.V.Rodricks, K.R.Henery-Logan, A.D.Campbell, L.Stoloff and M.J.Verrett, *Nature*, 217 (1968) 668.
470 A.C.Waiss, M.Wiley, D.R.Black and R.E.Lundin, *Tetrahedron Letters*, 25 (1968) 2975.
471 C.J.Mirocha, C.M.Christensen and G.H.Nelson, *Appl. Microbiol.*, 15 (1967) 497.
472 C.J.Mirocha, C.M.Christensen and G.H.Nelson, *Cancer Res.*, 28 (1968) 2319.
473 T.Tatsuno, *Cancer Res.*, 28 (1968) 2393.
474 T.Tatsuno, M.Saito, M.Enomoto and H.Tsunoda, *Chem. Pharm. Bull. (Tokyo)*, 16 (1968) 2519.
475 Y.Ueno, I.Ueno, T.Tatsuno, K.Ohokubo and H.Tsunoda, *Experientia*, 25 (1969) 1062.
476 C.W.Holzapfel, *Tetrahedron*, 24 (1968) 2101.
477 I.Ishikawa, Y.Ueno and H.Tsunoda, *J. Biochem. (Tokyo)*, 67 (1970) 753.
478 P.S.Steyn, *J. Chromatog.*, 45 (1969) 473.
479 R.M.Jones, *Anal. Chem.*, 38 (1966) 338.
480 C.Bighi and G.Saglietto, *J. Chromatog.*, 18 (1965) 297.
481 C.Bighi, G.Saglietto and A.Betti, *Ann. Univ. Ferrara, Sez. 5*, 11 (1967) 163.
482 C.Bighi, A.Betti and G.Saglietto, *Ann. Chim. (Rome)*, 57 (1967) 1142.
483 C.Bighi and G.Saglietto, *J. Gas Chromatog.*, 4 (1966) 303.
484 J.L.Spigarelli and C.E.Meloan, *J. Chromatog. Sci.*, 8 (1970) 420.
485 N.A.Kirshen and G.H.Olsen, *Anal. Chem.*, 40 (1968) 1341.
486 G.Neurath and W.Lüttich, *J. Chromatog.*, 34 (1968) 257.

487 R. L. HINMAN, *Anal. Chim. Acta*, 15 (1956) 125.
488 E. BOYLAND AND D. MANSON, *Biochem. J.*, 101 (1966) 84.
489 W. B. DEICHMAN AND J. L. RADOMSKI, *J. Natl. Cancer Inst.*, 43 (1969) 263.
490 H. UEHLEKE AND E. BRILL, *Biochem. Pharmacol.*, 17 (1968) 1459.
491 J. L. RADOMSKI AND E. BRILL, *Science*, 167 (1970) 992.
492 Y. MASUDA AND D. HOFFMANN, *Anal. Chem.*, 41 (1969) 650.
493 Y. MASUDA, K. MORI AND M. KURATSUNE, *Intern. J. Cancer*, 2 (1967) 489.
494 Q. QUICK, R. F. LAYTON, H. R. HARLESS AND O. R. HAYNES, *J. Gas Chromatog.*, 6 (1968) 46.
495 D. M. MARMION, R. G. WHITE, L. H. BILLE AND K. H. FERBER, *J. Gas Chromatog.*, 4 (1966) 190.
496 H. MATSUSHITA, *Ind. Health (Kawasaki)*, 5 (1967) 260.
497 K. SHIMOMURA AND H. F. WALTON, *Separation Sci.*, 3 (1968) 497.
498 J. LATINAK, *Collection Czech. Chem. Commun.*, 24 (1959) 2939.
499 E. BOYLAND AND R. NERY, *Biochem. J.*, 94 (1965) 198.
500 S. MIRVISH, *Biochim. Biophys. Acta*, 93 (1964) 673.
501 S. MIRVISH, *Biochim. Biophys. Acta*, 117 (1966) 1.
502 S. S. MIRVISH, *Analyst*, 90 (1965) 244.
503 R. NERY, *Analyst*, 91 (1966) 388.
504 R. NERY, *Analyst*, 94 (1969) 130.
505 W. L. ZIELINSKI, JR. AND L. FISHBEIN, *J. Gas Chromatog.*, 3 (1965) 142.
506 J. A. FRESEN, *Pharm. Weekblad*, 16 (1965) 532.
507 G. SZASZ, L. KHIN AND M. SZASZ, *Acta Pharm. Hung.*, 28 (1958) 55.
508 N. A. MILAS AND I. BELIČ, *J. Am. Chem. Soc.*, 81 (1959) 3358.
509 N. A. MILAS, R. S. HARRIS AND A. GOLUBOVIC, *Radiation Res. Suppl.*, 3 (1963) 71.
510 J. CARTLIDGE AND C. F. H. TIPPER, *Anal. Chim. Acta*, 22 (1960) 106.
511 A. RIECHE AND M. SCHUCZ, *Angew. Chem.*, 70 (1958) 694.
512 E. KNAPPE AND D. PETERI, *Z. Anal. Chem.*, 190 (1962) 386.
513 M. M. BUZLANOVA, V. F. STEPANOVSKAYA, A. F. NESTEROV AND V. L. ANTONOVSKY, *Zh. Analit. Khim.*, 21 (1966) 507.
514 J. A. BRAMMER, S. FROST AND V. W. REID, *Analyst*, 92 (1967) 91.
515 B. GLABIK AND W. WALCZYK, *Chem. Anal. (Warsaw)*, 12 (1967) 1299.
516 A. Y. VALENDO AND Y. D. NORIKOV, *Izv. Akad. Nauk Belorussk. SSR., Ser. Khim. Nauk,* 5 (1968) 120.
517 D. B. ADAMS, *Analyst*, 91 (1966) 397.
518 H. EWALD, G. ÖHLMANN AND W. SCHIRMER, *Z. Physik. Chem. (Leipzig)*, 234 (1967) 104.
519 N. R. GREINER, *J. Chromatog.*, 31 (1967) 525.

Chapter 3

ALKYLATING AGENTS

Alkylating agents represent the largest structural and functional class of environmental hazards categorized as carcinogens, teratogens, mutagens, and chromosome-breaking agents. They are found in industrial, pesticidal, drug and environmental pollutants, both synthetic and naturally occurring. The alkylating agents can be further subdivided on the basis of their functional groups. *e.g.* nitrosamines, aziridines, epoxides, lactones, aldehydes, dialkyl sulfates, etc. The reactivity of this class of agents (from a consideration of the myriad byproducts in their preparation, types of residues and cellular interaction) is stressed. In the latter instance it is important to note their reactivity with DNA either by an $S_N 1$ or $S_N 2$ mechanism (both appearing to react preferentially with the N-7 of guanine).

1. NITROSAMINES

The general considerations of nitrosamines that are relevant include toxicity[1-4], metabolism[5-8], mechanism of action[9-19], carcinogenic[20-27], mutagenic[28-33], and blastomogenic effects[23-25]. The analyses of N-nitrosamines have been achieved *via* polarographic[3, 34-40], colorimetric[41-43], iodometric[44], ultraviolet[42, 43, 45], infra-red[43], NMR[46, 47], and fluorescence[43] spectroscopy, as well as by acid–base and decomposition methods[48, 49].

Nitrosamines are representative of a category of compounds of great activity and considerable diversity of action, especially as carcinogens. (They have been shown to induce a great variety of tumors at different sites in many species.) Their occurrence, whether as synthetic derivatives, natural products, or accidental products in food processing or tobacco smoke, is of significant environmental concern.

A number of nitrosamines, *e.g.* dialkyl-, N-alkyl-N-aryl-, N-nitrosocyclic amines and dinitroso derivatives have been patented for use as gasoline and lubricant additives, antioxidants and pesticides. Dimethylnitrosamine (DMN) is used primarily in the electrolytic production of the hypergolic fuel 1,1-dimethylhydrazine. Other areas of utility include the control of nematodes, the inhibition of nitrification in soil, plasticizer for acrylonitrile polymers and its use in active metal anode–electrolyte systems for high-energy batteries.

It seems established that the dialkylnitrosamines are metabolized *in vivo* to yield an active alkylating agent which can be diazoalkane itself or an active alkene or an alkyl carbonium ion derived from it by further decomposition. Figures 1 and 2 illustrate the possible mechanism for the *in vivo* metabolism and reaction of dialkylnitrosamines as postulated by Druckrey *et al.*[50] and some postulated metabolic pathways of dimethylnitrosamine as proposed by Magee and Farber[19], respectively.

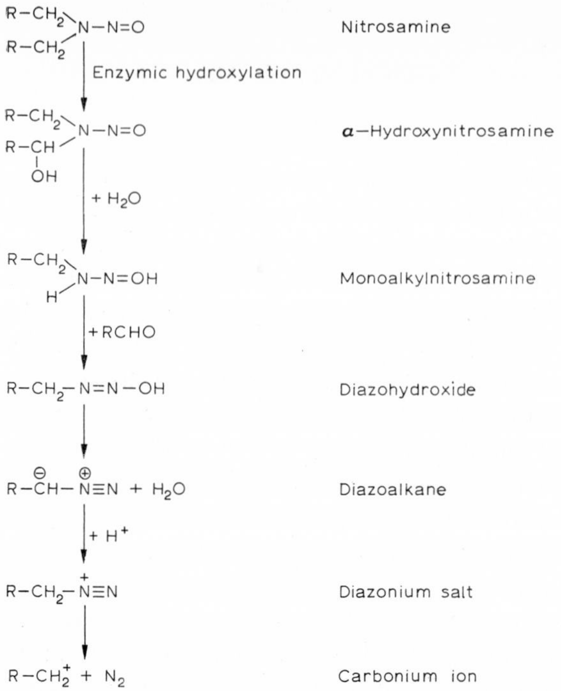

R-CH$_2$
R-CH$_2$⟩N-N=O   Nitrosamine

| Enzymic hydroxylation

R-CH$_2$
R-CH⟩N-N=O α-Hydroxynitrosamine
|
OH

| + H$_2$O

R-CH$_2$
H⟩N-N=OH   Monoalkylnitrosamine

|+RCHO

R-CH$_2$-N=N-OH   Diazohydroxide

↓

$\overset{\ominus}{R-CH}-\overset{\oplus}{N}≡N$ + H$_2$O   Diazoalkane

|+ H$^+$

R-CH$_2$-$\overset{+}{N}$≡N   Diazonium salt

↓

R-CH$_2^+$ + N$_2$ Carbonium ion

Fig. 1. Possible mechanism for the *in vivo* metabolism and reaction of dialkylnitrosamines[50].

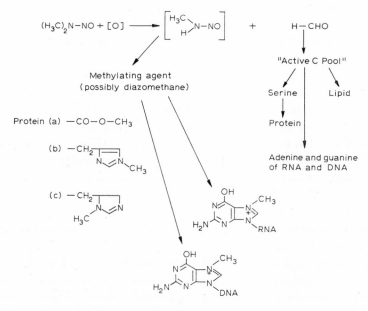

Fig. 2. Some postulated metabolic pathways of dimethylnitrosamine (ref. 19).

It is important to note that the hepatoxin dimethylnitrosamine has been shown to methylate both RNA and DNA *in vivo* to produce methylguanine[19]. The possible significance of nitrosamines as etiological factors in human carcinogenesis is based on the widespread occurrence of nitrosamine precursors (secondary amines and nitrites) in biological systems and the remarkable manner in which the nitrosamines can affect different organs in the same species of experimental animals.

It is possible that nitrosamines occur in foods since many foods contain large amounts of amines and small amounts of nitrite as preservative. The presence of nitrosamines has been suggested in tobacco, tobacco smoke, fish meals, wheat kernels and flour, dairy products (milk and cheese), and in various smoked fish, meat and mushrooms.

Preussmann *et al.*[51, 52] have described the thin-layer chromatography on Silica Gel G of a large variety of N-nitrosamines, *e.g.* symmetrical dialkyl-, methylalkyl-, cyclic, and aralkyl- and diarylnitrosamines. The solvent systems used were (*1*) for symmetrical dialkylnitrosamines and methylalkylnitrosamines: hexane–diethyl ether–methylene chloride (4:3:2); (*2*) for cyclic nitrosamines: hexane–diethyl ether–methylene chloride (5:7:10); and (*3*) for aralkyl- and diarylnitrosamines: hexane–diethyl ether–methylene chloride (10:3:2). Detection was achieved using (*a*) a 4:1 mixture of 1.5% diphenylamine in ethanol and 0.1% palladium(II) chloride in 0.2% sodium chloride; and (*b*) a 1:1 mixture of 1.0% sulfanilic acid in 30% acetic acid and 0.1% solution of α-naphthylamine in 30% acetic acid, preceded by UV irradiation. The diphenylamine–palladium chloride spray above was particularly sensitive for N-nitro compounds. After application of a light spray of this reagent, the moist plate was irradiated with UV light (λ_{max} = 240 nm) for several minutes, producing blue to violet spots on a colorless background. The limit of detection for nitrosamines was 0.5 µg. The mechanism of the reaction is believed to be a photochemical trans-nitrosation with transfer of the nitroso group to diphenylamine as with the palladium chloride reaction[53, 54]. Table 1 illustrates the TLC of various nitroso derivatives on Silica Gel G using the solvent systems and detection reagents depicted above.

Yasuda and Nakashima[55] described the TLC of a variety of aromatic nitroso compounds on silica gel. Detection was accomplished by heating the plates for 10–20 min at 105–110°C with an acidic 2-naphthol solution (1% 2-naphthol in a mixture of methanol–ethylene glycol–35% HCl (10:8:2), especially sensitive for *p*-nitrosoaniline derivatives. The detection limits for *p*-nitrosodimethylaniline, *p*-nitrosophenol, 1-nitroso-2-naphthol, and N-nitrosomethylaniline were 0.03, 0.2, 0.3, and 1 µg, respectively. R_F values using benzene–methanol (8:2) as solvent were: 4-nitrosodimethylaniline 0.62, 4-nitrosodiethylaniline 0.64, 4-nitrosomethylaniline 0.36, N-nitrosomethylaniline 0.76, N-nitrosoethylaniline 0.79, N-nitrosodiphenyl-amine 0.85, 4-nitrosophenol 0.32, 1-nitroso-2-naphthol 0.65, 2-nitroso-2-naphthol 0.47, 4-nitroso-1-naphthol 0.55, 5-nitroso-2-cresol 0.53, 4-phenylenediamine 0.35, N,N-dimethyl-4-phenylenediamine 0.42, 4-aminophenol 0.14, 4-nitrophenol 0.42, 4-nitrodimethylaniline 0.79, and 5-nitro-2-cresol 0.67. The R_F values for N-nitroso

TABLE 1

TLC OF VARIOUS NITROSO DERIVATIVES ON SILICA GEL G

Solvents: A, hexane–diethyl ether–methylene chloride (4:3:2); B, hexane–diethyl ether–methylene chloride (5:7:10); C, hexane–diethyl ether–methylene chloride (10:3:2). Detectors: 1, Diphenyl-amine–palladium chloride (a 4:1 mixture of a 1.5% solution of diphenylamine in ethanol and 0.1% palladium chloride in 0.2% sodium chloride). Application of detector 1 followed by uv irradiation of the plate. 2, Griess reagent (a 1:1 mixture of a 10% sulfanilic acid in 30% acetic acid and a 0.1% solution of α-naphthylamine in 30% acetic acid). Application of detector 2 preceded by uv irradiation of the plate.

Derivative	Solvent	R_F
Symmetrical dialkylnitrosamines	A	
Dimethylamine		0.24
Diethylamine		0.49
Di-*n*-propylamine		0.69
Di-isopropylamine		0.64
Di-*n*-butylamine		0.77
Diamylamine		0.82
Methylalkylnitrosamines	A	
Methylethylamine		0.36
Methylpropylamine		0.46
Methylbutylamine		0.51
Methylamylamine		0.57
Methylheptylamine		0.60
Methylbenzylamine		0.53
Cyclic nitrosamines	B	
N-Nitroso-N'-methylpiperazine		0.04
Dinitrosopiperazine		0.27
N-Nitroso-N'-carbethoxypiperazine		0.35
Morpholine		0.40
Pyrrolidine		0.41
Proline ethyl ester		0.52
Piperidine		0.63
Indoline		0.74
Aralkyl- and diarylnitrosamines	C	
p-Nitrosomethylaminobenzoic acid		0.00
p-Nitrosomethylaminobenzaldehyde		0.22
Methylphenylamine		0.63
Ethylphenylamine		0.72
Diphenylamine		0.80
Methyl(2-phenylethyl)amine		0.29

compounds were higher than those for the C-nitroso compounds and the R_F values were found to decrease in the order 4-nitro > 4-nitroso > 4-amino. 4-Nitroso-dimethylaniline was separated from related 4-amino and 4-nitro compounds with benzene–methanol (4:1) or benzene–dioxan–acetic acid (90:25:4) as solvent; and 4-nitrosophenol from 4-amino- and 4-nitrophenols with benzene–methanol (4:1) or cyclohexane–pyridine (7:3).

Sen *et al.*[56] described the determination of dimethyl-, diethyl-, di-*n*-propyl- and other nitrosamines in foods. The nitrosamines were determined semiquantitatively as follows: (*a*) initial extraction of the nitrosamines from foods into suitable solvents, (*b*) isolation of the nitrosamines from other interfering materials by steam distillation and ion-exchange clean-up or Amberlite CG-120 (strongly acidic resin, sodium form or REXYN101), (*c*) detection and semiquantitative estimation by TLC, and (*d*) final identification by GLC analysis of the separated compounds. The method was successfully applied to wheat flour, fish, cheese and spinach samples.

The detection limit for dimethylnitrosamine in foods was 0.15 p.p.m. and for diethyl- and di-*n*-propylnitrosamines, 50 p.p.b. For TLC, Silica Gel G-HR (Macherey, Nagel) was used with *n*-hexane–diethyl ether–methylene chloride developers, *e.g.* 4:3:2; 5:7:10 and 10:3:2. The spray reagents were (*a*) 1.5% diphenylamine in ethanol and 0.1% palladium(II) chloride in 0. 2% sodium chloride[52], (*b*) Griess reagent: 1% sulfanilic acid and 0.1% α-naphthylamine and (*c*) 0.3% ninhydrin in ethanol containing 2% pyridine.

Griess and ninhydrin reagents are generally used for detecting nitrous acid and compounds containing primary or secondary amino groups, respectively[57], and since N-nitrosamines produce secondary amines and nitrous acid upon irradiation with UV light and an acidic pH, a positive reaction to both the reagents can be considered specific for N-nitrosamines. Other compounds such as organic nitrites, C-nitroso compounds and free amines, although rarely present in the final food extracts, gave positive tests with these reagents usually *without* irradiation with UV light.

Table 2 lists the R_F values of a number of nitrosamines in various hexane–diethyl ether–methylene chloride developers. Gas chromatography of the nitrosamines

TABLE 2

R_F VALUES OF DIFFERENT NITROSAMINES IN VARIOUS SOLVENT SYSTEMS

Nitrosamines	*n-Hexane: diethyl ether: CH_2Cl_2*		
	4:3:2	*5:7:10*	*10:3:2*
Dimethylnitrosamine (DMN)	0.32	0.35	0.10
Diethylnitrosamine (DEN)	0.57	0.55	0.24
Di-*n*-propylnitrosamine (DPN)	0.77	0.63	0.38
Di-*n*-butylnitrosamine (DBN)	0.87	0.72	0.46
Phenylbenzylnitrosamine (PBN)	0.93	0.74	0.53
Nitrosopiperidine (NP)	0.56	0.52	0.23
N-Methyl-N'-nitrosopiperazine	0.09	0.28	0.64
Dinitrosopiperazine	0.09	0.31	0
Methyl-*n*-propylnitrosamine	0.48	0.76	0.27
Diisopropylnitrosamine	0.66	0.89	0.41
Ethyl-*n*-propylnitrosamine	0.61	0.87	0.37
Methyl-*n*-butylnitrosamine	0.57	0.83	0.33
Dibenzylnitrosamine	0.83	0.92	0.63
Methylbenzylnitrosamine	0.58	0.85	0.32

was achieved with a Varian Aerograph Model 1200 equipped with a hydrogen flame detector and $\frac{1}{8}$ in. × 5 ft. stainless steel columns containing either (*1*) 6 % Reoplex 400 on 60–80 mesh Chromosorb W or (*2*) 5 % SE-30 on acid-washed dimethylchlorosilane-treated 60–80 mesh Chromosorb W. The operating conditions were: for (*1*) column temperature 115°C, nitrogen flow 29 ml/min, range setting 10, and attenuation 8; for (*2*) column temperature 48°C, nitrogen flow 27 ml/min, range setting 10, and attenuation 1. Figure 3 illustrates the GLC separation of dimethyl-, diethyl- and di-*n*-propyl-nitrosamine standards.

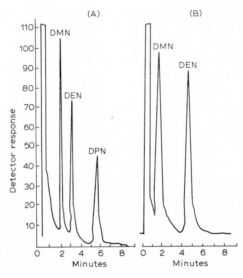

Fig.3. GLC separation of DMN, DEN and DPN standards. (A) 2 μg each of DMN, DEN and DPN in 2 μl ether, using GLC conditions 1; (B) 1 μg each of DMN and DEN, using GLC conditions 2.

Table 3 depicts the recovery of dimethyl-, diethyl- and di-*n*-propylnitrosamines added to various food samples (based on the TLC analysis of food extracts) using the Griess reagent and the ninhydrin solution as the location reagents. It was found possible to isolate DMN, DEN, and DPN from various foods spiked with suitable levels of different nitrosamines (0.3 p.p.m. for DMN and 0.1 p.p.m. for DEN and DPN) and identify them by GLC. However, examination of ten samples of flour (both whole wheat and white), six of Cheddar cheese, eight of frozen or pickled herring and four of spinach revealed that most of the samples did not contain DMN, DEN, and DPN. In a few cases, the extracts from Cheddar cheese and herring gave positive spots for nitrosamines (with the Griess reagent).

Hedlar and Marquardt[58] reported the presence of diethylnitrosamine (DEN) (as confirmed by TLC) in some samples of wheat plant, wheat grain, wheat flour, milk, and cheese. The presence of nitrosamines was established by the detection techniques of Preussmann *et al.*[51, 52] utilizing both the Griess and diphenylamine–palladium

TABLE 3

RECOVERY OF DMN, DEN, AND DPN ADDED TO VARIOUS FOOD SAMPLES (100 g)

Food	Compound added	Amount added (µg)	Approx. % rec.[a]
Cheese (Havarti)	DMN	15	65
	DEN	5	65
	DPN	5	65
Cheese (Cheddar)	DEN	10	65
	DPN	20	50
Fish (herring)	DMN	20	70
	DEN	6	70
	DPN	5	70
	DEN	10	80
	DPN	20	80
	DEN	25	80
	DEN	10	80
Corned beef	DMN	5	40
	DEN	5	50
	DPN	5	50
Canned luncheon meat	DMN	5	30
	DEN	5	70
	DPN	5	70
Wheat flour (white)	DEN	10	65
	DEN	400	90
	DEN	30	90
	DEN	25	65
Whole wheat flour	DEN	10	40
Dark rye flour	DEN	10	60

[a] Results based on visual estimation of spots on TLC plates.

chloride reagents described above (*e.g.* if both reactions are positive, the presence of nitrosamines can be accepted). The nitrosamines in flour were determined following the heating of samples at 175°C for 7 h. Figure 4 illustrates the TLC of 10 µg of diethylnitrosamine and 100 µg each of extracts of 5 commercial grades of flour treated as above. The R_F value and color of the depicted spots corresponded to those of DEN. Nitrosamines were found to be concentrated to an even greater extent than in the extracts by preparative TLC on silica gel. Figures 5 and 6 show that 15 µg of extract yielded spots approximately equivalent to those obtained with 2.5 µg DEN. It is of interest to note that nitrosamines were also detected in wheat stalk and leaves shortly after germination. In comparative tests, the grain extract (processed analogously as the flour extracts above) contained more nitrosamine than the stem of the same plant

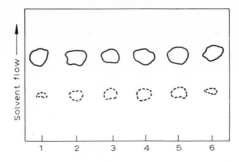

Fig.4. Thin-layer chromatogram of 10 µg diethylnitrosamine. 1, and 100 µg each of extracts from five commercial grades of flour: 2, Auroramehl; 3, Backerblume; 4, Goldpuder; 5, REWE; 6, Weizenkrone; reacted with diphenylamine–palladium chloride.

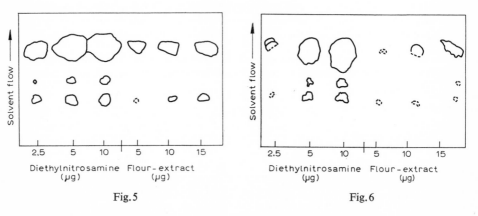

Fig. 5   Fig. 6

Fig.5. Thin-layer chromatogram of 5, 10 and 15 µg of extract of Edeka flour, concentrated by preparative thin-layer chromatography, and 2.5, 5 and 10 µg diethylnitrosamine; reacted with diphenylamine–palladium chloride.

Fig.6. Thin-layer chromatogram of 2.5, 5 and 10 µg diethylnitrosamine and 5, 10 and 15 µg of extract of Edeka flour, purified by preparative thin-layer chromatography; reacted with sulphanilic acid–naphthylamine.

(Figs. 7 and 8). As far as the nitrosamine content is concerned, wheat treated with artificial fertilizer did not differ from wheat fertilized exclusively with "organic material". It was considered that the nitrosamine occurrence depends upon the nitrogen content of the fertilizer, and not the kind of fertilizer used.

Nitrosamines were also detected in some samples of milk and cheese (Figs. 9 and 10). Tilsit cheese was selected for ether extraction and preparative TLC and sampled both immediately after manufacture and after ripening for 6 weeks. In both samples DEN was present, Fig. 10. Figure 10 also demonstrates that oxidative decomposition of DEN occurred during concentration of the extracts (the spots due to oxidation products in the extracts became larger than the spots of DEN). In further tests with Tilsit cheese after 3 weeks of ripening, both with and without the addition of nitrite,

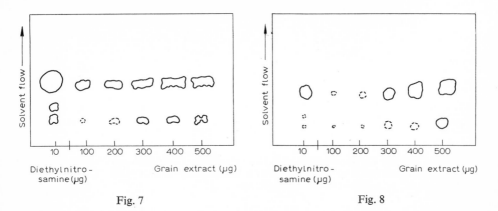

Fig. 7   Fig. 8

Fig. 7. Thin-layer chromatogram of 10 μg diethylnitrosamine and 100–500 μg of a grain extract purified by preparative thin-layer chromatography; reacted with diphenylamine–palladium chloride.

Fig. 8. Thin-layer chromatogram of 10 μg diethylnitrosamine and 100–500 μg of a grain extract purified by preparative thin-layer chromatography; reacted with sulphanilic acid–naphthylamine.

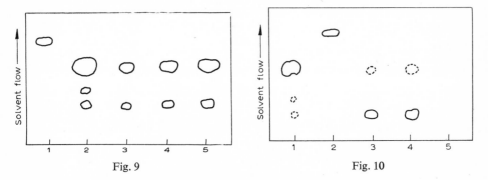

Fig. 9   Fig. 10

Fig. 9. Thin-layer chromatogram of 1, 10 μg dipropylnitrosamine; 2, 10 μg diethylnitrosamine; 3–5, 100, 200 and 300 μg of an ethereal extract of milk purified by preparative thin-layer chromatography; reacted with diphenylamine–palladium chloride.

Fig. 10. Thin-layer chromatogram of 1, 10 μg diethylnitrosamine; 2, 10 μg dipropylnitrosamine; 3 and 4, 20 and 40 μg of an extract of cheese ripened for 6 weeks; 5, 40 μg propionic acid; reacted with diphenylamine–palladium chloride.

the extract was found to contain comparable amounts of DEN, suggesting to the authors that the addition of nitrate plays no part in the synthesis of nitrosamine.

Neurath et al.[59] have described another test for the presence of nitrosamines. After reduction of the nitrosamine with lithium aluminum hydride, the hydrazine obtained is condensed with 5-nitro-2-hydroxybenzaldehyde to form a benzalhydrazine. When this test was used, the flour and cheese extracts yielded such small amounts of benzalhydrazine that its presence could only be demonstrated by TLC (Fig. 11). The spot obtained had the same R_F value as the benzalhydrazine (m.p. 98 °C) obtained from diethylnitrosamine (R_F 0.45) with the solvent system carbon tetrachloride–ethyl acetate (19:1).

Although the quantities of nitrosamine in the foods studied could not be specified, Hedlar and Marquardt[58] depicted the degree of concentration of the sample (*e.g.* 500 g of initial product) necessary for the *qualitative* determination of nitrosamine (of the order of 100 μg) (Table 4).

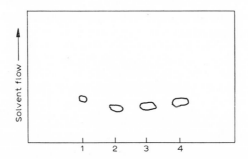

Fig. 11. Thin-layer chromatogram of 1, 5 μg diethylnitrosamine benzalhydrazine; 2–4, 100, 200 and 400 μg of an extract of cheese containing benzalhydrazine; reacted with 3 % KOH in 95 % ethanol.

TABLE 4

DEGREE OF CONCENTRATION OF 500 g INITIAL PRODUCT NECESSARY FOR THE QUALITATIVE DETERMINATION OF NITROSAMINE (OF THE ORDER OF 100 μg)

Initial product	Total no. of samples	No. of samples containing nitrosamine	Average degree of concentration (times)
Wheat plant	9	9	20,000
Wheat grain	2	2	9,000
Flour (unheated)	3	3	5,000
Flour (heated at 175 °C)	30	22	1,400[a]
Pasteurized milk	1	1	8,000
Cheese	2	2	250
			500

[a] The flour samples in which nitrosamines were not detected were from the harvest of 1967.

The presence of diethylnitrosamine in extracts from flour and cheese described in the above studies of Hedlar and Marquardt[58] were confirmed by TLC and GLC studies of Petrowitz[60]. The nitrosamines were separated from ether extracts of flour by TLC on silica gel with hexane–ether–methylene chloride (4:3:2) as solvents. The spots were identified by spraying with diphenylamine–palladium chloride or sulfanilic acid–α-naphthylamine reagents. As a further step, reduced diethylnitrosamine could be condensed with 5-nitro-2-hydroxybenzaldehyde and the product separated by TLC.

The product from preparative TLC (2.5 μg) was separated by GLC using an F&M Model 500 gas chromatograph equipped with a hot wire detector (at 100°C) and operated isothermally with helium as carrier gas at 100 ml/min. The columns and packings used were (a) stainless steel, 2 ft., containing silicone gum rubber on Chromosorb, (b) stainless steel, 6 ft., containing Apiezon L on Chromosorb, and (c) stainless steel, 6 ft., containing Carbowax 20 M on Chromosorb. Figure 12 and Table 5 illustrate the gas chromatograph and the retention times, respectively, of diethylnitrosamine when chromatographed as above. The GLC procedure using a 2-ft. column containing silicone oil at 60°C was quantitative with peak areas representing 50–500 μg of diethylnitrosamine.

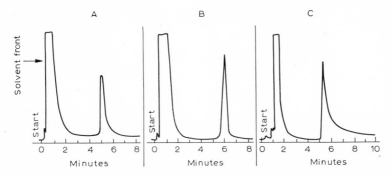

Fig. 12. Gas chromatography of diethylnitrosamine. Conditions: (A) 2-ft. column containing silicone gum rubber at 60°C; (B) 6-ft. column containing Carbowax 20M at 150°C; (C) 6-ft. column containing Apiezon L at 125°C.

TABLE 5

RETENTION TIMES OF DIETHYLNITROSAMINE

Stationary phase	Column length (ft.)	Retention times			
		60°C	100°C	125°C	150°C
Silicone gum rubber	2	5 min 10 sec	3 min 20 sec	60 sec	
Apiezon L	6			5 min 30 sec	3 min 40 sec
Carbowax 20 M	6		22 min 40 sec		6 min

Kröller[57] utilized both TLC and GLC procedures for the detection of nitrosamines in food and tobacco smoke. Silica Gel G (250 μ) chromatoplates were used with n-hexane–diethyl ether–dichloromethane (100:40:10) as developer[52] and both di-phenylamine–palladium chloride[52] and Griess reagents[41] as detectors. A ninhydrin reagent (0.2% ethanolic) was used for the further detection of the amines on the plates following preliminary treatment with 10% hydrochloric acid and 5 min UV irradiation. After drying the plates at 110°C for 1 h, the spot colors were recorded. Table 6 depicts

TABLE 6

TLC OF N-NITROSAMINES ON SILICA GEL G

N-Nitroso derivative	$R_F \times 100$	Spot color
Dimethylamine	18	Light pink
Diethylamine	31	Dark pink
Di-*n*-propylamine	51	Pink
Diisopropylamine	47	Ochre
Di-*n*-butylamine	59	Pink
Diisobutylamine	66	Pale pink
Di-*n*-pentylamine	69	Light pink
Di-*n*-hexylamine	75	Pink
Methyl-*n*-butylamine	32	Violet-pink
Ethyl-*n*-hexylamine	40	Dark pink
Diallylamine	54	Yellow-pink
Benzylmethylamine	34	Red
Dimethylaniline	12	Brown-violet
Diethylaniline	67	Green
Morpholine	14	Yellow
Piperidine	26	Yellow
Diphenylamine	75	Blue-green
Pyrrolidine	13	Grey-brown

Developing solvent: *n*-hexane–diethyl ether–dichloromethane (100:40:10).
Detector: 2% ethanolic ninhydrin following 10% hydrochloric acid spray and 5 min UV irradiation.

the R_F values as well as the spot colors of a variety of N-nitrosamines following TLC as delineated above. GLC analysis was performed on a Perkin-Elmer Model 116 chromatograph equipped with a flame-ionization counter and 2 m × 3 mm columns containing 5% Silicone Rubber SE-32 on 60–100 mesh Celite and 15% Silicone DCO on 60–100 mesh Celite. The column temperature was 200°C and nitrogen was the carrier gas at 36 ml/min. Table 7 depicts the retention times of various N-nitrosamines

TABLE 7

GAS CHROMATOGRAPHY OF N-NITROSODIALKYLAMINES ON 5% SE-32 AND 15% DCO ON 60–100 MESH CELITE

N-Nitroso compound	Retention time (sec)
Diethylamine	315
Di-*n*-propylamine	368
Diisopropylamine	323
Di-*n*-butylamine	559
Diisobutylamine	453
Di-*n*-pentylamine	1013

Column temperature 200°C.

following injection of 1 µl of a 0.5% solution of nitrosamine in hexane. The nitrosamines were also chromatographed on 2 m × 2.77 mm glass columns coated with
1% SE-30 on 80–100 mesh acid-washed DMCS-treated Chromosorb G. The column
and injection port temperatures were 165 and 260°C respectively, and the carrier gas
was nitrogen at 1.5 atm.

Table 8 illustrates the decreased retention times of N-nitrosamines on the
silicanized support following injection of 0.15 µl of a 0.5% solution of nitrosamine in
hexane.

TABLE 8

GAS CHROMATOGRAPHY OF N-NITROSODIALKYLAMINES ON 1% SE-30
ON 80–100 MESH ACID-WASHED DMCS-TREATED CHROMOSORB G

N-Nitroso compound	Retention time (sec)
Diethylamine	45
Di-n-propylamine	57
Diisopropylamine	62
Di-n-butylamine	92
Diisobutylamine	68
Di-n-pentylamine	169

Column temperature 165°C.

Möhler and Mayrhofer[43] investigated the presence of dimethyl- and diethylnitrosamines in meat, cheese, and flour using colorimetric, spectroscopic, and gas
chromatographic techniques. A Beckman Model 2 A gas chromatograph was used
equipped with a flame-ionization detector and 1.8 m column containing DEGS as
liquid phase operated at 160°C with hydrogen as carrier gas at 15 p.s.i.

Mayrhofer and Möhler[61] described the utility of combining GLC and TLC
techniques for the enhanced resolution and estimation of nitrosamines in foods. The
effluent from the GLC analysis[43] was adsorbed on silica gel thin-layer plates, developed
with hexane–ether–methylene chloride (4:3:2), and detected according to the method
of Preussmann et al.[52] using the diphenylamine–palladium chloride reagent or the
Griess reagent following UV irradiation of the plates.

Howard et al.[62] described the extraction and gas chromatographic determination of dimethylnitrosamine (DMN) in smoked fish. The fish sample was digested
in methanolic potassium hydroxide, the nitrosamine distilled from alkaline solution,
further isolated by extraction and column chromatography on Celite 545 and determined by GLC with a modified thermionic detector. A Barber-Colman Model 5000
instrument was used with the following: a Model 5000 electrometer, Model 8000 strip
chart recorder 0–5 mV, Model 5077 coiled-glass column cabinet, Model 5403-1 injector–detector temperature controller (SCR's–2 units), Model 5084 temperature
programmer, Model 5003 rear-chamber detector bath, and modified KCl thermionic

detector (Fig. 13). GLC conditions were: 9 ft. × 4 mm i.d. coiled-glass column packed with 10% Carbowax 1540 + 3% KOH on 100–120 mesh Gas Chrom P support. (Column conditioned at 180°C for 24 h, while maintaining argon carrier gas flow at 50 ml/min.) The injector and detector temperatures were 185 and 220°C, respectively,

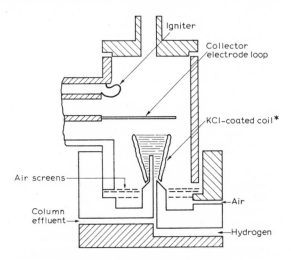

Fig. 13. Modified thermionic detector.
*7/16 in. continuous platinum–iridium helix.

and the flow rates (ml/min), argon 43, hydrogen 40 (producing a standing current of 3×10^{-9} A) and air 300, electrometer setting 100 (2×) equivalent to 2×10^{-9} A FSD, and temperature programmed from 80 to 120°C at 10°/min after initial 3 min hold at 80°C. Typical gas chromatograms of commercially processed smoked nitrite-treated chubs without and with DMN fortification are shown in Fig. 14. It was shown that the identity of DMN can be confirmed at levels as low as 10 p.p.b. by mass spectrometry. The use of a modified thermionic detector yielded increased sensitivity for DMN. Linearity studies up to 50 μg have indicated that a straight line (peak height response to weight) can be achieved.

The formation of nitrosamines in nitrite-treated fish was described by Sen et al.[63]. Eighteen samples of smoked and 5 of canned fish were analyzed after cooking with or without sodium nitrite (up to 200 p.p.m.). The results indicated that certain kinds of fish, particularly those rich in amines, can form dimethylnitrosamine (verified by both TLC and GLC). Analysis of uncooked and cooked samples was performed basically as described by Sen et al.[56] and modified to include a preliminary purification of samples over an alumina column (Woelm, basic) prior to the semi-quantitative determination of the nitrosamines present in the fish extracts by TLC using the method of Sen et al.[56]. When the test for nitrosamine was positive (Griess reagent) a duplicate extract was prepared and examined by the ninhydrin reagent. Two solvent systems were used for TLC analysis, e.g. n-hexane–diethyl ether–methylene chloride in the ratio

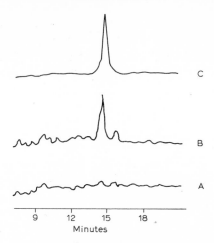

Fig.14. Gas chromatograms of (A) commercially processed nitrite-treated smoked chub; (B) commercially processed nitrite-treated smoked chub fortified with 10 p.p.b. DMN; (C) reference DMNA standard (0.5 µg/ml).

of 4:3:2 and 10:3:2. For GLC identification, the nitrosamine spots from TLC were scraped off and eluted with diethyl ether using a micro-Soxhlet apparatus, concentrated to *ca.* 0.3 ml and a 2.4 µl aliquot analyzed on two different columns. Figures 15 and 16 illustrate the GLC identification of DMN isolated from mackerel and hake using 6% Reoplex 400 and 10% Carbowax 20M columns, respectively. Table 9

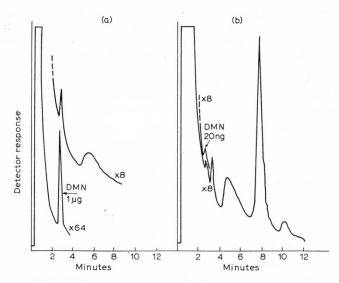

Fig.15. GLC identification of DMN. (a) Isolated from Mackerel (No.5); (b) from Hake (No.7). GLC conditions: 6% Reoplex 400 on 60–80 mesh Chromosorb W, 1/8 in. × 5 ft. stainless steel column, nitrogen flow 27 ml/min, column temperature 105 °C. Gas chromatograph as in Sen *et al.*[56].

depicts the amounts of nitrosamines formed after cooking various fish samples with or without sodium nitrite. The GLC and TLC results for samples 5 (smoked mackerel) and 7 (smoked hake) compared well with each other, *e.g.* the GLC estimations for DMN for the two samples were 8.2 μg and 4 μg, respectively, the respective TLC values were 7 μg and 4 μg. However, both sets of data are suggested by the authors to be very approximate values and hence should be considered semiquantitative only.

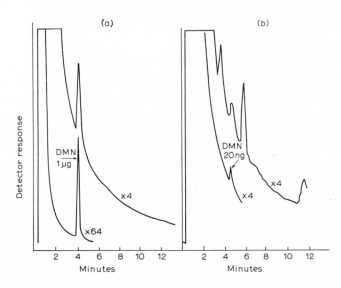

Fig. 16. GLC identification of DMN. (a) from Mackerel (No. 5); (b) from Hake (No. 7). GLC conditions: 10% Carbowax 20M on 60–80 mesh Chromosorb W (HMDS treated), 1/8 in. × 6 ft. stainless steel column, nitrogen flow 26 ml/min, column temperature 103–105 °C.

It is interesting to note that Ender and Ceh[64] found traces of DMN (1.15 μg/kg) in a few smoked fish samples but it was not stated whether or not the samples were preserved with nitrite.

Ender and Ceh[65] reported the occurrence of nitrosamines in foodstuffs for human and animal consumption. Table 10 shows the levels of nitrosamine in food-stuffs and indicates that the quantity of nitrosamines in smoked fish and meat exhibit great variation (the amount probably depending on the mode of the smoking procedure and the method of handling used). Mushrooms were also found to contain low levels of nitrosamines (Table 10). The nitrosamines detected in mushrooms and smoked fish and meat products appeared to be identical with dimethylnitrosamine as based on analytical methods described by Neurath and Doerk[66] and Neurath *et al.*[59] using TLC on silica gel with chloroform as solvent.

Ender *et al.*[45] previously reported the isolation and identification of dimethyl-nitrosamine in poisonous herring meal caused by the uncritical use of sodium nitrite as a preservative. The formation of dimethylnitrosamine in herring meal is caused by

TABLE 9

AMOUNT OF NITROSAMINES FORMED AFTER COOKING VARIOUS FISH SAMPLES
WITH OR WITHOUT SODIUM NITRITE

Code no.	Kind	Nitrite levels (p.p.m.)	Approx. quantity of nitrosamines detected (µg/200 g) (uncorrected for percentage recoveries)
1	Smoked haddock	0 (control)	N.D.
		200	(DEN) 1
2	Smoked cod	0	N.D.
		50	(DPN) 1
		200	(DPN) 2.5; (DMN) 0.7
		200ᵃ	(DPN) 0.7; (DMN) 2.5
3	Smoked cod	0	N.D.
		200	(DPN) 0.5; (DMN) 2
4	Canned mackerel, brand A	0	N.D.
		200	(DMN) 8
5	Canned mackerel, brand A	0	N.D.
		200	(DMN) 7
6	Smoked mackerel	0	N.D.
		200	(DMN) 4
7	Smoked hake	0	N.D.
		200	(DMN) 4
8	Canned mackerel, brand B	0	N.D.
		200	(DMN) 4
9	Canned salmon	0	N.D.
		200	(DMN) < 0.5
10	Canned mackerel[c], brand A	200 (control)	N.D.
		200	(DMN) 5
11	Smoked mackerel[c]	200 (control)	N.D.
		200	(DMN) 0.5
12	Smoked hake[b,c]	200 (control)	N.D.
		200	(DMN) 9

[a] Cooked at 210°C for 25 min.
[b] Cooked for 20 min under 20 p.s.i. All other samples were cooked for 60–70 min at 110°C.
[c] Analyzed by the methylene chloride extraction procedure. Sodium nitrite was added to the control samples after cooking the fish and cooling.

Abbreviations: N.D., not detectable; DMN, dimethylnitrosamine; DEN, diethylnitrosamine; DPN, di-*n*-propylnitrosamine.

various methylamines which occur normally in fish. Dimethylamine, the most potent producer of dimethylnitrosamine, reacts even at temperatures below 0°C[67], thus emphasizing the potential risk of using nitrite as a preserving agent.

Sakshaug *et al.*[68] investigated the possible occurrence of dimethylnitrosamine in herring meal as well as its hepatotoxicity in sheep. Samples of herring meal with

TABLE 10

LEVELS OF NITROSAMINE IN FOODSTUFFS

Type of food	Nitrosamine $(\mu g/kg)^a$	No. of samples analyzed
Fish		
Smoked herring	0.5–9.5	5
Kippers	0.5–2.4	4
Kippers	40	1
Smoked haddock (from Iceland)	15	1
Smoked mackerel	0.6	1
Meat		
Smoked sausage	0.8, 1.1, 2.4	3
Bacon	0.6, 1.2, 6.5	3
Smoked ham (from Iceland)	5.7	1
Mushrooms		
Polyporus ovinus	11.6	1
Boletus scaber	1.4	1
Amanita muscaria	30	1
Champignon	0.4, 5	2
Hydnum imbricatum	3, 15	2
Armillaria mellea	12	1
Lactarius trivialis	9.2	1
Russula emetica	10.2	1
A mixture of various edible mushrooms	14	1

^{a} Determined by the hydrazine method[67].

and without known toxic properties were examined for dimethyl- and diethylnitrosamine by a method where the final detection and semiquantitation were obtained by TLC[69] using silica gel with *n*-hexane–ether–dichloromethane (4:3:2) as developer and a 5:1 mixture of 1.5% diphenylamine in ethanol and palladium chloride (0.1% in 0.2% hydrochloric acid) used for detection. The R_F values of the pure substances were: dimethylnitrosamine 0.19 and diethylnitrosamine 0.18. Six sample batches known to have produced toxic hepatosis in ruminants all contained dimethylnitrosamine in concentrations ranging from 30 to 100 p.p.m. after prolonged storage at 2–4°C. In six samples from alleged non-toxic batches of herring meal, dimethylnitrosamine could not be detected by the method employed (detection limit 10–15 p.p.m.). Diethylnitrosamine was not found in any of the samples examined, but other unidentified nitroso compounds were observed in a few samples.

It was suggested that the detection of dimethylnitrosamine in samples from herring meal known to be toxic indicates an apparent connection between dimethylnitrosamine and the toxic hepatosis in ruminants caused by some batches of herring meal.

Devik[70] studied the possibility that the necessary precursors for N-nitrosamines in fish meals might be formed by side reactions occurring during the condensation of amino acids and aldoses (the Maillard reaction). This condensation is assumed to be an initial step for the subsequent formation of pigments responsible for the browning of various food products. Polarography as well as TLC and GLC were used to elaborate the products of the condensation of various amino acids and D-glucose adsorbed on potato starch. Silica gel was used with hexane–ether–methylene chloride (4:2:3) as developer[52]. Detection was achieved using UV irradiation followed by spraying with a diphenylamine–palladium chloride reagent. For gas chromatographic determination[71], the distillate from alkaline hydrolysis was extracted with methylene chloride, nitrosamine-containing extract, dried over anhydrous sodium sulfate, then concentrated to about 1 mg nitrosamine/ml and analyzed on a Pye Argon Chromatograph equipped with a standard β-ionization detection and a column of 10% PEGA on Celite 100 at 85°C.

Foreman *et al.*[72] described the GLC determination of a number of alkyl N-nitrosamines. The study was limited to simple nitrosamines which are volatile in steam; the separation procedure used was that of Heath and Jarvis[35]. At the 1 p.p.m. level, dimethyl- and diethylnitrosamines were recovered at 95% and 97% levels, respectively, when added to corned beef. The gas chromatographic method was found to be directly applicable to aqueous samples. (Using microporous polymer beads as a column packing, water is rapidly eluted in a narrow band which does not interfere with the determination of alkyl-N-nitrosamines.) A typical chromatogram is shown in Fig. 17. A lower limit of detection of 1 p.p.m. (corresponding to 10^{-2} μg in a 10 μl

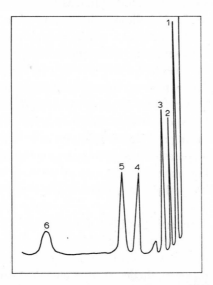

Fig. 17. Chromatogram of mixture of N-nitrosamines. 1, Dimethylnitrosamine; 2, methylethyl-nitrosamine; 3, diethylnitrosamine; 4, dipropylnitrosamine; 5, methylhydroxyethylnitrosamine; 6, dibutylnitrosamine. The gas chromatograph used was a Varian Aerograph 1200 with a 6 ft. × 1/8 in. (o.d.) stainless steel column packed with Chromosorb 101, at 200°C, with a flame-ionization detector. The flow rate was 25 ml/min, chart speed 4 in./h; 5 μl samples were taken.

injection on the column) was achieved using standard solutions, but the practical detection limit for sample extracts would depend upon the degree of purification achieved during the extraction stage. It was suggested that with suitable concentration techniques such as repeated distillation or evaporation of a suitable solvent, amounts of nitrosamine of the order of 0.01 p.p.m. should be detectable.

The GLC determination of dimethylnitrosamine as dimethylnitramine at picogram levels and its application to possible environmental application was described by Sen[73]. Crystalline DMNA (m.p. 52°C) was prepared by the nitrolysis of N,N-dimethylformamide according to the method of Robson[74]. The conversion of DMN to DMNA was achieved by the method of Emmons and Ferris[75] via the oxidation of DMN in methylene chloride by a mixture of trifluoracetic acid–50% hydrogen peroxide (5:4). GLC analysis was performed using a Varian Aerograph gas chromatograph, Model 1200, equipped with an electron-capture detector (^{3}H) and using 6 ft. × $\frac{1}{8}$ in. stainless steel column containing 10% Carbowax 20M on 60–80 mesh Chromosorb W (HMDS-treated). The column, injector and detector temperatures were 152, 225 and 210°C, respectively, and the nitrogen flow was 24 ml/min. DMNA, the oxidation product of DMN, was found to be extremely sensitive to electron capture, e.g. about 8 pg of DMNA (or 16 pg DMN) could be detected under the conditions used (Fig. 18). This is about a thousand times more sensitive than the detection limit of nitrosamines by the hydrogen-flame detector[63, 72]. The peak heights were proportional to the amount of DMNA at least up to 100 pg. Although the study demonstrated that DMN could be oxidized to DMNA on a micro scale and detected by GLC at picogram levels, Sen[73] indicated that proper clean-up procedures needed to be developed and applied to different foods before this procedure could be directly applied to food extracts. However, the technique as described could be used to confirm the identity of nitrosamines isolated from foods by the existing procedures. For example, the isolation of DMN from nitrite-treated fish was prepared and extracted as described by Sen et al.[63] but modified by using macro and micro Snyder columns for all concentration steps. The spot corresponding to DMN on the TLC plate was eluted with methylene chloride using a micro Soxhlet apparatus. Following conversion of DMN to DMNA, the preparation was chromatographed on an alumina column (Woelm, basic), interfering materials removed by washing with n-pentane, the adsorbed DMNA eluted with diethyl ether, eluate concentrated and finally analyzed by GLC (Fig. 18).

Thewlis[76] reported the testing of English wheat flour for the presence of nitrite and nitrosamine and found that neither could be detected in any sample of normal flour or bread examined, nor was nitrosamine detected in flour which had been deliberately exposed to nitrogen dioxide from a chemical generator or from vehicle exhaust gases. Only when a considerable excess of diethylamine was added under slightly acid conditions to the exposed flour and the mixture maintained at 70°C for 1 h could any diethylnitrosamine be detected. For detection, the method of Marquardt and Hedlar[77] was employed using ether extraction, followed by TLC as described by

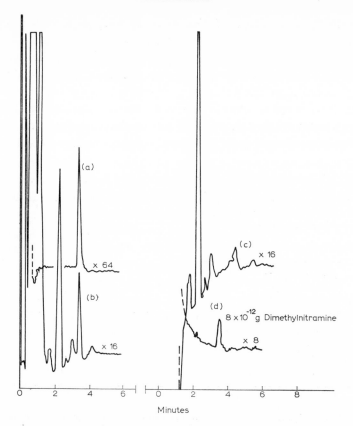

Fig. 18. GLC diagrams. (a) DMNA from nitrite-treated smoked hake (after clean-up through the alumina column); (b) DMNA prepared from DMN by pertrifluoroacetic acid oxidation; (c) pertri-fluoroacetic acid oxidation blank; (d) DMNA prepared by nitrolysis of dimethylformamide. Range setting, 1; attenuator setting as shown in the diagrams

Preussmann et al.[52]. The above findings of Thewlis[76] are contradictory with those of Marquardt and Hedlar[77] regarding the finding of nitrosamines in German flour.

Keybets et al.[78] investigated the possible presence of nitrosamines in nitrite-bearing spinach using the TLC procedure of Preussmann et al.[52] for separation and detection procedures of Daiber and Preussmann[42] (diphenylamine–palladium chloride reagent followed by UV irradiation) and Preussmann et al.[51] (sulfanilic acid–naphthylamine reagent). With this reagent, alkylnitrosamines give reddish-violet spots while aromatic nitrosamines yield green to blue spots with a red border. Within the limits of detection (0.1–0.5 p.p.m.) no nitrosamines were demonstrable in nitrite-bearing spinach. It was, however, possible to detect minute amounts of diethyl-nitrosamine under very extreme experimental conditions, e.g. extremely high diethyl-amine concentrations. It was also found that both the weakly alkaline reaction of spinach and the low concentration of secondary amines are important factors limiting the synthesis of nitrosamines even in spinach with a high nitrite concentration.

Sen *et al.*[79] studied the formation of N-nitrosamines from secondary amines and nitrite in human and animal gastric juice. The *in vitro* formation of diethylnitrosamine (DEN) was demonstrated when diethylamine and sodium nitrite were incubated with gastric juice from rats, rabbits, cats, dogs, and man. Human and rabbit juices (pH 1–2) produced more DEN than did rat gastric juice (pH 4–5); the nitrosation reaction has been shown to occur *in vivo* in cats and rabbits. The yields of DEN were low but the identity of the formed DEN was confirmed by both TLC[56] and GLC[56]. Table 11 depicts the amount of DEN detected in the stomach extracts of animals given

TABLE 11

AMOUNT OF DEN DETECTED IN THE STOMACH EXTRACTS OF ANIMALS GIVEN AN ORAL DOSE
OF DIETHYLAMINE AND SODIUM NITRITE

Species	*Oral dose*			*DEN (μg) detected in stomach extract by*	
	Diethylamine hydrochloride (mg)	*Sodium nitrite (mg)*	*pH of stomach extract*	TLC	GLC
Rabbit	1000	1000	4.5	2000	2000
	450	300	4.8	200	180
	450	300	3.8	100	
Cat	450	300	5.9	68	63

an oral dose of diethylamine and sodium nitrite. Figure 19 illustrates the gas chromatographic identification of DEN from an incubated (20 min at 37°C) mixture of human gastric juice with diethylamine (100 mg) and sodium nitrite (2 mg) at pH 1.9, compared with a reference sample of DEN and a sample of human gastric juice incubated without diethylamine and sodium nitrite.

The above studies of Sen *et al.*[79] confirm the observations of Sander[80] that secondary amines and nitrite, if present in ingested foods, can produce traces of nitrosamine in the human stomach. More recently, Sander *et al.*[81] concluded that the ease of formation of nitrosamines depended greatly on the basicity of the amines. The synthesis of nitrosamines from nitrite and secondary amines in the stomach of rats could be demonstrated only after feeding weakly basic amines such as diphenylamine, N-methylaniline and N-methylbenzylamine, but not with the strongly basic amines such as diethylamine. Sander[82] has also shown that nitrate-reducing bacteria, *e.g.* *Escherichia coli*, *Proteus vulgaris* and *Serratia marcescens* under suitable conditions could nitrosate amines such as diphenylamine, N-methylaniline and di-*n*-propylamine. The isolation and detection of the nitrosamines was accomplished by TLC using the technique of Preussmann *et al.*[52].

Lijinsky and Epstein[83] in their evaluation of nitrosamines as environmental carcinogens, suggest that human cancer might be caused by nitrosamines formed in the body from ingested nitrites and secondary amines and that cooking could be a secondary source of secondary amines. For example, pyrolysis of protein and cooking of protein food might produce free amino-acids such as proline, arginine and hydroxy-proline and nitrosatable secondary amines such as pyrrolidine and piperidine. (The latter amines could be formed when meat or fish are cooked.) The possibility also exists that proline ingested in food could be converted into nitrosoproline by nitrite

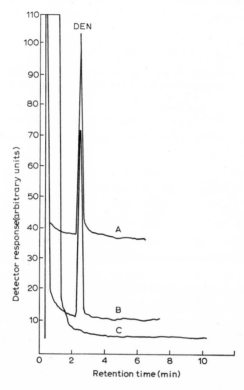

Fig. 19. Gas-chromatographic identification of DEN. (A) 1 μl final extract of incubation mixture containing 22 ml human gastric juice and suitable amounts of diethylamine and sodium nitrite; (B) 2 μg DEN; (C) 2 μl of extract from the gastric juice used for (A) but incubated without diethyl-amine and sodium nitrate. Conditions: 6% coating of Reoplex 400 on 60–80 mesh Chromosorb W in stainless steel column 6 ft. × 1/8 in. i.d.; nitrogen flow rate 29 ml/min; column temperature 125 °C; range setting at 10, attenuator at 4 or at 1 (sample C); Varian Aerograph gas chromatograph, Model 1200, equipped with hydrogen-flame detector and 1 mV recorder.

present in the stomach, and then decarboxylated, possibly bacterially, in the alkaline conditions prevailing in the duodenum and small intestine, yielding the highly carcinogenic nitrosopyrrolidine.

    The TLC and GLC isolation and identification of dimethylnitrosamine from the fruit of a solanaceous bush, *Solanum incanum* (the juice of which is used to curdle milk and provide the chief source of sustenance of the Bantu people in localized areas of the Transkei where there is a high incidence of esophageal cancer) has been reported by Du Plessis et al.[84]. Silica Gel G chromatoplates were used with carbon tetrachloride–dichloromethane (3:2) as developing solvent. Dimethylnitrosamine as well as other nitrosamines were detected by spraying with Preussmann's reagent[51]. For GLC, a Beckman GC-4 instrument was used equipped with dual flame-ionization detectors, a stream splitter and a glass column (1.19 m × 12.5 mm) packed with 15% Carbowax 1000 on Gas Chrom P, 80–100 mesh. The column temperature was 230°C and the carrier gas argon.

The effluent from GLC analysis was further analyzed by NMR and infrared spectroscopy and shown to be dimethylnitrosamine when compared with authentic samples.

It is also of interest to note that McGlashan et al.[85] reported the presence of dimethylnitrosamine (or a similar substance) in concentrations of 1–3 p.p.m. in samples of alcoholic spirits from Zambia. This concentration would be predicted to be carcinogenic in laboratory animals and it was pointed out that geographical studies have linked cancer of the esophagus in Africa to the drinking of locally distilled spirits (Kachasn). Polarography and TLC using Preussmann's reagents[52] were used to detect the presence of nitrosamine in the spirit samples.

Preussmann[16] studied the oxidative breakdown of nitrosamines by enzyme-free model systems. It had been previously shown that nitrosamines are degraded in vivo by drug-metabolizing enzyme, and that diazoalkanes (potent alkylating agents) resulted from this oxidative process. Presuming the first step of this oxidation to be a hydroxylation of nitrosamine at the α-carbon atom, Preussmann investigated whether similar reactions could be effected by simple enzyme-free oxidating model systems. The following systems were studied: (1) water + UV irradiation; (2) hydrogen peroxide + UV irradiation; (3) Fenton's reagent (Fe^{2+}/H_2O_2); (4) alkaline peroxydisulfate and (5) the hydroxylase model of Udenfriend et al.[86] consisting of the Fe^{2+} complex of EDTA, ascorbic acid, and molecular oxygen. It was shown that irradiation of aqueous nitrosamine solutions with UV light resulted in photochemical cleavage of the nitrosamine bond yielding the respective secondary amine and nitrite ion.

$$\begin{array}{c} R_1 \\ \diagdown \\ \diagup \\ R_2 \end{array} N\!-\!N\!=\!O + H_2O \xrightarrow{h\nu} \begin{array}{c} R_1 \\ \diagdown \\ \diagup \\ R_2 \end{array} \overset{+}{N}H_2 + NO_2^-$$

The hydroxylase-model system[86] was shown to exhibit a clear oxidative effect on the nitrosamines tested. Several newly formed reaction products were determined by TLC and characterized by the R_F values. Silica Gel G was used with hexane–ether–methylene chloride (4:3:2) as developer with detection achieved using the procedure of Preussmann et al.[52]. Ethyl-n-butyl- and di-n-butylnitrosamines each yielded five degradation products while ethyl-tert.-butylnitrosamines yielded three products when similarly exposed to the hydroxylase-model system.

The question of the occurrence of N-nitrosamines in tobacco and tobacco smoke is of increasing interest primarily due to the assumed existence of a causal relationship between cigarette smoke and lung cancer incidence. Although Neurath *et al.*[87] claimed that nitrosamines in tobacco smoke are formed in a time-dependent chemical reaction occurring in the vapor phase and after the combustion zone, the possibility cannot be entirely dismissed that the nitrosamines in tobacco smoke originate, at least in part, from the material of the tobacco plant (*e.g.* various secondary amines and nitrate) and because of the known carcinogenic properties of tobacco extracts[88].

Serfontein and Smit[89] reported the occurrence of N-nitrosamines in tobacco. An ether extract of 1 kg of Barley tobacco was extracted with 10% sodium hydroxide and 10% hydrochloric acid to remove acids and bases and the neutral solution concentrated to 10 ml, thence analyzed by GLC using various stationary phases, disclosing the presence of N-nitrosopiperidine. (Although no reliable estimate could be made of the quantities of nitrosamines involved in these experiments, the amounts were believed to be very low.) In subsequent studies, neutral fractions as obtained previously were reduced with lithium aluminum hydride[59, 89], the resultant hydrazines were reacted with 5-nitrosalicylic acid yielding red-colored hydrazones which were analyzed on silica gel columns and the mixture of nitrosamine derivatives then separated by means of TLC on Silica Gel G. In preliminary experiments, methanolic potassium hydroxide was used to reveal "nitrosamine spots". (Subsequently repeated one-dimensional development was used to separate these spots.) Although the nitrosamine zone occupied the same position on the chromatogram as an authentic sample of N-nitrosopiperidine, the spectra of the 5-nitroso-salicylaldehyde hydrazones were not identical. Serfontein and Smit[89] suggested that the results of this study appeared to confirm the earlier views reported by Druckrey[90] that nitrosamines are likely to occur in tobacco.

Evidence for the presence of nitrosamines in tobacco smoke condensate was reported by Serfontein and Hurter[91]. Trace amounts of N-nitrosamines were determined qualitatively and semi-quantitatively by conversion of the nitrosamines into the corresponding asymmetrical hydrazines. The hydrazines were concentrated by extraction techniques and identified by means of TLC after reaction with 4-nitroazo-benzene-4'-carboxylic acid chloride. The Silica Gel G plate (6 g/plate) was developed with chloroform–petroleum ether (b.p. 60–80°C) (9:5). Quantitative measurements were made colorimetrically after the removal and extraction of the yellow-brown colored spots. In multiple two-dimensional experiments, five to seven consecutive developments were required for the satisfactory separation of the colored components. GLC was used for the satisfactory resolution of nitrosamines in authentic synthetic mixtures containing N-nitrosopiperidine, diethylnitrosamine and dimethylnitrosamine, as well as in fractions from the neutral dichloromethane extraction of cigarette tar condensate. The column used was a $\frac{5}{16}$ in. × 150 in. copper spiral packed with 10% polyethylene glycol 4000 on Celite 545 (35–65 mesh, acid-washed); detection was achieved by flame ionization. The injection block, column and detector temperatures were 110, 100, and 105°C, respectively. The carrier gas was nitrogen at 20 ml/min (inlet pressure 1.5 atm, outlet pressure 1.0 atm). Figure 20 illustrates the scheme for the analytical separation and identification of nitrosamines in biologic material. The combined evidence from the above study indicated that at least three nitrosamines

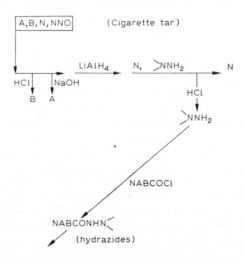

Fig. 20. Scheme for analytical separation and identification of nitrosamine in biologic materials. Separation by means of thin-layer chromatography. A = acids; B = bases, N = neutral compounds; NNO = nitrosamines; 4-nitroazobenzene-4'-carboxylic acid chloride.

may be present in cigarette smoke inhaled by smokers. A rough estimate indicated that approximately 1–5 µg of N-nitrosopiperidine was formed per cigarette under experimental conditions; the amounts of nitrosamines formed appeared to vary widely with different batches of cigarettes.

Neurath et al.[92] identified dibutylnitrosamine in cigarette smoke condensate in a 10–15% yield following its addition to a cigarette. Much of the loss was attributed to side stream smoke or to combustion processes. It was also found that the recovery of dimethylnitrosamine under comparable experimental conditions was much poorer than that of dibutylnitrosamine.

The presence of N-nitrosamine in tobacco smoke has been recently studied by Johnson et al.[93]. Examination of neutral smoke condensates utilizing a highly selective gas chromatographic system[94] (e.g. a detector selective for amine compounds including nitrosamines, and eliminating many classes of nitrogen compounds, in addition to standard hydrocarbons, chlorine and phosphorus-containing compounds) revealed no detectable nitrosamine peaks. From the available literature, it appears that if nitrosamines are present in tobacco smoke under conditions similar to those prevailing in cigarettes smoked by humans, they are present in concentrations not exceeding 0.01 µg/cigarette (for cigarettes not enriched with nitrates). This quantity would be increased somewhat for nitrate-enriched cigarettes.

Schoental and Gibbard[95] investigated the presence of carcinogens in Chinese incense smoke. Several polycyclic aromatic hydrocarbons were separated by thin-layer and column chromatography and identified by UV spectroscopy. A search for nitrosamines in the smoke condensate by the polarographic method of Heath and Jarvis[35] gave negative results, but the TLC procedure of Preussmann et al.[52] and application of the Griess reagent gave a salmon-pink spot, R_F 0.35. The same color was obtained by spraying the plate with sulfanilic acid (1.0% in 30% acetic acid) without the need of previous irradiation and of the second component of the Griess reagent, α-naphthylamine. Nitrosamines do not yield a color with the sulfanilic acid reagent, but aromatic aldehydes give chiefly yellow colors. Schoental and Gibbard suggested that the colors given by aromatic aldehydes with sulfanilic acid be considered when applying the Griess reagent for the detection of nitrosamines according to the procedure of Preussmann et al.[52].

Mohr et al.[96] studied the diaplacental effect of dimethylnitrosamine in the golden hamster. The passage of DMN through the placenta after intracardial injection of the drug was proven by TLC. The TLC facet of this study involved the analysis of ether extracts of liver placenta and fetus homogenates following administration of 0.010–0.030 g of DMN intracardially to hamsters on the 15th day of pregnancy. Silica gel (Merck 254) was used with a solvent system of hexane–ether–dichloromethane (4:3:2) with detection accomplished by spraying with a 5:1 mixture of 1.5% diphenylamine in ethanol and 0.1% palladium chloride in 0.2% sodium chloride; nitrosamine was also detected by UV light at 254 nm (blue fluorescent spot). Other metabolites were found on the chromatograms in addition to DMN.

In addition to the variety of TLC and GLC methods employed for the separation and identification of nitrosamines, other allied techniques have been tried with varying degrees of success.

The separation of N-nitrosamines on Sephadex LH-20 has been described by Eisenbrand et al.[97]. Figure 21 illustrates the gel chromatography of six symmetrical di-n-alkylnitrosamines on Sephadex LH-20 using methanol–water as eluent. The compounds were applied in a sample volume of 0.5 ml, elution was performed at a rate of 4.3 ml/h and fractions of 2 or 3 ml were collected. Nitrosamine concentrations in the fractions were determined by UV spectroscopy at 230–235 nm using a Zeiss DMS 21 recording spectrophotometer. The elution sequence of the methylalkyl-nitrosamine homologs follows the same principle, e.g. methylethyl-eluted before methylbutyl- which is followed by methylpentylnitrosamine. While dialkylnitros-amine homologs which differ in two methylene groups were well separated, the separation of nitrosamines with a difference of only one methylene group was incomplete, e.g. methylbutyl- from methylpentylnitrosamine. Apparently the elution behavior of symmetrical and unsymmetrical nitrosamines under the above conditions is determined primarily by absorption effects, not upon gel filtration separation. Thus the affinity of a given nitrosamine to the Sephadex LH-20 matrix increases with its lipo-

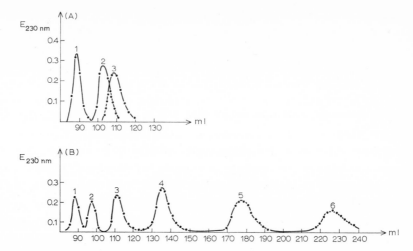

Fig.21. Gel chromatography on Sephadex LH-20. Column, Sephadex LH-20, 100 × 100 mm; eluent, methanol–water (1:1), 4.3 ml/h; sample volume, 0.5 ml. (A) 1 = 19.4 μg methylethylnitrosamine; 2 = 26.4 μg methylbutylnitrosamine; 3 = 25.3 μg methylpentylnitrosamine. (B) 1 = 10.5 μg dimethylnitrosamine; 2 = 14 μg diethylnitrosamine; 3 = 21.5 μg di-*n*-propylnitrosamine; 4 = 50 μg di-*n*-butylnitrosamine; 5 = 74 μg di-*n*-pentylnitrosamine; 6 = 52 μg di-*n*-hexylnitrosamine.

philic character and can be correlated with the distribution coefficient of the compounds in the system *n*-hexane–aqueous buffer[20].

Although mixtures of pure nitrosamines were found to be separable by gel chromatography on Sephadex LH-20 in the study of Eisenbrand *et al.*[97], the method could not be used for the separation of nitrosamines from biological contaminants due to their UV-absorbing nature. However, other analytical techniques for the determination of nitrosamines, *e.g.* GLC, could possibly circumvent this facet of interference.

Liquid–liquid distribution in acetonitrile–heptane as a clean-up method for possible applications for nitrosamine trace analysis has also been investigated by

TABLE 12

DISTRIBUTION COEFFICIENTS[a] FOR NITROSO DERIVATIVES

Compounds	K_D	Compounds	K_D
Dimethylnitrosamine	17.3	Methyl-2-hydroxyethylnitrosamine	32.0
Diethylnitrosamine	8.4	N-NO-morpholine	23.9
Di-*n*-pentylnitrosamine	2.6	N-NO-piperidine	8.8
Dicyclohexylnitrosamine	2.1	N-NO-pyrrolidine	17.5
Di-*n*-octylnitrosamine	0.5	N-NO-trimethylurea	19.6
Methylpentylnitrosamine	8.3		

[a] $K_D = \dfrac{\text{conc. in acetonitrile}}{\text{conc. in } n\text{-heptane}}$

Eisenbrand *et al.*[98]. Distribution coefficients K_D ($K_D = c_{acetonitrile}/c_{n-heptane}$) are given in Table 12. It can be seen that distribution into acetonitrile as lower phase is very favorable. Three distributions were found to extract all nitroso compounds shown to be more than 95% into the polar acetonitrile phase.

2. AZIRIDINE, AZIRIDINE PHOSPHINE OXIDES, SULFIDE, AND MELAMINE

2.1 Ethylenimine (aziridine)

Ethylenimine, available in commercial quantities, is an extremely reactive compound that it (*a*) reacts with many organic functional groups containing an active hydrogen to yield an aminoethyl derivative, or products derived from them and (*b*) undergoes ring-opening reactions similar to those undergone by ethylene oxide. Thus ethylenimine, because of its dual functionality and high degree of reactivity, exhibits actual or potential utility in a broad and expanding range of applications (the majority of which are in polymeric processes) including textiles (crease-, flame-, water- and shrink-proofing), agricultural chemicals (*e.g.* insecticides, chemosterilants (preparation of tepa, metepa, and apholate) and soil conditioners), chemotherapeutics (triethylene melamine), petroleum products and synthetic fuels. The physiological effects[99], toxicology[100], carcinogenicity[100], mutagenicity[101, 102] and metabolism[103] of ethylenimine have been described. The analysis of ethylenimine has been achieved by titrimetry[104, 105] and colorimetry[106-108].

The metabolic fate of ^{14}C-ethylenimine in the rat following intraperitoneal administration was studied by Wright and Rowe[103]. The purity of the labeled material was determined by GLC and colorimetric analysis according to the procedure of Epstein *et al.*[106]. Paper chromatography utilizing Whatman No. 1 paper with *n*-butanol–acetic acid–water (4:1:5) as well as liquid scintillation were used for the analysis of blood, total tissues, urine, and expired air. Radioactivity was widely distributed with some accumulation found in the liver, intestines, cecum, spleen, and kidneys. About half of the administered dose was excreted in the urine (a small amount of ethylenimine was excreted as such, but the major portion of the excreted radioactivity was found in a number of unidentified products). Three to five percent of the dose was expired as CO_2 and 1–3% was expired as a volatile basic material, probably ethylenimine. After 24 h, all the remaining radioactivity in the rat was firmly bound or incorporated into tissues with a very slow turnover and thus was essentially unavailable for further metabolism.

The determination of ethylenimine and its impurities by GLC has been described[109]. Up to 0.001% of monoethanolamine in ethylenimine was determined using an LKLM-7 gas chromatograph equipped with a flame-ionization detector, a 160 cm × 0.2 cm stainless steel column filled with 5% polyethylene glycol 4000 on Chromosorb P (0.15–0.23 mm grain size) at 100°C with nitrogen as carrier gas. Water (up to 0.005%) in ethylenimine was determined on the same apparatus equipped with a katharometer and a 160 cm × 0.2 cm column loaded with 12% polyethylene glycol 400 on Chromosorb W at 85°C.

DiLorenzo and Russo[110] described the gas chromatographic analysis of a number of aliphatic imines and amines. A Perkin-Elmer F-6 gas chromatograph was used equipped with a hot-wire detector and a 200 cm × 4 mm stainless steel column coated with 20% UCON LB-550-X on Chromosorb P (60–80 mesh) and subsequently alkanized with 20% potassium hydroxide. The injection block temperature was fixed at 200°C; the carrier gas was helium at a flow rate of 75 ml/min.

2.2 Aziridines (tepa, metepa, apholate and thiotepa)

Tepa (tris(1-aziridinyl)phosphine oxide) prepared by the reaction of ethylen-imine and phosphorus oxychloride in base, has been the most extensively employed of the aziridinyl phosphine oxides. Its industrial applications include the flame-proofing of textiles, preparation of water-repellant, wash and wear, crease-resistant fabrics, the treatment of paper and wood and in dyeing and printing, adhesives and binders. The above utility of tepa (as well as other aziridines) is illustrative of the ability of these compounds to act as cross-linking agents for polymers containing active hydrogen groups such as carboxyl, phenol, sulfhydryl, amide and hydroxyl. This can be represented schematically as follows with ZH representing the active hydrogen group in the polymer chain.

Interest in tepa and related compounds has been intensified in recent years as a result of the discovery of their activity as insect chemosterilants (antifertility agents) and their potential utility as new and powerful tools for insect control and eradication. To date the chemosterilants have proven effective in the control of houseflies, weevils, gypsy moths, Japanese beetles, Mexican fruit flies and codling moths in selected experimental areas. However, inherent in such programs of eradication is the dispersion of large numbers of treated insects into the environment. Hence, there is an obvious need in elaborating how much active chemosterilant remains on and in the insect that is to be released and how long it will persist.

Their chemistry[111-113], chemosterilant application[111, 112], toxicity[114-116], as well as their mutagenicity[117-119], has been described. The analyses of the aziridinyl chemosterilants is primarily performed using colorimetric techniques[106, 120] based on the reaction of the alkylating agent with 4-(p-nitrobenzyl)pyridine under acidic conditions and subsequent formation of the colored product with alkali.

The stability of ethylenimine and the chemosterilants tepa, apholate, metepa and tretamine was investigated by Beroza and Boŕkovec[121] using titrimetric, colorimetric, TLC and NMR spectroscopic techniques. A marked vulnerability of the aziridine chemosterilants to even mildly acidic or neutral solutions gave a stable degradation product which was identified as ethylenimine with the aid of NMR. Regulation of the pH of insect diets resulted in a marked increase in activity of added chemosterilants. TLC on Silica Gel G utilized buffer solutions (MacIlwaine's and Sorensen's) of pH 3.0, 5.0, and 7.5. Detection was accomplished with iodine crystals or, preferentially, 0.1 M aq. potassium hydrogen phthalate followed by 5% acetone solution of 4-(p-nitrobenzyl)pyridine (NBP); after heating the treated plates at 110°C for 30 min, they were allowed to cool and then sprayed with 1 M aqueous potassium carbonate (compounds containing the aziridine moiety are revealed as blue spots).

The fate of uniformly labeled ^{14}C-tepa in male houseflies was studied by Chang et al.[122]. The homogeneity of the tagged chemosterilant was indicated by a single radioactive spot on Whatman No. 1 paper developed with acetonitrile–water (4:1). Radiometric and colorimetric determinations indicated that flies treated with 1 μg ^{14}C-tepa/fly, retained about 50% of the dose 5 h after treatment and 9% over prolonged periods (48–72 h).

Collier and Tardif[123] described the determination of tepa (to 0.5 μg) in gypsy moth *Porthetria dispar* L using column chromatography of moth extracts on silica gel followed by colorimetric analysis according to the procedure of Chang and Boŕkovec[124]. TLC[121] performed on a typical clean-up sample indicated the complete absence of any alkylating type of compounds other than tepa. The studies indicated that the amount of tepa residues in moths depended on the holding time after treatment (the residues diminishing with increasing holding times). All residue data were determined on moths which had been exposed for a period of 8 h to residual films of 1 mg/dm^2 of tepa (this treatment is currently considered the most practical for producing sterile moths for field release).

The sterilizing activity of tepa and metepa solutions (buffered to pH 6–8) injected into male houseflies, *Musca domestica* L was determined by Boŕkovec et al.[125]. The degradation of aqueous and acidic solutions of tepa and metepa was followed by TLC and bioassay. At zero time, tepa and metepa solutions yielded single spots with R_F values of 0.53 and 0.60, respectively. After acidification (30 min), each compound showed an additional new spot. With tepa this new spot was ethylenimine resulting from hydrolytic cleavage of the P–N bond, and analogously with metepa the spot corresponded to 2-methylaziridine. In the course of degradation (30–180 min), the tepa and metepa spots gradually faded away and finally (90 min for tepa and 120 min for metepa) the chromatograms indicated the disappearance of the original compounds. The results indicated a similarity in the degradative pathways of the two chemosterilants. It was concluded that in partially degraded solutions, the sterilizing activity was proportional to the contents of *intact* tepa or metepa rather than to the total contents of aziridine function.

Maitlen and McDonough[126] investigated the residues of tepa on sterilized codling moths *Carpocapsa pomonella* L after aerolization at initial levels of 4.5 and 22.8 μg/moth. The purity of tepa was determined by TLC as described by Beroza and Bořkovec[121], except that Eastman Cromatogram sheets, type K301R (silica gel) were used. Residues of tepa were determined by a modification of the colorimetric method of Epstein *et al.*[106] as well as by TLC. The loss of tepa after 72 h was 88% and 93% for the high-level and low-level treatments, respectively.

The amounts and persistence of a number of chemosterilants in treated insects was investigated by Bowman and Beroza[127] utilizing GLC, TLC[121] and radiometric techniques. An F&M Model 700 gas chromatograph equipped with a Melpar flame photometric detector[128] was used with a 526 mμ interference filter to detect compounds containing phosphorus and a 394 mμ filter to detect those containing sulfur. A glass column (6 mm o.d., 4 mm i.d. × 45 cm) packed with 1% Carbowax 20M on 80–100 mesh Gas Chrom Q was used for the analysis of chemosterilants other than apholate while 5% DC-200 on 80–100 mesh Gas Chrom Q was used for apholate. Table 13 depicts the GLC analysis of 5 ng of the chemosterilants (tepa, metepa, thio-

TABLE 13

GAS CHROMATOGRAPHIC ANALYSIS OF 5 NANOGRAMS OF FIVE CHEMOSTERILANTS
WITH THE 526 mμ INTERFERENCE FILTER

Chemosterilant	Temperature (°C)			Retention time (min)	Peak height for 5 ng (A)
	Oven	Injection port	Detector[a]		
Hempa	100	170	140	1.40	8.0×10^{-8}
Methiotepa	110	170	140	1.20^b	4.0×10^{-8}
Metepa	110	170	140	2.70	2.9×10^{-8}
Tepa	110	170	140	4.10	2.1×10^{-8}
Tepa	125	170	140	1.90	4.4×10^{-8}
Apholate	200	220	200	2.00	4.0×10^{-8}

[a] External temperature.
[b] At 100 °C oven temperature, the retention time of methiotepa is 5.15 min.

tepa, methiotepa, apholate and hempa). The GLC method was found to be sensitive to as little as 0.1 ng, however, and also be directly applicable to the analysis of methanol extracts of insects fed tepa. Typical chromatograms of the sterilants are shown in Figs. 22 and 23. As shown in Fig. 23, 5 ng of tepa produces a substantial peak that is readily visible above the background interference of a raw methanol extract equivalent to 1.25 mg of the moth.

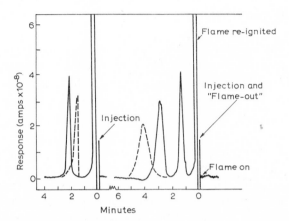

Fig. 22. Typical chromatograms of chemosterilants analyzed in 4 µl of absolute methanol with the 526 mµ filter. Right (column temp. = 110°C): solid line, 5 ng methiotepa (1.20 min) and 5 ng metepa (2.70 min); dotted line, 5 ng tepa (4.10 min). Left: solid line, 5 ng apholate (2.00 min, column temp. = 200°C); dotted line 2 ng hempa (1.40 min, column temp. = 100°C).

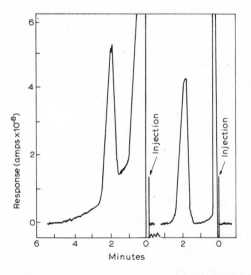

Fig. 23. Chromatograms of 5 ng of tepa with the oven temperature at 125°C. Right, injected in 5 µl of absolute methanol; left, injected in 5 µl of a methanol extract of female fall army worm moths without clean-up. The 5 µl sample contained 1.25 mg equivalents of the moths.

TLC was used to check the purity of the above chemosterilants as well as ^{14}C-tepa. Eastman TLC sheets (Type K301R, silica gel) were conditioned overnight at 110°C, cooled in a desiccator before use, and developed in absolute methanol (10 cm). Detection of the aziridines was accomplished with iodine crystals.

The persistence of tepa in male and female army worm moths, *Spodoptera frugiperda*, J.E.Smith, was determined by radiometric and GLC procedures[127] by Cox *et al.*[129]. The radiometric analyses showed considerably more tepa on and in the insects than did the specific GLC analyses, an indication that ^{14}C-labeled fragments of the tepa molecule were being analyzed as tepa. It was found that within 24 h, more than 90% of the tepa disappeared from moths that ingested as much as 100 µg/moth and that 95% disappeared within 48 h. However, it was determined that the recovery of ^{14}C (extractable tepa) diminished with time after 96 h. This phenomenon may have occurred because the material became polymerized or conjugated and remained in the homogenate of the insect tissue and/or excretor, or it may have been caused by volatilization.

The rates of absorption, degradation and excretion of ^{32}P-metepa (tris-(2-methyl-1-aziridinyl)phosphine oxide) in mosquitoes, houseflies and mice have been reported by Plapp *et al.*[130]. The purity of ^{32}P-metepa as well as its excretory products from both houseflies and mice were characterized by paper chromatography using untreated Whatman No. 1 paper with acetonitrile–water–ammonium hydroxide (80:20:2) in a 20-cm development. Known samples of metepa, as well as its suspected major metabolite phosphoric acid (sampled as potassium phosphate), were chromatographed along with all radioactive samples. In both larvae and adults of the mosquito *Culex tarsalis* Coquillet degradation of metepa was complete within 48 h after administration. Adult houseflies, *Musca domestica* L, degraded 50% of large doses of metepa within 2 h. The rates of degradation were similar in a susceptible fly strain and in two organophosphate resistant strains. This suggested that the degradation of metepa was carried out by enzymes different from those responsible for resistance to the organophosphate and suggested that phosphoramide chemosterilants would be equally effective against both resistant and susceptible strains. In mice, the observed rate of degradation was comparable to that in houseflies.

Chamberlain and Hamilton[131] studied the absorption, excretion and metabolism of ^{32}P-labeled metepa by screw-worm (*Cochliomyia hominivorax* Coquerel) and stable flies (*Stomoxys calcitrans* L) to determine whether a species difference could explain the great difference in dosage required for sterilization[132]. Sterilization of the screw-worm required 3.9–18 times more metepa/g of insect than the amounts needed for the stable fly. Similar differences were noted for apholate by Chamberlain[133] and Harris[134]. On the basis of data obtained 6 h after treatment, the screw-worm fly absorbed only half as much radio-labeled material in proportion to its size as the stable fly. Excretion by the screw-worm fly was twice that of the stable fly. The principal metabolite was phosphoric acid and it occurred in greater quantity in the screw-worm fly.

Three ascending paper chromatographic systems were used, *i.e.* (*1*) mobile phase of acetonitrile–water–ammonium hydroxide (80:20:2) and a stationary phase of uncoated Whatman No. 1 paper, (*2*) Whatman No. 1 paper coated with a solution of 5% silicone 550 in acetone used with a mobile phase of dimethyl formamide–water (75:25), and (*3*) the same stationary phase as (*2*), and water–ethanol–chloroform (60:40:2) as mobile phase. Radiochemical purity was determined by use of the systems (*1*) and (*2*) above. The first system was used primarily for the analysis of the homogenates, extracts and residues. The third system was used for ancillary confirmation of compound homogeneity. Localization of radioactive areas on the chromatograms and their quantitative evaluation were determined with X-ray film and by counting sections of the chromatogram as previously described by Chamberlain *et al.*[135].

Thiotepa (tris(1-aziridinyl)phosphine sulfide) is prepared *via* the reaction of ethylenimine and $PSCl_3$ in the presence of base. In addition to its use as a chemosterilant, it has demonstrated clinical utility in the temporary palliation of certain cancers (malignant breast cancer, Wilm's tumor, chronic lymphatic leukemia, ovarian carcinoma) and Hodgkin's disease. Thiotepa also has patented applicability in dyeing, flame-proofing and water-proofing of textiles as well as in the stabilization of polymers and photographic emulsion hardening.

Parish and Arthur[136] studied the metabolic fate of ^{32}P-thiotepa in rats and 4 species of insects, the German cockroach, *Blatella germanica* L.; housefly, *Musca domestica* L.; stable fly, *Stomoxys calcitrans* L. and boll weevil, *Anthonomus grandis* Boheman. Tepa, the oxygen analog of thiotepa, was the only chloroform-soluble metabolite recovered from insects. In German cockroaches, houseflies, and stable flies, the amount of thiotepa decreased with time after topical treatment; the opposite was true for boll weevils. During the 5-day experimental period, 24.3% of the administered thiotepa was eliminated in the urine of orally treated rats and 27.4% was eliminated in the urine of dermally treated rats. The orally and dermally treated rats eliminated 17.2 and 34.7% in the feces, respectively. From 54 to 91% of the ^{32}P materials eliminated in the urine and feces of rats were hydrolytic products. Chloroform-soluble ^{32}P materials recovered from the urine were 90% tepa from orally treated rats and about 50% tepa from dermally treated rats. Tepa was also the predominant product recovered in the chloroform extracts of feces from orally or dermally treated rats. Column chromatography on silica gel and anion-exchange chromatography were used to examine the urinary metabolites[137].

Benckhuijsen[138] studied the behavior of thiotepa (I) in acid medium and presented evidence of an intramolecular alkylation of the sulfur on protonation of the molecule. The 5-membered ring (III) formed in this manner was hydrolyzed slowly in neutral solution at room temperature liberating an SH group. In the SH derivative (IV), 2 ethylenimine rings were retained, which were demonstrated to be more reactive towards water and sulfhydryl than were those of thiotepa. The sequence of above

reactions can be depicted

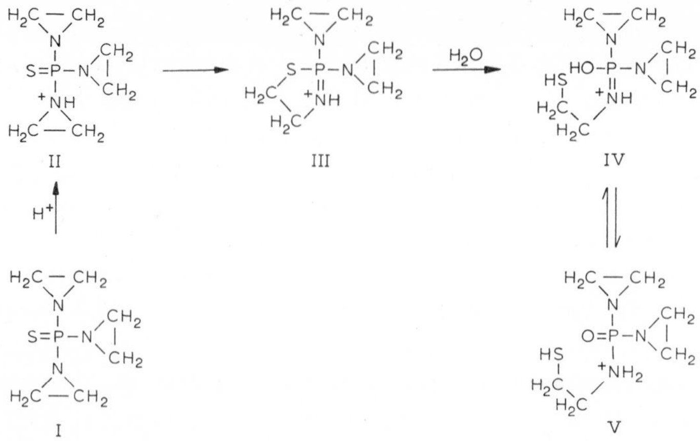

The isomerization of thiotepa *via* II to III was postulated to occur by the following mechanism.

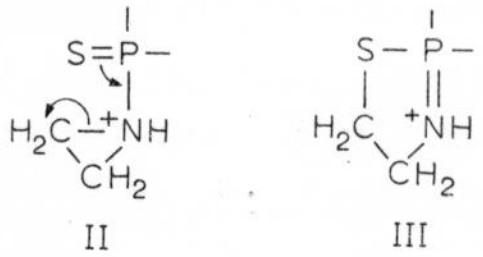

By thin-layer chromatography in three different solvent systems it was shown that a thiotepa solution kept at pH 1.1 for 5 min or for 24 h gave rise to a single spot with alkylating properties. The R_F values of these spots differed from those of thiotepa and ethylenimine. Table 14 depicts the thin-layer chromatography of alkylating products derived from thiotepa on solution in 0.1 N acid.

TABLE 14

TLC[a] OF ALKYLATING PRODUCTS DERIVED FROM THIOTEPA ON SOLUTION IN 0.1 N ACID

	R_F value			
	Thiotepa after 5 min at pH 1.1[b]	Thiotepa after 24 h at pH 1.1	Thiotepa	Ethylenimine
Propanol–water (70:30)	0.96	0.96	0.90	0.57
Butanol (water sat'd.)	0.97	0.95	0.89	0.09
Acetone–water (70:30)	0.97	0.97	0.95	0.63

[a] Layers of MN 300 cellulose powder (Macherey, Nagel) (0.25 nm); alkylating products detected with 5% solution of 4-(4-nitrobenzyl)pyridine in acetophenone, heating the plates for 15 min at 110°C, followed by alkalinization in ammonia vapor.
[b] Concentration of thiotepa, 6 mM, temperature 220°C.

Mellett and Woods[139] studied the comparative physiological disposition of thiotepa and tepa in the dog. After intravenous or oral administration of thiotepa to dogs, the presence of tepa in the urine was demonstrated by both spectrofluorometry and paper chromatography. Whatman No. 1 chromatograms were developed with two solvent systems, (a) n-butanol saturated with 1% ammonia and (b) acetone–water (80:20). Detection was achieved with 0.5% ninhydrin solution in n-butanol and heating for several hours at 45°C. The R_F values for thiotepa and tepa (the only spots obtained from urine extracts) were n-butanol and ammonia, thiotepa 0.84, tepa 0.68; acetone–water, thiotepa 0.80, tepa 0.90. Tissue distribution of thiotepa could not be determined because the drug was almost completely destroyed when in contact with tissue. Tepa, on the other hand, could be determined and displayed a marked affinity for bone marrow as compared with other tissues. The stability of thiotepa and tepa in various media was also investigated. No significant difference was found between the two drugs upon incubation in plasma, whole blood, or various buffers. Both drugs were stable in alkaline medium (pH 8.4) and were rapidly destroyed in acid medium (pH 4.2).

Mellett et al.[140] studied the fate of ^{14}C-labeled thiotepa in the dog after intravenous administration and the absorption and the fate of the drug in the human after oral and intravenous administration. Paper chromatography using n-butanol saturated with 1% ammonia in descending system was used for urinary analysis. Controls of labeled and unlabeled thiotepa treated in the same manner as the urine samples, were always determined on the same chromatogram.

Chromatographic studies of the urine of dogs receiving ^{14}C-thiotepa intravenously, confirmed the presence of tepa. (The chromatograms of the urine samples through the first 6 h displayed a distinct radioactive peak at an R_F value of 0.65 which corresponded to the ninhydrin-positive spot of pure unlabeled tepa on the same paper and a slight but distinct peak at an R_F value of 0.89 corresponding to pure unlabeled thiotepa.) The results in the dog described in the above study utilizing ^{14}C-labeled thiotepa agreed precisely with the results obtained by the fluorimetric method of Mellett and Woods[139] (where 8–13% of a 3.0 mg/kg dose of unlabeled thiotepa was excreted as tepa) and the isotope method of Craig et al.[141] using ^{32}P-labeled thiotepa.

The latter authors had shown that the manner in which thiotepa is metabolized was dependent upon the species studied. For example, thiotepa was metabolized in part to tepa in the rat, dog and rabbit, whereas the mouse converted the drug entirely to inorganic phosphate. The study of Mellett et al.[140] above, indicated that the human excretes only trace amounts of tepa after oral thiotepa administration.

The metabolic fate of ^{32}P-labeled thiotepa has been examined in the rat, mouse, dog, and rabbit by Craig et al.[141]. Single-dimension ascending paper chromatography on Whatman No. 1 paper was used for the examination of ^{32}P-thiotepa homogeneity as well as for the analysis of plasma and urine samples. The main solvent system employed was freshly prepared butanol–dioxan–2 N ammonia (4:1:1). The distribution of radioactivity on chromatograms was accurately determined with a multicounter system[142] as well as by contact autoradiography using Ilford X-ray film. The above studies revealed that thiotepa was metabolized rapidly *in vivo*. The primary stage was the rapid replacement of sulfur by oxygen with the formation of tepa. The salient differences in species metabolism were as follows: (*1*) in the rat, tepa was the main metabolite, (*2*) the dog and rabbit excreted three other metabolites, although the major product was tepa, and (*3*) the mouse was exceptional in that thiotepa was degraded so that the principal radioactive metabolite excreted was in organic phosphate, although the plasma analysis indicated that the first stage is the production of tepa. The phosphorus label provided no clue to the fate of the aziridine rings, although earlier work in the mouse with ^{14}C-ethylenimino-labeled tepa suggested a general breakdown rather than a detoxification by alkylation of specific substances[143].

It has been shown that alkylating compounds such as the aziridines, epoxides, and nitrogen mustards can react with nucleic acids and protein and that cellular growth, especially in rapidly growing tissue, is thus inhibited or altered. The nucleic acid synthesis in normal viable and non-viable housefly eggs deposited by flies chemosterilized by apholate and thiotepa has been studied by Painter and Kilgore[144]. During normal embryonic development there was a rapid many-fold increase in DNA after an initial long period. In the chemosterilized eggs there is almost no increase in DNA but some accumulation of the deoxyribosidic components of the acid-soluble extract. When ^{32}P-thiotepa was the chemosterilant, all components of the sodium RNA were labeled. This suggests that thiotepa may have been degraded and the resulting inorganic phosphate re-incorporated into the RNA. There was no evidence of alkylated guanine in the RNA. Apholate-sterilized eggs were found to be significantly lower in adenylic acid than normal fly-egg RNA. The RNA of thiotepa sterilized eggs was a little lower in guanylic acid and contained an unidentified compound not present in normal egg RNA. The presence of ^{32}P in the nucleotides suggests that when insects are sterilized with metepa, tepa and thiotepa, most of the sterilizing compound is rapidly excreted unchanged but one of the degradation products, probably as inorganic phosphate, is re-incorporated into the normal metabolites.

Chromatography utilizing Whatman 3M paper with (*a*) methanol–ethanol–conc. HCl–water (50:25:6:19) and (*b*) isopropanol–conc. HCl–water (170:44:25) as solvent systems and UV detection was used to elaborate acid-hydrolyzed RNA from thiotepa-fed flies. Radioaudiograms were also made of the chromatographs on Kodak Royal Blue Medical X-ray film.

2.3 *Triethylenemelamine*

Triethylenemelamine (2,4,6-tris-(1-aziridinyl)-*s*-triazine) (TEM) is prepared from ethylenimine and cyanuric acid, and is used in the manufacture of resinous products, as a cross-linking agent in textile technology, the finishing of rayon fabrics, and in the water-proofing of cellophane. Major interest, however, has been related to its medical utility as an antineoplastic agent and more recently its use as a chemosterilant for the housefly *Musca domestica*, screw-worm and oriental melon and Mediterranean fruit flies. The clinical and biological effects[145, 146], toxicity[145, 147] and mutagenicity[117, 148-151] have been described.

Similar to all aziridine chemosterilants, TEM is extremely susceptible to moisture and acidic conditions (at pH 3.0, degradation is complete and occurs almost immediately, whereas minor degradation occurs in buffered or unbuffered solutions at pH 7.5). Crystalline TEM stored at room temperatures polymerizes to an inactive material. The metabolism of both triazine-[14]C- and ethylenimino-^{14}C–TEM in normal and tumor-bearing mice[152, 153] and man[143] has been studied.

Following a single large dose (2 mg/kg) about 80% was excreted in 24 h. Chromatographic examination of the urine revealed only one radioactive metabolite, cyanuric acid, and no unchanged drug. The presumption was made that the ethylenimino groups were removed by a metabolic process and it was suggested that the biological activity of TEM might be related to a small portion of intact drug retained within the body. The rapid elimination of this drug is in agreement with previous experiments which suggested that its effect is short-lived, like that of nitrogen mustards. Paper and ion-exchange chromatography were utilized for the elaboration of radioactive homogeneity of the tri-^{14}C-ethyleniminotriazine as well as for the separation and identification of urinary metabolites following intraperitoneal or intravenous administration of the tagged drug[143]. Chromatography on Whatman No. 1 paper revealed a single spot (R_F 0.64) in *n*-butanol saturated with 1% ammonia and R_F 0.85 in acetone–water (3:2) solvent systems. These values corresponded to those obtained with triazine-labeled ^{14}C-TEM. For the separation and identification of urinary metabolites, Whatman No. 1 chromatograms were developed with the following solvent systems: (*1*) methylethyl ketone–propionic acid–water (15:5:6), (*b*) *n*-butanol–95% ethanol–water (4:1:1), (*c*) phenol saturated with an aq. solution containing 6.3% sodium citrate and 3.7% sodium dihydrogen phosphate, and (*d*) methanol–glacial acetic acid–water (19:2:1). The ion-exchange chromatographic procedure was adapted from Moore and Stein[154].

The radioactivity found in blood samples of mice and rats administered the ethylenimino-labeled ^{14}C-TEM intravenously indicated that over 90% of radioactivity was removed from the blood within a few minutes after injection. Between 68 and 73% of the injected radioactivity was excreted in the urine within 24 h and only 4–6% in the next 24 h, with chromatographic separation revealing at least 16 metabolites. There appeared to be some conversion of the ethylenimino moiety of TEM to two-carbon fragments as evidenced by the appearance of labeled creatinine in the urine. It was concluded that the unidentified urinary metabolites which accounted for the major portion of the excreted radioactivity were alkylated derivatives of normal break-down products resulting from metabolic cleavage after TEM combined with an as yet unidentified constituent of the cells or organs.

3. EPOXIDES (ETHYLENE AND PROPYLENE OXIDES AND THEIR DEGRADATION PRODUCTS)

3.1 Ethylene oxide

The general considerations of ethylene oxide that are germane include toxicity[155], pharmacological and physiological properties[156], mode of action[157], mutagenicity[158,159] and induction of chromosome aberrations[160]. Analyses of ethylene oxide have been achieved *via* titrimetric[161] and colorimetric[162, 163] methods.

Ethylene oxide is produced on an enormous industrial scale by the action of alkali on ethylene chlorohydrin (a mutagen)[164] or by catalytic oxidation of ethylene in air. Epoxides such as ethylene and propylene oxides owe their industrial importance to their high reactivity which is due to their ease of opening of the highly strained three-membered ring. A host of industrially important chemicals are synthesized from ethylene oxide, *e.g.* ethylene glycol, diethylene glycol, dioxan, carbowax, methyl carbital, ethylene chlorohydrin, monoethanolamine, acrylonitrile, and surface active agents. (Figure 24 shows some commercially important derivatives of ethylene oxide.)

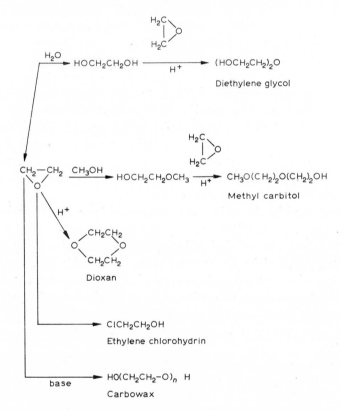

Fig. 24. Some commercially important derivatives of ethylene oxide.

Ethylene oxide also has utility as a solvent and a plasticizer in combination with other chemicals and in the production of high-energy fuels, diverse plastics, textile auxiliaries and hydroxyethylated cellulosis fibers and starch. The area of utility which is more important in terms of residues of toxic and mutagenic agents is that of gas sterilization and fumigation. All of the compounds most actively employed in gaseous sterilization, *e.g.* ethylene oxide, propylene oxide, β-propiolactone, formaldehyde, and methyl bromide, are alkylating agents. (The mode of action upon microbes appears to be a non-specific alkylation of such chemical groups as –OH, –NH– and –SH with a loss of a hydrogen atom and the production of an alkyl hydroxyethyl group.) The use of ethylene oxide for sterilization has thus raised a number of significant questions regarding (*a*) the possible entrapment and/or interaction of ethylene oxide in plastic, pharmaceuticals or food which may then exert a toxic effect when placed in contact with living tissue and (*b*) the effect of sorbed ethylene oxide on the possible changes in the physical and chemical properties of the medical plastics *per se*. The literature is replete with examples of not only trace amounts of ethylene oxide entrapped for extended periods of time in plastic devices, drugs, foods, undergoing sterilization and/ or fumigation, but chemical interactions of this alkylating agent with a number of the reactants as well, *e.g.* antibiotics, steroids, plasticizers and vulcanization accelerators of plastics and rubber devices. Residual ethylene oxide might produce a toxicity from the following sequences of oxidation: epoxide $\rightarrow$ glycol $\rightarrow$ glyoxal $\rightarrow$ glyoxalic acid $\rightarrow$ glycolic acid $\rightarrow$ oxalic acid. The area of greatest concern is both in residual epoxides and the formation of toxicants resulting in the fumigation of diverse food-stuffs (*e.g.* wheat, glazed fruits, gums, processed nut meats, dried prunes, and pro-cessed spices and starches) with ethylene and propylene oxides.

Heuser and Scudamore[165, 166] described methods for the cold solvent extraction and analyses of traces of the fumigants ethylene oxide and methyl bromide present in flour and wheat after treatment (as well as fumigant residues such as ethylene chloro-hydrin produced *in situ* by naturally occurring inorganic chloride with ethylene oxide); and for determining the efficiency of extraction in which a combination of gas chro-matographic and chemical techniques were used.

A Perkin-Elmer 452 gas chromatograph equipped with a flame-ionization de-tector and a stainless steel column (2 m × 4.6 mm i.d.) was used containing 15% poly(propylene glycol) ("Ucon" Oil LB-550-X) on 60–80 mesh Chromosorb W. The injection block, column and detector temperatures were 125, 85, and 125°C, respec-tively, and dry helium was the carrier gas at 80 ml/min. Initial exposure of flour to ethylene oxide for periods of 1–6 h, as well as with periods of preliminary aerations from $\frac{3}{4}$–2 h and main aeration from 3–4 h, resulted in recoveries of free ethylene oxide of 95.0–95.8% with a lower detection limit of about 0.3 p.p.m., hence suggesting *in situ* residues of ethylene oxide reaction products of approximately 4–5%.

The gas chromatographic determination of 34 fumigant gases including ethylene oxide has been described by Berck[167]. An F&M Model 500 linear temperature-programmed gas chromatograph with a four-filament tungsten thermal conductivity detector was used. The operating conditions were: 6 ft. × ¼ in. o.d. stainless steel column packed with 10% SE-30 on 60–80 mesh Diatoport S; the column temperatures were 50–180°C, depending on the gas investigated; the injection port and detector-block temperatures were 265 and 285°C, respectively. Helium was the carrier gas at a flow rate of 50 cc/min. The absolute retention time of ethylene oxide chromatographed

TABLE 15

ABSOLUTE AND RELATIVE RETENTION TIMES OF 34 FUMIGANTS[a]

No.	Fumigant	Formula	Column temp. (°C)	Abs. t_r (min)	Rel. t_r[b]
1	Methyl chloride	CH_3Cl	60	0.84	0.71
2	Methyl bromide	CH_3Br	60	1.00	0.85
3	Ethylene oxide	$(CH_2)_2O$	80	0.85	0.89
4	Phosphine	PH_3	75	0.70	0.71
5	Hydrogen cyanide	HCN	75	0.88	1.11
6	Methylene dichloride	CH_2Cl_2	60	1.32	1.12
7	Carbon disulfide	CS_2	60	1.52	1.29
8	Chloroform	$CHCl_3$	60	2.22	1.89
9	1,1,1-Trichloroethane	$CCl_3 \cdot CH_3$	60	2.40	2.04
10	Ethylene dichloride	$CH_2Cl \cdot CH_2Cl$	60	2.67	2.27
11	Carbon tetrachloride	CCl_4	60	3.17	2.70
12	Propylene dichloride	$CH_2Cl \cdot CHCl \cdot CH_3$	60	3.55	3.02
13	Ethyl bromide	C_2H_5Br	70	1.17	1.11
14	2-Bromopropane	$CH_3 \cdot CHBr \cdot CH_3$	70	1.55	1.48
15	3-Bromopropene	$CH_2:CH \cdot CH_2Br$	70	1.80	1.71
16	1-Bromopropane	C_3H_7Br	70	1.93	1.84
17	2-Bromobutane	$C_2H_5 \cdot CHBr \cdot CH_3$	70	2.63	2.51
18	Trichloroethylene	$CHCl:CCl_2$	70	2.74	2.61
19	Chloropicrin	$CCl_3 \cdot NO_2$	80	(3 peaks)	c
20	1-Bromobutane	$C_3H_7 \cdot CH_2Br$	80	2.58	2.71
21	1-Bromopentane	$C_4H_9 \cdot CH_2Br$	100	2.38	2.83
22	Ethylene dibromide	$CH_2Br \cdot CH_2Br$	100	2.57	3.05
23	Tetrachloroethylene	$CCl_2:CCl_2$	100	2.63	3.13
24	Acrylonitrile	$CH_2:CHCN$	110	0.92	1.21
25	2-Bromopentane	$C_3H_7 \cdot CHBr \cdot CH_3$	110	2.28	3.02
26	Chlorobenzene	C_6H_5Cl	125	1.88	2.64
27	Bromoform	$CHBr_3$	125	2.33	3.28
28	sym.-Tetrachloroethane	$CHCl_2 \cdot CHCl_2$	125	2.57	3.70
29	1,3-Dibromopropane	$CH_2Br \cdot CH_2 \cdot CH_2Br$	125	2.78	3.90
30	Bromobenzene	C_6H_5Br	125	2.85	4.00
31	β,β'-Dichloroethyl ether	$(C_2H_4Cl)_2O$	125	3.03	4.26
32	p-Dichlorobenzene	$C_6H_4Cl_2$	125	4.05	5.68
33	o-Dichlorobenzene	$C_6H_4Cl_2$	125	4.20	5.90
34	Hexachlorobutadiene	$CCl_2:CCl \cdot CCl:CCl_2$	180	2.68	4.32

[a] 2 μg of fumigant; instrument attenuation = 1 × ; flow rate = 50 ml helium/min.
[b] Relative to n-pentane = 1.00.
[c] Three peaks for chloropicrin at 0.90, 2.75, and 3.62 min abs. t_r.
t_r = Retention time.

at a column temperature of 80 °C was 0.85 min. Table 15 lists the absolute and relative retention times of the 34 fumigants and Fig. 25 illustrates the elution profiles of the fumigants at the parameters described in the text.

Ben-Yehoshua and Krinsky[168] described the gas chromatography of ethylene oxide and its toxic residues following fumigation of date fruits. The gas chromatographs used were (*1*) F & M Model 500 equipped with a thermal conductivity detector

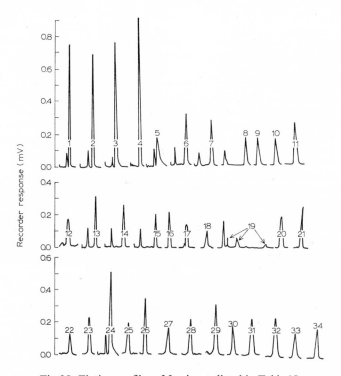

Fig. 25. Elution profiles of fumigants listed in Table 15.

and a 10 ft. × ¼ in. copper coil column packed with 4% SE-30 + 2% XE-60 on 100–200 mesh silanized Gas-Chrom P (helium was the carrier gas), and (*2*) Packard gas chromatograph with a flame-ionization detector containing a 6 ft. × ⅙ in. i.d. glass coil column packed with Porapak-R or 10% polypropylene glycol on Chromosorb W. Nitrogen was the carrier gas. Table 16 depicts the relative retention values of ethylene oxide and its residues on various columns.

Heuser and Scudamore[169] described a multi-detection scheme for the evaluation of fumigant residues in foodstuffs in which GLC using three types of detector is used to analyze the contents of processed solvent extracts. (The recovery of a range of 20 compounds as volatile fumigant residues in cereal and other foodstuffs, with sensitivities generally better than 0.1 p.p.m. was found feasible.) Table 17 lists the standard GLC conditions and sensitivity for multi-detection and individual analyses

TABLE 16

RELATIVE RETENTION VALUES OF ETHYLENE OXIDE AND ITS TOXIC RESIDUES
ON VARIOUS COLUMNS

Materials	B.p. (°C)	Relative retention time[a]			Column temp. (°C)[e]
		I[b]	II[c]	III[d]	
Ethylene glycol	197.6	30.0		3	120
Ethylene chlorohydrin	128.8	24.0	24.4		
Diethylene glycol	244.5	52.0		17	165
Ethylene oxide	10.7	1.0	1.0	1.0	40
Acetone	56.5	3.3	1.8	e	50–60

[a] Relative to ethylene oxide (EO). EO was diluted in 2 min on Porapak-R, in 0.96 min on 10% polypropylene glycol, and in 1.2 min on 4% SE-30 + 2% XE-60.
[b] Flame ionization detector. Column: Porapak-R 6 ft. × $\frac{1}{6}$ in. i.d. glass-coil column. Conditions: column temp. 140°C, inlet temp. 200°C, detector temp. 190°C, nitrogen carrier flow 80 ml/min S.T.P., air 550 ml/min at 70°F and 14.7 lb. p.s.i.a., column inlet pressure 16 p.s.i.a., hydrogen 80 ml/min S.T.P.
[c] Flame ionization detector. Column: 10% polypropylene glycol on Chromosorb W. 6 ft. × $\frac{1}{6}$ in. i.d. glass-coil column. Conditions: column temp. 65°C, inlet temp. 120°C, detector temp. 110°C, nitrogen carrier flow 50 ml/min S.T.P., column inlet pressure 15 p.s.i.a., air 600 ml/min at 70°F and 14.7 lb. p.s.i.a., hydrogen 80 ml/min S.T.P.
[d] Thermal conductivity detector. Column: 4% SE-30 + 2% XE-60 on 100–120 mesh silanized Gaschrom P. 10 ft. × $\frac{1}{4}$ in. i.d. copper-coil column. Conditions: column temp. 110°C, helium carrier flow rate 50 ml/min S.T.P. Separation between ethylene oxide and acetone could not be carried out on this column without programming.
[e] Separation by programming on the column described in (d). Program rate 5.6°C/min.

of 18 fumigants (including ethylene and propylene oxides). Table 18 depicts the operating parameters for the two GLC flow systems depicted in Table 17 above.

Gordon et al.[170] explored the effect of fumigating dried prunes with ^{14}C-ethylene oxide in order to provide data on the quantity and identity of these relatively unknown food constituents. Over 50% of the total radioactivity was found to be combined as insoluble hydroxyethyl cellulose in the prune skin, 30% as hydroxy-ethylated sugars in the pulp, and 3% as glycols (mostly diethylene glycol). The remainder was tentatively identified as hydroxyethylated amino acids and protein. Both one-dimensional paper chromatography and paper electrophoresis were used for the separation and identification of the prune constituents. The solvent systems used were (1) isopropanol–pyridine–water–acetic acid (8:8:4:1) and (2) isobutanol–pyridine–water–acetic acid (12:6:4:1). Table 19 illustrates the tentative identification of constituents of alkaline hydrolyzate of radioactive prune skin following the fumigation of dried prunes with ^{14}C-ethylene oxide.

Ethylene oxide has been used in the tobacco industry to both shorten the ageing process as well as reducing the nicotine content in tobacco leaves. The reaction products of nicotine with ethylene oxide as well as their pyrolysis products have been investigated using both paper and gas chromatography[171]. The reaction product of

STANDARD GLC CONDITIONS AND SENSITIVITY FOR MULTI-DETECTION AND INDIVIDUAL ANALYSES

Compound	4 m × 1/8 in. o.d. Carbowax 1540 column. Nitrogen: 20 p.s.i.						2 m × 1/8 in. o.d. Porapak Q column. Argon: 20 p.s.i.					
	Retention time at st. temp. (75°C) (min)	Order of elution	Detector response: Electron capture	Detector response: Flame	Pref. temp. (individual analysis) (°C)	Notes	Retention time at st. temp. (150°C) (min)	Order of elution	Detector response: β-Ion	Detector response: Flame	Pref. temp. (individual analysis) (°C)	Notes
Phosphine*	1	1		L	30	Poor sep. from air	2	1^b	H	L	80	Temp. above 100°C unsuitable
Hydrogen cyanide	2	2	M		40			3			130	Electron capture effective at low detector voltage
Methyl bromide	$1\,1/2^a$	3^a	M	M	40	Separated on Porapak Q	$3\,1/2$	5	H	M	130	
Ethylene oxide	$1\,1/2^a$	4^a	H	H	40	Separated on Porapak Q	3	4	H	H	130	
Carbon disulphide	$1\,3/4$	5			40	Separated on Porapak Q	$7\,1/2$	12	M^c		130	
Propylene oxide	2	6		M^c	40	Separated on Porapak Q	5	8	H^d	M^a	130	
Dimethyl sulphide	2	6		L^c	40	Separated on Porapak Q	$6\,1/2$	10	H^c	L^c	130	Sep. from dichloromethane on Carbowax 1540
Acetone	3	7		M	50		6	9	H	M	140	
Ethylidene chloride	$3\,1/4$	8	L^c	M^c	50		$10\,1/2$	13	H^c	M^c	160	
Carbon tetrachloride	4	9	H^c	L^c	50		22	18	M^c	L	130	β-Ion, anomalous response
Dichloromethane	5	10	M^c	M^c	50		$6\,1/2$	10	M	M^c	130	
Ethanol	6	11		M	60		4	6	M	M		
Acrylonitrile	$7\,1/2$	12	H^d	M^d	60	Separated on Porapak Q	7	11	M	M^c	130	
Trichloroethylene	$7\,1/2$	12	H^d	M^d	60	Separated on Porapak Q	24	19		M	180	
Chloroform	$8\,1/2$	13		M^d	60	Separated on Porapak Q	14	15	L	M	140	
Acetonitrile	$8\,1/2$	13		M	60	Separated on Porapak Q	$4\,3/4$	7	M	M		
Perchloroethylene	$9\,1/2$	14	H^d	M^d	60		52	22	L	M	180	
Chlorobromomethane	10	15	H^d	M^d	60		$11\,1/2$	14	H^c	M^c	140	
Dichloropropane	$10\,1/2$	16		M^d	60		31	20	H	M	180	
Ethylene dichloride	11	17	L	M^d	60	Not separated above 75°C	$16\,1/2$	16	H	M	160	
Water	$12\,1/2$	18	L		60	Not separated above 75°C	$11\,1/4$	2				β-Ion, suppressed
Chloropicrin	16	19	H	L	90		45	21		L	160	{ Partial decomp. above 160°C; Electron capture effective at low det. voltage
Ethylene dibromide	36	20	H	M	120	Use Porapak Q/β-ion	58	23	L	M	180	
Ethylene chlorohydrin	80	21		M	120	Use Porapak Q/β-ion	18	17	H	M	160	

Detector response: minimum detectable amount 2% f.s.d. at 0.5% f.s.d. noise level. H (High) = $< 10^{-10}$ g, M (Medium) = 10^{-10} g–2×10^{-9} g, L (Low) = $> 2 \times 10^{-9}$ g.

N.B. Injecting a 2.5 μl aliquot of 25 ml dehydrated solvent from 10 g commodity. 10^{-10} g ≡ 0.1 p.p.m., 2×10^{-9} g ≡ 2 p.p.m.

* Vapor phase only; standard solvent extraction impracticable.

^{a} Separated at 40°C. ^{b} At 80°C, retention time ~3 min. ^{c} Using acetonitrile. ^{d} Using acetone.

ALKYLATING AGENTS

TABLE 18

OPERATING PARAMETERS FOR THE TWO GAS CHROMATOGRAPHY FLOW SYSTEMS

	1	*2*
Column	4 m × 2.2 mm i.d. ($\frac{1}{8}$ in. o.d.) stainless steel	2 m × 2.2 mm i.d. ($\frac{1}{8}$ in. o.d.) stainless steel
Packing	10% by wt. Carbowax 1540 on Teflon 6	Porapak Q 50–80 mesh
Column temperature (isothermal)	75 °C	150 °C
Injection block temperature	160 °C	160 °C
Carrier gas (dried by mol. sieve 5 Å)	Nitrogen (20 p.s.i.)	Argon (20 p.s.i.)
Detectors (at column temperature)	(a) Perkin-Elmer 452, electron capture 100 mCi tritium	(a) Perkin-Elmer 452, dual purpose β-ionization/electron capture 100 mCi tritium
	(b) Flame ionization fed *via* 1:1 stream splitter Electron capture, ~0.5–50 V Flame, 200 V	(b) Flame ionization fed *via* 1:1 stream splitter β-ionization, 800–1200 V, electron capture, ~0.5–50 V Flame, 200 V
Amplifiers	1 and 2 Sensitivity 10^{-11}–10^{-6} A for f.s.d. Background elimination up to 10^{-6} A Detector supply voltages 0–1500 -ve Output 1 or 2.5 mV	
Recorders	1 mV or 2.5 mV f.s.d. 1 sec response Chart speed 0.5 or 1 cm/min	

nicotine dihydrochloride with ethylene oxide was N-hydroxyethyl nicotine as determined by paper chromatography using *n*-butanol–acetic acid–water (4:1:5) as developer and the Dragendorff reagent for detection. The reaction product of nicotine monoacetic acid with ethylene oxide yielded both N-hydroxyethyl nicotine (R_F 0.1) and N'-hydroxyethyl nicotine (R_F 0.48) as determined by paper chromatography using *n*-butanol–acetic acid–water (4:1:5) as developer and the Dragendorff and Kalnig reagent (1% alcoholic *p*-aminobenzoic acid followed by exposure of the chromatogram to gaseous cyanogen bromide) as detecting agent.

The pyrolysis of the above reaction products was studied utilizing gas chromatography. A Shimadzu GC-2 instrument was used equipped with a 3 m × 3 mm stainless steel column packed with Diabase B 60–80 mesh (79.7% polyethylene glycol 6000 (20%) and sodium hydroxide (0.3%). The temperature program was from 70 to 220 °C at 5°/min and the helium flow rate was 60 ml/min.

TABLE 19

TENTATIVE IDENTIFICATION OF CONSTITUENTS OF ALKALINE HYDROLYZATE
OF RADIOACTIVE PRUNE SKIN

Spot	Prune skin hydrolyzate		Ethylene oxide expected		Known compounds
	% of total radioactivity	$M_G{}^a$	%	M_G	
A	33	0.81	33	0.83	6-(2-Hydroxyethyl)-glucose
			3	0.71	3-(2-Hydroxyethyl)-glucose
B	21	0.67	22	0.67	6-(2-Hydroxyethoxyethyl)-glucose
C	5–15	0.46	2	0.63	3-(2-Hydroxyethoxyethyl)-glucose
			2	0.52	3,6-Di(2-hydroxyethyl)-glucose
			10	0.32	2-(2-Hydroxyethyl)-glucose
			7	0.24	2-(2-Hydroxyethoxyethyl)-glucose
D	30–40	0	5	0.22	2,6-Di(2-hydroxyethyl)-glucose
			15	0	Mixture of isomeric (2-hydroxyethyl)-cellobioses
			(15)	0	Mixture of degradation products of all of the above compounds, and also of (2-hydroxyethyl)-pentosan and tannins

^{a} M_G is the ionic mobility of a substance relative to that of glucose which is taken as 1.

TABLE 20

RELATIVE QUANTITY OF PYROLYZED COMPOUNDS OF NICOTINE, N-HYDROXYETHYL NICOTINE
AND THEIR HYDROCHLORIDES (3-CYANOPYRIDINE = 1)

Products	Free nicotine		Free N-hydroxy-ethyl nicotine		Nicotine dihydro-chloride	N-Hydroxy-ethyl nicotine dihydro-chloride	N′-Hydroxy-ethyl nicotine dihydro-chloride
	600°C	800°C	600°C	800°C	600°C	600°C	600°C
Low-boiling compounds	5	5	2	5	3	15	3
Pyridine	1.5	4	0.5	3	1	1	1
3-Picoline	4	1.5	0.3	1	4	5	5
3-Ethylpyridine	1.5	0.2	0.5	0.2	0.8	1	1
3-Vinylpyridine	10	5	5	2	4	5	5
3-Cyanopyridine	1	1	1	1	1	1	1
Nicotine	8	4	10	5	5	3	4
Quinoline	0.2	1	1	1.5	1.5	1.5	0.5
Isoquinoline	0.1	0.3	0.5	0.8	0.3	0.8	0.2
Myosmine	0.5	0.5	0.5	1.5	1.5	1	1

Table 20 lists the relative quantity of pyrolyzed compounds (as determined by
gas chromatography) of nicotine, N-hydroxyethyl nicotine and their hydrochlorides
following pyrolysis at 600°C for 15 sec.

Muramatsu[172] investigated the occurrence of ethylene oxide in cigarette smoke by GLC. The content of ethylene oxide in smoke from cigarettes whose tobacco leaves were originally sprayed with oxyethylene docosanol at their growing stage (at levels of 3×10^{-4} ml/cm^2/leaf) showed the same level as that from non-sprayed leaves. A slight amount of ethylene oxide was found in smoke from *non*-ethylene oxide-treated tobacco as well. GLC was performed isothermally at 40°C using 4 m × ⅛ in. stainless columns containing β,β-oxydipropionitrile or polypropylene glycol LB-X on Chromosorb W (60–80 mesh) with helium as carrier gas at 28 ml/min. Figure 26 illustrates a gas chromatogram of a mixture of ethylene oxide–freon (12:88) obtained on a β,β-oxydipropionitrile column at 40°C.

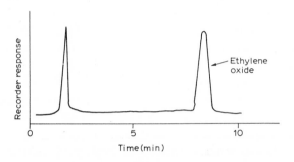

Fig. 26. Gas chromatograph of a mixture of ethylene oxide–freon.

In recent years ethylene oxide sterilization has become an important tool, primarily where sterilization by autoclaving is either impractical or, as in the case of many plastic devices, destructive to the product.

A quantitative method for determining residual ethylene oxide in plastic and rubber catheters using gas–liquid chromatography has been developed by Kulkarni *et al.*[173]. An F&M Model 700 gas chromatograph equipped with a dual hydrogen-flame-ionization detector was used. The column used was a 6 ft. × ¼ in. stainless steel tubing containing 10% polyphenyl ether on 80–100 mesh Diatoport S. Helium was used as the carrier gas at 25 p.s.i. and at a flow rate of 30 ml/min. The injection port, column and detector temperatures were 175, 45, and 220°C, respectively. Tests on ethylene oxide-stabilized catheters indicated that the amount of ethylene oxide in the catheters decreased as a fraction of aeration time up to 48 h. After this period, a steady concentration of ethylene oxide (0.4 mg ethylene oxide/g catheter for plastic, and 0.1 mg/g for rubber) in the catheters appeared to have been reached up to one week's testing. It was pointed out that ethylene oxide, if not removed, may be released later under use, and may cause blood hemolysis, erythema and edema of the tissues. Cunliffe and Wesley[174] demonstrated that ethylene chlorohydrin emanated from polyvinyl chloride tubing 6 days after ethylene oxide sterilization. It has also been demonstrated by Gunther[161] that high concentrations of ethylene oxide can be taken up by polyethylene, gum rubber, and plasticized polyvinyl chloride. Guess and Jones[175] have also described the solubility of ethylene oxide by a number of plasticizers commonly used in polyvinyl chloride and other medical plastics (*e.g.* dioctyl adipate, azelate, sebacate and phthalate and a variety of epoxidized soybean oils). These studies emphasize the fact that toxic concentrations of ethylene oxide are easily retained by plasticizers, plastics, etc., necessitating use of efficient degassing after ethylene oxide sterilization of plastics.

The determination of ethylene oxide and ethylene chlorohydrin in plastic and rubber surgical equipment stored with ethylene oxide was reported by Brown[176]. The procedure consisted of extracting the sample with *p*-xylene and passing the extract through three chromatographic columns in series. The top column (I) consists of Florisil and collects the ethylene chlorohydrin; the second column (II) mounted directly below the first column consists of Celite mixed with dilute HCl and converts the extracted ethylene oxide to ethylene chlorohydrin and the third column (III) placed directly below column II also contains Florisil and collects the ethylene chlorohydrin formed. The ethylene chlorohydrin collected on the columns is then eluted separately with ethyl ether and if further purification is necessary, sweep co-distillation is used prior to GLC. The equipment consisted of (*1*) Packard Model 76215 instrument equipped with hydrogen-flame-ionization detector and a coiled glass column, 6 ft. × 4 mm i.d., packed with 20% Carbowax 20M on 80–100 mesh Gas Chrom Q (conditioned for 24 h at 190°C with a nitrogen flow of 50 ml/min); (*2*) Hewlett-Packard F & M Model 5750 equipped with hydrogen-flame-ionization detector and coiled glass column 6 ft. × 3 mm i.d. packed with 18% Ucon 75-H-90,000 on 100–120 mesh Gas Chrom Q (conditioned for 24 h at 190°C with a nitrogen flow of 50 ml/min). The operating parameters were, for the Packard gas chromatograph, injection port 215°C, detector 250°C, voltage 100 V, gain 1×10^{-10} A, recorder chart speed 0.33 in./min, column temp. 120°C, nitrogen flow 100 ml/min and for the alternate GLC system (Hewlett-Packard), injection port 162°C, detector 200°C, range 10, attenuation 4, air flow 300 ml/min, hydrogen flow 30 ml/min, recorder chart speed 0.25 in./min, column temp. 110°C, nitrogen flow 55 ml/min.

The possible presence of residues resulting from the ethylene oxide sterilization of heat-labile pharmaceuticals is of scientific and legal concern. Adler[177] investigated residual ethylene oxide and ethylene glycol in ethylene oxide sterilized pharmaceuticals (including cortisone, hydrocortisone and prednisolone acetates). The methods used were (*a*) separation and concentration of ethylene oxide by distillation and measurement by internal standard GLC and (*b*) determination of total glycol as well as specific ethylene oxide by colorimetric determination of formaldehyde. The GLC assay utilized a Perkin-Elmer Model 154-B vapor fractometer and thermal-conductivity detector. The columns were (*a*) 2 m × 0.25 in. copper coils containing 25% dimethylsulfolane on 30–60 mesh firebrick at 30°C with helium as carrier gas at 45 ml/min and (*b*) 1 m of 25% 1,2,4-butanetriol in series with 2 m of 25% polyethylene glycol 1500 operated at 40°C (columns conditioned at 80–90°C, then further conditioned by injecting large samples of ethylene oxide in acetone). Figure 27 illustrates a gas chromatograph of both ethylene and propylene oxides in acetone and Table 21 depicts the gas chromatographic identification of ethylene oxide in the distillate from prednisolone–*tert.*-butyl acetate. The studies of Adler[177] indicated that ethylene oxide–carbon dioxide steril-

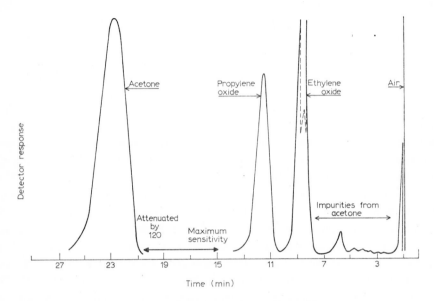

Fig. 27. GLC of ethylene and propylene oxides.

zation resulted in the formation of glycol residues in all the pharmaceuticals tested (in some cases, almost a 1 % residue was found).

Holmgren *et al.*[178] reported the content of ethylene chlorohydrin (in amounts of up to 1500 p.p.m.) in 21 different crude drugs which had been treated with ethylene

TABLE 21

GAS CHROMATOGRAPHIC IDENTIFICATION OF ETHYLENE OXIDE IN THE DISTILLATE
FROM PREDNISOLONE *tert.*-BUTYL ACETATE

Column substrate[a]	Length (m)	Corrected retention time (min) (ethylene oxide/propylene oxide)	
		Known mixture	Sample
Di-*n*-decyl phthalate	2	0.474	0.474
Dimethyl sulfolane	2	0.575	0.571
Polyethylene glycol	2	0.680	0.678
Polyethylene glycol 200 + 1,2,4-butanetriol	2 + 1	0.735	0.734
Polyethylene glycol 1500 + 1,2,4-butanetriol	2 + 1	0.701	0.703

[a] All substrates were on 30–60 mesh firebricks in 0.25 in. o.d. copper columns.

oxide. Ethylene chlorohydrin was determined by GLC[179] using a Varian Aerograph Model 200 with flame ionization and a column consisting of Celite 545 impregnated with 16.6% Carbowax 1500 (instead of Chromosorb W impregnated with 20% Carbowax 20M as originally prescribed[179]).

3.2 Propylene oxide

The general considerations of propylene oxide that are relevant include mode of action[180, 181], toxicity[182], carcinogenicity[183], and mutagenicity[184, 185].

Propylene oxide is synthesized commercially in a manner analogous to ethylene oxide, e.g. from propylene through the intermediate propylene chlorohydrin or by the direct oxidation of propylene with either air or oxygen. Although the bulk of production is used in the preparation of propylene glycols, mixed polyglycols and various propylene glycol ethers and esters, large amounts are also used in the synthesis of hydroxypropyl celluloses and sugars, surfactants, urethan elastomers, isopropanolamine and a variety of other derivatives that are useful in many applications including textiles, cosmetics, pharmaceuticals, agricultural chemicals, petroleum, plastics, rubber and paints. Propylene oxide which is highly reactive chemically (being intermediate between ethylene and butylene oxides) is also used as a fumigant, herbicide, preservative and, in some cases, as a solvent. (It is a powerful low-boiling solvent for hydrocarbons, cellulose nitrate and acetate, and vinyl chloride and acetate.)

Propylene oxide is commonly employed as a gaseous decontaminant to destroy specific groups of organisms such as fungi, yeast, coliform and salmonellae. Although its penetrating ability and microbiocidal properties are less than that of ethylene oxide, it is frequently used as a substitute for ethylene oxide and has been suggested for use as a soil sterilant. Propylene oxide has also been used on a variety of foods such as dried fruits, powdered or flaked foods (e.g. cocoa, yeast powder, and cereal flakes) to control spoilage. As has been discussed above, ethylene and propylene oxides may react with inorganic chloride and moisture in foodstuffs to form chlorohydrins and glycols which in themselves are toxic. Persistent residues of these are found because of their low volatility and relatively unreactive chemical nature.

Mestres and Barrois[186] described the gas chromatographic determination of propylene oxide and propylene glycol residues in fruits treated with propylene oxide. An Aerograph Model A600B Hy-Fi instrument was used equipped with a 10 ft. $\times \frac{1}{8}$ in. column containing 15% XF 1150 on Embacel (Merck). For the analysis of propylene oxide, the injection port and column temperatures were 125 and 70°C, respectively; nitrogen was the carrier gas at 25 ml/min. The analysis of propylene glycol required an injection port and column temperature of 145°C and nitrogen as the carrier gas at 37 ml/min.

Figure 28 shows a gas chromatogram of (A) a mixture of propylene oxide and ethanol and (B) a sample of prune pit treated with propylene oxide, and indicates retention of the oxide after 48 h. Figure 29 illustrates a gas chromatogram of (A) an

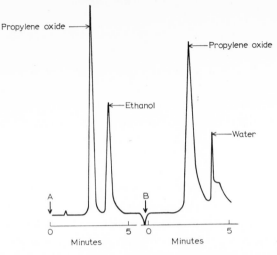

Fig.28. Gas chromatograms of (A) 1 µl of a mixture of propylene oxide (1 µg) and ethanol (1 µg), attenuation 1/80 (1 × 8); (B) 4.6 mg of prune pit treated with propylene oxide (after 48 h), attenuation 1/40 (1 × 4). Temperature of column 70°C, other operating conditions as indicated in text.

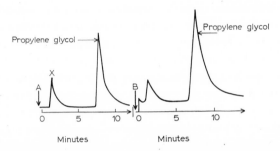

Fig.29. Gas chromatograms of (A) 0.5 µl of an aqueous solution of propylene glycol containing 5 µg propylene glycol. Attenuation 1/80 (1 × 8), × = unidentified peak, column temperature 145°C, other operating conditions as indicated in text. (B) 4.2 ng of prune pulp treated with propylene oxide (after 6 days). Attenuation 1/80 (1 × 8).

TABLE 22

GLC PARAMETERS FOR THE SEPARATION OF GLYCOLS

Stationary phase	TCEP	DEGS	Carbowax 20M	XF 1150
Percent	20%	20%	20%	15%
Support	(Chromosorb W (HMDS))			Embacel
	60/80 mesh			
Length column (ft.)	10	10	13	10
Width column (in.)	$\frac{1}{8}$	$\frac{1}{8}$	$\frac{1}{8}$	$\frac{1}{8}$
Temp. of column (°C)	123	145	145	145
Pressure of nitrogen carrier				
(input) (kg/cm^2)	2	1.1	2	1.5
Exit at ml/min	25	16.2	13.6	18
Retention time propylene glycol (min)	46	18	30	7.6
Relative retention time 2,3-butylene				
glycol	0.86	0.885	0.88	0.95
Propylene glycol	1	1	1	1
Ethylene glycol	1.33	1.26	1.25	1.065
1,3-Butylene glycol	2.17	1.89	1.875	1.98

aqueous solution of propylene glycol and (B) a sample of prune pulp treated with propylene oxide and indicates the retention of propylene glycol even after 6 days.

In addition to propylene glycol, Mestres and Barrois[186] described the GLC separation of 2,3-butylene, ethylene and 1,3-butylene glycols. Table 22 depicts the operating parameters for their separation as well as retention times.

3.3 Ethylene chlorohydrin and bromohydrin

As has been shown in a number of instances above, ethylene chlorohydrin is a byproduct in the gaseous fumigation and sterilization of a variety of foods, drugs and plastic devices. Ethylene chlorohydrin, however, has suggested utility in a number of areas, e.g. the separation of butadiene from hydrocarbon mixtures, in dewaxing and removing of naphthenes from mineral oil, in the refining of rosin, in the extraction of pine lignin, and as a solvent for cellulose acetate ethers and various resins. It has also been proposed as an effective agent in hastening the early sprouting of dormant potatoes and in the treatment of seeds for the inhibition of biological activity. The chemical properties[187], toxicity[188, 189], and tissue reactions[190] of ethylene chlorohydrin have been described.

In addition to the previously mentioned investigations involving the detection of ethylene chlorohydrin in fumigated materials, a number of additional studies need be cited.

The formation of persistent toxic chlorohydrins in foodstuffs by fumigation with ethylene oxide and with propylene oxide has been reported by Wesley et al.[191]. Chlorohydrin and ethylene oxide residues were examined both chemically and by gas chromatography. The apparatus utilizing a flame-ionization detector was essentially similar to that described by Smith[192]. The column was 5 ft. × ⅛ in. packed with 10% polyethylene glycol on Celite. Operation was isothermal at 100°C and the carrier gas was hydrogen containing 25% nitrogen at a flow rate of 30 ml/min. Concentrations of ethylene chlorohydrin up to 1000 p.p.m. were found in whole spices and ground spice mixtures after commercial fumigation with ethylene oxide.

The isolation and determination of chlorohydrins in foods fumigated with ethylene oxide or with propylene oxide has been described by Ragelis et al.[193]. Two isolative methods (steam distillation and sweep co-distillation in conjunction with ether extraction) were used with both gas chromatography and infrared spectroscopy for the isolation, determination and identification of residues and degradation products following fumigation. A Beckman Model GC-4 gas chromatograph was used equipped with a dual flame-ionization detector and stainless steel columns 10 ft. × 3 mm i.d., packed with 20% Carbowax 20M on 60–80 mesh acid-washed Chromosorb W. The column, detector, and injection port temperatures were 115, 250, and 200°C, respectively. The helium carrier gas flow rate was 100 ml/min.

Levels of ethylene chlorohydrin found in five food products commercially fumigated with ethylene oxide ranged from 45 p.p.m. for paprika to 110 p.p.m. for pepper. Levels of 1-chloro-2-propanol found in six food products commercially treated with propylene oxide ranged from 4 p.p.m. for cocoa to 47 p.p.m. for glazed citron. Figures 30 and 31 illustrate the sweep co-distillation and steam distillation apparatus, respectively, that were employed for the isolation of the chlorohydrin in foods fumigated with ethylene or propylene oxides.

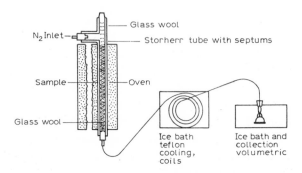

Fig. 30. Sweep co-distillation apparatus.

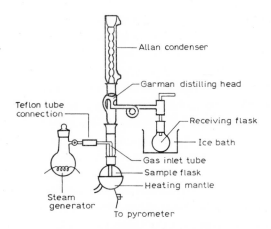

Fig. 31. Steam distillation apparatus.

Manchon and Buquet[194] described the determination and levels of ethylene oxide and ethylene chlorohydrin in bread. Gas chromatographic analysis of bread treated in air-tight packs with a mixture of ethylene oxide and carbon dioxide has demonstrated that the initial concentration of ethylene oxide in the atmosphere of the package is reduced by half immediately after the treatment and was virtually negligible after 3 days. GLC determination of ether extracts of the treated bread revealed an average concentration of 260 p.p.m. of ethylene chlorohydrin when treated with a mixture of 15% ethylene oxide in carbon dioxide. An F & M Model 700 gas chromatograph was used with a flame-ionization detector and a Moseley 7127A programmer,

6 ft. × ⅛ in. stainless steel columns containing 10% UCW 98-100S on 10% silicone oil on Diatoports (80–100 mesh). The injection port, column and detector temperatures were 150°, 85°, and 135°C, respectively, and the carrier gas was nitrogen at 30–50 ml/ min. The determination of ethylene chlorohydrin in bread was also effected using a 1.50-m stainless steel column containing 10% Carbowax 20M on Gas Crom Z.

The oral metabolism of ethylene chlorohydrin in the rat has been studied by Johnson[195] and the conversion of liver glutathione to S-carboxymethyl glutathione demonstrated. It was suggested that the toxicity of ethylene chlorohydrin is due to its conversion *in vivo* to chloroacetaldehyde (which reacts with glutathione). Both column and paper chromatography were used to identify the *in vitro* and *in vivo* metabolites. Descending chromatography was carried out on Whatman No. 1 paper for 17–24 h at 23–25°C with the following solvents: (*1*) *n*-butanol–acetic acid–water (11:4:5), (*2*) phenol–2 N ammonia (80:20), (*3*) *n*-propanol–pyridine–water (1:1:1), (*4*) methyl-ethyl ketone–pyridine–2 N ammonia (1:1:1), (*5*) 2-propanol–2 N ammonia (7:3), (*6*) 2-propanol–2 N HCl–20 mM EDTA (70:30:1), (*7*) 2-propanol–methylethyl ketone–water (1:1:1), and (*8*) *n*-butanol–formic acid–water (11:4:5). Both ninhydrin and platinic iodine reagents were used for detection[196].

Pagington[197] described a GLC method for the determination of chlorohydrins. The apparatus[198] incorporates an ozatron type J detector element from an A.E.I. leak detector type H.A. When used as a chromatographic detector it is only sensitive to halogenated compounds. This removes the necessity to clean up samples and allows extracts to be injected directly onto the column (2.2 m × ¼ in. o.d. U-shaped column packed with 80–100 mesh Porapak). The column was run at 175°C with a nitrogen flow of 50 ml/min and a dilution flow of 100 ml/min. Using samples of up to 25 µl, it was found possible to determine chlorohydrins at a level of 0.2 p.p.m. in aqueous solution.

In addition to ethylene chlorohydrin being formed during the fumigation of foods, Heuser and Scudamore[199] recently reported the formation of ethylene bromo-hydrin in flour and wheat during treatment with ethylene oxide, in which the required bromine is derived either from naturally occurring inorganic bromide or in larger amounts from bromide produced in prior fumigation with methyl bromide. Harvested crops, notably tobacco, which are treated with ethylene oxide for disinfestation pur-poses, may also contain inorganic bromide derived from soil treatment with methyl bromide or ethylene dibromide[200]. Ethylene bromohydrin, produced in wheat or flour by treatment with ethylene oxide vapor in air was determined by solvent extraction with acetone–water mixture and GLC using a flame-ionization detector, as described for ethylene chlorohydrin[166]. The retention time for ethylene bromohydrin under the conditions specified was about twice that for ethylene chlorohydrin. Its identity was confirmed by coincidence of retention times on polar and non-polar chromatographic columns with an authentic sample. It is important to note that ethylene bromohydrin as well as ethylene chlorohydrin is mutagenic[164] and that the use of methyl bromide or ethylene dibromide for fumigation may (especially in the latter case) give rise to ethylene bromohydrin *per se via* partial hydrolysis or leave residues of inorganic bromine which can react further with ethylene oxide as cited above to yield the mutagenic bromohydrin.

4. GLYCIDOL

Glycidol (2,3-epoxy-1-propanol) is made by the dehydrochlorination of glycerol and monochlorohydrin with caustic. It is used for the preparation of glycerol and glycidyl esters, ethers and amines which have utility as pharmaceutical intermediates, and in textile finishings, *e.g.* glycidol esters as water-repellant finishes. Glycidol has been used as an antibacterial and antimycotic agent for food products. Its toxicity[201,202], mutagenicity[184, 203] and chromosome-breaking properties[204] have been described.

The paper chromatographic determination of glycidol in brewery products has been described by Dyr and Mostek[205]. The solvents used were (*1*) *n*-butanol–acetic acid–water (4:1:5) and (*2*) *n*-butanol–ethanol–water (4:1.1:4.9) with detection accomplished with 7 reagents, the best of which were 10% benzidine in acetic acid and silver nitrate in acetone.

5. EPICHLOROHYDRIN

Epichlorohydrin (1-chloro-2,3-epoxypropane) is made commercially from propylene or the reactions of alkalies upon dichlorohydrins and is employed extensively as a solvent for natural and synthetic resins, gums, cellulose esters and ethers, paints, varnishes, nail enamels and lacquers, and as a raw material for the manufacture of a number of glycerol and glycidol derivatives, in the manufacture of epoxy resins, as a stabilizer in chlorine-containing materials, and as an intermediate in the preparation of condensates with polyfunctional substances. The patented uses of epichlorohydrin include its utility as an insecticide and nematocide, gaseous sterilant, a fungicide for seed treatment, as a cross-linking agent in the crease-proofing of textiles, paper processing, water-proofing of materials, fire-resistant epoxy resins and as a curing agent for aminoplast resins.

The toxicity[206], carcinogenic potential[207] and mutagenicity[20, 208, 209] of epichlorohydrin have been reported. The post-testicular antifertility effects of epichlorohydrin and glycidol in the rat have been recently described[210]. The antifertility effects of α-chlorohydrin (3-chloropropane-1,2-diol)[211-213], and the postulation of glycidol as the active intermediate in the antifertility action of α-chlorohydrin in male rats have also been reported. .

The paper chromatographic behavior of a number of epoxy compounds (*e.g.* epichlorohydrin, glycidol, and alkyl and aryl monoethers and diethers of glycidol) was described by Schaefer *et al.*[214, 215]. Epichlorohydrin and alkylmonoglycidyl ethers of C_1 to C_6 evaporate during treatment and could not be analyzed directly. Lower-boiling hydrophilic epoxy compounds were analyzed on paper chromatograms by using petroleum ether–propanol–water (2:1:1) as developer. The use of derivatives of the epoxy compounds prepared *via* reaction with benzenesulfonic acid, picric acid, picramide or N-methyl picramide were recommended[215]. Derivatives obtained by treatment of the epoxy derivative and picric acid were particularly suitable for paper chromatography using Whatman No. 1 paper impregnated with formamide and cyclohexane–toluene (2:1) as developer. Orange spots were obtained and after exposure of the chromatograms with ammonia, the stereoisomeric derivative appeared as a yellow spot fluorescing under UV light. Table 23 lists the R_F values of various epoxy compounds as their respective picrate derivatives.

TABLE 23

R_F VALUES[a] OF EPOXY COMPOUNDS
AS THEIR PICRATE DERIVATIVES

Substance	R_F
Epichlorohydrin	0.08 ± 0.02
Glycidol	0.00 ±
Allylglycidether	0.27 ± 0.05
Phenylglycidether	0.37 ± 0.07
Methylglycidether	0.09 ± 0.02
Ethylglycidether	0.23 ± 0.04
Propylglycidether	0.48 ± 0.08
Butylglycidether	0.69 ± 0.08
Hexylglycidether	0.89 ± 0.02

[a] Values obtained on Whatman No. 1 paper impregnated with formamide and developed with cyclohexane–toluene (2:1).

Copper and Roberts[216] described the qualitative and quantitative analysis of cotton finish based on epichlorohydrin or dichloropropanol. The quantitative method involved the GLC determination of isopropyl iodide liberated from the modified cellulose on treatment with boiling HI. Flame ionization was used for detection along with a column of dinonyl phthalate on 120–50 mesh glass beads at 27°C.

West *et al.*[217] described the GLC of a large number of possible organic air pollutants (including epichlorohydrin) on columns of tri-*m*-tolyl phthalate and β,β'-hydroxy-dipropionitrile.

6. DI(2,3-EPOXYPROPYL) ETHER

Di(2,3-epoxypropyl) ether (diglycidyl ether) is prepared through epoxidation of diallyl ether then chlorohydrination followed by dehydrochlorination with caustic. Its broad area of utility includes the preparation of trioxane copolymers, the curing of polysulfide polymers, preparation of thermoset resins, vulcanizable polyethers and acetals, anion exchangers, polymers as flocculating agents, diluent in aromatic amine-cured epoxy adhesives, as a hardener in photographic emulsions and for the removal of remnants of Ziegler–Natta catalysts in polymerization. Its toxicity[218] and mutagenicity[203, 219] have been described.

The paper chromatography of di(2,3-epoxypropyl) ether has been previously described above[214].

7. 1,2:3,4-DIEPOXYBUTANE

1,2:3,4-Diepoxybutane (butadiene diepoxide, DEB) is used in the prevention of microbial spoilage, the curing of polymers, as a cross-linking agent for textile fibers, and as an intermediate in the preparation of erythritol and pharmaceuticals. DEB has been evaluated for the treatment of Hodgkin's disease and lymphoreticulosarcoma. Experimentally it is an active radiomimetic agent producing skin cancers and sarcomas and depression of the hemopoietic system. Its toxicity[202, 220], effect on DNA[221], mutagenicity[222, 223], and induction of chromosome aberrations[224, 225] have been described.

The analysis of DEB has been achieved by titrimetry[226] as well as by GLC[227] (a Perkin-Elmer Model 154-C gas chromatograph was used with a 2-m glass column coated with silicone oil).

8. ALDEHYDES

8.1 Formaldehyde

Formaldehyde is used in considerable quantities, generally in the form of aqueous solutions (37–50% formaldehyde by weight). The major uses of formaldehyde and its polymers (hydrated linear polymer paraformaldehyde and the cyclic trimer s-trioxane) are in the synthetic resin industry (e.g. in the production of thermosetting and oil-soluble resins and adhesives which account for 50% of the total production. Large quantities are used in the manufacture of a broad spectrum of textiles, paper, fertilizer, miscellaneous products and specialty chemicals. The agricultural areas of utility of formaldehyde include disinfection of seeds, prevention of scab in potato, wheat, barley and oats, insecticidal fungicidal and food-preservative applications.

In the textile industry, formaldehyde alone and in the form of its N-methylol derivatives are extensively employed in the production of crease-proof, crush-proof, flame-resistant and shrink-proof fabrics. Formaldehyde has been found widely in man's environment, e.g. in tobacco leaf, tobacco smoke, incinerator effluents, automobile and diesel exhaust and as residues in treated textiles. Formaldehyde is extremely reactive and will react with practically every type of organic moiety, primarily involving the formation of methylol or methylene derivatives.

The chemistry[228], toxicity[229], carcinogenicity[230], and mutagenicity[231, 232] of formaldehyde have been described. Formaldehyde has been analyzed by a variety of

procedures including colorimetry[233, 234], titrimetry[235], and iodometry[236]. GLC procedures have been extensively used for the determination of formaldehyde in solutions
and in high-purity gas. Bombaugh and Bull[237] utilized a 5 m × ¼ in. copper coil
coated with Ethofat 60/25 to effect the separation of formaldehyde from water and
alcohols. Optimum separation of the components of formalin and butanol–formaldehyde solution was obtained with Ethofat 60/25 loaded on Columnpak T (Fig. 32).

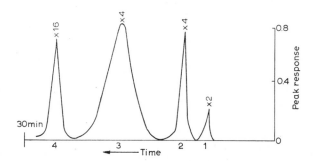

Fig. 32. GLC separation of the components of formation and butanol–formaldehyde solutions.
Operating conditions: top chromatogram, column, 5M Ethofat 60/25, 10 wt.% on Columnpak T
(5 m × 1/4 in. copper coil); temperature, column 115°C, injector 250°C; flow, 43 ml/min; bridge
current, 230 mA; sample size, 5 µl. 1, Methanol; 2, water; 3, formaldehyde; 4, butanol.

Figure 33 illustrates a chromatogram of products obtained from the degradation of
paraformaldehyde (as separated on an Ethofat 60/25 column), while Fig. 34 shows the
separation of formaldehyde from methanol, acetaldehyde, and butyraldehyde on
Ethofat 60/25. The major portion of the separations in the study of Bombaugh and
Bull[237] were obtained utilizing an Aerograph Model A-100-C gas chromatograph with
a thermistor detector.
     Commercial formaldehyde, which usually contains methanol, has presented one
of the more difficult analytical problems. Since water, formaldehyde and methanol are

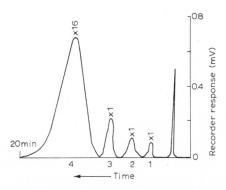

Fig. 33. GLC of high-purity formaldehyde. Composition: 0.08% methyl formate (1), 0.27% methanol (2), 0.41% water (3), 99.24% formaldehyde (4). The initial peak is argon sweep gas.

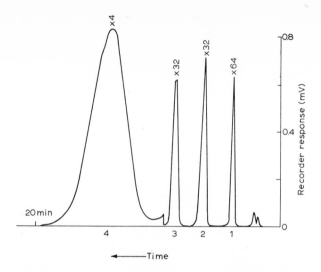

Fig. 34. GLC separation of formaldehyde, acetaldehyde, butyraldehyde and methanol on Ethofat 60/25. Operating conditions as in Fig. 32. 1, Acetaldehyde; 2, methanol; 3, butyraldehyde; 4, formaldehyde.

strongly adsorbed on conventional diatomite supports, the analysis of aqueous solutions of formaldehyde has generally been restricted to the polyfluorocarbon supports which are inert but have low loading characteristics, low efficiencies, and are difficult to handle. The evaluation of a number of solid supports (including a number of crosslinked polystyrene supports) in conjunction with several substrates was evaluated by Jones[238] for the analysis of aqueous formaldehyde solutions utilizing a Phase Separation LCI chromatograph fitted with stainless steel columns. Tables 24 and 25 show

TABLE 24

ELUTION ORDER AND RETENTION TIME FOR THE GLC OF AQUEOUS FORMALDEHYDE SOLUTIONS
ON COATED PHASE PAK SUPPORTS

Support material	Liquid phase	% w/w	Temp. (°C)	Elution order and time (min)
Phase Pak and Porapak P	None		100	W (1.5); M + F (2.28)
Phase Pak and Porapak Q	None		100	W (1.9); M + F (3.4)
Phase Pak Q	Squalane	5	100	W (1.9); M + F (3.4)
Phase Pak Q	Carbowax 20M	5	80	W (3.6); F (7.2)[a]; M (9.3)[a]
Phase Pak Q	Silicone Oil MS 550	10	100	W (1.8); M + F (5.9)
Phase Pak Q	Paraffin Oil	10	100	W (2.6); M + F (5.6)
Phase Pak Q	Carbowax 400	20	100	M + F (4.4); W (5.6)
Phase Pak Q	Ethofat 60/25	20	100	W (3.6); M (5.1); F (6.3)
Phase Pak Q	Sucroseoctaacetate	20	100	W (3.0); M (3.3); F (3.9)
Phase Pak Q	Pentaerythritol tetra-acetate	20	95	F (2.4); W (3.6); M (4.9)

[a] Indicates partially resolved peaks with heavy tailing.
Key: W = water, M = methanol, F = formaldehyde.

TABLE 25

ELUTION ORDER AND RETENTION TIMES FOR THE GLC OF AQUEOUS FORMALDEHYDE SOLUTIONS
ON COATED UNIVERSAL SUPPORTS

Support material	Liquid phase	% w/w	Temp. (°C)	Elution order and time (min)
Universal A	Carbowax 400	20	100	M + F (1.9); W (5.6)
Universal B	Carbowax 400	10	100	M + F (1.8); W (5.5)
Universal A	Ethofat 60/25	10	100	M (2.5); W (4.5); F (5.7)
Universal A	Sucrose octaacetate	10	100	M (2.3); W (3.9); F (5.1)
Universal A	Pentaerythritol tetraacetate	20	100	F (2.1); M (3.5); W (4.5)
Universal B	Pentaerythritol tetraacetate	10	100	F (2.0); M (3.7); W (4.7)

the elution order and retention time on coated phase supports. Figures 35 and 36 illustrate the separation obtained with pentaerythritol tetraacetate on Phase Pak Q and Universal A, respectively.

The study indicated that the analysis of commercial formaldehyde solution can be carried out quantitatively on a variety of support materials. The utility of GLC on porous polymer beads, Porapak N, for the analysis of formaldehyde–methanol–water mixtures has been reported[239]. A Carlo Erba Model C chromatograph was used equipped with a thermal conductivity detector and a glass column, 196 cm × 0.175 cm i.d. packed with Porapak N (100–120 mesh). The carrier gas was pure argon at a flow rate of 5.6 ml/min and an overpressure of about 57 cm mercury.

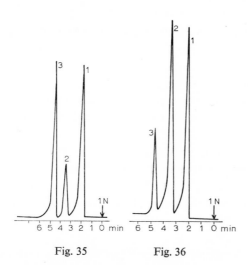

Fig. 35   Fig. 36

Fig. 35. 6 ft. × 0.125 in. o.d. column of 20% pentaerythritol tetraacetate on Universal A: temp. 100°C; He flow 22 ml/min. 1, Formaldehyde; 2, methanol; 3, water.

Fig. 36. 6 ft. × 0.125 in. o.d. column of 20% pentaerythritol tetraacetate on Phase Pak Q: temp. 100°C; He flow 22 ml/min. 1, Formaldehyde; 2, water; 3, methanol.

The column vaporizer and detector temperatures were 120, 200, and 240 °C, respectively. The identification of peaks was made by injecting pure compounds and making the test for the aldehyde with 2,4-dinitrophenyl hydrazine. Figure 37 illustrates a chromatogram of formaldehyde solution on Porapak N at 120°C. Although the

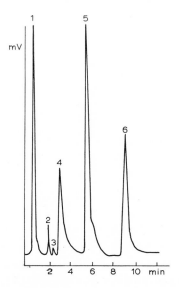

Fig. 37. A chromatogram of formaldehyde solution on Porapak N at 120 °C. 1, Air; 2, 3, not identified; 4, formaldehyde; 5, water; 6, methanol.

column when operated at optimum conditions had only 800 theoretical plates/196cm, it was found to separate rapidly formaldehyde, water and methanol (in that order in 12 min). This is in contrast to the finding of Mann and Hahn[240] that formaldehyde elutes after methanol and water on Porapak N.

The quantitative data summarized in Table 26 shows that gas chromatography yields results that are reproducible and comparable with those obtained by the classical titrimetric procedure[228].

It has been previously shown in the GLC analysis of formaldehyde that the polymerization of formaldehyde in the column has to be prevented by raising the column temperature to about 100°C. This requirement is thus a complicating factor with respect to the choice of stationary phase. The recommended packings have been Tide on Fluoropak 80[241], Ethofat 60/25[242], polyethylene glycol adipate[237], and sucrose octaacetate[242, 243].

GLC analysis of formaldehyde, methanol and water using acetylated polyesters as liquid stationary phases was investigated by Iguchi and Takiuchi[244]. Glycols such as ethylene glycol and diethylene glycol, and dibasic acids such as succinic, adipic and phthalic acids were used for preparing the polyesters. Acetylated polyethylene glycol adipate was particularly useful for the analyses of aldehyde–alcohol–ketone–

TABLE 26

QUANTITATIVE GAS CHROMATOGRAPHIC AND VOLUMETRIC ANALYSES
OF FORMALDEHYDE SOLUTIONS

	Formaldehyde		Methanol	
	GC	Titration	GC	Titration
1	0.06	0.08	0.10	0.00
2	0.58	0.59	0.16	0.00
3	1.20	1.30	0.67	0.55
4	3.65	3.60	1.06	0.90
5	12.4	12.3	16.6	16.7
6	16.0	16.1	0.06	0.00
7	34.4	34.4	14.1	14.1
8	35.5	35.6	11.6	11.5
9	36.1	36.2	6.59	7.10
10	46.1	46.1	1.42	0.96

water mixtures containing formaldehyde and of thermal products of formaldehyde
resins (Figs. 38 and 39). Acetylated polyethers such as polyethylene glycol 4000, 20M
and polypropylene glycol 2000 were also effective. Table 27 lists the retention ratios
and relative sensitivities of a variety of aldehydes and ketones determined on acetylated polyethylene glycol adipate.

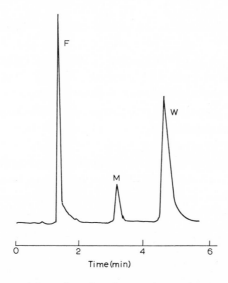

Fig. 38. Gas chromatogram of formalin using fluorocarbon resin support (Shimalite F). Liquid
phase: acetylated polyethylene glycol adipate. Column temperature: 110 °C; H_2 flow rate 26 ml/min.
F, Formaldehyde; M, methanol; W, water.

References pp. 152–160

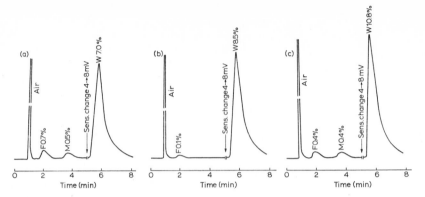

Fig. 39. Gas chromatogram of phenol- and urea-formaldehyde resins. Sample: (a) Phenol-formalde-hyde resin (resol type) impregnated paper; (b), (c) urea-formaldehyde molding powder 1 and 2. Heating condition: (a), 160 °C, 5 min, sample wt. 6.8 mg; (b), (c), 150 °C, 5 min, sample wt. 9.0 mg; column temp.: (a), (b), (c), 100 °C; support: Shimalite F; liquid phase: acetylated polyethylene-glycoladipate.

TABLE 27

RETENTION RATIOS AND RELATIVE SENSITIVITIES (WT.)

Substance	r (methyl alcohol 1.00)	R.S. (water 1.00)
Formaldehyde	0.45	0.88
Acetaldehyde	0.67	
Methyl formate	0.79	
Propionaldehyde	0.85	
Methyl alcohol	1.00	0.89
Acetone	1.12	
t-Butyl alcohol	1.20	
Ethyl alcohol	1.22	
Water	1.50	1.00
Methyl ethyl ketone	1.64	
Benzene	1.87	
n-Propyl alcohol	2.00	0.72
Isobutyl alcohol	2.55	
Toluene	3.03	
n-Butyl alcohol	3.50	
1,4-Dioxan	4.05	0.50

Liquid phase: acetylated polyethyleneglycol adipate.
Support: Shimalite F; column temp. 110 °C.

The determination of both free formaldehyde and phenol is important as a control test in the production of phenolic resins and in determining the effects of residual reactants on polymer properties. Kinetic studies of formaldehyde–phenol condensations also rely on the analysis of reactants. Stevens and Percival[245] described the GLC determination of free formaldehyde and free phenol in phenolic resins. A

Wilkins Aerograph A-350 dual column gas chromatograph was used with the following columns: (A) (for phenol) 12 ft. $\times \frac{1}{4}$ in. o.d. copper tubing packed with 10% G.E. silicone SF-96 on Fluoropak; (B) (for formaldehyde) 16 ft. $\times \frac{1}{4}$ in. o.d. copper tubing packed with 10% sucrose octaacetate on Teflon 6. The operating conditions were detection cell, 250°C; d.c. current, 200 mA; injection temp., 250°C; column temp., 130°C; helium flow rate at 60 p.s.i.g., 120 ml/min for column A and 59 ml/min for column B. Figures 40 and 41 illustrate the gas chromatograms of formaldehyde

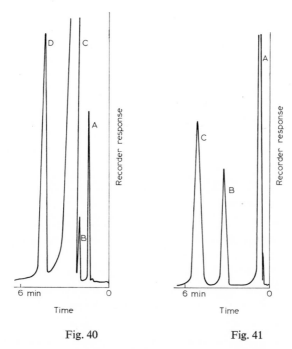

Fig. 40   Fig. 41

Fig. 40. Determination of formaldehyde on column B. (A) Formaldehyde; (B) methanol; (C) water; (D) 1-butanol.

Fig. 41. Determination of phenol on column A. (A) Ether; (B) phenol; (C) *m*-cresol.

on column B and phenol on column A, respectively. As internal standards, *n*-butanol and *m*-cresol were chosen for formaldehyde and phenol, respectively, because they gave peaks symmetric enough to allow peak height analysis. Both columns A and B eluted their respective solvents with little tailing. The retention times are shown in Table 28.

The determination of formaldehyde in acetaldehyde based on gas-chromatographic separations of mixtures before or after the reduction of formaldehyde to methanol with borohydride were described by Harrison[246]. The latter technique is capable of determining 400 p.p.m. of formaldehyde in acetaldehyde, but the presence of methanol and methyl formate were found to interfere. The former technique

TABLE 28

RETENTION TIMES

Compound	Retention time[a] (min)	
	Column A[b]	Column B[c]
Formaldehyde		1.2
Methanol[d]	0.4	1.9
Water	0.3	2.1
Ether	0.5	
1-Butanol	0.8	4.4
Phenol	3.0	
m-Cresol	5.0	

[a] From point of injection.
[b] Flow rate 120 ml He/min.
[c] Flow rate 59 ml He/min.
[d] Usually present as stabilizer in phenolic resins.

in which a borohydride reactor reduces formaldehyde after separation was found to determine formaldehyde in acetaldehyde down to 10 p.p.m. in the presence of methanol and methyl formate.

In the first method, the sample is dissolved in water and allowed to react with an aqueous solution of sodium borohydride–sodium hydroxide. The formaldehyde is quantitatively converted to methanol, which is then determined by GLC. In the second method the borohydride (KBH$_4$) is contained in a small reactor tube attached to the end of a gas–liquid chromatographic column and the sample is injected directly on the column. In this case separation of the components occurs before the reduction of the formaldehyde.

A Perkin-Elmer F-11 chromatograph equipped with a flame-ionization detector was used with a column of 10 ft. × $\frac{1}{4}$ in. o.d. Staybrite tubing packed with 5% Citroflex 1 + 8 (O-acetyl triethylhexyl citrate, Pfizer Ltd.) and 0.4% orthophosphoric acid on Chromosorb G. (This column required aging at 120°C for 12 h before use.) Attached to the end of this column was a reactor of 3 ft. × $\frac{1}{4}$ in. i.d. Staybrite tubing packed with evenly ground potassium borohydride. Column temperature was 60°C; the pre-heater about 200°C and the carrier gas nitrogen at 30 p.s.i.g. with a linear gas flow rate of 3 cm/sec. Using the above method it was possible to detect easily 10 p.p.m. of formaldehyde in acetaldehyde as shown in Fig. 42.

A variety of paper and thin-layer chromatographic techniques had been reported for the separation and detection of volatile aldehydes (as well as ketones) primarily as their respective 2,4-dinitrophenylhydrazine (2,4-DNPH) derivative. Meigh[247] separated the 2,4-dinitrophenylhydrazines of a number of aldehydes and ketones on Whatman No. 1 paper in a 20-cm development at 20°C using methanol–heptanes as developer. Yellow spots of the original dinitrophenylhydrazines were

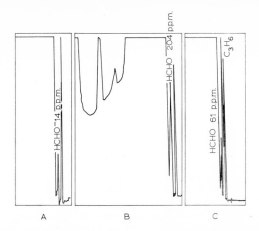

Fig. 42. GLC separations of formaldehydes in acetaldehyde. (A) Chromatogram of 14 p.p.m. of formaldehyde in acetaldehyde; (B) chromatogram of 204 p.p.m. of formaldehyde in acetaldehyde; (C) chromatogram of 61 p.p.m. of formaldehyde in acetaldehyde containing propylene.

revealed more clearly by spraying with aqueous 10% sodium hydroxide. The red-brown color obtained varies slightly with the compound and is sometimes an aid to identification. Table 29 lists the R_F values obtained for the dinitrophenylhydrazines as determined above.

TABLE 29

PAPER CHROMATOGRAPHY[a] OF 2,4-DINITROPHENYLHYDRAZINES ON WHATMAN NO. 1 PAPER

Compound	R_F	Compound	R_F
Formaldehyde	0.09	Furfural	0.06
Acetaldehyde	0.18	Acrolein	0.21
Propionaldehyde	0.32	Crotonaldehyde	0.22
n-Butyraldehyde	0.42	Acetone	0.30
n-Valeraldehyde	0.54	Methylethyl ketone	0.43
n-Caproaldehyde	0.57	Methyl n-propyl ketone	0.53
		Methyl n-butyl ketone	0.65

[a] Developer: methanol–heptane.

Other paper chromatographic separations of derivatives of aldehydes have been reported using normal filter paper[248] and paper impregnated with propylene glycol[249, 250].

The TLC of a number of 2,4-dinitrophenylhydrazines of aldehydes has been described by Bloem[251]. The 2,4-DNPH compounds were prepared by the method of Jones[252] using 2 N hydrochloric acid as solvent. The hydrazines were dissolved in 1,2-dichloroethane and applied to chromatoplates coated with a 0.3 mm layer of a suspension of 10 g Silica Gel G and a solution of 3 g Carbowax 4000 and 35 mg

Tinopal WG (Geigy) in 15 ml distilled water and 5 ml 1 N NaOH. (The plates were dried at 80°C for 30 min before use.) The plates were developed in benzene–heptane (65:35) (12 cm, 40 min) then air dried. The compounds showed up as colored areas on a white background; some of these change color within a few hours. Very small quantities could be detected under UV light using a Blacklight lamp, Philips type HPW.

Table 30 lists the R_F values and colors and sensitivities of a number of 2,4-DNPH compounds on Silica Gel G plates as prepared above. The temperature of drying and activation was found to be important. Higher temperatures increased the

TABLE 30

R_F VALUES AND COLORS OF CARBONYL COMPOUNDS

Compounds	R_F	Immediate color	Color after 6 h	Sensitivity (γ)
Formaldehyde	0.37	Yellow	Dark reddish brown	0.2
Acetaldehyde	0.52	Yellow	Ochre	1.0
Propionaldehyde	0.63	Yellow	Ochre	0.5
Butyraldehyde	0.69	Yellow	Brown	0.5
Valeraldehyde	0.75	Yellow	Brown	0.5
Caproaldehyde	0.81	Yellow	Brown	0.5
Oenanthal	0.85	Yellow	Brown	0.5
Acrolein dimer	0.36	Yellow	Reddish brown	0.2
Crotonaldehyde	0.57	Yellow	Greyish brown	0.2
Glyoxal	0.02	Pink	Purple	2.0
Malonaldehyde	0.29	Very faint yellow	Vivid orange-brown	1.0
Glutaraldehyde	0.07–0.14	Yellow	Dark green	0.2
Glyceraldehyde	0.10	Blue	Lilac	5.0
Pyruvaldehyde	0.08; 0.15; 0.36	Red; blue; red	Red; yellow; pink	2.0
2-Hydroxyadipaldehyde	0.00	Brown	Dark brown	0.5
Salicylaldehyde	0.04	Brick-red	Faint brown	0.5
Terephthalaldehyde	0.15	Faint brown	Purple	1.0
Cinnamaldehyde	0.46; 0.66	Yellow	Yellow	5.0
Furfural	0.25; 0.49	Purple	Dark red	0.5
Acetone	0.73	Blue	Yellow	2.0

R_F values but decreased the separation of the successive compounds. Higher temperatures also effect the colors, e.g. at 150°C formaldehyde turns green and at 80°C yellow-brown; at 150°C crotonaldehyde turns orange-brown and at 80°C greyish-brown. The R_F value combined with the color was found to be specific for each compound. The method as described was used for determining aldehydes in leather in which aldehyde or glyceraldehyde are used on a commercial scale. Of added interest, however, is the fact that malonaldehyde, glyoxal, glyceraldehyde, acrolein, crotonaldehyde and furfural are formed in situ during tanning[253, 254].

The separation of 2,4-dinitrophenylhydrazines of a variety of carbonyl compounds by TLC using both silica gel and alumina plates (in some cases with silver nitrate) was investigated by Denti and Luboz[255]. The most useful solvent mixtures employed were: (I) benzene–petroleum ether (b.p. 40–70°C) (60:40); (II) chloroform–petroleum ether (b.p. 40–70°C) (75:25); (III) benzene–n-hexane (50:50) and (IV) cyclohexane–nitrobenzene–petroleum ether (b.p. 40–70°C) (30:15:10). The plates were developed up to 14 cm at 18–22°C in a Shandon chamber (type 2842). Tables 31, 32 and 33 list the R_F values (relative to R_F of formaldehyde-2,4-DNPH)

TABLE 31

THIN-LAYER CHROMATOGRAPHY ON SILICA GEL G OF SOME 2,4-DNPH DERIVATIVES

2,4-DNPH from	R_F (average from 5 determinations)			
	I^a	II	III	IV
Formaldehyde	(1)	(1)	(1)	(1)
Propionaldehyde	1.38	1.36	1.37	1.72
Butyraldehyde	1.68	1.41	1.55	2.10
Isobutyraldehyde	1.90	1.63	2.10	2.24
Methylethyl ketone	1.50	1.46	1.62	1.95
n-Valeraldehyde	1.92	1.60	2.05	2.41
3-Pentanone	2.00	1.51	2.14	2.34
2-Pentanone	1.76	1.41	1.81	2.25
Cyclopentanone	0.88	1.03	0.98	1.53
n-Caproaldehyde	2.36	1.74	2.22	2.58
α-Methyl-n-valeraldehyde	0.84	0.98	0.90	1.43
4-Methyl-2-pentanone	1.95	1.57	2.05	2.61
2-Hexenal	2.18	1.49	2.08	2.28
5-Hexen-2-one	1.72	1.42	1.67	2.12
Cyclohexanone	1.14	1.44	1.16	2.35
Oenanthaldehyde	2.68	1.79	2.38	2.86
Benzaldehyde	1.85	1.39	1.90	1.65

^{a} The numbers refer to the solvents used as described in text.

of some 2,4-DNPH derivatives on Silica Gel G, Silica Gel G + 25% AgNO$_3$ and Alumina G + 25% AgNO$_3$, respectively. The TLC identification of formaldehyde in muscle tissues, fish and other aquatic animals was described by Harada et al.[256]. The formaldehyde in muscle was extracted with petroleum ether after converting it into its 2,4-DNPH derivative. The ether was then evaporated to dryness and the residue was re-extracted with a small volume of toluene, applied on activated Silica Gel G plates and developed with hexane–ethanol–acetic acid (90:10:3).

Other thin-layer chromatographic procedures for the analysis of aldehydes and their derivatives include silica gel impregnated with 3% polyvinyl alcohol[257], a mixture of magnesia and Celite[258], activated alumina and silica gel impregnated with

TABLE 32

THIN-LAYER CHROMATOGRAPHY ON SILICA GEL G + 25% AgNO₃ OF SOME 2,4-DNPH DERIVATIVES

2,4-DNPH from	R_F (average from 5 determinations)			
	I^a	II	III	IV
Formaldehyde	(1)	(1)	(1)	(1)
Propionaldehyde	1.46	1.23	1.73	1.49
Butyraldehyde	1.81	1.22	2.11	1.92
Isobutyraldehyde	1.96	1.60	2.48	2.09
Methylethyl ketone	1.43	1.01	1.77	1.51
n-Valeraldehyde	1.90	1.45	2.21	2.17
3-Pentanone	1.83	1.20	2.40	1.92
2-Pentanone	1.73	1.04	1.95	1.84
Cyclopentanone	0.90	0.47	1.03	1.11
n-Caproaldehyde	2.13	1.57	2.49	Solvent front
α-Methyl-n-valeraldehyde	0.78	0.64	0.99	0.86
4-Methyl-2-pentanone	2.12	1.07	2.17	2.09
2-Hexenal	1.90	1.16	2.25	1.98
5-Hexen-2-one	0.18	0.20	0.26	0.14
Cyclohexanone	0.98	0.35	1.24	1.15
Oenanthaldehyde	2.46	1.74	2.47	Solvent front
Benzaldehyde	1.75	1.10	1.74	1.46

ᵃ The numbers refer to the solvents used.

TABLE 33

THIN-LAYER CHROMATOGRAPHY ON ALUMINA G + 25% AgNO₃ OF SOME 2,4-DNPH DERIVATIVES

2,4-DNPH from	R_F (average from 5 determinations)			
	I^a	II	III	IV
Formaldehyde	(1)	(1)	(1)	(1)
Propionaldehyde	1.10	1.25	1.20	1.08
Butyraldehyde	1.14	1.10	1.26	1.12
Isobutyraldehyde	1.19	1.40	1.26	1.19
Methylethyl ketone	1.12	0.65	1.17	1.08
n-Valeraldehyde	1.18	1.05	1.47	1.17
3-Pentanone	1.22	0.74	1.53	1.10
2-Pentanone	1.19	0.56	1.41	1.06
Cyclopentanone	0.86	0.30	0.75	0.75
n-Caproaldehyde	1.19	1.08	1.37	1.20
α-Methyl-n-valeraldehyde	0.85	0.35	0.87	0.77
4-Methyl-2-pentanone	1.20	0.53	1.35	1.13
2-Hexenal	1.11	0.53	1.25	1.10
5-Hexen-2-one	0.20	0.03	0.13	0.03
Cyclohexanone	0.74	0.15	0.70	0.69
Oenanthaldehyde	1.19	1.20	1.33	1.23
Benzaldehyde	0.95	0.47	0.90	0.99

ᵃ The numbers refer to the solvents used.

2-phenoxyethanol[259]. Column chromatographic techniques for the separation of aldehydes include the use of polyethylene[260], magnesia[261], and column-partition systems using a stationary phase of ethanolamine[262] or acetonitrile[263].

8.2 Acetaldehyde

Acetaldehyde is produced commercially by (a) the vapor-phase dehydrogenation or partial oxidation of ethanol, (b) high-temperature oxidation of saturated hydrocarbons, (c) the liquid-phase hydration of acetylene, or (d) the liquid-phase oxidation of ethylene, and is used in considerable quantities in the manufacture of a host of important products including acetic acid, acetic anhydride, butanol, butyraldehyde, chloral, pentaerythritol, peroxyacetic acid, acrylonitrile, cellulose acetate, and vinyl acetate resins.

Acetaldehyde is a highly reactive compound exhibiting the general reactions of aldehydes (e.g. under suitable conditions the oxygen or any hydrogen may be replaced) and hence undergoes a number of condensation, addition or polymerization reactions. Acetaldehyde has a number of industrial uses; the condensation products with phenol or urea are thermosetting resins; the reaction with aliphatic and aromatic amines yield Schiff bases which are used as accelerators and antioxidants in the rubber industry and for the production of butadiene and polyvinyl acetal resins. It has been used as a preservative for fruit and fish, as a denaturant for alcohol, in fuel compositions, for hardening of gelatin, glue, and casein products, for the prevention of mold growth on leather and as a solvent in the rubber, tanning and paper industries. It is the product of most hydrocarbon oxidations, a normal intermediate product in the respiration of higher plants and occurs in traces in all ripe fruits and may also form in wine and other alcoholic beverages after exposure to air. Acetaldehyde is an intermediate product in the metabolism of sugars in the body and hence occurs in traces in blood, and also has been reported to cause significant effects on biotransformation and metabolism of catechol and indole amines[264, 265]. It has been reported in fresh leaf tobacco, as well as in tobacco smoke and along with propionaldehyde, butyraldehyde, isobutyraldehyde, isovaleraldehyde, and crotonaldehyde has been identified in automobile and diesel exhaust by paper[266] as well as gas chromatography[267-270]. The toxicity[271], pharmacology[272], metabolism[273,274], and mutagenicity[275] of acetaldehyde have been described. The analysis of acetaldehyde has been accomplished using polarographic[276], argentometric titration[277], mercurometric oxidation[278], and ultraviolet spectrophotometric[279] techniques.

The GLC determination of acetaldehyde as well as the by-products of its manufacture, e.g. water, acetone, propionaldehyde and crotonaldehyde, has been effected[280] using two serially joined columns, 4.5 mm diam. and 2.5 and 1.5 m long, respectively, filled with 1% PEGA and 1% PEG-200, respectively, on NaCl (grain size 0.16 to 0.25 mm). The column temperature was 102°C, the carrier gas was helium at 3 l/h and detection was by thermal conductivity. The time of analysis was 20–25 min with a relative error of 5–7%. The use of GLC to determine admixture (0.007–0.05%) of acetaldehyde and propylene oxide in ethylene oxide was described by Mokeeva and Tsarfin[281]. The chromatograph was equipped with a flame-ionization detector and the best separation was obtained with 30% E-301 silicone elastomer coated on Celite 545 (grain 0.1 mm). The column was a polyethylene tube 600×0.4 cm kept at 28°C in a Heppler ultrathermostat. The velocity of the carrier hydrogen was 48 ml/min, air 300 ml/min and the pressure at the column inlet was 0.6 atm. Under these conditions, the corrected retention volumes (V_r^0) were 237.7 ml for acetalde- hyde, 349.6 ml for ethylene oxide, and 672.7 ml for propylene oxide. Figure 43

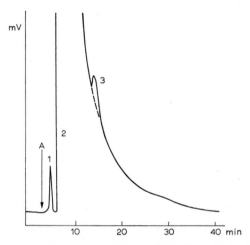

Fig. 43. Chromatogram of 0.035% acetaldehyde and 0.035% propylene oxide in ethylene oxide. 1, Acetaldehyde; 2, ethylene oxide; 3, propylene oxide; A, sensitivity 10 mV.

shows a chromatogram of 0.035% acetaldehyde and 0.035% propylene oxide in ethylene oxide. The determination required about 45 min and the relative error of determination for acetaldehyde and propylene oxide was below 7.3 and 8.5%, respectively.

The determination of small amounts of acetaldehyde in the presence of form- aldehyde, acetone, methanol, ethanol, isopropanol, propanol, ethyl, butyl and iso- amyl acetates, ethyl chloride, dichloroethane, chloroform, carbon tetrachloride, n-hexane, cyclohexane, n-pentane, toluene or petroleum ether was achieved by GLC[282] and a thermal-conductivity detector and 6×980 mm columns impregnated 3:1 with poly(vinyl butyl ether) (M.W. 2000) on Diatomite 1NZ-600 (0.25–50 mm) at 75°C with diethyl ether as an internal standard.

The chromatographic determination of acetaldehyde and ethanol in the alde- hyde fraction of acetone production was effected using a 2-m long column packed with 20% Reoplex 400 or Chromosorb W, at 73.5°C, a flame-ionization detector and air as carrier gas at 32 ml/min. The relative errors were 1.8–6% for acetaldehyde and 1.8–8.6% for ethanol[283].

Kyryacos *et al.*[284] studied the mechanism of hydrocarbon combustion through separation and identification of the intermediates formed in cool-flame oxidation by means of gas chromatography. Components previously not reported have been separated and identified including ethylene oxide, propionaldehyde, acetone, acrolein, *cis*- and *trans*-2,5-dimethyltetrahydrofuran, *n*-butyraldehyde, methylethyl ketone, methylvinyl ketone, 2-ethyltetrahydrofuran, methanol, and crotonaldehyde. In addition, the previously identified acetaldehyde, propylene oxide, formaldehyde, propanol, and ethanol were also separated. An 8 ft. × 5 mm i.d. column containing 40% polyethylene glycol-400 on 30–60 mesh Firebrick C-22 (Johns-Manville) maintained at 35°C with a helium flow rate of 50 ml/min was used for the separation of olefin oxides and carbonyl compounds. A 4-ft. column containing the same liquid

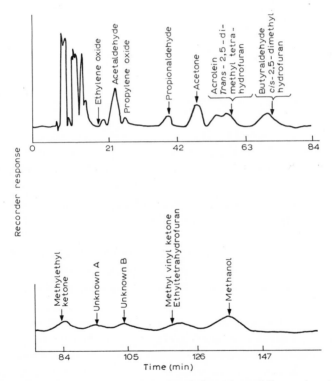

Fig. 44. Chromatogram of a 20 ml sample taken directly from the cool-flame exhaust gases. 8′-Polyethylene glycol-400 column; temperature, 30°C; flow rate, 50 ml/min.

phase and support was used for the separation of alcohols in cool-flame combustion products. This column was maintained at 60°C with a helium flow rate of 125 ml/min. A third column, 2 ft. in length (containing the same liquid phase and solid support as above for the first column) was maintained at 60°C with a flow rate of helium of 400 ml/min for the determination of water.

References pp. 152–160

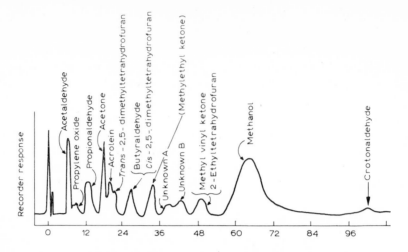

Fig. 45. Chromatogram of 0.1 ml sample exhaust liquid condensate. 8′-Polyethylene glycol-400 column; temperature, 30 °C; flow rate, 150 ml/min.

Figure 44 illustrates a chromatogram obtained from a 20 ml gas sample taken directly from the cool-flame exhaust gases and analyzed on the 8 ft. polyethylene glycol-400 column. Figure 45 is a chromatogram on an 8 ft. polyethylene glycol-400 column, of 0.1 ml of a non-homogeneous liquid condensate obtained by the dry-ice condensation of exhaust gases during the cool-flame combustion of *n*-hexane.

TABLE 34

OPERATING CONDITIONS FOR CHROMATOGRAPHIC COLUMNS

Cutter column
0.25 in. o.d. 0.18 in. i.d. 304 stainless steel 12 ft. long
20% 1,2,3-tris(2-cyanoethoxy) propane on 45–60 mesh
N.A.W. Chromosorb-W
Temperature: 22 °C
Carrier: helium
Flow rate: 100 ml/min
Analytical column
0.125 in. o.d. 0.105 in. i.d. 304 stainless steel 12 ft. long
50–80 mesh Porapak Q
Temperature: 156 °C
Carrier: helium
Flow rate: 50 ml/min
Soxhlet extracted with methanol for 2 h. Packed and conditioned at 175 °C for 60 days

A direct automatic gas chromatographic analysis of low molecular weight substituted organic compounds, *e.g.* C_2–C_4 aldehydes, C_3 and C_4 ketones, and various other oxygen-, nitrogen-, and halogen-containing organic compounds present in combustion effluents was reported by Bellar and Ligsby[285]. The method

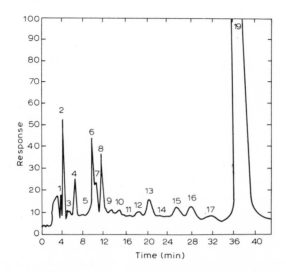

Fig.46. Chromatogram of auto exhaust diluted 13.7:1 with air.

	Compound	Full scale response (amps) equal to	Concentration (p.p.m.)
1	Methanol	1×10^{-11}	0.69
2	Acetaldehyde		
3	Unknown	5×10^{-12}	
4	Unknown	5×10^{-12}	
5	Unknown	5×10^{-12}	
6	Acrolein	5×10^{-12}	0.38
7	Propanol	5×10^{-12}	0.13
8	Acetone	5×10^{-12}	0.25
9	Unknown	5×10^{-12}	
10	Unknown	5×10^{-12}	
11	Unknown	5×10^{-12}	
12	Unknown	5×10^{-12}	
13	2-Methylpropanol	5×10^{-12}	0.13
14	Unknown	5×10^{-12}	
15	Butanol	5×10^{-12}	0.082
16	Methylethyl ketone	5×10^{-12}	0.090
17	Unknown	5×10^{-12}	
18	Benzene	5×10^{-12}	

emp!oys two columns: the first (the Cutter column) vents the hydrocarbons and retains the compounds of interest; the retained compounds are then transferred to the second (analytical) column analysis. Table 34 lists the operating conditions for the chromatographic columns. Figure 46 is a chromatogram of a sample obtained by variable dilution samples[286] from a car operated on a chassis dynamometer. Figure 47 shows chromatograms of the effluent from an incinerator burning wood under two conditions. (The samples were taken before and after forced air was intro-

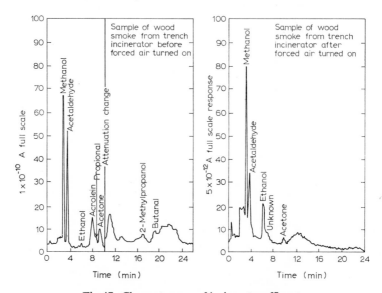

Fig. 47. Chromatogram of incinerator effluent.

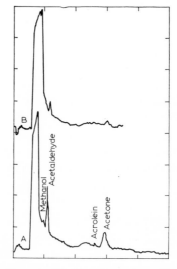

Fig. 48. A chromatogram of ambient air. (A) Air collected in traffic; (B) air collected in open field.

duced.) The incinerator samples show a considerably higher ratio of alcohols to aldehydes than the auto exhaust does, a reflection of the differences in type of combustion and source. Figure 48 shows chromatograms of ambient air collected at various sites. These chromatograms of samples from various sources suggest the broad applicability of this technique, *e.g.* since most oxygenated and nitrogen containing materials of approximately the same molecular weight are observed in complex mixture by thin layer technique, the difference in each source can be ascertained. Table 35 lists retention times for various compounds detected in trace quantities in the presence of relatively abundant hydrocarbons by the system described.

TABLE 35

RETENTION TIMES OF VARIOUS COMPOUNDS ELUTING FROM THE ANALYTICAL COLUMN–TRAP COMBINATION

Compound	Time (min)	Compound	Time (min)
Methanol	3.63	Allyl alcohol	15.50
Acetaldehyde	4.05	2,3-Butylene oxide	16.31
Ethylene oxide	4.21	Methyl acrolein	16.95
Methyl formate	5.11	2-Methylpropanal	17.82
Ethanol	7.17	Isobutylene oxide	17.88
Furan	8.05	Propionitrile	18.28
Propylene oxide	8.32	2-Methylfuran	18.78
Acetonitrile	8.51	1,2-Butylene oxide	20.11
Acrolein	8.92	2-Methylpropane-2-ol	20.68
Propanal	9.35	Tetrahydrofuran	22.23
Dichloromethane	9.56	Butanal	22.27
Methyl nitrate	9.71	Vinyl methyl ketone	22.38
Acetone	10.45	2,3-Butanedione	22.84
Ethyl formate	11.23	Methylethyl ketone	24.60
Acrylonitrile	11.78	Cyclobutanone	31.29
Methylal	12.42	2,2-Dimethylbutanal	31.53
2-Propanol	12.80	Benzene	31.60
Methyl acetate	12.89	Crotonaldehyde	34.50
Nitromethane	13.44	2-Methylpropanol	35.55

The identification of a number of oxygenates (*e.g.* acetaldehyde, propionaldehyde, isobutyraldehyde, n-butyraldehyde, acetone, methylethyl ketone, methanol and ethanol) in automobile exhausts has been accomplished by combined GLC and infrared techniques[268]. The oxygenates were separated from exhaust gases by scrubbing with a 1% solution of $NaHSO_3$. Oxygenates which eluted ahead of water were separated from the solution in a preparatory column. The carbonyls indicated in the chromatograms were derived from the thermal decomposition of the bisulfite complexes of these compounds in the chromatographic column. The eluted oxygenates were collected in a cold trapping needle and charged to an analytical GLC unit employing thermal conductivity detection. A preparatory unit consisting of a Beckman GC-2 Chromatograph with a column 2 ft. × ⅝ in. diameter containing 9% Carbowax 600 on unsized Teflon was used for separating oxygenates from water. A custom-built analytical GLC unit was used to separate the oxygenates obtained from the preparatory GLC unit. This unit contained a 20 ft. × ¼ in. column of 9% Carbowax on unsized Teflon. After identification of the oxygenates present, GLC analyses employing flame-ionization detection were made directly upon the scrubber solutions and also on the preparatory column effluents.

Figure 49 illustrates the trapping needle for preparatory GLC. The trapping assembly is connected to the inlet of the thermal conductivity GLC unit as shown in Fig. 50.

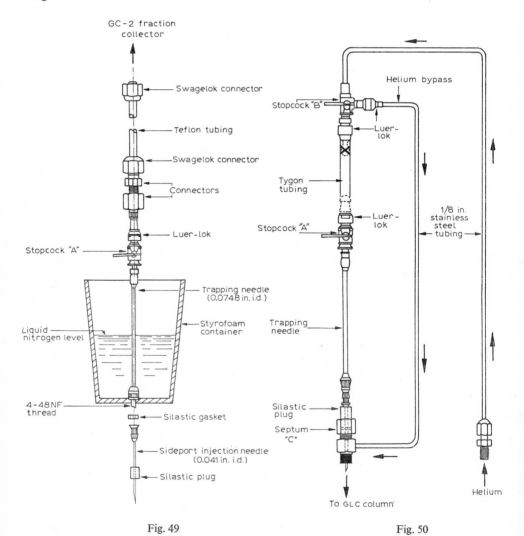

Fig. 49   Fig. 50

Fig. 49. Trapping needle for preparatory GLC unit.

Fig. 50. Injection system for thermal conductivity GLC unit.

Direct GLC of the 2,4-dinitrophenylhydrazines of a variety of carbonyl compounds, *e.g.* alkyl aldehydes and ketones up to C-13, alkyl phenyl ketones, some
terpene aldehydes and ketones, and aromatic aldehydes, has been accomplished[287].

An Aerograph Hy-Fi Model 600 gas chromatograph was used equipped with
a flame-ionization detector and modified with a Thermis Temp temperature controller, Model 63 RA (Yellow Springs Instrument Co.) and a glass insert for the injection
chamber. The column was a 2 m × ⅛ in. o.d., 0.085 in. i.d. stainless steel tube containing 10% SF-96 (100) silicone oil (General Electric) coated on Chromosorb W,
acid-washed. The column and injector temperatures were 250°C and 275°C, respectively, and helium was the carrier gas at 86 ml/min. The detectability of the 2,4-DNPH
derivatives was in the range of 10^{-3}–10^{-5} mg. Figure 51 shows a gas chromatographic separation of the 2,4-DNPH derivatives of *n*-alkyl aldehydes and Table 36
lists the retention times of the 2,4-DNPH derivatives of various carbonyl compounds.

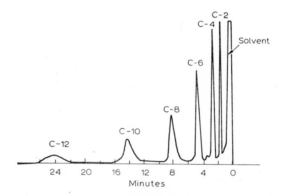

Fig. 51. Gas chromatographic separation of the 2,4-DNPH derivatives of *n*-alkylaldehydes. Sample
size, 3 μl containing 1% of each derivative in benzene; column temperature, 250°C; injector temperature, 267°C; attenuation, 64×; helium flow, 86 ml/min.

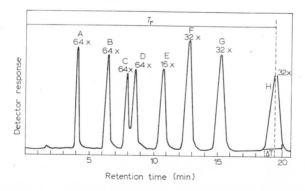

Fig. 52. Chromatogram with DMAB–oxalic acid (6.3 to 5.0 mg) as exchange reagents in flash
exchange of 6.0 mg of DNP standard mixture and 2 mg of Celite. A, Acetaldehyde; B, propionaldehyde; C, acetone; D, isobutyraldehyde; E, butyraldehyde; F, 2-butanone; G, isovaleraldehyde;
H, valeraldehyde.

TABLE 36

RETENTION TIMES OF 2,4-DNP DERIVATIVES OF VARIOUS CARBONYL COMPOUNDS[a]

Parent compound	Retention time (min)	Parent compound	Retention time (min)
Formaldehyde	1.3	Benzaldehyde	10.5
Acetaldehyde	1.7	Propiophenone	10.6
Acetone	2.2	Nonanal	10.9
Propionaldehyde	2.3	Acetophenone	11.4
Isobutyraldehyde	2.5	d-Citronellal	11.6
2-Butanone	2.8	dl-Menthone	11.9
Butyraldehyde	3.0	Butyrophenone	12.6
Diacetyl	3.0 (mono)	Hydrocinnamaldehyde	13.8
Isovaleraldehyde	3.3	Decanal	14.4
Crotonaldehyde	3.4	Valerophenone	16.2
2-Pentanone	3.5	1-Carvone	16.3
Valeraldehyde	3.6	2-Undecanone	16.6
2-Hexanone	4.4	Undecanal	19.4
Hexaldehyde	4.9	p-Anisaldehyde	22.8
2-Heptanone	5.7	Dodecanal	24.4
Furfural	5.8	Caprylophenone	36.6
Phenylacetaldehyde	5.8	o-, m-,p-Hydroxybenzaldehydes	N.R.[b]
Heptaldehyde	6.4	Cinnamic aldehyde	N.R.
2-Octanone	7.4	Citral	N.R.
Octanal	8.4		
2-Nonanone	9.5		

[a] Conditions as given in text.
[b] N.R. = No response.

Jones and Monroe[288] described a flash exchange method for the quantitative GLC analysis from aliphatic carbonyls as their 2,4-dinitrophenylhydrazines. A mixture of oxalic acid dihydrate and p-dimethylaminobenzaldehyde (DMAB) was used to effect maximum regeneration of the carbonyls from their 2,4-DNPH derivatives. An F&M Model 300 gas chromatograph was used equipped with a Model 1609 flame-ionization detector and a 16 ft. × ¼ in. column of 8% XE-60 and 12% QF-1 on Chromosorb P. Figure 52 shows a chromatogram obtained with DMAB–oxalic acid as exchange reagents in the flask exchange of an alkyl carbonyl-2,4-DNPH mixture.

Acetaldehyde is the first and the only specific oxidation product of ethanol in the intermediary metabolism and is present in the blood after the ingestion of alcohol. It has been postulated[289] that acute and chronic exposure of the brain to acetaldehyde may be responsible for brain damage incurred by chronic alcoholics. Gas chromatography has played an important role in the determination of acetaldehyde concentrations both in blood and alveolar air. For example, Freund and O'Hollaren[290] described a serial GLC determination of acetaldehyde, acetone, and alveolar air. An F & M Model 400 gas chromatograph with a hydrogen-flame detector was used with a 10 ft. × $\frac{1}{8}$ in. o.d. stainless steel column packed with 40% caster wax on 60–80 mesh Gas Chrom RP (firebrick). The operating temperatures were column 103°, flask heater 140°, detector 220°C and the carrier gas was nitrogen at a flow rate of 30 ml/min. The retention times in minutes were acetaldehyde 2.1, acetone 3.1, unknown (ethyl acetate) 3.8, and ethanol 5. Acetic acid was not detectable with the column used. The concentration of acetaldehyde in alveolar air was suggested to correspond to the rate of formation of acetaldehyde in the liver (acetaldehyde in the hepatic venous blood is carried directly to the lung, where it may be expected to equilibrate rapidly with the alveolar air since its boiling point is 21°C and it readily diffuses through tissues).

Fukui[291] also developed a GLC method for the determination of acetaldehydes in expired air after ingestion of alcohol and demonstrated the utility of a sampling bag made of Saran (polyvinylidene chloride film) for breath analysis. The concentration of acetaldehyde and ethanol in the expired air was obtained under the following conditions.

Apparatus	Shimadzu GC-1C gas chromatograph equipped with FID-1B hydrogen-flame-ionization detector and with gas sampler
Column	Dual stainless steel columns, each 3 mm i.d. × 1.9 m
Column packings	25% PEG (polyethylene glycol) 6000 on 60–80 mesh Shimalite
Column temp.	60°C
Detector temp.	90°C
Carrier gas flow	Nitrogen, 73 ml/min (at 20°C)
Hydrogen flow	45 ml/min
Air flow	0.8 l/min
Sensitivity	× 10^4
Range	0.2 V for AcH, 12.8 V for EtOH
Sample size	5 ml
Recorder	Shimadzu 250AO11 electronic self-balancing recorder set at 1 mV for full scale response.

Figure 53 illustrates typical chromatograms of (1) mixed standard gas of acetaldehyde and ethanol, (2) and (3) breath samples of subjects first hour after ethanol ingestion and control subject, respectively. In this study it was established that the concentrations of acetaldehyde were relatively well correlated with ethanol in the particular individuals tested but the existence of individual variation in ethanol metabolism is emphasized.

A gas chromatographic method for the determination of acetaldehyde and ethanol blood was developed that could be applied to 50 μl samples of blood and detect down to 4 μg/ml of acetaldehyde[292]. An F & M Model 402 gas chromatograph was used with a flame-ionization detector and a 6 ft. × $\frac{1}{4}$ in. i.d. U-shaped glass column packed with Polypak-2 (80–120 mesh, Hewlett-Packard) pretreated by washing in sequence with heptane, ethyl acetate and 10% HCl in methanol and drying at 110°C for 2 h. The temperature of the column and injection port was 110°C, the detector temperature was 135°C and helium was the carrier gas at 96 ml/min. A chromatogram of a mixture of acetaldehyde, ethanol, acetone, and *tert.*-butanol is shown in Fig. 54. The retention times for the four compounds are acetaldehyde 2.8, ethanol 4.6, acetone 8.9 and *tert.*-butanol 18.7 min.

Baker *et al.*[293] described the simultaneous determination of lower alcohols, acetaldehyde and acetone in blood by GLC. By using a column containing Porapak Q

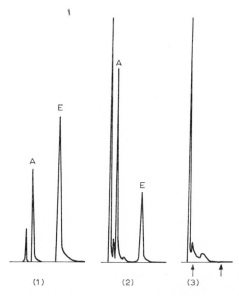

Fig. 53. Some examples of chromatograms. For operating conditions, see text. (1) Mixed standard gas of AcH (A) and EtOH (E); (2) breath sample from Subject No. 17 obtained on the first hour of drinking experiment. Retention time is 1.0 min for AcH (A) and 3.2 min for EtOH (E); (3) control breath of Subject No. 9. Note that neither AcH nor EtOH is detected, as indicated by the arrows.

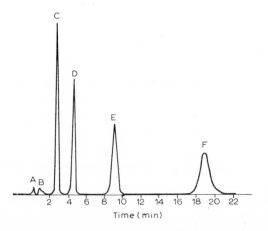

Fig. 54. Typical gas chromatogram obtained by the described procedure. The peaks are: A, air; B, water response; C, acetaldehyde; D, ethanol; E, acetone; F, *tert.*-butanol.

and interchangeable glass inlets, complete separation of these volatiles was achieved and whole blood may be directly injected without prior processing. A Model 220 Micro-Tek gas chromatograph was used with a flame-ionization detector and a 183 cm × 5 mm i.d. glass column filled with Porapak Q (60–80 mesh) preconditioned for 18 h at 225°C. The column temperature was 100°C and the inlet and detector

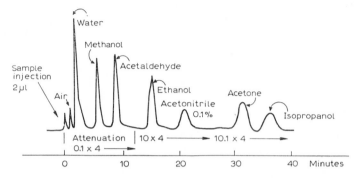

Fig. 55. Chromatogram of standard solution of methanol, acetaldehyde, ethanol, acetone and iso-propanol with acetonitrile as internal standard.

were maintained at 125°C. The flow rate of nitrogen carrier gas was 35 ml/min. The chromatogram of stock standards (Fig. 55) shows the complete separation of all of the test compounds. The retention times of these substances relative to acetonitrile are shown in Table 37.

TABLE 37

RELATIVE RETENTION TIMES OF BLOOD VOLATILES AS COMPARED TO ACETONITRILE

Water	0.12	Acetonitrile	1.0
Methanol	0.29	Acetone	1.5
Acetaldehyde	0.44	Isopropanol	1.7
Ethanol	0.73	n-Propanol	2.4

Figure 56 shows the relationship of the concentrations of blood volatiles to that of the internal standard acetonitrile and illustrates linearity over a wide range of concentrations. The chromatogram of a patient suffering from acute alcoholism is shown in Fig. 57; the presence of methanol, isopropanol and possibly acetaldehyde is of interest.

A variety of paper, thin-layer, column, and ion-exchange chromatographic techniques are also of utility for the separation and identification of acetaldehyde and a variety of carbonyls. The chromatography of the 2,4-dinitrophenylhydrazones of some aldehydes and ketones in tobacco smoke has been described by Buyske et al.[294]. Whatman No. 1 paper coated with dimethylformamide and developed with n-hexane effected the separation of the 2,4-DNPH derivatives of furfural, formaldehyde, acetaldehyde, propionaldehyde, acetone, methylethyl ketone, diethyl ketone, and butyraldehyde. (These compounds were identified by comparative paper chromatography and by their absorption spectra.) The free aldehydes and ketones of low molecular weights totaled 3–3.5 mg/cigarette and comprised about 10% of

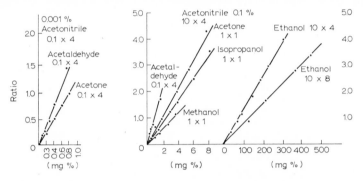

Fig. 56. Relationship of the concentrations of blood volatiles to that of the internal standard acetonitrile.

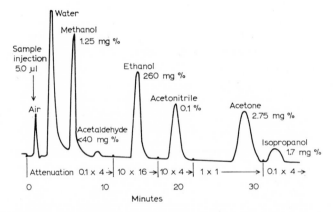

Fig. 57. Chromatogram of whole blood of an acute alcoholic. Note the presence of methanol, isopropanol and possibly acetaldehyde.

the total weight of the smoke with acetaldehyde present in the highest concentration (accounting for approximately ⅓ of the total amount). Table 38 lists the R_F values for the 2,4-DNPH derivatives on dimethylformamide treated paper developed with hexane.

Severin[295] studied the TLC of 2,4-dinitrophenylhydrazones of a number of aldehydes on Silica Gel G and Aluminum Oxide G (Merck) (silica gel was the absorbent of choice). Table 39 shows the $R_F \times 100$ values obtained on silica gel with 6 solvent systems.

Hunt[296] described the TLC on silica gel of aldehyde hydrazones of 2-hydrazino-benzothiazole. The reactions involved are firstly between an aldehyde and 2-hydra-

zinobenzothiazole to form a hydrazone which then reacts with *p*-nitrobenzene
diazonium fluoborate to form a colored dye which reacts in turn with a base to form
a blue or green anion. Silica Gel F_{254} plates were activated at 110 °C for 1 h and then
chloroform or ethyl acetate solutions of the hydrazones were applied and after 1 h
equilibration of the spotted plate, developed (15 cm) with a solvent system of light
petroleum (b.p. 30–40 °C)–ethyl acetate–acetic acid (88:10:2). Detection of the
separated hydrazones was accomplished by spraying firstly with an ethanol solution
of 0.1 % *p*-nitrobenzene diazonium fluoborate until the colored spots appeared then

TABLE 38

R_F VALUES FOR 2,4-DINITROPHENYLHYDRAZONES IN N,N-DIMETHYLFORMAMIDE–HEXANE PAPER
CHROMATOGRAPHY SOLVENT SYSTEM

Compound	$R_F \times 100$
Furfuraldehyde	
trans form	11
cis form	19
Formaldehyde	22
Acetaldehyde	30
Crotonaldehyde	36
Acetone	45
Propionaldehyde	48
Methylethyl ketone	64
Diethyl ketone	82
Butyraldehyde	89

TABLE 39

$R_F \times 100$ VALUES OF 2,4-DINITROPHENYLHYDRAZONES OF $C_1 - C_8$ ALDEHYDES
ON SILICA GEL G

2,4-DNPH	R_F values $\times 100^a$					
	Solvents					
	A	B	C	D	E	F
C_1	27	40	18	19	19	19
C_2	31	45	18	19	21	21
C_3	41	56	24	25	32	31
C_4	48	64	28	28	38	38
iso-C_4	48	64	30	30	40	41
C_6	55	66	33	35	45	48
C_8	58	69	37	39	48	51

[a] Average of 4 or 5 determinations.
Solvents: A, cyclohexane–cyclohexene–ethyl acetate (50:30:20); B, benzene–cyclohexane–pyridine
(70:20:10); C, benzene–cyclohexane (90:10); D, benzene–cyclohexene (90:10); E, cyclohexane–
methyl acetate (85:15); F, cyclohexane–ethyl acetate (85:15).

with a solution of 10% alcoholic KOH to form the final blue or green anions. (Alternatively, the final colors can be developed by exposure of the diazonium reagent detected spots to ammonia vapor.) Table 40 lists the R_F values and colors obtained with spray reagents for aldehyde hydrazones of 2-hydrazinobenzothiazole.

The separation of 2,4-dinitrophenylhydrazone derivatives of aliphatic monocarbonyls into classes on magnesia–Celite columns was reported by Schwartz et al.[297]. The classes elute in the sequence methyl ketones, saturated aldehydes, 2-enals, and 2,4-dienals. Characteristic colors for each class were shown on the adsorbent and

TABLE 40

R_F VALUES AND COLORS OBTAINED WITH SPRAY REAGENTS FOR ALDEHYDE HYDRAZONES
OF 2-HYDRAZINOBENZOTHIAZOLE

Aldehyde hydrazone	R_F value[a]	Color	
		With p-nitro-benzenediazonium fluoborate	With p-nitro-benzenediazonium fluoborate, then alcoholic KOH or NH₃ vapor
Formaldehyde	0.35	Orange	Blue
Acetaldehyde	0.40	Orange	Blue
Propionaldehyde	0.44	Orange	Blue
Butyraldehyde	0.50	Orange	Blue
Isobutyraldehyde	0.48	Purple	Blue
2-Methylbutyraldehyde	0.52	Purple	Blue
Hexyl aldehyde	0.55	Orange	Blue
Heptyl aldehyde	0.56	Orange	Blue
Octyl aldehyde	0.57	Orange	Blue
Nonyl aldehyde	0.58	Orange	Blue
Benzaldehyde	0.38	Purple	Blue
p-Tolualdehyde	0.36	Purple	Blue
Cuminyl aldehyde	0.44	Purple	Blue
Phenylacetaldehyde	0.40	Orange	Blue
Hydrotropaldehyde	0.39	Orange	Blue
Cinnamaldehyde	0.38	Orange	Green
Hydrocinnamaldehyde	0.36	Orange	Blue
α-Amylcinnamaldehyde	0.52	Orange	Blue
α-Hexylcinnamaldehyde	0.51	Orange	Blue
Citral	0.49, 0.57	Orange	Blue
Citronellal	0.54	Orange	Blue
Hydroxycitronellal	0.09	Orange	Blue
Salicylaldehyde	0.25	Purple	Blue
Acrolein	0.35	Red	Blue
Isovaleraldehyde	0.45	Orange	Blue
Piperonal	0.23	Pale green	Blue
Anisaldehyde	0.26	Olive	Blue

[a] R_F values obtained on Silica Gel F_{254} plates developed with light petroleum (b.p. 30–40 °C) – ethyl acetate–acetic acid (88:10:2).

aid in their identification, *e.g.* methyl ketones are gray, saturated aldehydes tan, 2-enals rust-red, and 2,4-dienals lavendar. Separation of the 4 classes of 2,4-dinitro-phenylhydrazones was effected on Magnesia 2665 (iodine number 19, Fisher Scientific)–Celite 545, 3 cm × 30 cm columns by use of the following sequence of solvents: 150 ml of 15% $CHCl_3$ in hexane, 150 ml of 30% $CHCl_3$ in hexane, 100 ml of 60% $CHCl_3$ in hexane, and finally $CHCl_3$. Figure 58 shows the separation of a mixture

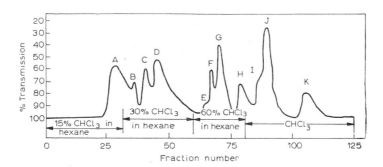

Fig. 58. Separation of mixture of 2,4-dinitrophenylhydrazone derivatives of methyl ketones, saturated aldehydes, 2-enals, and 2,4-dienals into classes of magnesia. Time, 5.0 h. A, Higher methyl ketones; B, 2-butanone; C, acetone; D, higher saturated aldehydes; E, acetaldehyde; F, formaldehyde; G, higher 2-enals; H, crotonal; I, acrolein; J, higher 2,4-dienals; K, penta-2,4-dienal.

of 2,4-dinitrophenylhydrazones of methyl ketones, saturated aldehydes, 2-enals, and 2,4-dienals into classes in magnesia. The members of the individual peaks were identified by column partition chromatography using the acetonitrile–hexane system of Corbin *et al.*[298] and by paper chromatography[299, 300]. The liquid–liquid partition chromatographic separation of the 2,4-dinitrophenylhydrazones of saturated aldehydes (including acetaldehyde), methyl ketones, 2-enals, and 2,4-dienals was reported[298]. Both Celite–acetonitrile and 2-chloroethanol were found effective in accomplishing separations of the above classes. Figure 59 shows a chromatogram obtained with the acetonitrile column.

       Schwartz *et al.*[301] described the analysis of a variety of 2,4-dinitrophenyl-hydrazones (including acetaldehyde, formaldehyde and acrolein) on 1 cm i.d. × 15 cm columns of the cation exchangers AG50W-X4 (200–400 mesh Bio Rad Labs) and Dowex 1-X4 (50–100 mesh).

       Metaldehyde, a polymer of acetaldehyde $(C_2H_4O)_n$, is prepared by the poly-merization of acetaldehyde in the presence of HCl or H_2SO_4 at low temperature and decomposes with the partial regeneration of acetaldehyde. It is used in compressed form as a fuel instead of alcohol and also as a slug and snail poison. Mays *et al.*[302] described the TLC of metaldehyde in plant extracts. Adsorbosil-1 (250 μ) was used with 4 solvent systems that afforded good separations of metaldehyde from plant constituents, chloroform, acetone–benzene (20:80), ethyl acetate, and methanol–toluene (20:80). Metaldehyde was located after development by spraying the plates

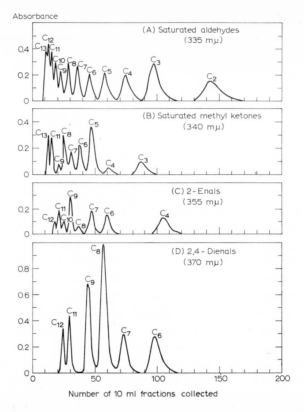

Fig. 59. Liquid–liquid partition chromatographic separation of saturated aldehydes, saturated methyl ketones, 2-enals, and 2,4-dienals on acetonitrile column.

TABLE 41

R_F VALUES FOR METALDEHYDE AND PLANT CONSTITUENTS IN VARIOUS SOLVENT SYSTEMS

Solvent system	Metaldehyde red[a]	R_F	
		Plant constituent 1 green[a]	Plant constituent 2 orange[a]
Chloroform (aged 1 week)[b]	0.34	0.27	0.22
Acetone–benzene (20:80)	0.77	<0.69	<0.69
Ethyl acetate	0.74	0.81	0.69
Methanol–toluene (20:80)	0.55	0.67	0.26

[a] Spot color.
[b] Fresh chloroform gives R_F of 0.56 for metaldehyde.

heavily with 5% guiacol in chloroform followed by spraying with hot (90°C) conc. sulfuric acid. Table 41 shows the R_F values of metaldehyde and plant constituents in various solvent systems.

8.3 Acrolein

Acrolein (acrylic aldehyde) is prepared on a commercial scale by (a) the direct oxidation of propylene utilizing catalysts such as mixed oxides of bismuth and molybdenum, molybdenum and cobalt and molybdenum and cuprous oxide and (b) cross-condensation of acetaldehyde with formaldehyde using lithium phosphate on activated alumina or sodium silicate on silica gel as catalyst. The extreme reactivity of acrolein is attributed to the conjugation of a carbonylic group with the vinyl group within its structure. Acrolein is largely used in the manufacture of acrolein dimer (3,4-dihydro-2-formyl-2H-pyran) which is a valuable starting point for the synthesis of a variety of chemicals (e.g. hexanetriol, hydroxyadipaldehyde, and glutaldehyde) which are useful in textile finishing, paper treatment, and the manufacture of rubber chemicals, pharmaceuticals, plasticizers, and synthetic resins. Other important reactions of acrolein involve its ability to undergo a variety of polymerization reactions (homo-, co-, and graft polymerizations) as well as to interact with ammonia and formaldehyde to yield the industrially important derivatives acrylonitrile and pentaerythritol, respectively. Other major products of acrolein synthesis include methionine (useful in supplementing food, swine and ruminant feeds) and glycidaldehyde which is extensively used as a cross-linking agent for textile treatment and leather tanning. Acrolein has been identified in both tobacco leaf and tobacco smoke and it has also been found that the use of humectants such as glycerol in tobacco can serve as precursors of volatile aldehydes upon combustion[303]. For example, the pyrolysis of humectants at 600°C yielded 10–15 g of volatile carbonyls (acrolein, acetaldehyde and acetone, calculated as acetaldehyde)/100 g of tested polyol. Acrolein (as well as formaldehyde, hydrocarbons, organic peroxides, formic acid, sulfur dioxide, ammonia, and nitrogen oxide) has been identified as a volatile contaminant in smog[304]. The toxicity [229, 305] as well as the mutagenicity[275] of acrolein have been described.

The GLC determination of the products from the oxidation of propylene to acrolein over bismuth/molybdenum catalysts has been described by Verhaar and Lankhuijzen[306]. The compounds determined—oxygen, carbon monoxide, propylene, water, acetaldehyde and acrolein—together constitute more than 99% by weight of the total reactants and reaction products present in the effluent of the reactor. Both qualitative and quantitative analyses were performed with a heated-katharometer-programmed chromatograph model 44 of Philips/Pye, equipped with a Hitachi Perkin-Elmer 196 recorder. The coiled columns were AISI 321 stainless steel with ⅛ in. o.d. The carrier gas, helium, was dried over silica gel and had a flow rate of about 25 ml/min controlled by a Becker Delft Dc GC-2 constant-flow regulator. For the qualitative analysis, 5 columns were used (a) 15% polyethylene glycol 400 on Chromosorb P-NAW (60–80 mesh), length 4.0 m, (*b*) 6% DEGS on Chromosorb W (60–80 mesh), length 1.8 m, (*c*) Porapak Q (80–100 mesh), length 2.0 m, (*d*) Porapak T (80–100 mesh), length 1.5 m and (*e*) molecular sieve 13X (Union Carbide) (40–70 mesh), length 1.8 m.

Figure 60 shows a diagram of the apparatus used for the investigation of the oxidation of propylene to acrolein at atmospheric pressure and in a temperature range of 325–450 °C. In order to prevent condensation of the reaction products, the sample stream from the reactor to the sample valves and from the sample valves to the chromatographic columns was kept at about 110 °C. Table 42 illustrates the

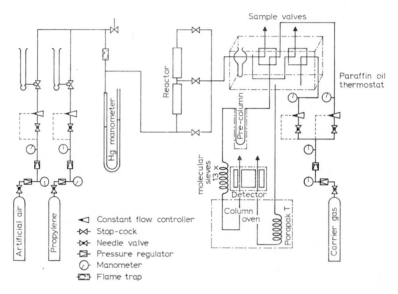

Fig. 60. Diagram of the apparatus used for the investigation of the kinetics of the oxidation of propylene to acrolein.

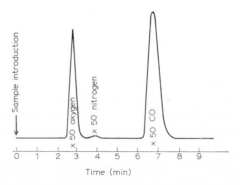

Fig. 61. Analysis of the oxidation products of propylene on the molecular sieve 13X column at a column temperature of 20 °C. The oxidation experiments were carried out with an oxygen/helium mixture consisting of 25% oxygen and 75% helium, so that only a trace of nitrogen was present.

TABLE 42

QUALITATIVE ANALYSIS OF THE REACTION EFFLUENT OBTAINED IN THE OXIDATION
OF PROPYLENE TO ACROLEIN

Column	PEG 400	DEGS	Porapak Q	Porapak T	Mol sieve 13X
Initial temp. (°C)	60	125	65	110	20
Isotherm period (min)			4	0	
Program (°C/min)			12	12	
Final temp. (°C)	60	125	160	160	20
Compound					
Oxygen					+
Nitrogen					+
Carbon monoxide					+
Carbon dioxide			+		
Water		+	+	+	
Ethylene			+		
Propylene			+	+	
Formaldehyde				+	
Acetaldehyde	+		+	+	
Propionaldehyde	−				
Acrolein	+			+	
Acetone	+			+	
Formic acid	−				
Acetic acid		+			
Propionic acid	−				
Acrylic acid		+			

+ = eluted separately and identified as a component of the effluent.
− = identifiable on this column but not present in the effluent.

qualitative analysis of the reaction effluents obtained in the oxidation of propylene to acrolein and Figs. 61–64 illustrate chromatograms of the oxidation products on three different columns. The main compounds, *e.g.* oxygen, carbon monoxide, carbon dioxide, propylene, water, acetaldehyde, and acrolein, could be determined (within 20 min) by means of the following combinations of columns (*1*) molecular sieve 13X parallel with Porapak Q and (*2*) molecular sieve 13X parallel with Porapak T (Figs. 61–64).

Stevens[307] described the GLC determination of acrolein and furfural in phenolic resins. A Wilkins Aerograph A-350 dual-column gas chromatograph was used with the following columns: (A) 12 ft. × ¼ in. copper tubing packed with 10% GE silicone SF-96 on Fluoropak, helium flow rate 105 ml/min and (B) 16 ft. × ¼ in. copper tubing packed with 10% sucrose octaacetate on Teflon 6, helium flow rate 36 ml/min. Furfural and phenol were determined on column A with *m*-cresol as internal standard. Acrolein as well as formaldehyde was determined on column B using *n*-butanol as internal standard. (The retention times are shown in Table 43 and Fig. 64 shows a chromatogram of the analysis of acrolein on column B.)

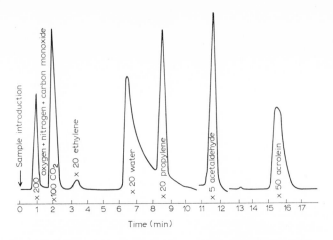

Fig. 62. Analysis of the oxidation products of propylene on the Porapak Q column. Initial temperature 65 °C; isotherm period 4 min; program 12 °C/min; final temperature 160 °C.

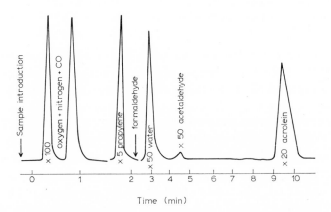

Fig. 63. Analysis of the oxidation products of propylene on the Porapak T column. Initial temperature 110 °C; program 12 °C/min directly after injection; final temperature 160 °C.

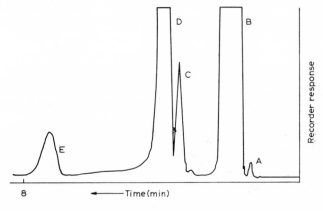

Fig. 64. Analysis of acrolein on column B. A, Air; B, ether; C, acrolein; D, water; E, 1-butanol

TABLE 43

RETENTION TIMES OF RESIN INGREDIENTS, SOLVENTS, AND INTERNAL STANDARDS

Compound	Retention times (min)[a]	
	Column A[b]	Column B[c]
Formaldehyde		0.4
Ether	0.3	0.6
Methanol	0.1	1.5
Acrolein	0.2	1.9
Water	<0.1	2.1
1-Butanol	0.6	5.5
Furfural	1.8	
Phenol	3.6	
m-Cresol	6.3	

[a] From air peak.
[b] Flow rate 105 ml He/min.
[c] Flow rate 36 ml He/min.

Bevilacqua et al.[308] described the utility of a number of substrates for the GLC of a variety of carbonyl derivatives including acrolein and acetaldehyde. Analyses were made with an F & M Model 202B gas chromatograph equipped with 10 ft. × ¼ in. copper tubing, 4.6 mm i.d. containing (a) 10% P84 (block polymer of ethylene oxide and propylene oxide, Wyandotte Chemicals), (b) 10% CET (polymer of acrylonitrile and trimethylol propane), (c) 10% XF1150 (siloxane polymer in which methyl groups have been replaced by β-propionitrile groups), and (d) 20% TEGD (tetramethylene glycol dimethyl ether). The liquid phases (a, b, and c) were supported on 20–80 mesh Haloport F and TEGD was coated on 30–60 mesh Chromosorb P. The columns were at 45°C (70°C for the TEGD) and helium was the carrier gas at 30 ml/min, inlet pressure to flow controller 30 p.s.i., outlet pressure atmospheric. Table 44 lists the retention times of a number of polar compounds on the above 4 columns.

The GLC separation of a mixture of seven carbonyl compounds with two to four carbon atoms (e.g. acrolein, acetaldehyde, propionaldehyde, acetone, isobutyraldehyde, butyraldehyde) and 2,3-butanedione, was achieved using nitrobenzene or 2-nitrobiphenyl as stationary phase[309]. Analyses were performed on an Aerograph Model A-100 equipped with 5 ft. × ¼ in. o.d. columns packed with 33% nitrobenzene or 2-nitrophenyl on 40–60 mesh firebrick. (The columns were operated isothermally at 50° and 70°C, respectively, with helium as carrier gas at 60 and 50 ml/min, respectively.) As shown in Fig. 65, nitrobenzene resolved the mixture into 7 distinct peaks with almost complete baseline separation of the four components most frequently involved in inseparable combinations. However, even at 50°C, nitrobenzene is volatile enough that the column bleeds and deteriorates in a few days. The 2-nitrodiphenyl columns operated at 70°C (Fig. 66) gave 7 distinct peaks which were less well separated

TABLE 44

REPRESENTATIVE RETENTION TIMES (MINUTES FROM AIR PEAK)

Compound	P84	CET	XF 1150	TEGD
Acetaldehyde	4.1	3.6		6.9
Furan	9.9	8.0	6.1	13.2
Propionaldehyde	9.7	12.8	10.7	14.0
Acrolein		16.1	12.2	18.4
Acetone	10.4	18.6	14.2	17.6
Sylvan	20.7	16.0	12.1	25.4
Tetrahydrofuran	22.1	23.1	19.4	
Methanol	17.3	15.1	9.5	19.1
Methacrolein	19.3	22.8	18.1	27.0
Butyraldehyde	21.2	26.4	22.8	28.1
Ethanol	26.4	22.6	14.6	30.2
Butanone	23.7	36.6	29.1	36.5
Butenone	28.7	43.1	33.6	42.7
Propanol	67.1	51.6	33.6	67.6
Water	43.1	30.1	16.1	45.6
Valeraldehyde	51.6	57.6	52.1	

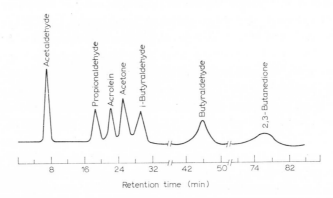

Fig. 65. Separation of volatile carbonyls on nitrobenzene. Temperature 50 °C; flow rate, 60 ml/min of He.

for the 4 critical compounds (acrolein, propionaldehyde, acetone, and 1-butyralde-hyde), but this column has the advantage of being considerably more stable than nitrobenzene and allows a more rapid analysis because of lower retention times.

The analysis of aldehydes by GLC using methylolphthalimide for regeneration of their Girard-T derivatives has been reported by Godbois et al.[310]. An F & M gas chromatograph Model 810 was used with dual hydrogen-flame detectors, an automatic column oven temperature programmer, a −0.2–1.0 mV Honeywell recorder and a Disc Chart Integrator Model 201-B. The retention column consisted of a 10 ft. × ⅛ in. stainless steel coiled tube packed with 10% Carbowax 20M coated on

90–100 mesh Diatoport-S. Initially the column temperature was set at 90°C and then programmed at 4°C/min until 175°C. The injection port and detector temperatures were 260° and 255°C, respectively, nitrogen carrier gas at 15 ml/min, hydrogen flow 40 ml/min, air flow 400 ml/min, range 10^2, attenuation 2. The methylolphthalimide

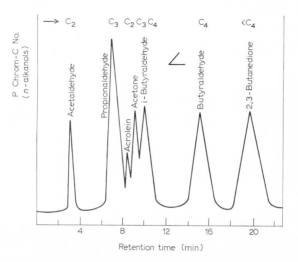

Fig. 66. Separation of volatile carbonyls on 2-nitrobiphenyl. Temperature 70°C; flow rate, 50 ml/min of He.

TABLE 45

GAS CHROMATOGRAPHIC ANALYSIS OF ALDEHYDES, AFTER REGENERATION
FROM THEIR GIRARD-T DERIVATIVES

Peak no.	Aldehyde compound	Retention time (min)	Recovery obtained (%)
1	Propionaldehyde	1.8	80.3
2	Isobutyraldehyde	2.0	85.7
3	Acrolein	2.4	33.4
4	Butyraldehyde	2.8	a
5	Isovaleraldehyde	3.5	93.5
6	Valeraldehyde	4.7	95.4
7	Hexanal	7.5	96.7
8	Heptanal	10.9	100.0
9	Octanal	14.6	100.0
10	Nonanal	18.5	100.0
11	Decanal	22.3	100.0
12	Undecanal	26.8	100.0
13a	Neral	31.9	87.2
13b	Geranial	33.0	87.7

[a] Not calculated because the solvent peak (tert.-butanol) and the butyraldehyde peak were not sufficiently separated from each other on the chromatogram.

regenerant was prepared *via* the reaction of 0.28 mole of phthalimide with 0.29 mole of formaldehyde[311]. The formation of Girard-T complexes was obtained by treatment of a mixture of the aldehydes in *tert.*-butanol azeotrope with Girard-T reagent in the presence of Rexyn 102 (H). Table 45 depicts the retention times and the percent recovery obtained following GLC analysis of aldehydes after regeneration from their Girard-T derivatives and shows that the procedure is quantitative for most aldehydes studied. Figure 67 illustrates the efficiency of the GLC procedure for resolving the

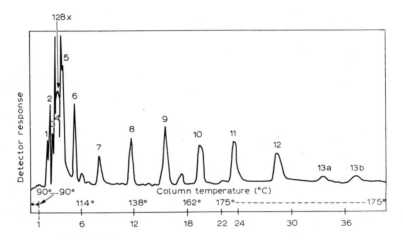

Fig. 67. Standard aldehyde mixture in *tert.*-butanol diluted 1:10.

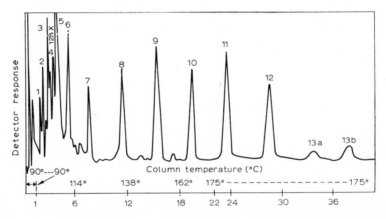

Fig. 68. Girard-T derivatives of standard aldehyde mixture liberated by formaldehyde.

components of a mixture of standard aldehydes while Fig. 68 shows a similar efficiency after treatment of this mixture with Girard-T reagent and the methylol regenerant. Comparison of Figs. 67 and 68 shows that the derivatization–regeneration procedure did not introduce extraneous peaks on the chromatogram.

Ralls[312] described the flask exchange gas chromatography for the semiquantitative determination of a number of volatile aldehydes, ketones and acids. The derivatives of volatile compounds were heated with reagents in glass capillary tubes and the products volatilized directly into a gas chromatograph. 2,4-Dinitrophenylhydrazones are exchanged with excess α-ketoglutaric acid by heating at 250°C for 10 sec. The technique of Ralls[312] was suggested to have particular application to the analysis of trace amounts of compounds in large volumes of water; it has a precision of 10–20% and requires 10 mg of mixed derivatives for multiple analysis. The columns used were $\frac{1}{4}$ in. × 5 or 10 ft. stainless steel packed with Dow-Corning 550 silicone and 85% stearic acid on firebrick (30–60 mesh) (75 parts) or LAC 446 (glycol–adipate polymer) (30 parts on firebrick, 70 parts); or Carbowax 1540 (30 parts on firebrick, 70 parts). Temperature was controlled to ±1°C, helium was the carrier gas at 32, 46, or 50 ml/min with an inlet pressure of 8–11 p.s.i.g. and outlet pressure atmospheric. Detection was by a four-filament thermal-conductivity cell. Table 46 shows the retention times

TABLE 46

RETENTION TIMES FOR CARBONYL COMPOUNDS REGENERATED FROM 2,4-DNPH
BY EXCHANGE WITH α-KETOGLUTARIC ACID
10-foot Carbowax, flow rate 32 ml/min, theoretical plates = 1150 (propionaldehyde)

Carbonyl compound	Column temperature (°C)	Retention time (min)	Remarks
Aldehydes			
Formaldehyde	90		Polymerizes?
Acetaldehyde	90	4.5	
Propionaldehyde	90	6.8	
Isobutyraldehyde	90	7.7	
n-Butyraldehyde	90	10.4	
Isovaleraldehyde	90	13.1	
n-Valeraldehyde	90	17.0	
2-Methyl-1-butanal	90	12.0	
Acrolein	90	9.2	
Crotonaldehyde	90, 150	26.9, 6.9	
2,4-Pentadienal	150	12.2	
2-Hexenal	150	12.7	
Methional	150		Decomposes
Benzaldehyde	150, 210	>32, >20	210° on LAC
Ketones			
Acetone	90	7.6	
Butanone	90	11.7	
2-Pentanone	90, 150	17.8, 4.6	
3-Pentanone	90, 150	19.1, 5.3	
3-Methyl-2-butanone	90	14.0	
2-Hexanone	90, 150	30.0, 6.6	
4-Methyl-3-penten-2-one	150	12.3	
Cyclohexanone	150	24.6	
Biacetyl(bis)	90	>25	Not regenerated

for carbonyl compounds regenerated from 2,4-DNPH by exchange with α-ketoglutaric acid.

Kelker[313] reported the GLC of a large number of organic compounds including a number of mutagenic aldehydes and epoxides and related derivatives. Table 47 lists the retention volumes of a number of these compounds on β,β-hydroxydipropionitrile and hexaethylene glycol dimethyl ether.

TABLE 47

RETENTION VOLUMES OF SOME ORGANIC COMPOUNDS

Compound	β,β-Hydroxydipropionitrile V_t^0 benzene	Hexaethylene glycol dimethyl ether V_t^0 benzene
Acrolein	0.96	0.46
Acetaldehyde	0.43	0.16
Ethylene oxide	0.37	0.16
Propylene oxide	0.48	0.24
Isobutylene oxide	0.52	0.31
Cis-2,3-butylene oxide	0.78	0.46
Trans-2,3-butylene oxide	0.57	0.35
1,2-Butylene oxide	0.78	0.52

9. LACTONES

9.1 β-Propiolactone

β-Propiolactone (β-hydroxypropionic acid lactone, BPL) is produced commercially from formaldehyde and ketene in the presence of a catalyst such as zinc chloride, hydrated aluminum silicate, or zinc trifluoroacetate. The commercial grade BPL contains about 18% impurities, e.g. acrylic acid, acetic anhydride, and polymers. BPL possesses a broad spectrum of current and suggested industrial uses, e.g. in wood processing, protective coatings, and impregnation of textiles, modification of flax cellulose, urethan foam manufacture, intermediate in the preparation of insecticides, pesticides, medicinals, additive for leaded gasolines, and in the modification of tobacco flavors. BPL has been widely used as a sterilizing agent and as a solvent. The chemical properties and reactions[314], mechanisms of action[315], toxicity[316], and mutagenicity[317, 318] have been described.

Procedures for the analysis of BPL include titrimetry[319] and infrared spectroscopy[320]. Chemically, BPL, an alkylating carcinogen[321], is highly reactive due to the strained four-membered ring and undergoes rapid electrophilic reactions, e.g. with thiols forming thioethers. The reaction of BPL with cysteine was followed by paper chromatography[322] using two solvents: aq. phenol–ammonia (1:1) and n-butanol–

pyridine–water (1:1:1). The reaction mixture contained a product believed to be the thioether S-(2-carboxyethyl)-cysteine. The *in vivo* binding of tritiated BPL to mouse skin DNA, RNA and protein was demonstrated by Colburn and Boutwell[323]. Hydrolysis and two-dimensional paper chromatography[324] of skin RNA and DNA after reaction with tritiated BPL showed that the major covalent binding product was 7-(2-carboxyethyl)-guanine and much smaller amounts of the pyrimidine nucleotides and adenine, respectively. Tritiated alkylating agents (BPL, 3-iodopropionic acid, 3-chloropropionic acid and iodoacetic acid) were compared as initiators of mouse skin tumors[315, 325]. The *in vivo* binding of the labeled compounds to mouse skin DNA as well as the hydrolysis, chromatography and autoradiography of RNA was elaborated as described previously[323].

9.2 β- and γ-Butyrolactones

Of the two lactones, γ-butyrolactone is by far the most important commercially. It is manufactured by the dehydrogenation of 1,4-butanediol or by the hydrogenation of maleic anhydride and its utility includes vehicle for pigments in the dyeing of nylon, protein and polyester fibers, plasticizer for epoxy resins, wool treatment with sulfonyl

TABLE 48

$R_F \times 100$ VALUES OF LACTONES AS THEIR HYDROXAMIC ACID DERIVATIVES
ON WHATMAN NO. 1 PAPER

Lactone	Solvent system		
	1	*2*	*3*
γ-Butyrolactone	0	2	8
γ-Valerolactone	0	7	15
γ-Hexalactone	3	15	33
γ-Heptalactone	6	28	54
γ-Octalactone	14	43	68
γ-Nonalactone	32	54	77
γ-Decalactone	64	64	83
γ-Hendecalactone	74	71	87
γ-Dodecalactone	84	83	92
δ-Nonalactone	18	50	73
δ-Decalactone	33	60	79
δ-Hendecalactone	59	69	84
δ-Dodecalactone	70	78	89

Solvent systems: (1) Equilibrating solvent: aq. phase from a 5:5:1 by vol. mixture of benzene–water–glacial acetic acid. Developing solvent: 10% solution of glacial acetic acid in benzene. (2) Equilibrating solvent: aq. phase from a 5:2:5:1 by vol. mixture of benzene–isopropanol–water–glacial acetic acid. Developing solvent: upper phase of the above. (3) Same as system (2) except that three parts of isopropanol were used.

chlorides and chlorophosphates, alkylations of olefins, preparation of polymers with s-trioxane, and acrylonitrile polymer fibers, activator in the alkaline polymerization of caprolactones as well as 2-pyrrolidinone, and manufacture of pressure-sensitive adhesives from cross-linked glycol. γ-Butyrolactone has also been used as an anti-helminthic, a nematocidal synergist, and in the prevention of plant frost damage.

β-Butyrolactone is made from ketene dimer and has been utilized in various polymerization reactions. The carcinogenicity[326-329], mechanism of action[315, 330], denaturation of DNA[331], and pharmacology[332] of the butyrolactones have been described.

The paper chromatographic separation of a number of aliphatic lactones via their respective hydroxamic acid derivatives has been reported by Keeney[333]. The developing solvent systems were (1) 10% solution of glacial acetic acid in benzene, (2) upper phase of benzene–isopropanol–water–glacial acetic acid (5:2:5:1), and (3) upper phase of benzene–isopropanol–water–glacial acetic acid (5:3:5:1). A 5% ferric chloride–5% HCl (1:1) reagent was used for detection. For lactones containing 8–12 carbon atoms, solvents (1) and (2) were recommended and for lactones of lower molecular weight, solvent (3) was the solvent of choice. Table 48 lists the R_F values × 100 for a variety of lactones as their hydroxamic acid derivatives. Van Duuren et al.[329] described the utility of GLC for purity determinations of a variety of lactones, epoxides, and peroxides. A Perkin-Elmer Model 154-C gas chromatograph was used equipped with flame-ionization detector and 2-m glass columns coated with either silicone oil or polypropylene glycol. Table 49 lists the retention times for a number of lactones, epoxides and peroxides.

The presence of γ-butyrolactone in the reaction products of the manufacture of polyvinyl pyrrolidone and methyl pyrrolidone has been confirmed by GLC analysis[334].

A 2.25-m column filled with Kieselguhr containing 12% polyester of 1,4-butanediol and adipic acid was used at 180°C. The relative retention volumes of the various products were ethylene glycol 1, butanediol 2.83, butenediol 6.90, methyl pyrrolidinone 502, γ-butyrolactone 1.64, α-pyrrolidinone 5.02, and vinyl pyrrolidinone 2.47. The relative error of determination by the peak area method was 1–3%.

γ-Butyrolactone has been isolated in coffee aroma. Viani et al.[335] obtained and separated the most volatile portion of the aroma of coffee by GLC on four different stationary phases, SE-30, UCON-LAB-550-X; UCON HB-2000 and DEGS. Thirty-four substances known to be present in coffee aroma were identified and their known Kovats retention times listed, while the new compounds discovered included 2,3-dimethylpyrazine, γ-butyrolactone, and hexane-2,3-dione.

Reymond et al.[336] described the GLC of aroma constituents of coffee, tea, and cocoa performed at 80°C on concentrated methylene chloride solutions, using flame-ionization detectors. The following columns were used: (a) capillary UCON-LB-550-X, (b) conventional 20% UCON-LB-550-X on Embacel, 60–80 mesh, and (c) conventional 35% silicone oil DC-200 on Celite 30–80 mesh. The separation and identification by

TABLE 49

GAS CHROMATOGRAPHY OF EPOXIDES, LACTONES AND PEROXIDES

Compound	Temperature (°C)	Retention time (min)
Ethylene oxide	23	4.0
1-Ethyl-3,4-epoxycyclohexane	148	4.0
1-Vinyl-3,4-epoxycyclohexane	148	3.5
Epoxycyclohexane	70	8.1
1,2,4,5-Diepoxypentane	90	8.5
1,2,5,6-Diepoxyhexane	90	18.5
1,2,6,7-Diepoxyheptane	90	10.5
1,2,5,6-Diepoxycyclooctane	168	2.6, 8.0
Glycidaldehyde	163	1.75
β-Butyrolactone	97	2.3
4,5-Epoxy-3-hydroxyvaleric acid β-lactone	118	6.5
2,2,4-Trimethyl-3-hydroxy-3-pentenoic acid β-lactone	130	1.6
β-Angelicalactone	130	1.4
α-Angelicalactone	130	1.0
1-Hydroperoxycyclohex-3-ene[a]	140	8.6
Cumene hydroperoxide[a]	126	5.0
Ascaridole	125	5.0

A 2-m silicone column was used for these analyses except for glycidaldehyde, β-butyrolactone, and 1-hydroperoxycyclohex-3-ene, for which a 2-m polypropyleneglycol column was used.

[a] Both these hydroperoxides show decomposition products on the chromatograms, which was expected in view of low heat stability of these materials.

GLC was followed by identification by IR spectroscopy. The positions of the peaks were given by the Kovats indices, with an error margin of ± 2 units for the capillary columns and ± 10 units for the conventional columns.

γ-Butyrolactone has also been detected with 86 volatile constituents in roasted coffee by vapor-phase chromatography[337] and further identified by IR and mass spectroscopy. The volatile components of pineapple also were found to contain γ-butyrolactone, γ- and δ-octalactones and 9 other constituents as identified by GLC, IR, NMR and mass spectroscopy[338].

10. TRIMETHYLPHOSPHATE

Trimethylphosphate (TMP) is a methylating agent largely employed as a low-cost gasoline additive (at a concentration of approximately 0.25 g/gallon) and in the preparation of organophosphorus insecticides, polymethyl polyphosphates, as a flame-retardant solvent for paints and polymers, as a catalyst in the preparation of polymers and resins and has recently been proposed as a food additive for stabilizing egg white. TMP is a chemosterilant in rodents[339] (probably related to *in vivo* alkylation) and is mutagenic in mice[340] and *Neurospora*[341]. Short-term toxicity[342] and the metabolism[343] of trialkyl phosphates have been described. In this latter regard, Scott et al.[344] has extensively studied the biological monodealkylation of TMP and triethyl phosphate (TEP) and it appears that it is analogous to their chemical reactivity in which further dealkylation occurs only under extremely severe conditions.

Chromatograms of rat and mouse urine from either oral or intraperitoneal administration of ^{32}P-TMP revealed one radioactive area corresponding to ^{32}P-dimethylphosphate. With uniformly-labeled ^{14}C-TMP two additional urinary metabolites were detected and identified as S-methyl cysteine and its N-acetate indicating that TMP, at least as far as its detoxification is concerned, acts in an alkylating capacity. Table 50 indicates the R_F values for trialkyl phosphate metabolites on both

TABLE 50

R_F VALUES* FOR TRIALKYL PHOSPHATE METABOLITES

	TLC[a]	Paper
S-Alkyl cysteine[a]		
Methyl-	0.38	0.43
Ethyl-	0.44	0.54
n-Propyl-	0.53	0.65
Isopropyl-	0.54	0.73
n-Butyl-	0.52	0.72
S-Methylglutathione	0.21	0.27
Phosphate[b]		
Dimethyl-		0.46
Diethyl-		0.54
Di-n-propyl-		0.74
Diisopropyl-		0.50
Di-n-butyl-		0.79
Diethylthio-		0.65

* R_F values for metabolites are on Whatman No. 17 papers from which the phosphate metabolites were isolated from untreated urine and the cysteine conjugate from acid-hydrolyzed urine, and Silica Gel G plates (250 μ).
[a] Ascending paper chromatograms were developed in n-butanol–glacial acetic acid–water (4:2:1) and detected by ninhydrin reagent.
[b] Ascending paper chromatograms were developed in isopropanol–880 ammonia–water (8:1:1) and detected by molybdate reagent.

Silica Gel G plates and Whatman No. 17 papers. The solvent systems used were (a) n-butanol–glacial acetic acid–water (4:2:1) and (b) isopropanol–880 ammonia–water (8:1:1) and detection being by ninhydrin and molybdate[345] reagents, respectively.

The TLC of organophosphorus compounds of the groups phosphates, phosphites, phosphines, phosphonates, phosphinates, and phosphine oxides has been reported by

Lamotte *et al.*[346-348]. The phosphate esters $(RO)_3PO$ were most conveniently separated on silica gel using hexane–acetone–ethyl acetate $(60:20:20)$[346]. Table 51 shows the TLC of phosphate esters on four silica absorbents using the above developer.

Klement and Wild[349] used TLC on silica gel to separate various mixtures of phosphoric esters (*e.g.* $OP(OR_3)$, $OP(OR)_2NH_2$, $OP(OR)(NH_2)_2$, $SP(OR)_3$, $SP(OR)_2NH_2$, $SP(OR)(NH_2)_2$, $P(OR)_3$, $P_2ON(OR)_5$).

TABLE 51

TLC OF PHOSPHATE ESTERS ON SILICA GEL ABSORBENTS WITH HEXANE–ACETONE–ETHYL
ACETATE $(60:20:20)$ AS DEVELOPING SOLVENT

Compound	R_F			
	Silica Gel HR	Silica Gel H	Silica Gel G	Silica Gel DC
Trimethyl phosphate	0.27	0.28	0.29	0.21
Triethyl phosphate	0.45	0.51	0.53	0.37
Triallyl phosphate	0.66	0.68	0.70	0.54
Tri(*n*)-propyl phosphate	0.73	0.74	0.76	0.59
Tri(*n*)-butyl phosphate	0.81	0.87	0.86	0.68
Tribenzyl phosphate	0.70	0.69	0.73	0.55
Triphenyl phosphate	0.86	0.89	0.90	0.73
Diphenylmethyl phosphate	0.77	0.76	0.82	0.64

A gas chromatographic screening technique based on the detection of methylated hydrolysis products was described by Askew *et al.*[350]. Most of the organophosphorus pesticides now in use, together with many of their oxidation products, can be hydrolyzed (by methanolic sodium hydroxide) to produce dialkyl phosphorothionates which on methylation yield products that can be separated by gas chromatography. The products of hydrolysis normally found are dimethyl and diethyl phosphates and the corresponding O,O-dialkyl phosphorothionates. Methylation of the hydrolysis products was carried out with diazomethane using the procedure of Schlenk and Gellerman[351]. A Versamid 900 column was selected to separate the methylated derivatives yielding better resolution than the Ucon Polar[352] or Carbowax 20M[353] stationary phase used by other workers. Table 52 shows the relative retention times of the methylated hydrolysis products derived from organophosphorus pesticides. Using a thermionic detector (Aerograph phosphorus detector) the derivatives could be detected at a level down to 0.1 ng.

Figure 69 shows chromatograms of trialkyl phosphates and thiophosphates. Pesticides which give trimethyl phosphate on hydrolysis include demeton-S-methyl, dichlorovos (DDVP), mevinphos, oxydemetonmethyl, phosphamidon, trichlorophon and vamidothion.

Thin-layer chromatography was used in this study to obtain pure methylated derivative reference standards. Silica gel plates were developed in diethyl ether to yield

TABLE 52

Chromatographic conditions: column, 150 cm glass, 3 mm o.d.; packing, 4% Versamid 900 on Chromosorb G, acid washed, dimethylchlorosilane treated 80–100 mesh; temperature, oven 120°C, injector 160°C, detector 180°C; instrument, Varian Aerograph 205-B with phosphorus detector; gas flow rates, nitrogen 25 ml/min, hydrogen 22 ml/min, air 200 ml/min.

Pesticide derivative	Relative retention time
Trimethyl phosphate (TMP)	41
Diethyl methyl phosphate (DEMP)	75
Trimethyl phosphorothionate (TMPT)	100 (2.30 min)
Diethyl methyl phosphorothionate (DEMPT)	175
Di-2-chloroethyl methyl phosphate	200
Di-dimethylamino methyl phosphate	240

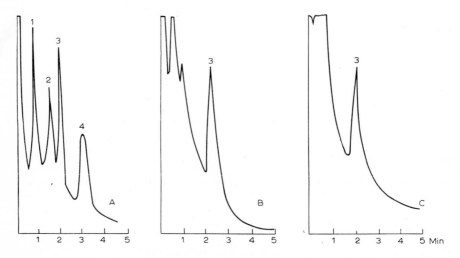

Fig. 69. Chromatograms of trialkyl phosphates. 1, TMP; 2, DEMP; 3, TMPT; 4, DEMPT. (A) Mixture of derivatives. Each peak = 0.25 ng material. (B) Fish extract (=10 g) spiked with 1 μg menazon, hydrolyzed, extracted and methylated. (C) Runner bean extract (=10 g) spiked with 1 μg menazon, hydrolyzed, extracted and methylated.

the following R_F values: trimethyl phosphate 0.15, diethyl methyl phosphate 0.25, trimethyl phosphorothionate 0.50, and diethyl methyl phosphorothionate 0.70.

The metabolism and urinary hydrolysis of organophosphorus pesticides in mammals results in the excretion of a variety of alkyl phosphates. Shafik and Enos[354] described the determination of these metabolic and hydrolytic products in human blood and urine via the isothermal GLC of their methyl and ethyl derivatives. A Micro-tek Model MT 220 gas chromatograph was used equipped with 394 mμ and 526 mμ filters to determine sulfur and phosphorus compounds, respectively. The use of a 12 ft. × ¼ in. o.d. aluminum column coated with 20% Versamid 900 on 60–80 mesh

Gas Chrom Q allowed for the simultaneous determination of the 6 major metabolites of organophosphorus and thiophosphorus insecticides. Table 53 lists the relative retention times of the trialkyl phosphates obtained with a phosphorus flame photometric detection system.

TABLE 53

ISOTHERMAL GAS CHROMATOGRAPHY OF TRIALKYL PHOSPHATES WITH PHOSPHORUS FLAME PHOTOMETRIC DETECTION SYSTEM

Trialkyl phosphate	Relative retention time	Detector sensitivity (4/1 signal-to-noise ratio) (ng)	Limit of detectability (p.p.m.)
TMTP	0.70		
TMP	1.00		
DEMMP	1.72	0.4	0.01
TMDTP	2.09	15.0	0.20
DEMMDTP	2.98	1.0	0.02
DEMMTP	4.12	1.0	0.02
DMMETP	0.71	1.5	0.02
DMMEP	1.00	1.5	0.02
TEP	1.75		
DMMEDTP	2.18		
TEDTP	3.08		
TETP	3.96		

TMTP = trimethyl thiophosphate; TMP = trimethyl phosphate; DEMMP = diethyl monomethyl phosphate; TMDTP = trimethyl dithiophosphate; DEMMDTP = diethyl monomethyl dithiophosphate; DEMMTP = diethyl monomethyl thiophosphate; DMMETP = dimethyl monoethyl thiophosphate; DMMEP = dimethyl monoethyl phosphate; TEP = triethyl phosphate; DMMEDTP = dimethyl monoethyl dithiophosphate; TEDTP = triethyl dithiophosphate; TETP = triethyl thiophosphate.

11. PYRROLIZIDINE ALKALOIDS

Alkaloids of the pyrrolizidine class (e.g., heliotrine, retrosine, lasiocarpine, monocrotaline) are found in members of the Senecio, Crotolaria, Amsinkia and other genera which are widely distributed throughout rangelands and are hepatoxic to livestock. Many of the plants containing such alkaloids have been and are being traditionally used as herbal folk medicines for various disorders. Pyrrolizidine alkaloids, of which the carcinogenic retrosine is an example, were among the first to be suggested as possible natural etiological factors in the high incidence of liver diseases (kwashiorkor, cirrhosis, and primary liver tumors) in the tropics and subtropics.

These compounds were found to possess alkylating activity in vitro against nucleophilic reagents[355]. Only the pyrrolizidine alkaloids with a sterically hindered allylic ester group ($>C=C-CH_2OCOR$) have alkylating activity and hepatoxic properties. For example, the alkaloids represented by retrosine may function by a mechanism involving alkyl–oxygen fission of the ester linkage. This reaction which is known to occur with carboxylic esters of allylic alcohols but not with those of saturated alcohols would result in displacement of the anion RCO_2^- by a nucleophilic agent X^-. Since combination may occur at either end of the allylic system, the product may be of type I or II. The alkylation mechanism may be depicted as follows.

$R = CH_3(CH_2)_2-$ or $CH_3(CH_2)_3-$
$R' = H, OH$ or acyl

Thus, the hypothesis was advanced that the fundamental biochemical lesion under-lying the action of pyrrolizidine alkaloids on cell nuclei results from alkylation.

It has also been suggested[356] that the pyrrolizidine alkaloids are not hepatoxic *per se*, but are transformed in the liver by ring dehydrogenation to pyrrole-like deriv-atives to form soluble "bound pyrroles" which are highly reactive and exert cytotoxic and other biological actions suggestive of attack on nucleic acids. The chemistry and distribution[357], pharmacology[358], hepatotoxicity[359, 360], teratogenicity[361], carcino-genicity[359, 362], and mutagenicity[363, 364] of the pyrrolizidine alkaloids have been reported.

The presence of the pyrrolizidine alkaloids in the *Crotalaria* species has been elaborated utilizing TLC[365]. The absorbent consisted of silica gel–$CaSO_4 \cdot \frac{1}{2}H_2O$ (9:1) with chloroform–methanol–ammonia (85:14:1) as developing solvent. Dragendorff reagent[366] detected the alkaloids as orange spots. Table 54 shows the TLC of the crude bases from seeds of several *Crotalaria* species and Table 55 lists the R_F values of authentic samples of *Crotalaria* alkaloids.

The characterization of pyrrolizidine alkaloids by gas, thin-layer and paper chromatography has been described by Chalmers *et al.*[367]. For gas chromatography, an argon-ionization detector was employed, and a glass column 6 ft. × 6 mm i.d. packed with 4% SE-30 siloxane polymer on Gas Chrom P was prepared and silanized with dimethyldichlorosilane according to the procedure of Fales and Pisano[368]. The interior surfaces of the column, inlet and outlet were similarly silanized and the column was operated with inlet pressure 650 mm and at a temperature of 140°C for the lower molecular weight non-ester bases and at 960 mm and 205°C for the higher molecular weight bases. The inlet was maintained at 250°C and included a short length (*ca.* 2 cm) of silanized Gas Chrom P which collects non-volatile constituents of the sample or carbonaceous pyrolysis products and was replaced at intervals of one week. Thin-layer chromatography was carried out on plates prepared from a slurry of Silica Gel G (Merck, 30 g) and sodium hydroxide (0.1 N, 60 ml) and kept for at least a day before use. Methanol was the developing solvent and spots were detected

TABLE 54

THIN-LAYER CHROMATOGRAPHY[a] OF THE CRUDE BASES FROM SEEDS OF SEVERAL
Crotalaria SPECIES

No.	Species	Number of alkaloids detected and their R_F values on TLC	Number of alkaloids detected by paper chromatography
1	Crotalaria intermedia	0.15, 0.43, 0.64 (3)	2
2	C. sericea	0.00, 0.46, 0.58 (3)	2
3	C. mucronata	0.00, 0.07, 0.33, 0.54 (4)	1
4	C. agatiflora	0.00, 0.21, 0.59 (3)	1
5	C. semperflorens	0.04, 0.24, 0.34 (3)	1
6	C. wightiana	0.00, 0.64, 0.71 (3)	1
7	C. barbata	0.03, 0.18, 0.26, 0.36, 0.46 (5)	1
8	C. walkerii	0.06, 0.22, 0.36, 0.47 (4)	2

[a] Absorbent: Silica Gel–CaSO$_4 \cdot \frac{1}{2}$ H$_2$O (9:1); solvent: chloroform–methanol–ammonia (85:14:1).

TABLE 55

THIN-LAYER CHROMATOGRAPHY[a] OF AUTHENTIC SAMPLES OF *CROTALARIA* ALKALOIDS
ON SILICA GEL–CaSO$_4 \cdot \frac{1}{2}$ H$_2$O (9:1)

No.	Alkaloid	R_F value of the pure alkaloid
1	Supinine	0.17
2	Heliotrine	0.30
3	Monocrotaline	0.44
4	Heleurine	0.49
5	Jacobine	0.50
6	Seniciphylline	0.56
7	C. wightiana alk. III	0.71
8	C. grahmiana alk. IV	0.76
9	C. agatiflora alk. IV (leaf)	0.80

[a] Solvent: chloroform–methanol–ammonia (85:14:1).

in iodine vapor. Paper chromatography was effected on Whatman No. 1 paper with ascending solvent which was the top layer resulting from shaking *n*-butanol with an equal volume of 5% acetic acid.

The relative retention times of the alkaloids are shown in Tables 56–58 along with R_F values measured in the thin-layer and paper chromatographic systems. Non-ester alkaloids and derivatives are run at a column temperature of 140°C and the ester alkaloids at 205°C. Table 59 illustrates the effects of structural changes on relative retention times. Minor structural changes were shown to produce additive effects on the logarithm of the relative retention times as has been found with other classes of compounds.

TABLE 56

RELATIVE RETENTION TIMES (AT 140 °C) AND R_F VALUES OF NON-ESTER PYRROLIZIDINE
ALKALOIDS AND DERIVATIVES
For conditions, see text

No.	Base	B.p. (°C/mm)	M.p. (°C)	R_T (min)	R_F (TLC)	R_F (paper)
1	1-Methylenepyrrolizidine	115/150		1.6	0.05	0.39
2	Heliotridane (1-β-methyl-8-α-pyrrolizidine)	169/760		1.8	0.01	0.39
3	Anhydroplatynecine	194/750		2.4	0.05	0.22
4	7-β-Hydroxy-1-methylene-8-α-pyrrolizidine	41/0.1	35–36	3.3	0.07	0.27
5	Desoxyretronecine (7-β-hydroxy-1-methyl-1,2-dehydro-8-α-pyrrolizidine)		79–80	3.4	0.07	0.31
6	7-β-Hydroxy-1-methylene-8-β-pyrrolizidine	62/0.03	34–36	4.4	0.14	0.28
7	Retronecanol (7-β-hydroxy-1-β-methyl-8-α-pyrrolizidine)	140/30	98–98.5	4.5	0.01	0.40
8	Hydroxyheliotridane (7-α-hydroxy-1-β-methyl-8-α-pyrrolizidine)	92/0.5	61.5–62.5	4.5	0.02	0.29
9	7-α-Hydroxy-1-methyl-1,2-dehydro-8-α-pyrrolizidine	114/3.5	67–68	4.5	0.13	0.29
10	1-Methoxymethyl-1,2-dehydro-8-α-pyrrolizidine	100/10		5.4	0.08	0.36
11	1-Methoxymethyl-1,2-epoxy-pyrrolizidine	53/0.1		6.4	0.23	0.32
12	Isoretronecanol (1-β-hydroxymethyl-8-α-pyrrolizidine)	115–16/1.5	39–40	7.4	0.02	0.24
13	Supinidine (1-hydroxymethyl-1,2-dehydro-8-α-pyrrolizidine)	90/0.1	29–30	7.4	0.04	0.24
14	1-Hydroxymethyl-1,2-epoxy-8-α-pyrrolizidine	80/0.04		8.3	0.18	0.20
15	7-β-Hydroxy-1-methoxymethyl-1,2-dehydro-8-α-pyrrolizidine	77/0.4	36–38	9.0	0.12	0.33
16	7-β-Acetoxy-1-methoxymethyl-1,2-dehydro-8-α-pyrrolizidine			15.0	0.30	0.54
17	Retronecine (7-β-hydroxy-1-hydroxymethyl-1,2-dehydro-8-α-pyrrolizidine)		117–118	15.0	0.07	0.20
18	Heliotridine (7-α-hydroxy-1-hydroxymethyl-1,2-dehydro-8-α-pyrrolizidine)		117–118	15.0	0.14	0.20
19	Platynecine (7-β-hydroxy-1-β-hydroxymethyl-8-α-pyrrolizidine)		148–148.5	15.0	0.01	0.21

The detection of pyrrolizidine alkaloids on thin-layer chromatograms has been described by Mattocks[369, 370]. The method used is an extension of that of Dann[371] for the detection of N-oxides of pyrrolizidine alkaloids. After chromatography, the alkaloids on the plate are converted to N-oxides by a hydrogen peroxide spray followed by heating. Treatment with acetic anhydride then converts the N-oxides to pyrroles which with Ehrlich reagent (4-dimethylaminobenzaldehyde) yield blue or mauve spots.

TABLE 57

RELATIVE RETENTION TIMES (AT 205 °C) AND R_F VALUES OF PYRROLIZIDINE ESTERS
WITH MONOCARBOXYLIC ACIDS

No.	Base	M.p. (°C)	R_T (min)	R_F (TLC)	R_F (paper)
20	7-Angelylretronecine	76–77	4.2	0.33	0.49
21	7-Angelylheliotridine	116–117	4.8	0.45	0.52
22	Heleurine	67–68	8.0	0.11	0.50
23	Supinine	148–149	8.8	0.10	0.40
24	Heliotrine	128	12.4	0.30	0.43
25	Indicine	97–98	14.3	0.19	0.38
26	Retronecine trachelanthate		14.3	0.19	0.38
27	Retronecine viridiflorate		14.3	0.19	0.38
28	Rinderine	100–101	15.6	0.29	0.34
29	Echinatine	109–110	15.6	0.30	0.36
30	Europine	(N-Oxide 171)	18.2	0.29	0.34
31	Sarracine	45–46	29.1	0.16	0.56
33	Echiumine	99–100	35.6	0.47	0.65
34	Lasiocarpine	96.5–97	46.4	0.54	0.59
35	Echimidine	(Picrate 142–143)	47.6	0.45	0.57
36	Heliosupine	(Picrate 103–106)	50.6	0.53	0.51
37	Latifoline	102–103	55.0	0.53	0.54

TABLE 58

RELATIVE RETENTION TIMES (AT 205 °C) AND R_F VALUES FOR MACROCYCLIC
DIESTER ALKALOIDS

No.	Base	M.p. (°C)	R_T (min)	R_F (TLC)	R_F (paper)
38	Retusine	174–175	13.9	0.16	0.44
39	Fulvine	213.5–214	15.5	0.33	0.43
40	Crispatine	137–138	15.6	0.29	0.39
41	Monocrotaline	202–203	19.5	0.29	0.39
42	Senecionine	245	20.6	0.40	0.56
43	Seneciphylline	217	21.1	0.38	0.50
44	Platyphylline	129	24.0	0.18	0.50
45	Integerrimine	172.5	24.3	0.39	0.57
46	Spectabiline	185.5–186	27.2	0.34	0.43
47	Senkirkine	198	31.6	0.29	0.47
48	Jacobine	228	34.0	0.38	0.36
49	Sceleratine	178	34.6	0.34	0.39
50	Jacozine	228	35.5	0.37	0.32
51	Jacoline	221	36.6	0.37	0.23
52	Rosmarinine	209	37.1	0.35	0.40
53	Jaconine	147	40.4	0.47	0.47
54	Retrorsine	219–220	40.7	0.35	0.36
55	Riddelliine	198	41.9	0.32	0.28
56	Retusamine	174.5	50.3	0.30	0.51
57	Otosenine	221	51.0	0.23	0.34
58	Grantianine	209–209.5	57.0	0.31	0.33

TABLE 59 EFFECTS OF STRUCTURAL CHANGES ON RELATIVE RETENTION TIMES

Structural change	Examples[a]	Ratio $\dfrac{R_T \text{ modified compd.}}{R_T \text{ initial compd.}}$	Structural factor (mean of example ratios)
Non-ester bases			
1-CH$_3$ → 1-CH$_2$OH	2 → 12	4.1	3.8
	5 → 17	4.4	
	7 → 19	3.3	
	9 → 18	3.3	
7-H → 7-α-OH	1 → 6[b]	2.75	2.43
	2 → 8	2.50	
	13 → 18	2.03	
7-H → 7-β-OH	1 → 4	2.06	2.06
	2 → 7	2.50	
	10 → 15	1.67	
	13 → 17	2.03	
	12 → 19	2.03	
7-β-OH → 7-α-OH	4 → 6[b]	1.33	1.16
	7 → 8	1.0	
	5 → 9	1.32	
	17 → 18	1.0	
1,2-dehydro → 1,2-epoxy	10 → 11	1.18	1.15
	13 → 14	1.12	
1-CH$_2$OH → 1-CH$_2$OMe	13 → 10	0.73	0.70
	14 → 11	0.77	
	17 → 15	0.60	
Esters of monocarboxylic acids[c]			
7-α-OH → 7-α-Angeloxy	26 → 33	2.49	2.51
	30 → 34	2.54	
7-H → 7-α-OH	22 → 24	1.55	1.66
	23 → 28	1.77	
4′-H → 4′-OH	24 → 30	1.47	1.40
	33 → 35	1.34	
7-β-Angeloxy → 7-α-Angeloxy	20 → 21	1.14	1.10
	26 → 28	1.09	
	27 → 29	1.09	
	35 → 36	1.06	
2′-OH → 2′-OMe	23 → 22	0.91	0.85
	28 → 24	0.80	
Macrocyclic diesters[c]			
1′-H → 1′-OH	42 → 54	1.98	1.98
	43 → 55	1.98	
5′,6′-double bond → 5′,6′-epoxy	42 → 48	1.65	1.67
	43 → 50	1.68	
Retronecine nucleus → otonecine nucleus	42 → 47	1.53	1.53
	48 → 57	1.50	
cis → trans 5′,6′-double bond	42 → 45	1.18	1.18
3′(H,CH$_3$) → 3′(=CH$_2$)	42 → 43	1.02	1.03
	48 → 50	1.04	
	54 → 55	1.03	

[a]Alkaloid numbers as given in Tables 56 and 57.
[b]Base 6 is the enantiomer of 7-α-hydroxy-1-methylene-8-α-pyrrolizidine.
[c]Dashed numbers indicate carbon atoms of the esterifying acids; see diagrams I and II.

(Only alkaloids having an unsaturated (3-pyrroline) ring in the basic moiety will respond to this procedure.) The basic composition of the above reagents was 30% hydrogen peroxide in which was dissolved sodium pyrophosphate (2–4 mg/ml). Acetic anhydride reagent consisted of acetic anhydride–light petroleum (b.p. 80–100 °C) and benzene (1:4:5) and Ehrlich reagent consisting of dimethylaminobenzaldehyde (1 g) dissolved in absolute ethanol (70 ml), carbitol 30 ml and hydrochloric acid (1.5 ml). The alkaloids, applied as methanol solutions, were chromatographed on Silica Gel G (Merck) and developed by either chloroform–acetone–ethanol–0.88 ammonia (5:3:1:1) or for N-oxides the non-aqueous phase from n–butanol–acetic acid–water (4:1:5) was also used.

Table 60 depicts the TLC of retrosine N-oxide and the relative strength of spots following detection by 3 techniques. Table 61 shows the TLC of retrosine and the relative strength of spots shown by platinum iodide, exposure to iodine vapor, modified Dragendorff reagent, and peroxide–anhydride–Ehrlich reagent. Table 62 describes the TLC of pyrrolizidine alkaloids and other bases developed with chloroform–acetone–ethanol–ammonia (5:3:1:1).

TABLE 60

THIN-LAYER CHROMATOGRAPHY OF RETRORSINE N-OXIDE (ISATIDINE)

Amount applied (μg)	Relative strengths of spots (R_F 0.20)[a]		
	A	B	C
26	+ +	+ + +	+ + +
12.8	+	+ +	+ + +
6.4	+	+ +	+ +
3.2	+	+	+ +
1.6	−	Trace	+
0.8	−	Trace	+
0.4	−	Faint trace	+
0.2	−	−	+
0.1	−	−	Trace

Solvent system: n-butanol–acetic acid–water, (4:1:5).
A, visible (brown) spots after spraying with acetic anhydride reagent and heating for 15 min.
B, yellow fluorescent spots visible under long wavelength ultraviolet illumination after treatment A.
C, blue spots visible after treatment A followed by Ehrlich's reagent and heating for 10–15 min.
[a] + Means a distinct spot; − means no spot visible.

The excretion of the pyrrolizidine alkaloid heliotrine in the urine and bile of sheep has been studied by Jago et al.[372]. Analysis of extracts of bile, blood and urine by column chromatography (Florisil, and elution with methanol and methanol–1% ammonia) and thin-layer chromatography (Silica Gel G, and chloroform–methanol–ammonia, 84:15:1) revealed the presence of three metabolites that gave a positive reaction with Dragendorff's reagent. (The R_F values of these metabolites corresponded

TABLE 61

THIN-LAYER CHROMATOGRAPHY OF RETRORSINE

Amount applied (μg)	Relative strengths of spots (R_F 0.35) shown by reagents			
	Platinum iodide spray	Exposure to iodine vapor	Modified Dragendorff reagent	Present method
128	+++	+++	+++	++++
64	++	++	+++	+++
32	++	++	++	++
16	+	+	+	++
8	+	+	+	+
4	Trace	+	Trace	+
2	−	Trace	Faint trace	+
1	−	−	−	+
0.5	−	−	−	+
0.25	−	−	−	Trace

Solvent system: chloroform–acetone–ethanol–ammonia, (5:3:1:1).

TABLE 62

THIN-LAYER CHROMATOGRAPHY OF PYRROLIZIDINE ALKALOIDS AND OTHER BASES

Group	Compound	$R_F{}^a$	Color of spot
I	Retrorsine	0.35	Blue
	Diacetylretrorsine	0.70	Blue
	Monocrotaline	0.43	Blue
	Senecionine	0.63	Blue
	Anacrotine	0.33	Blue
	Lasiocarpine	0.74	Blue
	Heliotrine	0.33	Blue
	Supinine	0.28	Mauve
II	Rosmarinine	0.33	Blue
III	Retronecine	0.09	Blue
	Heliotridine	0.05	Blue
IV	Platynecine	0.00	Weak yellow-brown
	Retronecanol		None
V	Strigosine		None
VI	Brucine		None
	Strychnine		None
	Arecoline	0.78	Weak yellow-brown
	Quinine		None
	Nicotine		None
VII	Benzylamine		None
	Pyrrolidine		None
VIII	Indole	0.80	Very weak brown
	Pyrrole		None

Solvent systems: chloroform–acetone–ethanol–ammonia, (5:3:1:1).

[a] R_F values varied slightly with different batches of solvent and adsorbent, thus these values should be regarded as relative, not absolute.

to those of heliotridine–trachelanthate, heliotrine N-oxide and heliotridine.) The chemical relationship of these metabolites to heliotrine are illustrated in Fig. 70.

The paper electrophoretic mobilities of 27 pyrrolizidine alkaloids and related compounds were determined for seven electrolytes[373]. In non-complexing electrolyte,

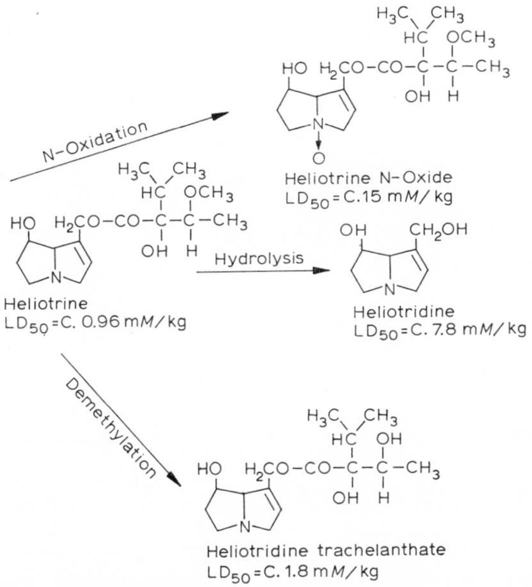

Fig. 70. Chemical relationships and respective toxicities of heliotrine and the 3 metabolites found in sheep[372].

the compounds migrate as cations, and electrophoresis at controlled pH afforded separations of mixtures dependent on differences in molecular weight and base strength of the alkaloids. Electrophoresis in complex-forming electrolytes (*e.g.* sodium borate and sodium arsenite) permitted separation and determination of some alkaloids which are chromatographically indistinguishable such as esters of trachelanthic and viridifloric acids and alkaloids of *Heliotropsium europacum* L. The paper electrophoresis was conducted using Whatman No. 4 papers in strips 13.5 × 1 cm with 45 cm under pressure and cooled in an enclosed strip apparatus previously described[374]. The composition of the electrolytes and reagents for the detection of the alkaloids on paper are as follows.

Electrolytes

(*A*) Acetate buffer (pH 4.6) containing 6.39 g $CH_3COONa \cdot 3H_2O$ and 3.2 g glacial acetic acid in 1 l of water. The solution was 0.1 M with respect to total acetate.

(*B*) Tris(hydroxymethyl)aminomethane–hydrochloric acid buffer (pH 7), prepared from stock solutions as prescribed by Gomori[375]. The electrolyte then referred to as "Tris buffer" or "electrolyte (*B*)".

TABLE 63

RELATIVE RATES OF MIGRATION OF PYRROLIZIDINE ALKALOIDS AND RELATED COMPOUNDS IN SEVEN ELECTROLYTES

For description of electrolytes (A)–(G), see text. Compounds were detected after paper electrophoresis in each electrolyte at approx. 20 V/cm and 26°C for 1–1.5 h. For each compound, the mol. wt. and pK_a value are included. Mobilities are relative to heliotridine (M_H values).

No.	Compound	Mol. wt.	pK_a[a]	$10^2 M_H$ values[b]						
				(A)	(B)	(C)	(D)	(E)	(F)	(G)
1	7-α-Hydroxy-1-methylene-8α-pyrrolizidine	139	9.3	102	103	125	153	151	137	122
2	Heliotridine	155	9.0	100	100	100	100	100	100	100
				(11.7)[c]	(10.5)	(6.6)	(4.4)	(4.9)	(4.9)	(5.9)
3	Retronecine	155	8.9	99	99	98	99	45	94	87
4	Platynecine	157	10.2	96	100	154	197	44	159	107
5	7-Angelylheliotridine	237	8.2	77	72	38	27	28	28	84
6	Crispatine	309	7.9	69	58	16	9	8	13	32
7	Fulvine	309	7.9	69	58	17	9	8	11	32
8	Supinine[d]	283	9.6	68	68	93	113	60	100	72
9	Supinidine viridiflorate[d]	283	9.6	68	67	92	113	12	75	71
10	Heleurine	297	9.6	67	68	92	110	113	102	76
11	Echinatine[d]	299	8.4	66	65	48	33	-48	15	61
12	Monocrotaline[d]	325	7.9	65	56	19	6	-100	-38	32
13	Heliotrine	313	8.5	65	64	48	30	35	35	62
14	Intermedine[d]	299	8.5	64	64	48	35	-6	34	56
15	Lycopsamine[d]	299	8.5	64	64	48	35	-48	15	56
16	Seneciphylline	333	7.9	64	54	11	5	10	4	26
17	Cynaustraline[d]	285	10.5	64	68	101	131	23	93	73
18	Jacobine	351	7.9	63	46	6	4	7	3	14
19	Spectabiline	367	7.9	63	54	12	5	8	8	31
20	Senecionine	335	7.9	61	54	7	4	6	8	28
21	Jaconine	387.5	7.9	59	46	6	4	7	3	17
22	Latifoline	393	7.9	59	50	8	5	3	4	38
23	Lasiocarpine[d]	411	7.6	56	46	9	4	-73	-10	54
24	Supinine N-oxide[d]	299	4.5	30	8	5	4	-39	3	2
25	Heliotrine N-oxide	329	4.5	22	6	3	3	8	2	2
26	Lasiocarpine N-oxide[d]	427	4.5	10	4	3	2	-76	-15	4
27	Heliotric acid	176		-69	-86	-131	-188	-164	-166	-130

[a] Values of pK_a in italics are calculated values; others are the experimental values (in aqueous medium) of Culvenor and Willette[377].

[b] Positive values represent cationic, negative values anionic migration.

[c] Values in parentheses are the absolute mobilities of heliotridine, in cm/h/kV of total potential applied.

[d] Vicinal glycol groups present.

TABLE 64

ELECTROPHORETIC DATA FOR CONSTITUENT ALKALOIDS OF SOME PLANT SPECIES

Plant species	Alkaloids	$10^2 M_H$			
		(C)	(D)	(E)	(F)
Heliotropium europaeum L.	Supinine	93	113	60	100
	Heleurine	92	110	113	102
	Heliotrine	48	30	35	35
	Lasiocarpine	9	4	−73	−10
	Europine[a]			−50	
Heliotropium lasiocarpum Fisch. & Mey.	Heliotrine	48	30	35	35
	Lasiocarpine	9	4	−73	−10
Heliotropium supinum L.	Supinine	93	113	60	100
	Echinatine	48	33	−48	15
	Heliosupine[a]	10	9	−78	−13
Cynoglossum australe R. Br.	Cynaustraline		131	23	93
	Cynaustine[b]		113	12	75
Cynoglossum amabile Stapf & Drummond	Echinatine	48	33	−48	15
	Amabiline[b]	92	113	12	75
Crotalaria crispata F. Muell. ex Benth.	Monocrotaline			−100	−38
	Fulvine			8	11
	Crispatine			8	13
Crotalaria spectabilis Roth.	Monocrotaline			−100	−38
	Spectabiline			8	8
Amsinckia hispida (Ruiz. & Pav.) Johnst.	Intermedine			−6	34
and *A. lycopsoides* Lehm.	Lycopsamine			−48	15
	Echiumine				

Alkaloid N-oxides occurring in the plants are not detailed separately; the list includes them as the corresponding tertiary bases. Values of relative rates of migration, taken from Table 63, are included, where appropriate to indicate potentially useful analytical applications for each electrolyte. Electrolytes are designated by letters as described in the text.

[a] Relative mobilities for europine and heliosupine, pure samples of which were not available for testing, were derived from experiments on the respective plant extracts. Data for these alkaloids are not included in Table 63.
[b] Cynaustine and amabiline are respectively, the (+)-supinidine and (−)-supinidine esters of viridifloric acid. The data entered here for the bases correspond with data for supinidine viridiflorate given in Table 63.

(C) Sodium hydrogen carbonate solution (0.1 M, pH 8.4). Because carbon dioxide tends to volatilize from this electrolyte, papers impregnated with it were enclosed in the apparatus without delay.

(D) Sodium hydrogen carbonate–sodium carbonate solution (pH 9.2), containing 6.72 g $NaHCO_3$ and 1.06 g anhydrous Na_2CO_3 in 1 l of water.

(E) Sodium borate buffer (pH 9.2) containing 0.2 g-atom of boron/l[376].

(*F*) Sodium arsenite solution containing 0.2 *M* arsenious acid adjusted to pH 9.2 with sodium hydroxide[376].

(*G*) Sodium phosphate buffer (pH 7.0) containing 6.24 g $NaH_2PO_4 \cdot 2H_2O$ and 10.68 g $Na_2HPO_4 \cdot 2H_2O/l$. The solution was 0.1 *M* with respect to total phosphate.

Reagents for the detection of the alkaloids on paper

(*1*) Chromium trioxide–permanganate–sulfuric acid[377].

(*2*) Hydrogen peroxide (B.D.H. AnalaR, 30% w/v), diluted tenfold immediately before use.

(*3*) Acetic acid solution (10% v/v) in hexane, used as a dip reagent

(*4*) Acetic anhydride–benzene–hexane. A mixture, 1:4:5 by volume used as a dip reagent.

(*5*) Ehrlich's reagent: *p*-dimethylaminobenzaldehyde (1 g) dissolved in 100 ml ethanol containing 1.5% (v/v) conc. hydrochloric acid.

(*6*) Iodine vapor.

Table 63 lists the relative rates of migration of pyrrolizidine alkaloids and related compounds in seven electrolytes and Table 64 describes the electrophoretic data for constituent alkaloids of some plant species.

REFERENCES

1   P.N. MAGEE, *Biochem. J.*, 70 (1958) 606.
2   J.M. BARNES AND P.N. MAGEE, *Brit. J. Ind. Med.*, 11 (1954) 167.
3   D.H. HEATH, *Biochem. J.*, 85 (1962) 72.
4   D.F. SWANN AND A.E.M. McLEAN, *Biochem. J.*, 107 (1968) 14P.
5   D.F. HEATH, *Biochem. Pharmacol.*, 16 (1967) 1517.
6   P.N. MAGEE AND M. VAN DE KAR, *Biochem. J.*, 70 (1958) 600.
7   D.C. HEATH AND A. DUTTON, *Biochem. J.*, 70 (1958) 619.
8   P.N. MAGEE, *Biochem. J.*, 64 (1956) 676.
9   H. DRUCKREY, R. PREUSSMANN, G. BLUM AND S. IVANKOVIC, *Naturwissenschaften*, 50 (1963) 99.
10 P.N. MAGEE, *1st Conf. on N-Nitroso Compounds and Lactones, Hamburg, 1963*, p.23.
11 W. LIJINSKY, J. LOO AND A.E. ROSS, *Nature*, 218 (1968) 1174.
12 S. VILLA-TREVINO, *Biochem. J.*, 105 (1967) 625.
13 R.C. SHANK, *Biochem. J.*, 108 (1968) 625.
14 P.N. MAGEE AND J.M. BARNES, *Advan. Cancer Res.*, 10 (1967) 163.
15 P. BROOKES, *Cancer Res.*, 26 (1966) 1994.
16 R. PREUSSMANN, *Arzneimittel-Forsch.*, 14 (1964) 769.
17 A.H. DUTTON AND D.F. HEATH, *Nature*, 178 (1956) 644.
18 D.F. HEATH, *Nature*, 192 (1961) 170.
19 P.N. MAGEE AND E. FARBER, *Biochem. J.*, 83 (1962) 114.
20 H. DRUCKREY, R. PREUSSMANN, S. IVANKOVIC AND D. SCHMÄHL, *Z. Krebsforsch.*, 69 (1967) 103.
21 H. DRUCKREY AND R. PREUSSMANN, *Naturwissenschaften*, 49 (1962) 111.
22 H. DRUCKREY, R. PREUSSMANN AND D. SCHMÄHL, *Acta Unio. Intern. Contra Cancrum*, 19 (1963) 510.
23 H. DRUCKREY AND D. STEINHOFF, *Naturwissenschaften*, 49 (1962) 497.

24   M.I.Argus and C.Hoch-Ligeti, *J. Natl. Cancer Inst.*, 27 (1961) 695.
25   D.Schmähl and H.Osswald, *Experientia*, 23 (1967) 497.
26   D.P.Griswold, Jr., A.E.Casey, E.K.Weisburger, J.H.Weisburger and F.M.Schabel, Jr., *Cancer Res.*, 26 (1966) 619.
27   P.N.Magee, *17th Colloquium of the Society for Physiological Chemistry, Mosbach/Baden, 1966*, p.79.
28   O.G.Fahmy, M.J.Fahmy, J.Massasso and M.Ondrej, *Mutation Res.*, 3 (1966) 201.
29   I.Pasternak, *Arzneimittel-Forsch.*, 14 (1964) 802.
30   J.Veleminsky and T.Gichner, *Mutation Res.*, 5 (1968) 429.
31   O.N.Pogodina, *Citologya*, 8 (1966) 503.
32   H.V.Malling, *Mutation Res.*, 3 (1966) 537.
33   O.G.Fahmy and M.J.Fahmy, *Mutation Res.*, 6 (1968) 139.
34   P.J.Elving and E.C.Olson, *J. Am. Chem. Soc.*, 79 (1957) 2697.
35   D.F.Heath and J.A.E.Jarvis, *Analyst*, 80 (1955) 613.
36   D.L.Lydersen and K.Nagy, *Z. Anal. Chem.*, 230 (1967) 277.
37   H.Hellmann, *Z. Physiol. Chem.*, 326 (1961) 15.
38   C.L.Walters, *Analyst*, 95 (1970) 485.
39   F.L.English, *Anal. Chem.*, 23 (1951) 344.
40   A.A.Smales and H.N.Wilson, *J. Soc. Chem. Ind.*, 67 (1948) 210.
41   P.Griess, *Ber.*, 12 (1879) 427.
42   D.Daiber and R.Preussmann, *Z. Anal. Chem.*, 206 (1964) 344.
43   K.Möhler and O.L.Mayrhofer, *Z. Lebensm. Untersuch. Forsch.*, 135 (1968) 313.
44   I.J.Gal, E.R.Stedrowsky and S.I.Miller, *Anal. Chem.*, 40 (1968) 168.
45   F.Ender, G.Havre, A.Helgebostad, N.Koppang, R.Madsden and L.Ceh, *Naturwissenschaften*, 51 (1964) 637.
46   R.K.Harris and R.A.Spragg, *Chem. Commun.*, 7 (1967) 362.
47   Ş.I.Chu and W.C.Lin, *J. Chinese Chem. Soc.*, 16 (1969) 55.
48   W.W.Becker and W.E.Shaefer, *Organic Analysis*, Interscience, New York, 1954, Vol. 2, p.93.
49   D.Tewari and J.P.Sharma, *Indian J. Chem.*, 2 (1964) 173.
50   H.Druckrey, A.Schildbach, D.Schmähl, R.Preussmann and S.Ivankovic, *Arzneimittel-Forsch.*, 83 (1963) 841.
51   R.Preussmann, D.Daiber and H.Hengy, *Nature*, 201 (1964) 502.
52   R.Preussmann, G.Neurath, G.Wulf-Lorentzen, D.Daiber and H.Hengy, *Z. Anal. Chem.*, 202 (1964) 187.
53   Y.H.Yoe and L.G.Overholser, *J. Am. Chem. Soc.*, 61 (1939) 2058.
54   Y.H.Yoe and L.G.Overholser, *J. Am. Chem. Soc.*, 63 (1941) 3224.
55   K.Yasuda and K.Nakashima, *Japan Analyst*, 17 (1968) 732.
56   N.P.Sen, D.C.Smith, L.Schwinghamer and J.J.Marleau, *J. Assoc. Offic. Anal. Chemists*, 52 (1969) 47.
57   E.Kröller, *Deut. Lebensm. Rundschau*, 10 (1967) 303.
58   L.Hedler and P.Marquardt, *Food Cosmet. Toxicol.*, 6 (1968) 341.
59   G.Neurath, B.Pirmann and M.Dünger, *Chem. Ber.*, 97 (1964) 1631.
60   H.J.Petrowitz, *Arzneimittel-Forsch.*, 18 (1968) 1486.
61   O.L.Mayrhofer and K.Möhler, *Z. Lebensm. Untersuch. Forsch.*, 134 (1967) 246.
62   J.W.Howard, T.Fazio and J.O.Watts, *J. Assoc. Offic. Anal. Chemists*, 53 (1970) 269.
63   N.P.Sen, D.C.Smith, L.Schwinghamer and B.Howsman, *Can. Inst. Food Technol. J.*, 3 (1970) 66.
64   F.Ender and L.Ceh, Alkylierend wirkende Verbindungen, *Second Conference on Tobacco Research, Freiberg, 1967*, p.83.
65   F.Ender and L.Ceh, *Food Cosmet. Toxicol.*, 6 (1968) 569.
66   G.Neurath and E.Doerk, *Chem. Ber.*, 97 (1964) 172.
67   F.Ender et al., *Z. Tierphysiol. Tierernaehr. Futtermittelk.*, 22 (1967) 189.
68   J.Sakshaug, E.Sognen, M.A.Hansen and N.Koppang, *Nature*, 206 (1965) 1261.
69   J.Sakshaug, personal communication.
70   O.G.Devik, *Acta Chem. Scand.*, 21 (1967) 2302.

71   O. DEVIK, personal communication.
72   J. K. FOREMAN, J. F. PALFRAMAN AND E. A. WALKER, Nature, 225 (1970) 554.
73   N. P. SEN, J. Chromatog., 51 (1970) 301.
74   J. H. ROBSON, J. Am. Chem. Soc., 77 (1955) 107.
75   W. D. EMMONS AND A. F. FERRIS, J. Am. Chem. Soc., 75 (1953) 4623.
76   B. H. THEWLIS, Food Cosmet. Toxicol., 5 (1967) 333.
77   P. MARQUARDT AND L. HEDLER, Arzneimittel-Forsch., 16 (1966) 778.
78   M. J. H. KEYBETS, E. H. GROOT AND G. H. M. KELLER, Food Cosmet. Toxicol., 8 (1970) 167.
79   N. P. SEN, D. C. SMITH AND L. SCHWINGHAMER, Food Cosmet. Toxicol., 7 (1969) 301.
80   J. SANDER, Arch. Hyg. Bakteriol., 151 (1967) 22.
81   J. SANDER, F. SCHWEINSBERG AND H. P. MENZ, Z. Physiol. Chem., 349 (1968) 1691.
82   J. SANDER, Z. Physiol. Chem., 349 (1968) 429.
83   W. LIJINSKY AND S. EPSTEIN, Nature, 225 (1970) 21.
84   L. S. DU PLESSIS, J. R. NUNN AND W. A. ROACH, Nature, 222 (1969) 1198.
85   N. D. MCGLASHAN, C. L. WALTERS AND A. E. M. MCLEAN, Lancet, II (1968) 1017.
86   S. UDENFRIEND, C. T. CLARK, J. AXELROD AND B. B. BRODIE, J. Biol. Chem., 208 (1954) 731.
87   G. NEURATH, B. PIRMANN, W. LUTTICH AND H. WICHERN, Beitr. Tabakforsch., 3 (1965) 251.
88   H. DRUCKREY, Advan. Cancer Res., 8 (1964) 356.
89   W. J. SERFONTEIN AND H. SMIT, Nature, 214 (1967) 169.
90   H. DRUCKREY, Acta Med. Scand., 170 (Suppl. 369)'(1961) 24.
91   W. J. SERFONTEIN AND P. HURTER, Cancer Res., 26 (1966) 575.
92   G. NEURATH, B. PIRMANN AND H. WICHERN, Beitr. Tabakforsch., 2 (1964) 311.
93   D. E. JOHNSON, J. D. MILLAR AND J. W. RHOADES, Natl. Cancer Inst. Monograph, 28 (1968) 181.
94   D. E. JOHNSON, J. W. RHOADES AND R. J. WHEELER, 154th American Chemical Society Meeting,
     Chicago, Ill., Sept. 14, 1967.
95   R. SCHOENTAL AND S. GIBBARD, Nature, 216 (1967) 612.
96   U. MOHR, J. ALTHOFF AND A. AUTHALER, Cancer Res., 26 (1966) 2349.
97   G. EISENBRAND, K. SPACZYNSKI AND R. PREUSSMANN, J. Chromatog., 51 (1970) 304.
98   G. EISENBRAND, P. MARQUARDT AND R. PREUSSMANN, Z. Anal. Chem., 247 (1969) 54.
99   H. C. HODGE AND S. H. STERNER, Am. Ind. Hyg. Assoc. Quart., 10 (1949) 93.
100 A. L. WALPOLE, D. C. ROBERTS, F. L. ROSE, J. A. HENDRY AND R. F. HOMER, Brit. J. Pharmacol.,
     9 (1954) 306.
101 H. LÜERS AND G. RÖHRBORN, Mutation Res., 2 (1965) 29.
102 F. K. ZIMMERMAN AND U. VON LAER, Mutation Res., 4 (1967) 377.
103 G. F. WRIGHT AND V. K. ROWE, Toxicol. Appl. Pharmacol., 11 (1967) 575.
104 S. D. SILVER AND F. P. MCGRATH, J. Ind. Hyg. Toxicol., 30 (1948) 7.
105 E. ALLEN AND W. SEAMAN, Anal. Chem., 27 (1955) 540.
106 J. EPSTEIN, R. W. ROSENTHAL AND P. J. ESS, Anal. Chem., 27 (1955) 1435.
107 D. H. ROSENBLATT, P. HLINKA AND J. EPSTEIN, Anal. Chem., 27 (1955) 1290.
108 T. R. CROMPTON, Analyst, 90 (1965) 107.
109 V. S. TATARINSKII, N. N. MOSKVITIN, V. G. BEREZKIN AND V. N. ANDRONOV, Sovrem. Metody
     Khim. Spektraln. Anal. Mater., (1967) 253; Chem. Abstr., 68 (1968) 46024T.
110 A. DILORENZO AND G. RUSSO, J. Gas Chromatog., 6 (1968) 509.
111 A. B. BOŘKOVEC, Residue Rev., 6 (1964) 87.
112 A. B. BOŘKOVEC, Insect Chemosterilants, Interscience, New York, 1966.
113 W. W. KILGORE AND R. C. DOUTT (Eds.), Pest Control, Academic Press, New York, 1967,
     p. 197.
114 W. J. HAYES, Bull. World Health Organ., 31 (1964) 721.
115 K. R. S. ASCHER, 5th Intern. Pesticide Congr., London, 1963, p. 7; World Review of Pest Control,
     1964, Vol. III, Part 1, p. 7.
116 J. M. BARNES, Roy. Soc. Trop. Med. Hyg. Trans., 58 (1964) 327.
117 S. S. EPSTEIN AND H. SHAFNER, Nature, 219 (1968) 385.
118 J. PALMQUIST AND L. E. LACHANCE, Science, 154 (1966) 915.
119 A. R. KANEY AND K. C. ATWOOD, Nature, 201 (1964) 1006.
120 Y. L. TAN AND D. R. COLE, Clin. Chem., 11 (1965) 50.
121 M. BEROZA AND A. B. BOŘKOVEC, J. Med. Chem., 7 (1964) 44.

122 S. C. CHANG, A. B. BOŘKOVEC AND C. W. WOODS, *J. Econ. Entomol.*, 59 (1966) 937.

123 C. W. COLLIER AND R. TARDIF, *J. Econ. Entomol.*, 60 (1967) 28.

124 S. C. CHANG AND A. B. BOŘKOVEC, *J. Econ. Entomol.*, 59 (1966) 102.

125 A. B. BOŘKOVEC, S. C. CHANG AND A. M. LIMBURG, *J. Econ. Entomol.*, 57 (1964) 815.

126 J. E. MAITLEN AND L. M. MCDONOUGH, *J. Econ. Entomol.*, 60 (1967) 1391.

127 M. C. BOWMAN AND M. BEROZA, *J. Assoc. Offic. Anal. Chemists*, 49 (1966) 1046.

128 S. S. BRODY AND J. E. CHANEY, *J. Gas Chromatog.*, 4 (1966) 42.

129 H. C. COX, J. R. YOUNG AND M. C. BOWMAN, *J. Econ. Entomol.*, 60 (1967) 1111.

130 F. W. PLAPP, JR., W. S. BIGLEY, G. A. CHAPMAN AND G. W. EDDY, *J. Econ. Entomol.*, 55 (1962) 607.

131 W. F. CHAMBERLAIN AND E. W. HAMILTON, *J. Econ. Entomol.*, 57 (1964) 800.

132 W. F. CHAMBERLAIN AND C. C. BARRETT, *J. Econ. Entomol.*, 57 (1964) 267.

133 W. F. CHAMBERLAIN, *J. Econ. Entomol.*, 55 (1962) 240.

134 R. C. HARRIS, *J. Econ. Entomol.*, 55 (1962) 882.

135 W. F. CHAMBERLAIN, P. E. GATTERMAN AND D. E. HOPKINS, *J. Econ. Entomol.*, 54 (1961) 733.

136 J. C. PARISH AND B. W. ARTHUR, *J. Econ. Entomol.*, 58 (1965) 976.

137 F. W. PLAPP, JR. AND J. E. CASIDA, *Anal. Chem.*, 30 (1958) 1622.

138 C. BENCKHUIJSEN, *Biochem. Pharmacol.*, 17 (1968) 55.

139 L. B. MELLETT AND L. A. WOODS, *Cancer Res.*, 20 (1960) 524.

140 L. B. MELLETT, P. E. HODGSON AND L. A. WOODS, *J. Lab. Clin. Med.*, 60 (1962) 818.

141 A. W. CRAIG, B. W. FOX AND H. JACKSON, *Biochem. Pharmacol.*, 3 (1959) 42.

142 C. W. GILBERT AND J. P. KEENE, *Radioisotopes in Scientific Research*, Pergamon Press, London, 1958, Vol. 1, p. 698.

143 E. I. GOLDENTHAL, M. V. NADKARNI AND P. K. SMITH, *J. Pharmacol. Exptl. Therap.*, 122 (1958) 431.

144 R. R. PAINTER AND W. W. KILGORE, *J. Insect Physiol.*, 13 (1967) 1105.

145 L. H. SCHMIDT, *Ann. N.Y. Acad. Sci.*, 68 (1958) 652.

146 H. B. MANDEL, *Pharmacol. Rev.*, 11 (1959) 743.

147 H. JACKSON, B. W. FOX AND A. W. CRAIG, *Brit. J. Pharmacol.*, 14 (1959) 149.

148 A. J. BATEMAN, *Nature*, 210 (1966) 205.

149 W. E. RATNAYAKE, *Mutation Res.*, 5 (1968) 271.

150 M. WESTERGAARD, *Experientia*, 13 (1957) 224.

151 D. T. NORTH, *Mutation Res.*, 4 (1967) 225.

152 M. V. NADKARNI, E. I. GOLDENTHAL AND P. K. SMITH, *Cancer Res.*, 14 (1954) 599.

153 M. V. NADKARNI, E. G. TRAMS AND P. K. SMITH, *Proc. Am. Assoc. Cancer Res.*, 2 (1956) 136.

154 S. MOORE AND H. J. STEIN, *J. Biol. Chem.*, 192 (1951) 663.

155 K. H. JACOBSON, E. B. HACKLEY AND L. FEINSILVER, *Arch. Ind. Health*, 13 (1956) 237.

156 A. HADDOW, in F. HAMBURGER AND H. FISHMAN (Eds.), *The Physiopathology of Cancer*, Harper, New York, 1953, Vol. II, p. 411.

157 P. ALEKANDER AND K. A. STACEY, *Ann. N.Y. Acad. Sci.*, 68 (1958) 1225.

158 M. J. BIRD, *J. Genet.*, 50 (1952) 480.

159 H. G. KOLMARK AND B. J. KILBEY, *Mol. Gen. Genet.*, 101 (1960) 89.

160 A. C. FABERGE, *Genetics*, 40 (1955) 571.

161 D. GUNTHER, *Anal. Chem.*, 37 (1965) 1172.

162 F. E. CRITCHFIELD AND J. B. JOHNSON, *Anal. Chem.*, 29 (1957) 797.

163 J. C. GAGE, *Analyst*, 82 (1957) 587.

164 C. E. VOOGD AND P. V. D. VET, *Experientia*, 25 (1969) 85.

165 S. G. HEUSER AND K. A. SCUDAMORE, *Analyst*, 93 (1968) 252.

166 S. G. HEUSER AND K. A. SCUDAMORE, *Chem. Ind. (London)*, (1966) 1557.

167 B. BERCK, *J. Agr. Food Chem.*, 13 (1965) 375.

168 S. BEN-YEHOSHUA AND P. KRINSKY, *J. Gas Chromatog.*, 6 (1968) 350.

169 S. G. HEUSER AND K. A. SCUDAMORE, *J. Sci. Food Agr.*, 20 (1969) 566.

170 J. T. GORDON, W. W. THORNBURG AND L. N. WERUM, *J. Agr. Food Chem.*, 7 (1959) 196.

171 Y. OBI, Y. SHIMADA, K. TAKAHASHI, K. NISHIDA AND T. KISAKI, *Tobacco*, 166 (1968) 26.

172 M. MURAMATSU, Y. OBI, Y. SHIMADAA, K. TAKAHASHI AND K. NISHIDA, *Sci. Papers Central Res. Inst. Japan Monopoly Corp.*, 110 (1968) 217.

173 R.K.KULKARNI, D.BARTAK, D.K.OUSTERHOUT AND F.LEONARD, *J. Biomed. Mater. Res.*, 2 (1968) 165.
174 A.C.CUNLIFFE AND F.WESLEY, *Brit. Med. J.*, 2 (1967) 575.
175 W.L.GUESS AND A.B.JONES, *Am. J. Hosp. Pharm.*, 26 (1969) 180.
176 D.J.BROWN, *J. Assoc. Offic. Anal. Chemists*, 53 (1970) 263.
177 N.ADLER, *J. Pharm. Sci.*, 54 (1965) 735.
178 A.HOLMGREN, N.DIDING AND G.SAMUELSSON, *Acta Pharm. Suecica*, 6 (1969) 33.
179 E.P.RAGELIS, B.S.FISHER AND B.A.KLIMECK, *J. Assoc. Offic. Agr. Chemists*, 49 (1966) 963.
180 P.ALEXANDER, *Advan. Cancer Res.*, 2 (1954) 4.
181 W.C.J.ROSS, *Advan. Cancer Res.*, 1 (1953) 429.
182 V.K.ROWE, R.L.HOLLINGSWORTH, F.OYEN, D.D.McCOLLISTER AND H.C.SPENSER, *Arch. Ind. Health*, 13 (1956) 228.
183 A.L.WALPOLE, *Ann. N.Y. Acad. Sci.*, 68 (1958) 750.
184 I.A.RAPOPORT, *Dokl. Akad. Nauk SSSR*, 60 (1948) 469.
185 A.SCHALET, *Drosophila. Inform. Serv.*, 28 (1954) 155.
186 R.MESTRES AND C.BARROIS, *Soc. Pharm. Montpellier*, 24 (1964) 47.
187 E.T.HUNTRESS, *The Preparation, Properties, Chemical Behavior and Identification of Organic Chlorine Compounds*, Wiley, New York, 1948, p.705.
188 S.CARSON AND B.L.OSER, *Toxicol. Appl. Pharmacol.*, 14 (1969) 633.
189 A.M.AMBROSE, *Arch. Ind. Health*, 2 (1950) 591.
190 W.L.GUESS, *Toxicol. Appl. Pharmacol.*, 14 (1969) 659.
191 F.WESLEY, B.ROURKE AND O.DARBISHIRE, *J. Food Sci.*, 30 (1965) 1037.
192 J.F.SMITH, in R.P.W.SCOTT (Ed.), *Gas Chromatography, Proceedings of the Third Symposium*, Butterworths, London, 1960, p.114.
193 E.P.RAGELIS, B.S.FISHER, B.A.KLIMECK AND C.JOHNSON, *J. Assoc. Offic. Anal. Chemists*, 51 (1968) 709.
194 P.MANCHON AND A.BUQUET, *Food Cosmet. Toxicol.*, 8 (1970) 9.
195 M.K.JOHNSON, *Biochem. Pharmacol.*, 16 (1967) 185.
196 M.K.JOHNSON, *Biochem. J.*, 98 (1966) 38.
197 J.S.PAGINGTON, *J. Chromatog.*, 36 (1968) 528.
198 T.W.LARKHAM AND J.S.PAGINGTON, *J. Chromatog.*, 28 (1967) 422.
199 S.G.HEUSER AND K.A.SCUDAMORE, *Chem. Ind. (London)*, (1969) 1054.
200 S.G.HEUSER AND G.GOODSHIP, *Pest Infestation Research*, 41 (1960) London, HMSO.
201 C.H.HINE, J.K.KODAMA, J.S.WELLINGTON, M.K.DUNLAP AND H.H.ANDERSON, *Arch. Ind. Health*, 14 (1956) 250.
202 J.K.KODAMA, R.J.GUZMAN, M.K.DUNLAP, G.S.LOQUUAM, R.LIMA AND C.H.HINE, *Arch. Environ. Health*, 2 (1961) 50.
203 G.KØLMARK AND N.H.GILES, *Genetics*, 40 (1955) 890.
204 J.T.BIESELE, F.S.PHILIPS, J.B.THIERSCH, J.H.BURCHENAL, S.M.BUCKLEY AND C.C.STOCK, *Nature*, 166 (1950) 1112.
205 J.DYR AND J.MOSTEK, *Knasny Prumsyl*, 4 (1958) 121; *Chem. Abstr.*, 52 (1958) 19011.
206 C.SCHULTZ, *Deut. Med. Wochschr.*, 89 (1964) 1342.
207 ANON, *Food Cosmet. Toxicol.*, 2 (1964) 1342.
208 A.LOVELESS AND S.HOWARTH, *Nature*, 184 (1959) 1780.
209 B.S.STRAUSS AND S.OKUBO, *J. Bacteriol.*, 79 (1960) 464.
210 J.D.HAHN, *Nature*, 226 (1970) 87.
211 R.J.ERICSSON, *J. Reprod. Fertility*, 21 (1970) 267.
212 A.R.JONES, P.DAVIES, K.EDWARDS AND H.JACKSON, *Nature*, 224 (1969) 83.
213 H.JACKSON, I.S.G.CAMPBELL AND A.R.JONES, *Nature*, 226 (1970) 86.
214 W.SCHÄFER, W.NUCK AND H.JAHN, *J. Prakt. Chem.*, 11 (1960) 1.
215 W.SCHÄFER, W.NUCK AND H.JAHN, *J. Prakt. Chem.*, 11 (1960) 10.
216 H.R.COPPER AND J.G.ROBERTS, *J. Soc. Dyers Colourists*, 80 (1964) 428.
217 P.W.WEST, P.SEN, B.R.SANT, K.L.MALLIK AND J.G.S.GUPTA, *J. Chromatog.*, 6 (1961) 220.
218 C.J.McCAMMON, P.KOTIN AND H.L.FALK, *Proc. Am. Assoc. Cancer Res.*, 2 (1957) 229.
219 A.LOVELESS AND C.C.STOCK, *Proc. Roy. Soc. (London)*, Ser.B, 150 (1959) 497.

220 C.H.HINE AND V.K.ROWE, *Industrial Hygiene and Toxicology*, Wiley, New York, 1967, Vol.II, p.1601.
221 M.S.MELZER, *Biochim. Biophys. Acta*, 138 (1967) 613.
222 Y.NAKAO AND C.AUERBACH, *Z. Vererbungslehre*, 92 (1961) 457.
223 C.H.CLARKE, *Mutation Res.*, 8 (1969) 35.
224 R.MATAGNE, *Radiation Botany*, 8 (1968) 489.
225 N.S.COHN, *Exptl. Cell Res.*, 24 (1961) 569.
226 D.J.SWANN, *Anal. Chem.*, 26 (1954) 878.
227 B.L. VAN DUUREN, N.NELSON, L.ORRIS, E.D.PALMES AND F.L.SCHMITT, *J. Natl. Cancer Inst.*, 31 (1963) 41.
228 J.F.WALKER, *Formaldehyde*, 3rd edn., Reinhold, New York, 1964.
229 E.SKOG, *Acta Pharmacol. Toxicol.*, 6 (1950) 299.
230 T.MATSUNAGA, T.SOEJIMA, Y.IWATA AND F.WATANABE, *Gann*, 45 (1954) 451.
231 C.AUERBACH, *Nature*, 210 (1966) 104.
232 T.ALDERSON, *Nature*, 207 (1965) 164.
233 A.P.ALTSHULLER AND S.P.MCPHERSON, *J. Air Pollution Control Assoc.*, 13 (1963) 109.
234 E.SAWICKI, T.R.HAUSER AND S.MCPHERSON, *Anal. Chem.*, 34 (1962) 1460.
235 S.EBEL, M.BRUEGGEMANN AND E.ROSSWOG, *Deut. Apotheker-Ztg.*, 107 (1967) 1718.
236 M.B.JACOBS, *The Analytical Chemistry of Industrial Poisons, Hazards and Solvents*, 2nd edn., Wiley, New York, 1949, Vol.I, p.524.
237 K.J.BOMBAUGH AND W.C.BULL, *Anal. Chem.*, 34 (1962) 1237.
238 K.JONES, *J. Gas Chromatog.*, (1967) 432.
239 F.ONUSKA, J.JANKA, S.DURAS AND M.KRCMAROVA, *J. Chromatog.*, 40 (1969) 209.
240 R.S.MANN AND K.W.HAHN, *Anal. Chem.*, 39 (1967) 1314.
241 S.SANDLER AND R.STROM, *Anal. Chem.*, 32 (1960) 1890.
242 H.CHERDON, L.HÖHR AND W.KERN, *Angew. Chem.*, 73 (1961) 215.
243 W.O.MCREYNOLDS, Pittsburgh Conf. Anal. Chem. Appl. Spectry. 1961, *Anal. Chem.*, 33 [2] (1961) 78A.
244 T.IGUCHI AND T.TAKIUCHI, *Bunseki Kagaku*, 17 (1968) 1080.
245 M.P.STEVENS AND D.F.PERCIVAL, *Anal. Chem.*, 36 (1964) 1023.
246 S.HARRISON, *Analyst*, 92 (1967) 773.
247 D.F.MEIGH, *Nature*, 170 (1952) 579.
248 R.G.RICE, *Anal. Chem.*, 23 (1951) 194.
249 R.ELLIS, *Anal. Chem.*, 32 (1959) 1997.
250 A.M.GADDIS AND R.ELLIS, *Anal. Chem.*, 31 (1958) 870.
251 E.BLOEM, *J. Chromatog.*, 35 (1968) 108.
252 L.A.JONES, *J. Org. Chem.*, 26 (1961) 228.
253 L.SELISBERGER, *J. Am. Leather Chemists' Assoc.*, 51 (1956) 2.
254 A.KÜNTZEL, *Leder*, 2 (1951) 233.
255 E.DENTI AND M.P.LUBOZ, *J. Chromatog.*, 18 (1965) 325.
256 K.HARADA, S.SHIGETSUGU, Y.SHINODA AND K.YAMADA, *Nippon Suisan Gakkaishi*, 36 (1970) 188.
257 D.J.PIETRZYK AND E.P.CHAN, *Anal. Chem.*, 42 (1970) 41.
258 K.ONOE, *J. Chem. Soc. Japan*, 73 (1952) 337.
259 D.P.SCHWARTZ, *Microchem. J.*, 7 (1963) 403.
260 J.RUSMUS, *J. Chromatog.*, 6 (1961) 187.
261 W.FREYTAG, *Fette, Seifen, Anstrichmittel*, 65 (1963) 603.
262 D.P.SCHWARTZ, *Anal. Chem.*, 35 (1963) 2191.
263 D.P.SCHWARTZ, *J. Chromatog.*, 9 (1962) 187.
264 E.B.TRUITT, JR., *Federation Proc.*, 28 (1969) 543.
265 R.A.LAHTI AND E.MAJCHROWICZ, *Biochem. Pharmacol.*, 18 (1969) 535.
266 E.D.BARBER AND J.P.LODGE, *Anal. Chem.*, 35 (1963) 348.
267 E.J.HUGHES AND R.W.HURN, *J. Air Pollution Control Assoc.*, 10 (1960) 367.
268 C.F.ELLIS, R.F.KENDALL AND B.H.ECCLESTON, *Anal. Chem.*, 37 (1965) 511.
269 R.H.LINNEL AND W.E.SCOTT, *Arch. Environ. Health*, 5 (1962) 616.
270 P.K.MUELLER, M.F.FRACCIA AND F.J.SCHUETTE, *152nd Natl. Meeting Am. Chem. Soc., New York, 1966*, pp.4–63.

271 H.F.Smith, Jr., *Am. Ind. Hyg. Assoc. Quart.*, 17 (1956) 311.
272 E.Asmussen, J.Hald and V.Larsen, *Acta Pharmacol.*, 4 (1948) 311.
273 E.Jacobsen, *Pharmacol. Rev.*, 4 (1952) 107.
274 R.T.Williams, *Detoxification Mechanisms*, 2nd edn., Wiley, New York, 1959.
275 I.A.Rapoport, *Dokl. Akad. Nauk SSSR*, 61 (1948) 713.
276 S.Sandler and Y.H.Chung, *Anal. Chem.*, 30 (1958) 1252.
277 H.Siegel and F.T.Weiss, *Anal. Chem.*, 26 (1954) 917.
278 J.E.Ruch and J.B.Johnson, *Anal. Chem.*, 28 (1956) 69.
279 J.H.Ross, *Anal. Chem.*, 25 (1953) 1288.
280 E.A.Demina, R.E.Podnebesnaya and Y.P.Grishutin, *Tr. Khim. Met. Inst. Akad. Nauk Kaz. SSR*, 2 (1968) 165; *Chem. Abstr.*, 70 (1969) 8677S.
281 R.N.Mokeeva and Y.A.Tsarafin, *Ind. Lab.*, 31 (1965) 1306.
282 A.A.Koldaev, *Sudebno-Med. Ekspertiza, Min. Zdravookhr. SSSR*, 11 (1968) 24.
283 O.L.Lapitskaya and L.V.Ivanova, *Ferment. Spirt. Prom.*, 34 (1968) 24; *Chem. Abstr.*, 69 (1968) 49069T.
284 G.Kyryacos, H.R.Menapace and C.E.Boord, *Anal. Chem.*, 31 (1959) 222.
285 T.A.Bellar and J.E.Sigsby, Jr., *Environ. Sci. Technol.*, 4 (1970) 150.
286 L.E.Broering, W.J.Werner and A.H.Rose, Jr., *Annual Meeting of Air Pollution Control Association, Cleveland, Ohio, June* 1967.
287 R.J.Soukup, R.J.Scarpellino and E.Danielczik, *Anal. Chem.*, 36 (1964) 2255.
288 L.A.Jones and R.J.Monroe, *Anal. Chem.*, 37 (1965) 935.
289 E.B.Truitt, Jr., F.K.Bell and J.C.Krantz, Jr., *Quart. J. Studies Alc.*, 17 (1956) 594.
290 G.Freund and P.O'Hollaren, *J. Lipid Res.*, 6 (1965) 471.
291 Y.Fukui, *Jap. J. Legal Med.*, 23 (1969) 22.
292 M.K.Roach and P.J.Creaven, *Clin. Chim. Acta*, 21 (1968) 275.
293 R.N.Baker, A.L.Acenty and J.F.Zack, Jr., *J. Chromatog. Sci.*, 7 (1969) 312.
294 D.A.Buyske, L.H.Owen, P.Wilder, Jr. and M.E.Hobbs, *Anal. Chem.*, 28 (1956) 910.
295 M.Severin, *Bull. Inst. Agron. Sta. Rech. Gembloux*, 32 (1964) 122.
296 F.C.Hunt, *J. Chromatog.*, 40 (1969) 465.
297 D.P.Schwartz, O.W.Parks and M.Keeney, *Anal. Chem.*, 34 (1962) 669.
298 E.A.Corbin, D.P.Schwartz and M.Keeney, *J. Chromatog.*, 3 (1960) 322.
299 F.Klein and K.DeJong, *Rec. Trav. Chim.*, 75 (1956) 1285.
300 F.E.Huelin, *Australian J. Sci. Res.*, 5B (1952) 328.
301 D.P.Schwartz, A.R.Johnson and O.W.Parks, *Microchem. J.*, 6 (1962) 37.
302 D.L.Mays, G.S.Born and J.E.Christian, *Bull. Environ. Contam. Toxicol.*, 3 (1968) 366.
303 T.Doihara, U.Kobashi, S.Sugawara and Y.Kaburaki, *Sci. Papers Central Res. Inst. Japan Monopoly Corp.*, No. 106 (1964) 129, 141.
304 R.D.Cadle and H.S.Johnston, *Proc. 2nd Natl. Air Pollution Symp., Pasadena, Calif.*, 1952, p. 28.
305 H.F.Smyth, Jr., C.P.Carpenter and C.S.Weil, *Arch. Ind. Hyg. Occup. Med.*, 4 (1951) 119.
306 L.A.Th.Verhaar and S.P.Lankhuijzen, *J. Chromatog. Sci.*, 8 (1970) 457.
307 M.P.Stevens, *Anal. Chem.*, 37 (1965) 167.
308 E.M.Bevilacqua, E.S.English and J.S.Gall, *Anal. Chem.*, 34 (1962) 861.
309 G.Mizuno, E.McMeans and J.R.Chipault, *Anal. Chem.*, 37 (1965) 151.
310 D.F.Gadbois, P.G.Scheurer and F.J.King, *Anal. Chem.*, 40 (1968) 1362.
311 L.W.Kissinger and H.E.Ungnade, *J. Org. Chem.*, 23 (1958) 815.
312 J.W.Ralls, *Anal. Chem.*, 32 (1960) 332.
313 H.Kelker, *Angew. Chem.*, 71 (1959) 218.
314 G.Mackell, *Ind. Chemist*, 36 (1960) 13.
315 R.K.Boutwell and N.H.Colburn, *Ann. N.Y. Acad. Sci.*, 163 (1969) 751.
316 D.W.Fassett, *Ind. Hyg. Toxicol.*, 4 (1967) 1823.
317 H.H.Smith and A.B.Srb, *Science*, 114 (1951) 490.
318 L.J.Lilly, *Nature*, 207 (1965) 433.
319 H.Roth, *Mikrochim. Acta*, 6 (1958) 766.
320 H.K.Hall, Jr. and R.Zbinden, *J. Am. Chem. Soc.*, 80 (1958) 6428.
321 F.Dickens and H.E.H.Jones, *Brit. J. Cancer*, 15 (1961) 85.

322 C.E.SEARLE, *Brit. J. Cancer*, 15 (1961) 804.
323 N.H.COLBURN AND R.K.BOUTWELL, *Cancer Res.*, 28 (1968) 642.
324 N.H.COLBURN AND R.K.BOUTWELL, *Cancer Res.*, 26 (1966) 1701.
325 N.H.COLBURN AND R.K.BOUTWELL, *Cancer Res.*, 28 (1968) 653.
326 F.DICKENS, *Brit. Med. Bull.*, 20 (1964) 96.
327 F.DICKENS, in R.TRUHAUT (Ed.), *Potential Carcinogenic Hazards from Drugs*, UICC Monograph Series, Vol.7, Springer-Verlag, Berlin, 1967, p.144.
328 F.DICKENS, H.E.H.JONES AND H.B.WAYNFORTH, *Brit. J. Cancer*, 20 (1966) 134.
329 B.L.VAN DUUREN, L.ORRIS AND N.NELSON, *J. Natl. Cancer Inst.*, 35 (1965) 707.
330 B.L.VAN DUUREN AND B.M.GOLDSCHMIDT, *J. Med. Chem.*, 9 (1966) 77.
331 L.LEVINE, J.A.GORDON AND N.P.JENCKS, *Biochem.*, 2 (1963) 168.
332 H.HAMPEL AND H.J.HAPKE, *Arch. Int. Pharmacodyn.*, 171 (1968) 306.
333 P.G.KEENEY, *J. Am. Oil Chemists' Soc.*, 34 (1957) 356.
334 G.M.GAL'PERN, G.A.GUDKOVA, E.Y.SHAPOSHNIKOVA AND E.B.YABUBSKII, *Zavodsk. Lab.*, 32 (1966) 931; *Chem. Abstr.*, 65 (1966) 17049f.
335 R.VIANI, F.MUEGGLER-CHAVAN, D.REYMOND AND R.H.EGLI, *Helv. Chim. Acta*, 48 (1965) 1809.
336 D.REYMOND, F.MUEGGLER-CHAVAN, R.VIANI, L.VAUTA AND R.H.EGLI, *Advan. Gas Chromatog., Proc. 3rd Intern. Symp., Houston, Texas, 1965*, Univ. of Houston, Houston, Texas, 1966, p.126.
337 V.M.GIANTURCO, A.S.GIAMARINO AND P.FRIEDEL, *Nature*, 210 (1966) 1358.
338 R.K.CREVELING, R.M.SILVERSTEIN AND W.G.JENNINGS, *J. Food Sci.*, 33 (1968) 284.
339 H.JACKSON AND A.R.JONES, *Nature*, 230 (1968) 591.
340 S.S.EPSTEIN, W.BASS, E.ARNOLD AND Y.BISHOP, *Science*, 168 (1970) 584.
341 G.KOLMARK, *Compt. Rend. Trav. Lab. Carlsberg Ser. Physiol.*, 26 (1956) 205.
342 M.R.GUMBMANN, W.E.GAGNE AND S.N.WILLIAMS, *Toxicol. Appl. Pharmacol.*, 12 (1968) 360.
343 A.B.JONES, *Experientia*, 26 (1970) 492.
344 F.L.SCOTT, R.RIORDAN AND P.D.MORTON, *J. Org. Chem.*, 27 (1962) 4255.
345 C.S.HANES AND F.A.ISHERWOOD, *Nature*, 164 (1949) 1107.
346 A.LAMOTTE AND J.C.MERLIN, *J. Chromatog.*, 45 (1969) 432.
347 A.LAMOTTE AND J.C.MERLIN, *J. Chromatog.*, 38 (1968) 296.
348 A.LAMOTTE, A.FRANCINA AND J.C.MERLIN, *J. Chromatog.*, 44 (1969) 75.
349 R.KLEMENT AND A.WILD, *Z. Anal. Chem.*, 195 (1963) 180.
350 J.ASKEW, J.H.RUZICKA AND B.B.WHEALS, *J. Chromatog.*, 41 (1969) 180.
351 H.SCHLENK AND J.L.GELLERMAN, *Anal. Chem.*, 32 (1960) 1412.
352 L.E.ST.JOHN, JR. AND D.J.LISK, *J. Agr. Food Chem.*, 16 (1968) 408.
353 C.W.STANLEY, *J. Agr. Food Chem.*, 14 (1966) 321.
354 M.J.SHAFIK AND H.F.ENOS, *J. Agr. Food Chem.*, 17 (1969) 1186.
355 C.C.J.CULVENOR, A.T.DANN AND A.T.DICK, *Nature*, 195 (1962) 570.
356 A.R.MATTOCKS, *Nature*, 217 (1968) 723.
357 F.L.WARREN, *Proc. Chem. Org. Nat. Prod.*, 12 (1955) 198.
358 J.S.MCKENZIE, *J. Exptl. Biol. Med. Sci.*, 36 (1958) 11.
359 R.SCHOENTAL, *Cancer Res.*, 28 (1968) 2237.
360 L.B.BULL AND A.T.DICK, *Australian J. Bull. Med. Sci.*, 38 (1960) 515.
361 C.R.GREEN AND G.S.CHRISTIE, *Brit. J. Exptl. Pathol.*, 42 (1961) 369.
362 R.SCHOENTAL, *Nature*, 227 (1970) 401.
363 N.G.BRINK, *Mutation Res.*, 8 (1969) 139.
364 A.M.CLARK, *Z. Vererbungslehre*, 94 (1963) 115.
365 R.K.SHARMA, G.S.KHAJURIA AND C.K.ATAL, *J. Chromatog.*, 19 (1965) 433.
366 R.MUNIER, *Bull. Soc. Chim. Biol.*, 35 (1953) 1225.
367 A.H.CHALMERS, C.C.J.CULVENOR AND L.W.SMITH, *J. Chromatog.*, 20 (1965) 270.
368 H.M.FALES AND J.J.PISANO, *Anal. Biochem.*, 3 (1962) 337.
369 A.R.MATTOCKS, *J. Chromatog.*, 27 (1967) 505.
370 A.R.MATTOCKS, *Anal. Chem.*, 39 (1967) 312.
371 A.T.DANN, *Nature*, 186 (1960) 1051.
372 M.V.JAGO, G.W.LANIGAN, J.B.BINGLEY, D.W.T.PIERCY, J.H.WHITTEM AND D.A.TITCHEN, *J. Pathol.*, 98 (1969) 115.

373   J.L. FRAHN, *Australian J. Chem.*, 22 (1969) 1655.
374   J.L. FRAHN AND J.A. MILLS, *Australian J. Chem.*, 17 (1964) 256.
375   G. GOMORI, in S.P. COLOWICK AND N.O. KAPLAN (Eds.), *Methods in Enzymology*, Academic Press, New York, 1955, Vol.I, p.144.
376   J.L. FRAHN AND J.A. MILLS, *Australian J. Chem.*, 12 (1959) 65.
377   C.C.J. CULVENOR AND R.E. WILLETTE, *Australian J. Chem.*, 19 (1966) 885.

Chapter 4

PESTICIDES

The pesticidal environmental hazards constitute a prime area of human concern. Included in this category are herbicides, fungicides, insecticides, fumigants, seed sterilants, and chemosterilants. A number of these agents are of major importance and are used in considerable amounts in such forms as granular formulations, dusts, powders, sprays, foams, aerosols, impregnated strips, etc., and hence man can be exposed to these agents *via* consumption of toxicant residues in food, handling or exposure of the agents *per se* or through ecological distribution. All organic pesticides, to a varying degree, are metabolized in living organisms and/or are degraded environmentally (*e.g.* photolytically, thermally). The extent and nature of these transformations vary with the agent causing them and with the pesticidal chemical structure and time being important factors, *e.g.* the transformation of some of these agents occur in a matter of minutes, while that of others requires months or even years. The chemical reactions involved hydrolysis, oxidation, reduction, dehalogenation, desulfurization, ring opening, isomerization, and/or conjugation. In addition, cognizance of the possibility of the existence of synthetic precursors and degradation products incorporated in the pesticidal agent all are of fundamental importance for consideration in chromatographic investigation of residues, metabolic and degradation products, and the nature of the agent *per se*.

1. MALEIC HYDRAZIDE

Maleic hydrazide (1,2-dihydro-3,6-pyridazinedione, MH) is prepared commercially *via* the reaction of hydrazine with maleic anhydride, and it has extensive application as a herbicide, fungicide growth inhibitor, and growth regulator. Its utility includes the prevention of such production in tobacco plants, growth control of weeds, grass and foliage, inhibition of sprouting of potatoes, onions, and stored root crops, and the protection of citrus seedlings against frost damage. The diethanolamine and sodium salts of maleic hydrazide are used in commercial formulations because of their enhanced stability (*e.g.* the commercial formulation MH-30 is a water-soluble liquid containing the diethanolamine salt of MH equivalent to 30% maleic hydrazide). The growth-restricting effects of MH consist of destroying the apical dominance of shoots, stopping root growth, and destroying germ cells. The action of MH is analogous to that of X-rays (it is specifically directed to growing tissues and therefore mitotic cells). Studies on the effect of MH on the enzymes of intact cells suggest that MH affects the enzymes requiring sulfhydryl groups for activity[1]. Maleic hydrazide is frequently applied at the rate of 2.25 lb./acre to control adventitious buds when tobacco is topped. Following this treatment, the green commercial leaves and the green sucker leaves contained 37 and 482 p.p.m. of MH, respectively[2]; cigarettes made from tobacco so treated contained 10–30 p.p.m. of MH[3]. Information on the fate of MH in cigarette smoke is sparse. Experimental cigarettes containing 10–30 p.p.m. of MH gave 0 and $\lesssim 2$ p.p.m. of unchanged MH in the smoke, respectively[3]. Using ^{14}C-labeled maleic hydrazide, 23.4% of the added radioactivity was found in CO_2, CO and "tars" of the smoke and 31% was calculated or found in the butt and ash. The remainder was assumed to have been lost in the sidestream smoke[4].

The chemistry and mode of action[5, 6], carcinogenicity[7, 8], cytochemical effects on cultured mammalian cells[9], and mutagenicity[10, 11] of maleic hydrazide have been described. The analysis of maleic hydrazide has been performed by conductometric titration[12], polarography[13], ultraviolet[14, 15], infrared[16] and NMR spectroscopy[17], and colorimetry[15, 18, 19].

A technique for the estimation of MH by paper chromatography using two different solvents, n-butanol–acetic acid–water (5:1:4) and isopropanol–ammonium hydroxide–water (7:1:2), was reported by Andreae[20]. A freshly prepared 1:1 mixture of 1% aq. ferric chloride–1% potassium ferricyanide detected MH as a blue color (lower limit of detection was 0.2 µg). The metabolism of ^{14}C-maleic hydrazide in ten plants has been described by Biswas et al.[21]. Based on two-dimensional paper chromatography, autoradiography and IR spectroscopy, the possible metabolic products were identified as lactic acid, succinic acid, maleimide, and hydrazine. Among the possible routes of metabolism and degradation for maleic hydrazide are the following.

A glycoside of maleic hydrazide from plants has also been identified as a metabolite[22]. The two-dimensional paper chromatography of Biswas et al.[21] utilized Whatman No. 1 paper with the first development with phenol saturated with water (descending, 25.4 cm) followed by n-butanol–acetic acid–water (4:1:5) (90° development, 38.1 cm).

The metabolism of ^{14}C-maleic hydrazide in the rat following a single oral administration has been described by Mays et al.[23]. Less than 0.001% of radioactivity was detected in tissues or blood samples after 3 days and expired [14]CO_2 accounted for 0.2% of the radioactivity administered. Recovery of administered radioactivity amounted to 77% in the urine and 12% in the feces in 6 days. The products excreted in the urine were MH (92–94%) and a conjugate of MH (6–8%). Aqueous ^{14}C-maleic hydrazide and urine samples (along with reference ^{14}C-MH) were examined on Silica Gel G (Brinkmann, 250 μ) or Adsorbosil (Applied Science, 1000 μ) plates and developed with ethanol–acetone–6 N HCl (120:30:10) or water–isopropanol (30:120). Autoradiograms were made from each plate by exposing medical X-ray film for a period of time calculated to allow 10^8 radioactive disintegrations in order to observe the presence of radioactive spots as small as 0.5% of the total radioactivity placed on the chromatogram.

The photolysis of maleic hydrazide has been studied by Stoessl[24, 25] and evidence for an intermolecular hydrogen transfer was presented[25]. It was shown that photolysis of aqueous MH under nitrogen leads to a complex mixture from which succinic acid was isolated in 26% yield[24]. A pathway for photolysis was suggested as follows.

The intermediate II could decompose in a number of ways (*e.g.* a, b or c) in which dicimide could, but need not be involved. The products could be succinic acid (*via* the ketene) and nitrogen, together with either hydrogen, or the products of a hydrogen transfer to one or several acceptors. The expected product, cyclic succinhydrazide (II), was obtained (in about 3% yield of photolyzed MH or 13% of the succinic acid formed) from the irradiation of aqueous MH in a sealed, evacuated vessel. Ionexchange, paper, thin-layer, and gas chromatographic procedures were used for the elaboration of reaction products. Semi-quantitative paper chromatography used the procedure of Andreae[20]. TLC was performed on silica gel using n-butanol–water for development and quantitative vapor phase chromatography was carried out using a 6 ft. $\times \frac{1}{4}$ in. column containing 30% Carbowax at 185°C with helium as carrier gas at 70 ml/min.

The metabolism and binding of ^{14}C-maleic hydrazide in corn coleoptile sections, tobacco pith explants, and pea seedlings was studied by Nooden[26]. The radiochemical purity of the 1-position-labeled MH was checked by descending chromatography using three solvent systems: (*1*) n-butanol–conc. acetic acid–water (6:1:2), (*2*) isopropanol–conc. ammonia–water (60:15:25), and (*3*) phenol–water (72:28). In the above solvents used, the R_F values of the ^{14}C-MH were 0.7–0.8. MH was taken up by corn and pea-seedling roots and bound to some material which is insoluble in 80% ethanol or 5% trichloracetic acid. ^{14}C-MH is stable metabolically, *e.g.* chromatography in butanol–acetic acid–water (6:1:2) gave no indication of degradation. Very little ^{14}C-MH was bound in the zone of cell division (where MH acts to inhibit root elongation) or even in the region of cell enlargement in corn roots and most was bound ≪1 cm behind the tip. Similarly, a small amount of MH was bound in corn coleoptile and tobacco root sections. Approximately 90% of ^{14}C-MH bound in corn roots was associated with large particles believed to be cell-wall fragments. The physiological role of the binding process may be to provide a mechanism for detoxifying or inactivating MH. The apparent stability of the bound and unbound MH is of decided interest since MH is carcinogenic (*e.g.* in the mouse[8]), and persistence of MH in treated food materials could constitute a potential hazard. This is compounded by the ability of roots and possibly other plants to accumulate MH actively.

The GLC analysis of MH *per se* is difficult owing to its high melting point (296–298°C). However, the use of derivatives of MH, *e.g.* alkyl carbonates[27] and trimethylsilyl[28], has permitted its facile analysis.

The alkyl carbonate derivatives of MH were prepared by treating equimolar quantities of MH and the respective alkyl chloroformate in aqueous potassium

TABLE 1

GAS CHROMATOGRAPHY OF ALKYL CARBONATE DERIVATIVES

R	M.p. (°C)	Found M.p.[b] (°C)	Relative elution[a]	
			Versilube F-50[c]	XE-60[c]
Methyl	105–107	121–122	0.60	0.47
Ethyl	106–108	101–103	0.69	0.67
Propyl	79–81	79–81	1.0	1.0
Butyl	84–85	83.5–85	1.6	1.4
Amyl	49–51	48–50	2.5	2.0
3-Chloropropyl	99–100	98–99	11.75[d]	11.1[d]

[a] Relative to the propyl carbonate derivative (propyl-2,3-dihydro-3-oxo-6-pyridazinyl carbonate) as 1.0. Elution of the propyl carbonate derivative was 7.2 min on Versilube F-50 and 4.7 min on XE-60.
[b] Maleic hydrazide melted at 296–298°C.
[c] 15% w/w on 60–80 mesh Chromosorb-W (HMDS pretreated), 4 ft. × 0.125 in. o.d. copper column. Operating conditions: column 175°C, injection port 270°C, detector 200°C, filament current 150 mA, wire detector, helium 1.7 ml/min for Versilube F-50 and 16.4 ml/min for XE-60.
[d] Determined at a column temperature of 205°C.

hydroxide at 0 to $+5°C$ according to the procedure of Stefanyl and Howard[29]. An
F & M Model 500 gas chromatograph was used equipped with a hot-wire detector and
4 ft. × $\frac{1}{8}$ in. o.d. copper columns containing 15% Versilube or 15% SE-60 on 60–
80 mesh HMDS-treated Chromosorb W. Columns were housed in an F & M Model
720 oven. Table 1 lists the relative elution of a number of alkyl carbonates of maleic
hydrazide determined on Versilube F-50 and XE-60 columns. The effect of acylation
on the lowering of the melting point of maleic hydrazide can be seen. This fact greatly
enhances the feasibility of GLC analysis of maleic hydrazide *via* the route of an appropriate derivative. Facile separations were observed for the methyl through amyl carbonate derivatives, a composite chromatogram of which is presented in Fig. 1.

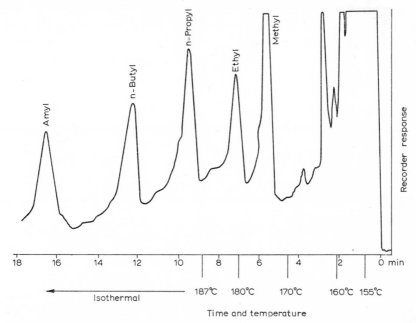

Fig. 1. Composite chromatogram of maleic hydrazide alkyl carbonates. Column: 4 ft. × 0.125 in.
o.d. copper, packed with 15% Versilube F-50 coated on 60–80 mesh HMDS-treated Chromosorb W.
Conditions: column programmed at 4°/min, 155–187°C; detector 250°C; injection port 265°C;
hot-wire detector; filament current 170 mA; helium 16 ml/min.

The preparation of the trimethylsilyl derivative of MH was carried out *via* the
reaction of MH in pyridine with hexamethyldisilazane[30] in the presence of trimethyl-
chlorosilane (TMCS) as shown below.

Chromatography was carried out on an F&M Model 1609 flame-ionization instrument utilizing a 6 ft. × ¼ in. o.d. coiled Pyrex glass column containing 3% Carbowax 20M on 60–80 mesh acid-washed DMCS-treated Chromosorb G and 4% SE-30 on 80–100 mesh HMDS-treated Chromosorb W. On-column injections were made using a conversion kit to provide an all-glass flow system (Applied Science Labs.). The specific operating conditions are shown in Table 2 and Fig. 2 illustrates the elution of the TMS–maleic hydrazide on SE-30 at 100°C.

TABLE 2

ELUTION OF TMS–MALEIC HYDRAZIDE

	Minutes	*Conditions*			
		Column temp. (°C)	*N$_2$ (ml/min)*	*H$_2$ (ml/min)*	*Air (ml/min)*	*Range*
Carbowax 20M	6.11	190	89	77	450	1000
SE-30	4.72	100	104	81	500	1000

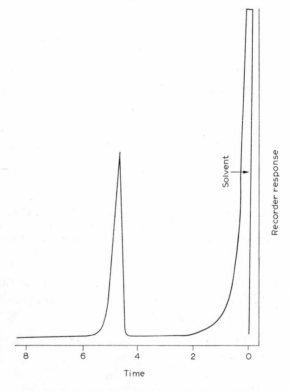

Fig. 2. TMS–maleic hydrazide on SE-30 at 100°C.

2. CAPTAN

Captan (N-(trichloromethylthio)-4-cyclohexene-1,2-dicarboximide) is produced by the reaction of perchloromethyl mercaptan with tetrahydrophthalimide. It is a general fungicide used for the treatment of folia and soil and against seedborne diseases including apple scab, grape mildews, corn seed infections, and many fruit, vegetable, and ornamental plant diseases. Captan is stable when dry but decomposes at or near its melting point (178 °C) to yield a variety of products. The rate of hydrolytic attack increases with increasing pH and temperature. The reaction of captan with thiols has been extensively investigated[31-35]. For example, the decomposition of captan by thiols is known to produce thiocarbonyl chloride which can react in turn with cell thiols[31]. Thiocarbonyl chloride is also rapidly hydrolyzed by water to carbonyl sulfide[32], so that the extent of the thiocarbonyl chloride–SH reaction depends on this rate of hydrolysis. A scheme for the reaction of captan with glutathione (GSH) has been proposed[31, 33] to proceed as follows.

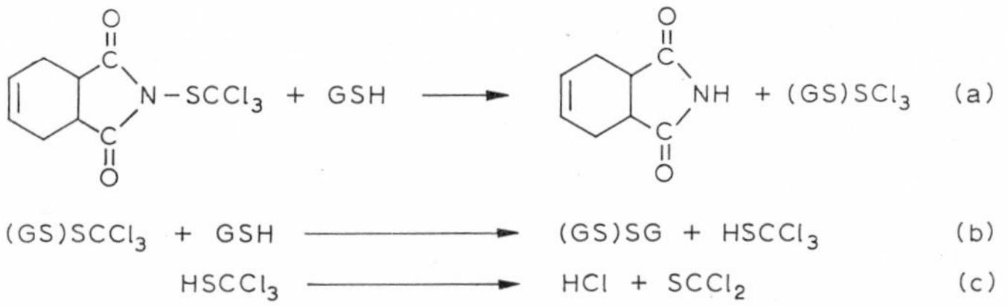

$$(GS)SCCl_3 \; + \; GSH \longrightarrow (GS)SG \; + \; HSCCl_3 \qquad (b)$$

$$HSCCl_3 \longrightarrow HCl \; + \; SCCl_2 \qquad (c)$$

Reactions with *Saccharomyces pastorianus*[31] suggested that thiophosgene, formed in the reaction of captan with –SH groups, is the ultimate toxicant, being free to combine not only with additional –SH groups, but also with –OH, $-NH_2-$, and –COOH. Owens and Blaak[33] regard the fungitoxicity of captan to be due to the intact molecule and not to its decomposition products. Figure 3 illustrates the mode of action of captan and depicts the initial condensation of captan with thiol followed by spontaneous decomposition to products of the type characterized by Lukens and Sisler[31].

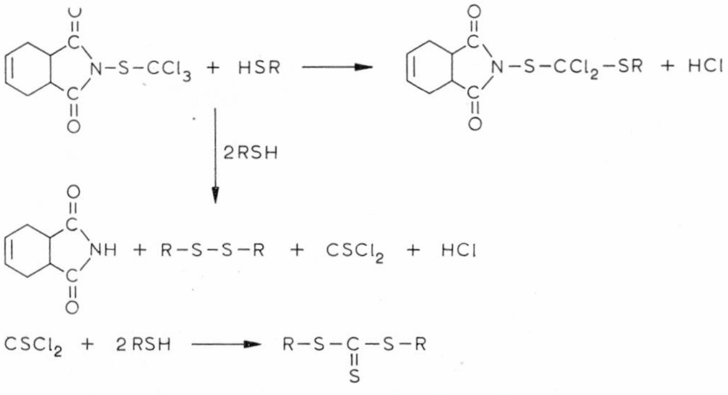

$$CSCl_2 \; + \; 2\,RSH \longrightarrow R-S-\underset{\underset{S}{\|}}{C}-S-R$$

Fig. 3. Mode of action of captan.

References pp. 267–272

The toxicity[36], teratogenicity[37, 38], and mutagenicity[38] of captan have been demonstrated.

The decomposition of ^{35}S-labeled captan by *Neurospora crassa conidia* has been described by Richmond and Somers[34] *via* an examination of the cellular distribution of ^{35}S. Nearly all the ^{35}S in the spores was bound to the water-soluble and protein fractions. Thin-layer chromatography of the hot-water extract of spores has shown that ^{35}S occurs largely in oxidized glutathione (GSSG) and in a product tentatively identified as a thiazolidine derivative of glutathione (I). It was suggested that this derivative, which only forms above pH 6.5, is produced by reaction between glutathione (GSH) and the thiocarbonyl chloride liberated on the decomposition of captan. Eastman chromagram sheets, type K301R$_2$ (100 μ silica gel) developed with *n*-propanol–water (70:30) was used to separate ^{35}S-labeled compounds.

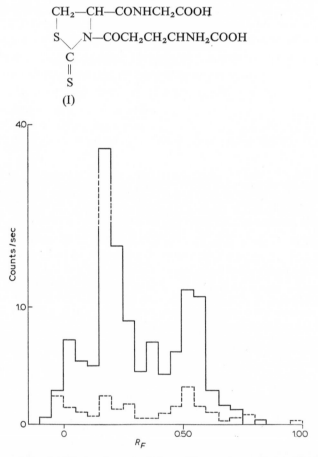

(I)

Fig.4. Radio-chromatogram of hot-water extract from *Neurospora crassa conidia* (20 million/ml) incubated with 2.5 × 10^{-5} *M* ^{35}S-captan. The thin-layer film was developed with *n*-propanol–water. ———, Spores incubated with ^{35}S-capton; – – –, spores incubated with 5 × 10^{-4} *M* iodoacetic acid for 30 min at 25 °C, washed twice, then incubated with ^{35}S-captan.

For TLC, Silica Gel HR (Merck) or Cellulose Powder 300 (Macherey, Nagel; 250 μ) and for preparative separations with silica gel, the layer was 500 μ thick. Cellulose layers were prepared by a modification of the procedure of Jones and Heathcote[39]. Samples were applied as 4 μl spots and developed at 25°C by the ascending technique with n-propanol–water (70:30). The spray reagents were platinic iodide for sulfur compounds[40] and ninhydrin for amino acids[41]. Figure 4 depicts a radiochromatogram of hot-water extract from *Neurospora crassa conidia* incubated with ^{35}S-captan and shows that nearly all the labeled sulfur was confined to two components at R_F 0.18 and 0.55.

The major component, at R_F 0.18, gave a positive reaction with platinic iodide and had the same R_F as GSSG. The second component (R_F 0.55) also gave a positive reaction with platinic iodide and is suggested to possess the structure of the thiazolidine derivative of glutathione (I).

A GLC determination of captan residues on apricots, peaches, tomatoes, and cottonseed has been reported by Kilgore *et al.*[42]. The residues were extracted with benzene or acetonitrile (residues as low as 0.01 p.p.m. could be detected). The overall average recovery of captan residues obtained from fortified control samples was 92%. An Aerograph Model 204 gas chromatograph equipped with an electron-capture detector was used for the analysis. The electrometer was operated at an output sensitivity setting of 3×10^{-11} afs. The chromatograph was equipped with a 6 ft. × $\frac{1}{8}$ in. o.d. spiral borosilicate-glass column packed with 10% DC-200 on 110–120 mesh Anachrom ABS. The column and glass-lined injection port temperatures were 185° and 210°C, respectively, and nitrogen carrier gas which was filtered through a molecular sieve was at a flow rate of 50 ml/min. The peak height, symmetry, band width, and column retention time of a 3 nanogram sample determined on DC-200 at 185°C are illustrated in Fig. 5. Figure 6 shows a gas chromatogram of apricot extract (untreated control, and untreated control fortified with 1.0 p.p.m. captan) determined on DC-200 at 185°C. Table 3 shows data obtained with apricots fortified with captan when both a resorcinol–colorimetric procedure[43] and GLC analysis was used.

The examination of mixtures of the fungicides captan and folpet (N-(trichloromethylthio)-phthalimide, phaltan) has been described by Bevenue and Ogath[44]. The gas chromatographs used were (*1*) an Aerograph Model 204-B equipped with an electron-capture detector (^{3}H 250 mc); column temp. 195°C, injection temp. 200°C, detector temp. 200°C, range 10, attenuation 2, and nitrogen carrier gas flow rate 30 ml/min; (*2*) an F&M Model 810 with an electron-capture detector (200 mc), column temp. 200°C, injection temp. 215°C, detector temp. 200°C, pulse interval 50 μsec, range 10, attenuation 32, and argon–methane (90:10) carrier gas flow rate 75 ml/min. The gas chromatographic columns consisted of 3% or 5% silicone GE XE-60 (nitrile)-coated Chromosorb W, acid washed (AW) (treated with dimethylchlorosilane, DMCS), and packed into 5 ft. × $\frac{1}{8}$ in. borosilicate glass columns for use in the Aerograph gas chromatograph and 4 ft. × $\frac{1}{4}$ in. columns for use in the F&M

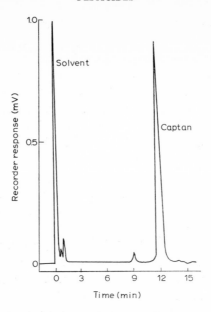

Fig. 5. Gas chromatogram of captan. The curve represents 3 ng of captan.

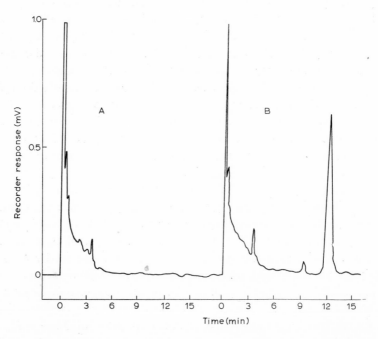

Fig. 6. Gas chromatogram of apricot extract. A, untreated control; B, untreated control fortified with 1.0 p.p.m. captan.

CAPTAN

TABLE 3

DETERMINATION OF CAPTAN RESIDUES ON APRICOTS USING RESORCINOL–COLORIMETRIC
AND GAS CHROMATOGRAPHIC PROCEDURES

Sample no.	Captan residue found (p.p.m.)[a]	
	Resorcinol–colorimetric procedure[b]	Gas chromatographic procedure[c]
1[d]	16.47	17.21
2[d]	16.47	17.21
3[e]	2.54	3.10
4[d]	7.64	7.60

[a] Application rate 2 lb. captan/100 gall. water, trees saturated to drip stage with hand sprayer.
[b] Average values of 3 determinations; figures uncorrected for an 83% recovery value.
[c] Average values of 3 determinations; figures uncorrected for a 91% recovery value.
[d] Fruit harvested 3 days after the last of 3 spray applications.
[e] Fruit harvested 6 days after the last of 2 spray applications.

instrument (all columns were conditioned for at least 18 h prior to use at the temperature (195–200°C) used for the analysis of the compounds).

One nanogram quantities each of captan (92% assay) and folpet (88% assay) were analyzed and linearity curves were obtained in the range of 0.2–5.0 µg. The retention time data for captan and folpet using GS XE-60 columns are given in Table 4. The effect of the column temperature (195–200°C) on the performance characteristics of the columns over an extended period (18–330 h) is also illustrated in the data in Table 4. With increased time, the retention time decreased. With the $\frac{1}{8}$ in., 5% column and the $\frac{1}{4}$ in., 3% column, the daily changes in retention time were minor after the columns had been aged for about 160 h. However, in all instances the relative position of the curves of the two compounds was practically constant (Table 4). Preliminary studies in which fresh papayas were fortified with captan and folpet at the 1.0 p.p.m. level, extracted with benzene, cleaned up over Nuchar C-190 carbon and analyzed as above, indicated recoveries of 85–95% of the fungicides from the extracts.

An extraction, clean-up and GLC method for the analysis of captan, folpet, and difolatan (N-(1,1,2,2-tetrachloroethylthio)-4-cyclohexene-1,2-dicarboximide) in crops was described by Pomerantz et al.[45]. The fungicides were determined by a procedure using electron-capture GLC. Either QF-1 or XE-60 columns were used after acetonitrile extraction from fortified crops, partitioning into methylene chloride–petroleum ether, and clean-up on Florisil. A Barber–Colman Model 5360 was used under the following conditions: (1) 6 ft. × 4 mm i.d. Pyrex coiled tube containing 5% QF-1 on 80–100 mesh Chromosorb W (HP) (conditioned 3 days at 250°C) with nitrogen flow at 120 ml/min, injection port, column, and detector temperatures 185, 155, and 175°C, respectively; (2) 3% XE-60 on 80–100 mesh Chromosorb W (HP) contained in a 6 ft. × 4 mm i.d. Pyrex coiled tube with nitrogen flow of 100 ml/min, injection port, column, and de-

TABLE 4

GAS CHROMATOGRAPH DATA ON PHALTAN AND CAPTAN

Column[a]	Age of column (h)	Retention time (min)		Ratio B/A
		Phaltan A	Captan B	
5% silicone GE XE-60 (nitrile);				
⅛ in. × 5 ft. glass spiral[b]	18	8.3	10.3	1.24
	42		9.5	
	66	7.1	8.9	1.25
	162	6.1	7.6	1.24
	186	6.0	7.4	1.23
	210	5.9	7.3	1.24
	234	5.7	7.1	1.25
	258	5.6	7.0	1.25
	330	5.2	6.6	1.26
5% silicone GE XE-60 (nitrile);				
¼ in. × 4 ft. glass spiral[c]	42	12.4	15.6	1.26
	186	11.6	14.6	1.26
	282	11.2	14.1	1.26
	306	10.3	12.9	1.25
	330	10.4	13.0	1.25
3% silicone GE XE-60 (nitrile);				
¼ in. × 4 ft. glass spiral[c]	18	4.2	5.2	1.24
	90	4.0	4.9	1.23
	114	3.7	4.6	1.24
	138	3.7	4.5	1.22
	162	3.3	4.1	1.24
	186	3.3	4.1	1.24
	258	3.2	3.9	1.22

a Solid support: Chromosorb W (AW, DMCS), high-performance grade, 80–100 mesh.
b Aerograph, Model 204, gas chromatograph.
c F & M, Model 810, gas chromatograph.

tector temperatures were 210, 178, and 210°C, respectively. The retention times and response data for captan, folpet and difolatan are given in Table 5 and Table 6 and indicates the percent recoveries of admixed fungicides from fortified crops (2 p.p.m.). The wider range of calculated recoveries on the QF-1 column were believed to be due to difficulties arising from partial overlap of the captan and folpet peaks. However, the complete separation of these two peaks and the more uniform recoveries were obtained using the XE-60 column.

Pomerantz and Ross[46] described the TLC and GLC determination of captan, folpet, and difolatan and their respective epoxides. The TLC method based upon chromatography on Silica Gel G with 1% methanol as developer followed by sequential color development with N,N-dimethyl-p-phenylenediamine, potassium permanganate, and chromic acid, permitted the differentiation of captan, folpet, difolatan,

TABLE 5

GLC RETENTION TIMES (RELATIVE TO CAPTAN) AND RESPONSE DATA FOR CAPTAN,
FOLPET AND DIFOLATAN[a]

Fungicide	R_t	Wt. for approx. 50% fsd (ng)
QF-1 column		
Captan[b]	1.00	4
Folpet	0.90	4
Difolatan	2.8	16
3% XE-60 column		
Captan[c]	1.0	2
Folpet	0.81	3
Difolatan	3.2	12

[a] See text for operating parameters.
[b] Actual R_t, *ca.* 13.6 min.
[c] Actual R_t, *ca.* 4.4 min.

TABLE 6

PERCENT RECOVERIES OF ADMIXED FUNGICIDES FROM FORTIFIED CROPS (2 p.p.m.)

Crop	Captan	Folpet	Difolatan
QF-1 column			
Carrots	103	118	89
	118	128	121
Cabbage	94	108	96
	84	111	95
Soybeans	113	119	88
	103	109	89
XE-60 column			
Carrots	102	95	103
	106	97	98
	98	97	99
Cabbage	99	98	97
	95	95	97
	92	95	96
Soybeans	95	96	95
	96	94	94
	96	96	94

tetrahydrophthalimide, phthalimide, captan epoxide, difolatan epoxide, and tetra-hydrophthalimide epoxide. By analogy with the heptachlor epoxide and dieldrin, the epoxides of captan and difolatan may be considered to be possible alteration products derived from the parent compounds by oxidative metabolism or by non-metabolic oxidation. The epoxide formation from the three fungicidal precursors is shown below.

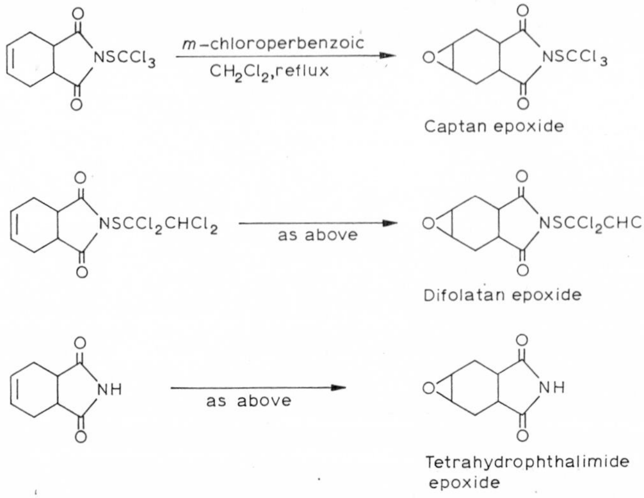

Captan epoxide

Difolatan epoxide

Tetrahydrophthalimide epoxide

The response of the 3 parent fungicides and 5 derivatives to several chromogenic sprays as well as the lower limits of their detection is indicated in Table 7. The DMPD reagent (N,N-dimethyl-p-phenylenediamine) appeared specific for difolatan and its epoxide. Table 8 shows the relative R_F values for captan-type compounds obtained on Silica Gel G using 1% methanol in chloroform as developer.

TABLE 7

RESPONSE[a] OF CAPTAN-TYPE COMPOUNDS TO CHROMOGENIC REAGENTS

Compound	Color response and lower limit of detection (μg) with reagents used			
	Resorcinol[b]	KMnO$_4$	DMPD[c]	Chromic acid[d]
Captan	Y/W ($\leqslant$1)	Y/P (0.2)	—	
Folpet	Y/W ($\leqslant$1)	Y/P (1)	—	
Difolatan	—	Y/P (0.2)	OB/W (0.2)	
Captan epoxide	Y/W ($\leqslant$1)	Y/P (1)	—	
Difolatan epoxide	—	Y/P (1)	OB/W (0.2)	
Tetrahydrophthalimide	—	Y/P (0.5)	—	BG/Y (1)
Phthalimide	—	—	—	BG/Y (1–2)
Tetrahydrophthalimide epoxide	—	—	—	BG/Y (1–2)

[a] Y/W = yellow spot on white background; Y/P = yellow spot on pink background; OB/W = orange-brown spot on white background; BG/Y = blue-green spot on yellow background; and — indicates no spot color development.
[b] Resorcinol followed by heating 10–15 min at 110°C.
[c] N,N-dimethyl-p-phenylenediamine.
[d] Chromic acid followed by heating 20–30 min at 130°C.

TABLE 8

RELATIVE R_F VALUES FOR CAPTAN-TYPE COMPOUNDS IN 1% METHANOL IN CHLOROFORM

Compound	R_F relative to captan
Captan	1.00[a]
Folpet	1.14 ± 0.04
Difolatan	1.02 ± 0.02
Captan epoxide	0.70 ± 0.02
Difolatan epoxide	0.68 ± 0.03
Tetrahydrophthalimide	0.19 ± 0.01
Phthalimide	0.35 ± 0.01
Tetrahydrophthalimide epoxide	0.08 ± 0.01

[a] R_F of captan relative to 154 mm solvent front distance was 0.61 ± 0.01 in eight determinations.

The GLC analysis of captan, folpet, difolatan, and their derivatives was carried out on a Barber–Colman Model 5360 gas chromatograph using columns of 10% DC-200 on 80–100 mesh Gas Chrom Q or a mixed column of equal portions of 15% QF-1 and 10% DC-200, each on 100–120 mesh Gas Chrom Q. Half-scale deflection data were obtained with an electron-capture detector under conditions which gave half-scale deflection (on a 5 mV recorder) for 1 ng heptachlor epoxide. (The conditions used in this work find wide applicability in pesticide residue analysis[47-49].) The retention time and response data for the above fungicides and their derivatives are shown in Table 9. Examination of the retention time data shows that the change to

TABLE 9

CAPTAN-TYPE COMPOUNDS: GLC[a] RELATIVE R_t AND RESPONSE DATA

Compound	Retention time relative to captan[b]		Wt. for half-scale deflection on mixed column (ng)
	DC-200 column	"Mixed" column	
Captan	1.00	1.00	3.5–4.0
Folpet	1.06	0.97	3.5–4.0
Difolatan	2.60	2.53	15–17
Captan epoxide	2.20	3.32	28–32
Difolatan epoxide	5.95	8.64	300–500
Tetrahydrophthalimide	0.13	0.15	450–550
Phthalimide	0.13	0.16	1.6
Tetrahydrophthalimide epoxide	0.31	0.58	2000–2400

[a] Barber–Colman 5360 gas chromatograph with concentric-type electron-capture detector. Conditions for DC-200 column: 10% DC-200 (12,500 cSt) on 80–100 mesh Gas Chrom Q; 6 ft. × 4 mm i.d.; conditioned 3 days at 250°C; N_2 flow 120 ml/min, injection temp. 225°C, column temp. 210°C, detector temp. 220°C.
Conditions for "mixed" column: 1:1 ratio of 15% QF-1 and 10% DC-200 each on 100–120 mesh Gas Chrom Q; 6 ft. × 4 mm i.d.; conditioned 3 days at 250°C; N_2 flow 120 ml/min, injection temp. 225°C, column temp. 200°C, detector temp. 200°C.
[b] Actual retention time of captan was 6.1 min on DC-200 column and 7.5 min on "mixed" column.

the more polar "mixed" column reverses the elution order of captan and folpet. The epoxides show considerably increased relative retention values on the more polar column (consistent with an increase in molecule polarity when the parent compound is epoxidized).

The site and fate of captan residues from the treatment of prunes prior to commercial dehydration was elaborated using GLC analysis[50] following a modified extraction and clean-up procedure[42]. An Aerograph Hy-Fi, Model 600-B chromatograph equipped with an electron-capture detector and a Leeds Northrup 1 mV recorder (Model G) with a chart speed of $\frac{1}{4}$ in./min was used under the following conditions: 5 ft. × $\frac{1}{8}$ in. Pyrex column packed with 60–80 mesh acid-washed Chromosorb W, DMCS treated, coated with 5% Dow 11 silicone grease; column 185°C, detector 200°C, injector 250°C (the injector port contained a glass insert), electrometer setting of 10 and attenuator setting of 2. Two to twenty nanograms of captan was an acceptable working standard range under these conditions and the sensitivity of the method was accepted as 0.05 p.p.m.

Archer and Corbin[51] described a TLC method as a rapid screening procedure for the detection of captan residues in the presence of difolatan residues in prune fruits and blossoms. Four grams or less of plant extractives containing 1 µg or more of captan were chromatographed on Silica Gel H (Brinkmann, 500 µ) and developed with isopropanol–benzene (4:96) (15 cm, 15 min). Two spray reagents were used in detecting captan. The preferred reagents are resorcinol (7.25 g) in glacial acetic acid (35 ml) and tetraethyl ammonium hydroxide (3 ml of 40% added to pyridine–water (24:3)). Both sprays produced an intense yellow captan spot on a white background. The color for the tetraethyl ammonium hydroxide–pyridine spray reagent developed almost immediately at room temperature but soon faded. The color for the resorcinol–glacial acetic acid spray developed only after the plate was sprayed and heated (difolatan does not react). Approximately 1 µg of captan was detected with either spray reagent, and the R_F value was 0.44 in the above developing solvent.

The determination of captan and folpet residues in foods by TLC was described by Engst and Schnaak[52]. Silica Gel G (Merck) was used with the following developers: (1) methylene chloride–toluene (5:2), (2) n-hexane–ether (1:1.5), and (3) n-hexane–ether (1:1.3). The spray reagent consisted of silver nitrate (1.7 g in 10 ml water and 5 ml conc. ammonium hydroxide diluted to 200 ml with acetone). The detection limits were 0.1 µg for captan and 0.15 µg for folpet. The procedure permits the detection of residual amounts from 1–10 p.p.m. after extraction from foods with chloroform without additional clean-up of the extract, and also of amounts below 1 p.p.m. when using acetonitrile–hexane partition. Table 10 lists the R_F values for captan and folpet obtained following their extraction from foods and development in solvents 1–3 above.

The TLC of captan as well as captax (2-mercaptobenzothiazole) has been described by Fishbein et al.[53]. Silica gel (TLC-7G, Mallinckrodt) was used with chloroform as developer for the analysis of captan and Silica Gel DF-5 (Camag) with isopropanol–ammonium hydroxide–carbon tetrachloride (50:10:40) as solvent for captax

TABLE 10

TLC DETERMINATION OF CAPTAN AND FOLPET IN FOODS

Food	Solvent	R_F values	
		Captan	Folpet
Applesauce	Methylene chloride–toluene (5:2)	0.46	0.70
Sweet cherries	Methylene chloride–toluene (5:2)	0.46	0.70
Carrots	*n*-Hexane–ether (1:1.5)	0.43	0.58
Tomatoes	*n*-Hexane–ether (1:1.5)	0.43	0.58
Strawberries	*n*-Hexane–ether (1:1.3)	0.38	0.55

analysis. Captan was detected with 25% resorcinol in glacial acetic acid[43] and captax with (a) cupric chloride reagent[54] mixed 1:2 prior to use with 20% aq. hydroxylamine and (b) conc. sulfuric acid–*n*-butanol (1:1). Densitometry and the technique of Purdy and Truter[55] (plot of the log sample weight *vs.* the square root of the spot area) were evaluated for the quantitation of both captan and captax on channel chromatoplates. For densitometry, a Model 52-C densitometer with a motorized TLC stage and a Model 520A-photomultiplier (Photovolt) were used. Figure 7 compares the quantita-

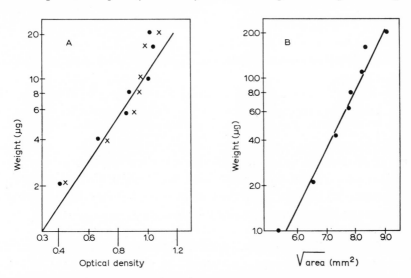

Fig. 7. Quantitation of captan. A, By densitometry; B, by channel technique.

tion of captan by densitometry and the channel technique. Linearity from 1–10 μg was obtained with densitometry compared to 1–20 μg range utilizing the channel technique which is dependent only upon spot area and not color density.

The determination of difolatan residues (as well as their separation from captan and botran) in fruits by electron-capture gas chromatography was described by Kilgore and White[56]. An Aerograph Hy-Fi gas chromatograph Model 600B was fitted with a 2 ft. × ⅛ in. spiral borosilicate glass column containing 5% purified Dow-11 silicone oil on acid-washed 70–80 mesh DMCS treated Chromosorb G. The column was conditioned for 3 days at 225°C. The glass-lined injection port and column temperatures were 212 and 197°C, respectively. The nitrogen carrier gas was passed through a small molecular-sieve filter (Varian) and was regulated to provide a flow rate of 75 ml/min. The peak height, band width, and column retention time for a 1 ng sample of difolatan are shown in Fig. 8. With a sensitivity setting of 1 × and an attenuation setting of 2,

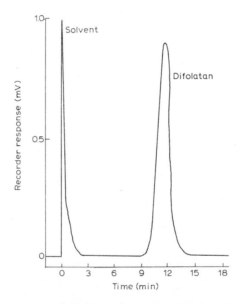

Fig. 8. Gas chromatogram of difolatan. Curve represents 1 nanogram of difolatan.

the linearity of response was from 2–10 ng. At a more sensitive setting (10 ×) the response also was linear and ranged from 0.2–1 ng. Difolatan on cherries was detected with an average recovery of 83% (std. dev. ±1.0) (Fig. 9). Actual field-treated samples containing captan, difolatan, and botran (2,6-dichloro-4-nitroaniline) were separated readily (Fig. 10).

The decomposition of captan by the conidia of *Neurospora crassa* has been examined by gas chromatography[32]. A Microtek gas chromatograph (Model GC 2000) was used with a dual thermal conductivity detector and a stainless steel column filled with Porapak Q (80–100 mesh). Helium at 90 ml/min was the carrier gas and the column, inlet, and detector were at 165°C. Figures 11 and 12 show gas chromatograms of products of reaction between captan and *Neurospora crassa* conidia and glutathione, respectively, and show the formation of carbonyl sulphide (COS) but no carbon disulfide. The formation of carbonyl sulfide from captan has biological as well as chemical interest because it has been shown that vapors of this compound can be appreciably fungitoxic[57].

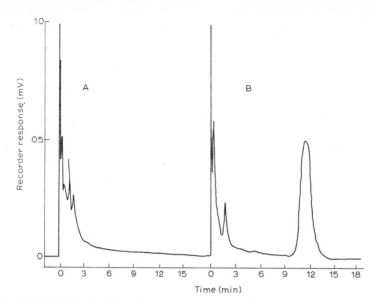

Fig. 9. Gas chromatogram of difolatan in cherry extract. A, Untreated control sample; B, untreated control sample fortified with 0.01 p.p.m. difolatan.

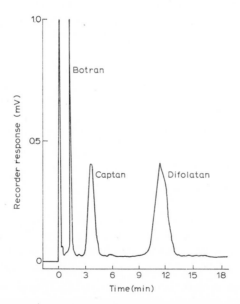

Fig. 10. Chromatogram of field-treated cherries. Sample was treated with botran, captan, and difolatan. A 2-μl aliquot of the benzene extract, equivalent to 1 mg of fruit, was chromatographed. Response curve of difolatan represents 500 picograms or 0.5 p.p.m.

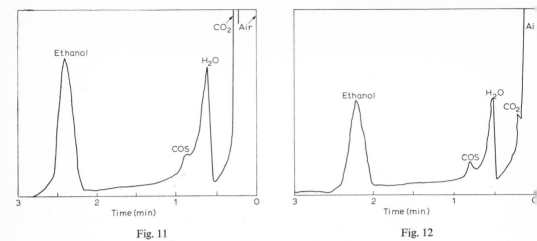

Fig. 11

Fig. 12

Fig. 11. Gas chromatogram of products of reaction between captan and *Neurospora crassa conidia*.

Fig. 12. Gas chromatogram of products of reaction between captan and glutathione.

3. HEMEL AND HEMPA

Hemel (2,4,6-tris-dimethylamino)-1-triazine, hexamethyl melamine) and hempa (hexamethylphosphoric triamide) are thermally stable non-alkylating analogs of tretamine and tepa, respectively, developed by Chang *et al.*[58] (see below).

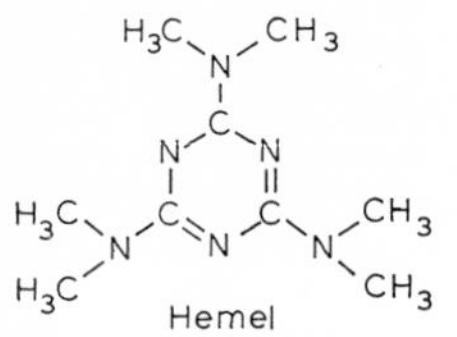

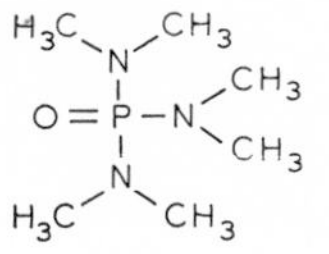

Both hemel (prepared from dimethylamine and cyanuric chloride) and hempa (prepared by the reaction of phosphorus oxychloride and triethylamine) are effective chemosterilants for houseflies *(Musca domestica)*. Although the activity of hempa in male flies is much lower than that of tepa, both compounds produce strikingly similar physiological and cytological effects in several different organisms. Both hempa and tepa are mutagenic in the wasp, *Bracon hebeta*[59], induce testicular atrophy in rats[60], and hempa induces a marked antispermatogenic effect in rats and mice[60]. The toxicity of hempa has also been described[58, 60, 61].

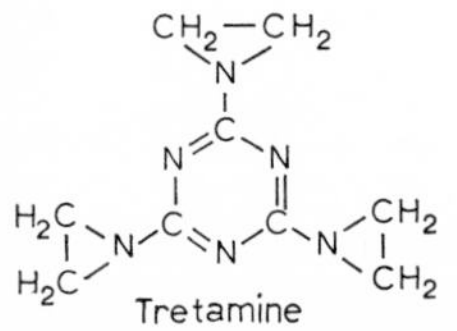

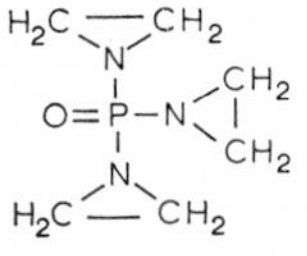

The metabolism of hempa uniformly labeled with ^{14}C in male houseflies has been described by Chang *et al.*[62]. The only major metabolite of hempa found in treated flies and in their excreta was pentamethyl phosphoric triamide (PMPT). Since PMPT is not active as a chemosterilant, and because it is not metabolized by male houseflies, it was concluded that the biologically active species was either hempa itself or an intermediate metabolite in the demethylation of hempa to PMPT. Unchanged hempa and its metabolite were separated from the extracts of treated flies and their excreta analyzed by TLC, GLC, and radiochromatography and confirmed by IR spectroscopy. Silica Gel G was used in an ascending development of 100 mm with 25% absolute ethanol in chloroform as developer. A Packard Model 7201 scanner with disc chart

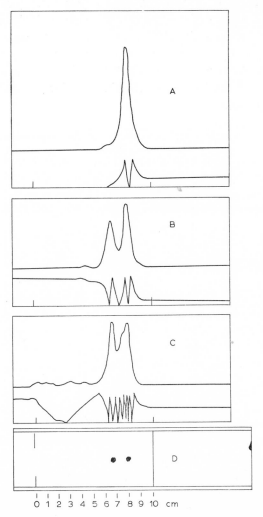

Fig. 12A. Radiochromatograms. A, ^{14}C-hempa; B, extract of flies; C, extracts of excreta of flies injected with ^{14}C-hempa; non-radioactive reference compounds (colorimetric); D, hempa (R_F 0.78) and pentamethylphosphoric triamide (R_F 0.65).

integrator was used to obtain the radiochromatograms of the extract from flies treated with ^{14}C-hempa and their excreta. The non-radioactive hempa and its meta-bolite were detected on the plates by the reagents of Haines and Isherwood used for the detection of phosphate esters[63].

Gas–liquid chromatography was performed on an F & M Model 720 thermal conductivity gas chromatograph equipped with a 2 ft. × ¼ in. o.d. (61 × 0.62 cm) stainless steel column containing 5 % Carbowax 20M on Haloport F. The column temperature was 190 °C, the helium flow rate was 60 ml/min, and the injection port and detector block temperatures were 225° and 235 °C, respectively. Figure 12A illus-trates radiochromatograms of pure ^{14}C-hempa (A); extract of flies treated with ^{14}C-hempa, (B), and extract of their excreta (C) compared with a TLC plate (D) on which the spots of hempa and PMPT were made visible colorimetrically. Radio-chromatograms B and C indicated that there were only 2 major radioactive peaks corresponding to the R_F values of hempa (0.78) and its pentamethyl homolog (0.55). Additional evidence of the identity of the single major metabolite of hempa was obtained by GLC (Fig. 13).

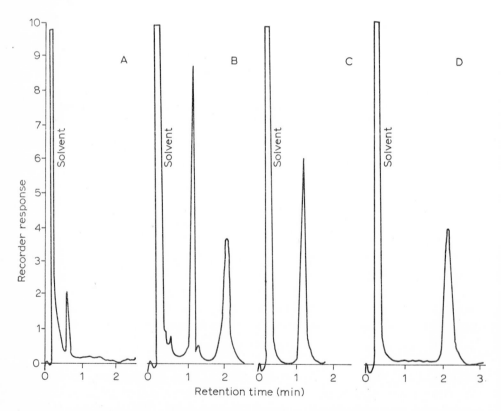

Fig. 13. Gas chromatograms. A, extract of untreated flies; B, extract of flies injected with hempa; C, authentic hempa; D, authentic pentamethyl phosphoric triamide (PMPT).

The comparative metabolism of ^{14}C-labeled hempa by houseflies susceptible or resistant to isolan (1-isopropyl-3-methyl-s-pyrozolyl dimethyl carbamate) was described by Chang and Borkovec[64]. Twice as much unchanged hempa was recovered from the susceptible than from the resistant flies at 4 h post-treatment. Metabolites of hempa in resistant flies contained a larger proportion of trimethyl phosphoric and tetramethyl phosphoric triamides.

The techniques for TLC, radiochromatography and IR spectroscopy of the metabolites were as described by Chang et al.[62]. GLC analyses were performed as described by Akov et al.[65]. Figure 14 shows the structural formulas and names of methyl phosphoric triamides isolated from hempa-treated flies and their excreta.

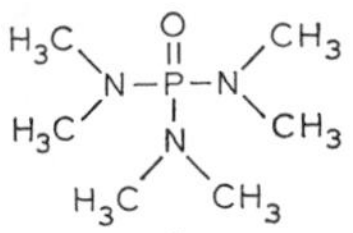

Fig. 14. Structural formulas and names of methylphosphoric triamides isolated from hempa-treated flies and their excreta. I, Hexamethylphosphoric triamide (hempa); II, pentamethylphosphoric triamide (PMPT); III, N,N,N',N''-terramethylphosphoric triamide; IV, N,N',N''-trimethylphosphoric triamide.

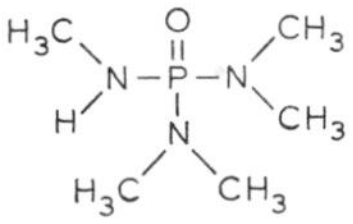

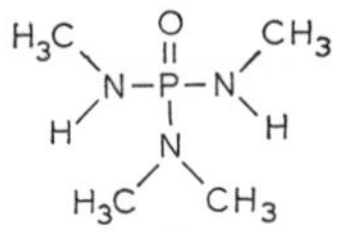

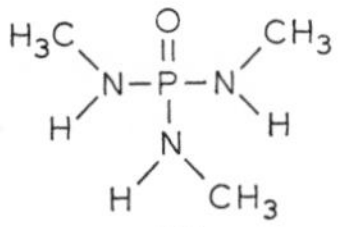

Figures 15 and 16 illustrate radiochromatograms of extracts of resistant and susceptible male houseflies and extracts of excreta from resistant and susceptible male houseflies, respectively, 4 h after injection of ^{14}C-hempa.

When flies of either strain were treated topically with the synergist tropital [piperonal bis-(2-(2-butoxyethoxy)ethyl)acetal] and then injected with hempa, the metabolism of the sterilant and the excretion of radioactivity were substantially reduced, but the metabolic pattern of the hempa was different from that of the untreated flies. Figures 17 and 18 show results of TLC analyses of the extracts of flies treated with tropital and then with labeled hempa indicating the suppression of the demethylation of hempa. Compounds I and II (peaks a and b in Fig. 16) were absent, but a new unidentified metabolite appeared (peak designated by an arrow in Fig. 18) could not be detected in the extract of synergized-resistant flies (Fig. 17), and the pattern in their excreta (Fig. 18) was also different from that shown in Fig. 16.

In rats and mice, hempa is degraded by a stepwise loss of methyl groups[66] to pentamethyl, tetramethyl, and trimethyl phosphoramide. Both liver slices and permanganate oxidation have shown that the demethylation reaction involves the loss

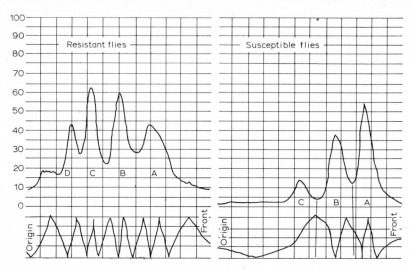

Fig. 15. Radiochromatograms (TLC) of extracts of resistant and susceptible male houseflies 4 h after injection of hempa-^{14}C ($6\,\mu g/\male$). Disc chart integrator units are shown at the bottom of each chromatogram. Peaks A–D correspond to compounds I–IV shown in Fig. 14.

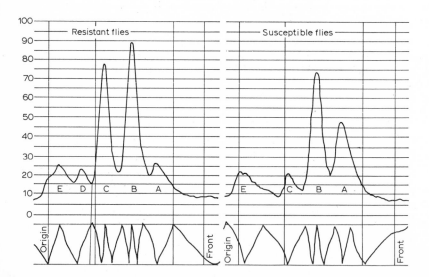

Fig. 16. Radiochromatograms (TLC) of extracts of excreta obtained from resistant and susceptible male houseflies 4 h after injection of hempa-^{14}C ($6\,\mu g/\male$). Disc chart integrator units are shown at the bottom of each chromatogram. Peaks A–D correspond to compounds I–IV shown in Fig. 14. The material indicated by peak C was not identified.

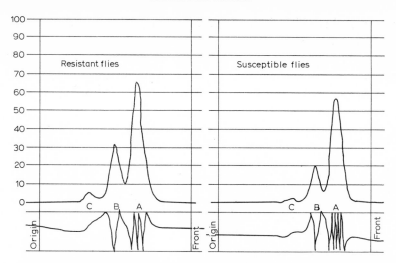

Fig. 17. Radiochromatograms (TLC) of extracts of resistant and susceptible male houseflies 4 h after injection of hempa-^{14}C (6.11 μg/♂) when each fly was treated topically with 31.4 μg of Tropital 1 h before injection. Peaks A–C correspond to compounds I–III in Fig. 14. Disc chart integrator units are shown at the bottom of each chromatogram.

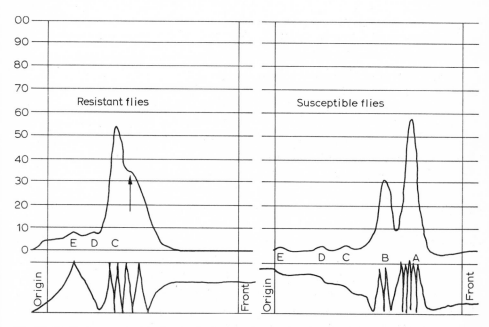

Fig. 18. Radiochromatograms (TLC) of extracts of excreta obtained from susceptible and resistant male houseflies 4 h after injection of hempa-^{14}C (6.11 μg/♂) when each fly was treated topically with 31.4 μg of Tropital 1 h before injection. Peaks A–D correspond to compounds I–IV in Fig. 14. Compounds pertaining to peak E and to the peak designated by an arrow were not identified. Disc chart integrator units are shown at the bottom of each chromatogram.

of the carbon as formaldehyde and the methylol (II) (Fig. 19) has been proposed as the intermediate[66].

Terry and Borkovec[67] demonstrated pentamethyl phosphoramide (III), N-formyl-pentamethyl phosphoramide (VI) and formaldehyde to be oxidation products of hempa by reaction with molar permanganate confirming that the methylol (II)

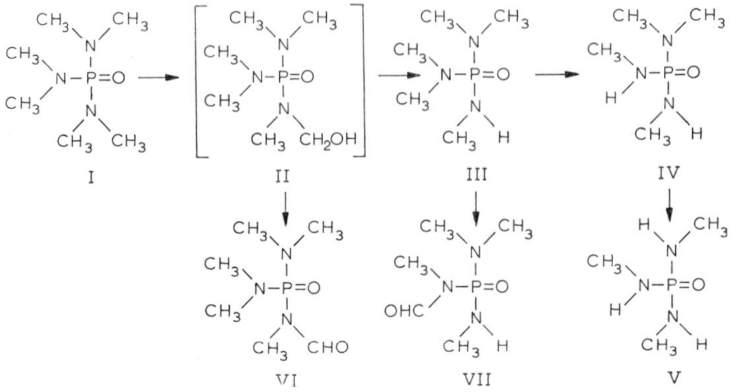

Fig. 19. Metabolic pathways of hempa. I, Hempa; II, methylol intermediate; III, pentamethylphosphoramide; IV, N,N,N',N''-tetramethyl phosphoramide; V, N,N',N''-trimethylphosphoramide; VI, N-formyl-pentamethyl phosphoramide; VII, N'-formyl-N,N,N',N''-tetramethyl phosphoramide.

is the most logical *in vitro* intermediate. Further investigations by Jones[68] into the metabolism of hempa in the rat has led to the isolation of (VI) as a minor metabolite so that both *in vitro* and *in vivo* oxidation appears to occur by a similar mechanism. Long-term administration of hempa to rats yielded N-formyl-N,N,N',N''-tetramethyl phosphoramide (VII) as a further metabolite thus suggesting that analogous methylol intermediates are probably involved at each demethylation stage. Figure 19 illustrates the *in vitro* and *in vivo* metabolic pathways of hempa.

Both column and TLC were used to elaborate the urinary metabolites following administration of hempa and pentamethyl phosphoramide to rats. The urine was extracted and chromatographed on Whatman SG-31 Chromedia[66] to remove I, III, and IV; the columns were then eluted with methanol and the extracts chromatographed on freshly activated[66] TLC plates (1.5 mm) of Silica Gel G and developed with chloroform–ethanol (3:1). From hempa-administered urine, the area R_F 0.44–0.55 (detected yellow with molybdate reagent[69]) was identified as N-formyl-pentamethyl phosphoramide (VI). From pentamethyl phosphoramide (III)-administered urine, the area R_F 0.30–0.40 was similarly extracted from preparative TLC and characterized as N'-formyl-N,N',N',N''-tetramethyl phosphoramide (VII) by reference to an authentic sample and IR spectroscopy.

Purification and analysis of ^{14}C-hempa by paper, thin-layer, and gas chromatographic techniques has been reported by Terranova and Schmidt[70]. The efficiency of paper partition chromatography was ascertained utilizing Whatman No. 1 paper with 27 solvent systems. After development, the resultant spots were visualized with iodine vapor and/or a Packard Chromatogram Scanner. Chromatograms were also attached to 8 × 8 in. glass plates and exposed to No-Screen X-ray film for 24 h to prepare autoradiograms. TLC utilized Silica Gel G with 13 solvent systems. Detection was achieved with the systems described above. Tables 11 and 12 list the results of

TABLE 11

TLC OF ^{14}C-HEMPA ON SILICA GEL G

Solvent	R_F value[a]
Ethanol	0.35, 0.51, 0.62
N,N-Dimethylformamide	0.85, 0.94
Methanol	0.50, 0.65, 0.75
Isopropyl alcohol	0.35, 0.65, 0.79
Acetone–chloroform (1:1)	0.08, 0.25
Acetone–chloroform–water (7:7:2)	0.08, 0.16, 0.29
Phenol	0.98
n-Butanol–water (3:1)	0.20, 0.33, 0.51
Benzene–ethanol (8:2)	0.00, 0.08, 0.14, 0.20, 0.27, 0.42
Benzene–ethanol (9:1)	0.00, 0.02, 0.05, 0.12
Benzene–ethanol (1:1)	0.50, 0.65, 0.77
Benzene	0.00
Ether	0.00

[a] R_F value of the main spot (corresponding to hempa) is underlined.

thin-layer and paper chromatography, respectively, of ^{14}C-labeled hempa. For gas–liquid chromatography, a Barber–Coleman Series 5000 gas chromatograph equipped with a radioactive monitoring system (RAM) that had a 10:1 splitter and a sodium thermionic detector was used to detect labeled hempa and its contaminants. The column was 6% DEGS (diethylene glycol succinate) on 80–100 mesh Diatoport S in a 5 mm × 8 ft. U-shaped glass column conditioned at 225°C for 48 h. Injector and detector temperatures were 230 and 280°C, respectively. The column was operated isothermally at 210°C. Helium at a column flow rate of 120 ml/min was used as the carrier gas.

The overall results of this study indicated that TLC on Silica Gel G with benzene–ethanol (8:2) was a satisfactory system for completely separating ^{14}C-labeled hempa from the contaminating materials present in the samples studied (confirmed by gas chromatography) and hence this system can be used as a preparative or purification method for hempa. The gas chromatographic system was also a satisfactory analytical means of determining the purity of low concentrations of hempa.

TABLE 12

SOLVENT SYSTEMS USED FOR THE PAPER CHROMATOGRAPHY OF ^{14}C-LABELED HEMPA
ON WHATMAN NO. 1 PAPER

Solvent	R_F [a]
Water	0.97
n-Butanol–H$_2$O (3:1)	0.90
N,N-Dimethyl formamide	0.92
Methanol	0.92
Isopropyl alcohol	0.90
Acetone	0.88
Acetone–chloroform (9:1)	0.88
Acetone–chloroform (1:1)	0, 0.90
Acetone–chloroform–water (7:7:2)	0.66, 0.89, 0.97
Tert.-butyl alcohol	0, 0.84
Phenol	0.95
n-Butyl alcohol	0, 0.90
Ethyl acetate	0.97
n-Butyl acetate	0, 0.87
Isopropyl ether	0, 0.10, 0.25, 0.66
Chloroform	0, 0.83
Benzene	0, 0.16, 0.45
Benzene–ethanol (99.6:0.4)	0, 0.43, 0.68
Benzene–ethanol (99:1)	0, 0.62, 0.80
Benzene–ethanol (97:3)	0, 0.90
Benzene–ethanol (1:9)	0.87
Toluene	0, 0.08, 0.34
Hexane	0, 0.06
Hexane–ether (9:1)	0, 0.11
Hexane–ether (7:3)	0, 0.08, 0.30
Hexane–ether (3:7)	0, 0.09, 0.30, 0.62
Kerosene	0, 0.21

[a] R_F of the main spot (corresponding to hempa) is underlined.

Figures 20 and 21 show GLC analysis of ^{14}C-labeled hempa from eluates of paper and thin-layer chromatograms and Fig. 22 illustrates flame-ionization response to 3 samples of hempa.

A comparison of the rate of conversion of hempa to pentamethylphosphoric triamide (PMPT) by microsomal preparations from an insecticide-susceptible strain with that of preparations from two carbamate-resistant strains of houseflies known to have high levels of N-demethylating enzymes was described by Akov et al.[65, 71].

Both TLC and GLC procedures[65] were used to elaborate the metabolic products. A Micro-Tek Model 220 gas chromatograph was used with a Melpar Model 65-34A

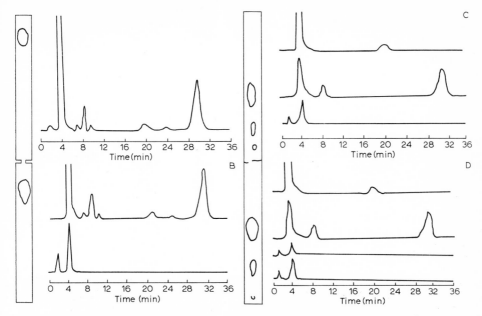

Fig. 20. Gas chromatographic analysis of ^{14}C-hempa from eluates of paper chromatography devel oped in solvent systems that resolved the material into A, 1 spot; B, 2 spots; C, 3 spots; D, 4 spots The column was 6% DEGS on 80–100 mesh Diatoport S, run isothermally at 210 °C. The material was detected by the radioactive monitoring system.

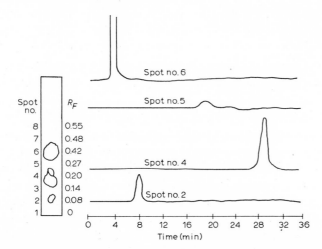

Fig. 21. Gas chromatographic analysis of ^{14}C-labeled hempa from eluate of thin layer chromato- graph developed in benzene–ethanol (8:2).

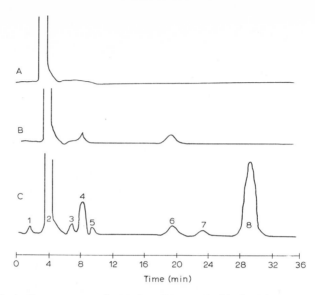

Fig. 22. Flame-ionization response to 3 samples of hempa. A, Obtained from Borden Chemical Co., Philadelphia, Penn.; B, cold hempa; C, ^{14}C-labeled hempa obtained from Dr. A.B. Borkovec and tested on a 6% DEGS on 80–100 mesh Diatoport and operated isothermally at 210°C.

flame-photometric detector and a 1.8 m × 6 mm o.d. glass column packed with 0.5% polyethylene glycol 20,000 on 60–80 mesh glass beads. Nitrogen was the carrier gas at 80 ml/min. Injection port, column, and detector temperatures were 200°, 160° and 150°C, respectively. The retention time of hempa was 2.0 min and that of PMPT 4.3 min. One nanogram of PMPT produced a deflection of *ca.* 15% of the total scale at an electrometer setting of 32×10^{-8} A full scale. As little as 0.3 ng of PMPT could be detected in a given injection in the presence of 450 ng of hempa, corresponding to 90 μg of hempa (0.5 μmole) and 60 ng of PMPT (0.36 nmole) in a reaction mixture.

The toxicity[72, 73], mutagenicity[74], and metabolism[75] of hemel have been described. The metabolism of C^{14}-hemel in male houseflies, *Musca domestica* was studied by Chang *et al.*[75] who isolated a number of lower methylmelamines, *e.g.* penta-, tetra-, and trimethylmelamines in addition to other acyclic products from the treated flies and from their excreta. Mono- and dimethylmelamines were not detected and unchanged hemel was found only in the flies. Column chromatography (basic aluminum oxide, Woelm, and benzene eluents) was used for the isolation of synthetic ^{14}C-hemel and TLC on Silica Gel G with 5% ethanol in chloroform as developer was employed for the elaboration of radiopurity of the tagged chemosterilant. The radiometric determinations were made with a Packard Tricarb Model 3003 liquid scintillation spectrometer. TLC and GLC were used to separate and identify the metabolites. Silica Gel G and 10% ethanol in chloroform (for extracts of fly), Silica Gel G and 50% ethanol in chloroform (for extracts of excreta) and Aluminum

Oxide G and 5% ethanol in chloroform (for resolution of fractions which co-chro-matographed on silica gel) were the TLC systems employed. A Packard Model 7201 scanner with a disc integrator was used to detect and to quantify the radioactive spots on the TLC plates. A comparison of the isolated metabolites with reference compounds was made by GLC with an Aerograph Model 204B instrument equipped with a flame-ionization detector and a 30 cm × 0.32 cm stainless steel column packed with 10% diethylene glycol succinate (LAC-728) on Chromosorb W (acid-washed, 60–80 mesh). Figure 23 illustrates the structural formulas of methylmelamines with 3–6 methyl groups.

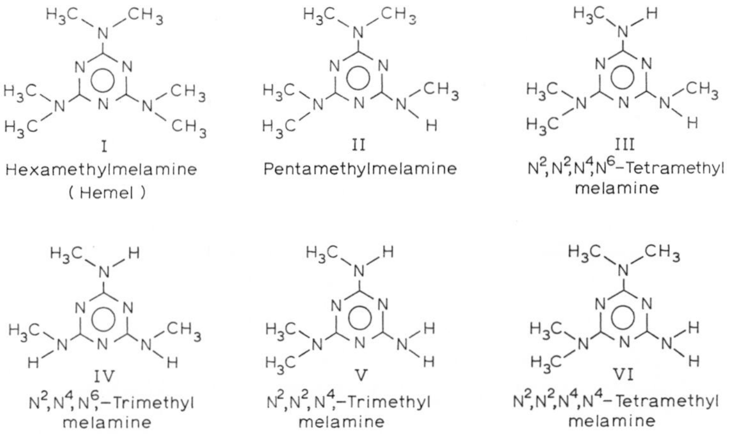

Fig. 23. Structural formulas of methylmelamines with 3–6 methyl groups. Compounds III and VI and compounds IV and V are isomeric.

Figure 24 depicts a radiochromatogram of fly extracts following administra-tion of ^{14}C-hemel and shows the formation of 5 radioactive peaks (A–E) (cor-responding to compounds I–V in Fig. 19), and is illustrative of the type of separa-tions and radiochemical separations effected in this study. Table 13 shows the UV absorption maximum λ_{max} and GLC retention times of metabolites and of correspond-ing reference compounds in this study.

^{14}C-labeled N^2,N^2,N',N'-tetramethylenemelamine (TMM) has been shown to be metabolized in male *Musca domestica* to N^2,N^2,N^4-trimethylmelamine and N^2,N^2-dimethylmelamine[76]. The absence of other isomeric trimethyl- and dimethylmelami-nes among the metabolites indicated that the dimethylamino group was demethylated in preference to the methylamino group. Since more than 80% of the originally in-jected TMM was recovered as metabolites containing the *s*-triazine ring, the main metabolic pathway was a demethylation and did not involve ring cleavage. Column and TLC and radiochemical procedures[75] were used for the isolation, purification, and determination of radiochemical purity and with GLC for the isolation and identi-fication of metabolites of ^{14}C-tetramethylenemelamine.

TABLE 13

UV ABSORPTION MAXIMA λ_{max} AND GLC RETENTION TIMES OF METABOLITES
AND OF CORRESPONDING REFERENCE COMPOUNDS

Fraction of fly extract	Reference compound	λ_{max} (mμ)				GLC retention time (sec)	
		Water		0.01 N HCl			
		Meta-bolites	Stand-ards	Meta-bolites	Stand-ards	Meta-bolites	Stand-ards
A	I	228	228	242	242	40	40[a]
B	II	223	223	236	236	83	83[a]
C	III	217	215	226	226	193	193[a]
D$_1$	IV	213	217	220	220	374	379[b]
D$_2$	V		219	221	221	227	225[b]
a	II	223	223	236	236	40	41[c]
b	III	220	215	226	226	90	90[c]
c$_1$	IV	215	217	220	220	261	261[d]
c$_2$	V		219	221	221	220	220[d]

[a] Column temp. 185 °C, injector temp. 240 °C, detector temp. 250 °C, flow rate of nitrogen 27 ml/min, flow rate of hydrogen 15 ml/min, flow rate of air 250 ml/min.
[b] Column temp. 200 °C, injector temp. 235 °C, detector temp. 265 °C, flow rate of nitrogen 30 ml/min, flow rate of hydrogen 15 ml/min, flow rate of air 250 ml/min.
[c] Column temp. 200 °C, injector temp. 250 °C, detector temp. 250 °C, flow rate of nitrogen 30 ml/min, flow rate of hydrogen 15 ml/min, flow rate of air 250 ml/min.
[d] The GLC was performed at conditions given in footnote b, but several hours elapsed between analyzing samples D and c.

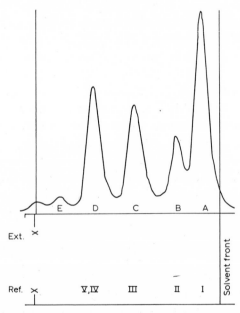

Fig. 24. Radio scanning and TLC of a fly extract containing 2 male equivalents. Each fly was injected with 10 μg of ^{14}C-hemel hydrochloride and held 5 h post treatment. A mixture of the reference compounds I–V was chromatographed on the same plate. The TLC plate was coated with Silica Gel G containing 1 % Cu–Sn sulfide phosphor and developed with 10 % ethanol in chloroform.

Hemel has demonstrated moderate but reproducible antineoplastic activity in animals[77] and significant regressions in human neoplasms have been recorded following therapy with hemel[78]. Bryan and Gorske[79] described a method for the determination of hemel in human whole blood, plasma, and urine based upon preliminary column chromatography on anion- and cation-exchange resin followed by UV absorption spectroscopy. Dowex 1 (Cl$^-$), 10% cross-linkage, 200–400 mesh and Dowex 50 W (H$^+$), 12% cross-linkage, 200–400 mesh were prepared as described by Price[80]. Recoveries of approximately 91–93% for hemel added to human whole blood (over the range of 10–200 µg of hemel) and urine samples were effected after passage through Dowex 1 (Cl$^-$) and Dowex 50 W (H$^+$). The ultraviolet absorption spectra were obtained in 0.1 N HCl at 242 nm.

4. ARAMITE

Aramite [2-(*p-tert.*-butylphenoxy)-1-methylethyl-2'-chloroethyl sulfite] is prepared by the reaction of 2-chloroethanol with sulfuryl chloride giving 2-chloroethylchlorosulfinate which is then reacted with the condensation product of propylene glycol and *p-tert.*-butylphenol. It is a non-systemic acaricide, formulated as an emulsifiable concentrate, wettable powder or dust and has been effective in the control of a variety of mites. Technical aramite is stable under ordinary storage, but does decompose when exposed to strong acid or alkali yielding sulfur dioxide and/or ethylene oxide.

The carcinogenicity of aramite has been established in chronic feeding studies in the rat[81] and dog[82, 83]. Aramite is also mutagenic in *Drosophila*[84]. It is of note to compare the structural similarities of aramite (I) with 2-chloroethylmethane sulfonate (II), an alkylating agent with both tumor-inhibiting activity and distinctive mutagenic effects in *Drosophila*[85, 86].

Aramite has been determined colorimetrically *via* the initial liberation of ethylene oxide[87], then conversion to formaldehyde[88].

Archer[89] described the quantitative measurement of combinations of aramite, DDT, toxaphene, and endrin in crop residues. Benzene extracts of alfalfa chaff, straw, and seed screenings were washed into an activated Florisil–anhydrous sodium sulfate column with pentane. DDT and endrin were eluted in that order with ethyl ether–pentane (1:9) then aramite was eluted with isopropanol–pentane (1:49).

TLC was employed for screening purposes and in combination with gas–liquid chromatography as an analytical tool. Silica Gel H (0.5 mm) was used with the following solvent systems: (*a*) benzene, (*b*) pentane, and (*c*) 5% diethyl ether–95% benzene. Detection was accomplished with indophenol reagent[90]. Table 14 lists the R_F values of DDE, DDT, DDD, toxaphene, endrin, and aramite in several solvents. DDT, its analogs, and toxaphene were separated (together) using pentane. Aramite was separated using 5% diethyl ether–95% benzene, and endrin was separated using benzene.

TABLE 14

R_F VALUES OF SEVERAL PESTICIDES ON SILICA GEL H

Compound	R_F values in solvent systems[a]		
	A	B	C
DDE	0.95	0.41	0.98
DDT	0.95	0.26	0.98
DDD	0.95	0.14	0.98
Toxaphene	0.95	0.10–0.20	0.98
Endrin	0.39	0	0.98
Aramite	0.76	0	0.81

[a] Solvents: A, benzene; B, pentane; C, 5% diethyl ether–95% benzene.

Gas chromatography of the above pesticidal extracts from TLC was carried out using a Varian Aerograph (Model 1200) equipped with an electron-capture detector. The column was 8 ft. × ⅛ in. stainless steel packed with 60–80 mesh DMCS-treated Chromosorb W, acid washed, coated with 5% Dow 710 silicone fluid and 5% SE-30 silicone gum rubber. The column temperature was 220°C and nitrogen was the carrier gas (50 p.s.i.; 20 ml/min). For DDT, toxaphene, and endrin, an electrometer setting of 10 and an attenuator setting of 1 were used. For aramite, an electrometer setting of 1 and an attenuator setting of 4 were used.

Blinn and Gunther[91] described TLC and GLC procedures for distinguishing aramite from closely related acaricide (OW-9) residues on citrus. The acaricide OW-9 is a mixture of 2 organosulfites the general structure of which is

$$CH_3-\underset{\underset{CH_3}{|}}{\overset{\overset{CH_3}{|}}{C}}-\langle\ \rangle-\left(O-CH_2-\underset{\underset{CH_3}{|}}{CH}\right)_m-O-\overset{\overset{O}{||}}{S}-O-CH_2CH_2Cl$$

When $m = 1$, the compound is aramite; OW-9 is a mixture of $m = 2$ and $m = 3$. TLC on silica gel with 3.5% ethyl acetate in benzene resulted in the separation of aramite and the two OW-9 compounds, with solvent front comparative values of 0.58 ± 0.02, 0.46 ± 0.02 and 0.30 ± 0.02, respectively, for $m = 1$, 2, and 3. All "m" compounds on the layer were detected by ready hydrolysis to sulfite ion, which complexes with buffered trinitrobenzene[92] to give a red di-adduct in aqueous methanol, or which reduces buffered malachite green to the bleached leuco form against a blue background. Vapor chromatography of aramite at 240°C resulted in a single sharp peak with a relative elution time of 1.8 compared to aldrin as 1.0. However, GLC of OW-9 under analogous conditions yielded evidence of major degradation based on elution behavior (Figs. 25 and 26). Despite the unpredictibility of the OW-9 in this copper-free system, the presence or absence of a sharp aramite elution peak was thought to help characterize a residue.

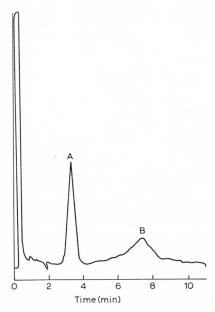

Fig. 25. Gas chromatographic separation of aramite A, from OW-9 compounds; B, by thermal conductivity detection. Conditions: 4 ft. × 1/4 in. stainless steel column of 18.5% high-vacuum silicone grease on acid-washed GC-22 fireback, 240°C, helium flow of 200 ml/min, 100 μg each of aramite and OW-9 compounds mixture.

When selective microcolorimetric gas chromatography[93, 94] was used, only partial clean-up was required to effect concentration and reduce non-volatile contamination. A useful technique that was found for the analysis of aramite and OW-9 was first to hydrolyze these organo-sulfites to their carbinol precursors, which in turn were separated by GLC as shown in Fig. 27.

Figure 28 illustrates a flow diagram of the procedure used to distinguish aramite from OW-9. Since aramite is no longer allowed as a residue in foodstuffs, the availability of a procedure to distinguish OW-9 residues from those resulting from illegal uses of aramite is of importance.

Other chromatographic procedures for the analysis of aramite have been reported. Bowman and Beroza[95] used an electron-affinity GLC to study the extraction

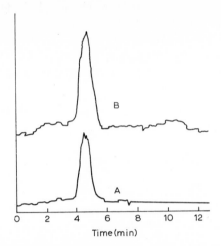

Fig. 26. Gas chromatographic behavior of A, aramite and B, aramite/OW-9 compounds mixture by microcolorimetric detection. Conditions: 4 ft. × 1/4 in. aluminum column of 5% Dow 200 silicone oil on acid-washed Chromosorb P, 220 °C, nitrogen flow of 100 ml/min, 20 μg each of aramite and OW-9 compounds mixture.

Fig. 27. Gas chromatographic separation of the carbinols resulting from A, the hydrolysis of aramite and B, C, the two OW-9 compounds (B) and (C). Conditions exactly the same as described in Fig. 26

p-values of several pesticides and related compounds including aramite in six binary solvent systems. Bosin[96] described the analysis of pesticide residues (including aramite) using microcoulometric temperature gas chromatography.

Mitchell[97] described the paper-chromatographic separation of 114 pesticides including aramite. Two solvent systems were used, (*1*) aq. immobile phase, soybean

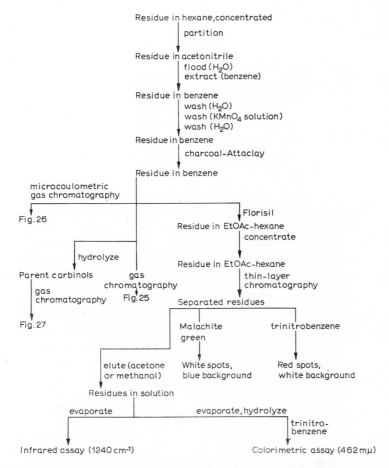

Fig. 28. Flow diagram of procedure for the separation of aramite from OW-9.

oil–ether (1:20) and mobile phase, water–2-methoxyethanol (1:3) and (*2*) non-aqueous immobile phase, 2-phenoxyethanol–ethyl ether (1:10) and mobile phase, 2,2,4-trimethylpentane. Detection was accomplished using a solution of silver nitrate and 2-phenoxyethanol in acetone followed by exposure to ultraviolet light.

5. DDVP

DDVP (2,2-dichlorovinyl dimethyl phosphate) (Dichlorovos, Vapona) is prepared by the reaction of trimethyl phosphate and chloral and is a widely employed insecticide used in baits, emulsifiable concentrates, space sprays, resin strips, aerosols, mechanical vapor generators, etc. Using the resin strips as recommended, the air concentration will be between 0.5 and 1 mg DDVP/m³ during the first week and above 0.1 mg/m³ during the first 5–6 weeks of the total usage period[98, 99]. DDVP is a contact and stomach poison and acts also as a fumigant and has also been formulated as an antihelminthic for swine. Saturated aqueous solutions of DDVP at room temperature convert to dimethyl phosphoric acid and dichloroacetaldehyde at a rate of about 3%/day; enhanced hydrolysis occurs in alkaline media.

Löfroth *et al.*[100, 101] have recently described the alkylating and biological properties of DDVP. DDVP was shown to alkylate calf thymus DNA yielding N-7-methyl guanine[101], cause chromosome aberrations in *Vicia faba*[101, 102] and induce mutagenicity in *E. coli*[100].

A variety of chromatographic procedures have been described for the determination of DDVP residues in food and tissues. A GLC determination of DDVP in milk, eggs and various body tissues of cattle and chickens has been described by Ivey and Claborn[103]. Extraction with acetonitrile and dichloromethane–hexane and chromatographic clean-up over silicic acid prior to GLC allowed the detection of 0.003 p.p.m. of DDVP in milk and 0.002 p.p.m. in body tissues and eggs. A Micro-Tek Model 160 was used equipped with a Melpar flame-photometric detector, and a 4 m × 1.22 m Pyrex glass column packed with 10% DC-200 coated on Gas Chrom Q (80–100 mesh) and conditioned 24 h at 190°C. The carrier gas was prepurified nitrogen at 120 ml/min, and the column, injector and detector temperatures were 135, 175, and 170°C, respectively. The detector gases were hydrogen 200 ml/min and oxygen 26 ml/min. (Under these conditions the retention time of DDVP is about 1.6 min.) Figure 29 illustrates chromatograms of extracts from milk, omental fat and blood of control and spiked samples. With an electrometer attenuation at input 10³, output 16 and bucking range 10⁻⁸, the noise level did not exceed 0.5% and 0.4 ng. DDVP gave a recorder response of 20% FSD. At the condition described, 0.05 ng DDVP is readily detected since it gives a response of 2–3%. Recoveries of 79–87% were obtained from fat, muscle and chicken skin, 93–97% from milk and 71% from eggs.

A gas chromatographic method for determining DDVP in plants and milk was reported by Draeger[104]. Plant material (banana, tobacco) was extracted by macerating with methanol and water, the macerate centrifuged and the supernatant extracted with diethyl ether–petroleum ether, the residue transferred to ethanol and the DDVP determined by GLC using a phosphorus detector. With milk, the active ingredient was partitioned between acetonitrile and diethyl ether–petroleum ether to separate milk fat. The lower limit of detection for DDVP is 0.2 ng corresponding to a limit of detection of 0.1 p.p.m. for tobacco and 0.01 p.p.m. for milk. The gas-chromato-

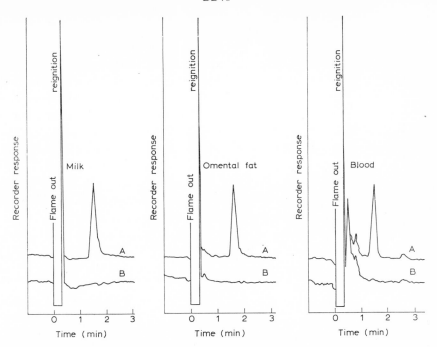

Fig. 29. Chromatograms of extracts from milk, omental fat, and blood. A, Control sample contain_ing 0.4 ng dichlorovos; B, untreated sample.

graphic conditions used are as follows: a 100 cm × 3.2 mm o.d. glass tube containing 10% Reoplex 400 on Embacel (60–100 mesh, acid-washed); the carrier gas was nitrogen at 25 ml/min and the gas mixture for the phosphorus detector was air at 170 ml/min and hydrogen at 16 ml/min; the column temperature was 150–160°C and the injection port and detector temperatures were 150 and 200°C, respectively. The sensitivity was $\frac{1}{1}$ or $\frac{1}{4}$ for an Aerograph 204 and $\frac{1}{16}$ for an Aerograph 1200, and the recorder chart speed was 17 mm/min (0.67 in./min). Under these conditions the retention time of DDVP is approximately 3 min. Figures 30 and 31 show gas chromatograms of DDVP in milk and tobacco, respectively.

    The determination of the vapor concentration of DDVP in air by trapping in a hydrocarbon solvent and direct analysis by GLC has been described by Miles[105, 106]. A gas chromatograph equipped with a phosphorus-sensitive flame-photometric detector (Micro-Tek MT-220 with Melpar detector) or an electron-capture detector were used. The latter was more sensitive, but more subject to interferences. A 6 ft. × 4 mm o.d. aluminum column was used with either (a) 3% Carbowax M 20 terephthalic acid (Suppelco #1033) on 100–200 Chromosorb G at 118°C and a nitrogen carrier flow of 60 ml/min (the retention time of DDVP is ca. 2 min)[106] or (b) 4 ft. × 0.25 in. o.d. aluminum column containing 3% SE-30 on 100–120 mesh Chromosorb W, DMCS treated, acid-washed; column temperature 150°C; nitrogen carrier gas at 133 ml/min gives a retention time of DDVP of ca. 1 min.

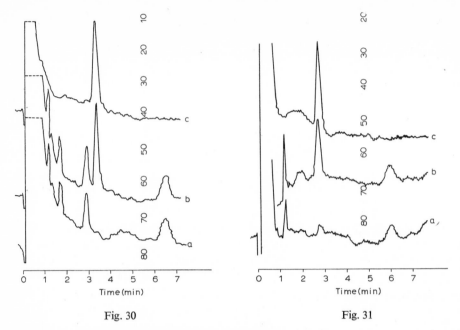

Fig. 30

Fig. 31

Fig. 30. Gas chromatographic determination of dichlorovos in milk. a, Control (2 µl); b, milk +0.01 p.p.m. dichlorovos (2 µl); c, standard 0.4 ng dichlorovos (2 µl).

Fig. 31. Gas chromatographic determination of dichlorovos in tobacco. a, Control (2 µl); b, tobacco +0.1 p.p.m. dichlorovos (2 µl); c, standard 0.4 ng dichlorovos (2 µl).

The hydrolysis of commercially important organophosphorus pesticides has been extensively studied[107–109]. The hydrolysis of DDVP in rat liver[107] was pictured to proceed as follows.

$$(CH_3O)_2P(O)OCH{=}CCl_2 \xrightarrow{\text{Fast}} (CH_3O)_2P(O)OH$$

DDVP   Dimethyl phosphate

↓ Slow

$$\underset{HO}{\overset{CH_3O}{>}}P(O)OCH{=}CCl_2 \xrightarrow{\text{Fast}} \underset{HO}{\overset{CH_3O}{>}}P(O)OH$$

Demethyldichlorovos   Methyl phosphate

↙| Slow

$$(OH)_3P(O)$$

Phosphoric acid

Ruzicka et al.[110] described the utility of GLC using a specific phosphorus detector in the study of the hydrolysis of organophosphorus pesticides. A Varian Aerograph

Model 205-B gas chromatograph was used with a Varian phosphorus detector and a 150 cm × 5 mm o.d. coiled column packed with an Apiezon L – Epikote 1001-coated Chromosorb G[111]. Table 15 lists the optimum column temperature and retention time relative to parathion for the pesticides studied and Table 16 lists the hydrolysis

TABLE 15

GAS CHROMATOGRAPHIC DETERMINATION OF PESTICIDES ON AN APIEZON L –
EPIKOTE 1001-COATED CHROMOSORB G

Pesticide	Optimum column temp. (°C)	Relative retention time	Pesticide	Optimum column temp. (°C)	Relative retention time
Azinphos ethyl	190	1375	Malathion-O-analogue	165	62
Azinphos methyl	190	1160	Mecarbam	190	122
Carbophenothion	190	450	Mevinphos	165	8.5, (11.5)
Chlorfenvinphos	190	140	Morphothion	190	285
Demeton-S-methyl	165	21	Parathion	165	100
Diazinon	165	32	Parathion-O-analogue	165	89
Dichlorovos	125	2.6	Parathion methyl	165	67
Dimefox	125	1.6	Phenkapton	190	825
Dimethoate	165	50	Phorate	165	25
Disulfoton	165	42	Phorate-O-analogue	165	19
Ethion	190	265	Phosphamidon	165	46, (66)
Fenchlorphos	165	81	Schradan	165	106
Fenitrothion	165	82	Thiometon	165	28
Fenthion	165	103	Thionazin	165	16
Formothion	165	67	Thionazin-O-analogue	165	18
Malathion	165	67	Vamidothion	190	318

rates of organophosphorus pesticides at 70 °C in ethanol–pH 6.0 buffer solution (20:80). All the pesticides studied above decomposed by a pseudo-first-order reaction. The hydrolysis rates of the pesticides are dependent upon their chemical structure and in pH 6.0 buffered solutions at 70 °C range from half-lives of 0.5 h to 4.0 days and larger still for phosphoramidates (at 20 °C the rates were several hundred times slower).

The persistence of a number of organophosphorus insecticides in non-sterile, heat-sterilized and gamma radiation-sterilized soils was determined by GLC of the extracts[112]. All of the pesticides degraded fastest in non-sterile soils and several compounds (e.g. DDVP, malathion, mevinphos and crodrin) decomposed much faster in irradiated soil than in autoclaved soil. The insecticide residues were measured by GLC with a phosphorus detector attached to a Varian Aerograph Model 600-D gas chromatograph containing a 5 ft. × ⅛ in. borosilicate glass column packed with QF-1 on 60–80 mesh Gas Chrom Q. Helium was the carrier gas at a flow rate

TABLE 16

HYDROLYSIS RATES OF PESTICIDES AT 70 °C IN ETHANOL–pH 6.0 BUFFER SOLUTION (20:80)

Pesticide	Half-life (h)	Pesticide	Half-life (h)
Phorate-O-analogue	0.5	Phosphamidon peak II	14.0
Dichlorovos	1.35	Thiometon	17.0
Phorate	1.75	Oxy-demeton methyl[a]	17.1
Trichlorphon[a]	3.2	Demeton-S	18.0
Mevinphos peak I	3.7	Morphothion	18.4
Mevinphos peak II	4.5	Fenthion	22.4
Demeton-S-methyl sulfone[a]	5.1	Vamidothion	25.4
Mecarbam	5.9	Menazon[a]	25.4
Malathion-O-analogue	7.0	Parathion-O-analogue	28.0
Demeton-S-methyl	7.6	Thionazin	29.2
Malathion	7.8	Disulfoton	32.0
Thionazin-O-analogue	8.2	Diazinon	37.0
Parathion methyl	8.4	Ethion	37.5
Fenchlorphos	10.4	Parathion	43.0
Azinphos methyl	10.4	Phenkapton	92.0
Phosphamidon peak I	10.5	Chlorfenvinphos	93.0
Fenitrothion	11.2	Carbophenothion	110.0
Dimethoate	12.0	Dimefox	212
		Schradan	Neglible hydrolysis in 96 h

[a] These compounds were studied using a colorimetric determination.

of 18 ml/min and the hydrogen flow varied from 20 to 25 ml/min. The injector column temperature was set at 230–250°C and oven temperatures were adjusted between 155 and 240°C to obtain insecticide retention periods within 2–4 min. With an electrometer setting of 1.6×10^{-10} afs, 1–20 ng of insecticide were required to produce 25% full scale deflection of a 1 mV Leeds & Northrup Model W recorder.

The correlation of structure of phosphate pesticides with response in electron affinity detectors was described by Cook et al.[113], utilizing GLC operating conditions previously described[114]. Three different detectors were used: (1) the standard Lovelock detector (Micro-Tek Model GC-1600-074-1)[114], (2) an adjustable parallel-plate detector (Barber-Colman Model 5120), and (3) the pin-cup detector (Micro-Tek).

The utility of electron-capture gas chromatography in the analysis of a number of phosphoric ester pesticides was described by Salame[115]. An Aerograph Hy-Fi Model 600 C with tritium detector and stainless steel columns (5 and 8 ft. × ⅛ in.) containing 5% SE-30 or 2.5% Carbowax 20M. The columns were run at 180 and 185°C, respectively, with nitrogen as carrier gas at 450–500 ml/min.

Burke and Holswade[47] listed the retention times and response data for over 50 pesticides (including DDVP) obtained with electron-capture and microcoulometric GLC systems. The column packing consisted of equal portions of previously coated 80–100 mesh Gas Chrom Q: one portion with 10% DC-200, the other with 15% QF-1. The operating conditions were as follows: column and injection temperatures 200 and 225°C, respectively, and nitrogen flow rate 120 ml/min. The apparatus used were (a) Barber–Colman Model 500 gas chromatograph equipped with a concentric-type ^{3}H electron-capture detector and a 6 ft. × 4 mm i.d. pyrex glass column, (b) Micro-Tek Model 2503 R gas chromatograph equipped with a 6 ft. × 4.5 mm i.d. aluminum column and (c) Dohrmann Model C-100 coulometer equipped with Models T-200 S and T-200 P titration cells.

Zweig and Devine[116] outlined GLC procedures for the determination of a variety of organophosphorus pesticides in water. A Varian Aerograph Hy-Fi gas chromatograph Model 600-D equipped with a Varian Aerograph phosphorus detector was used (Fig. 32). A 6 ft. × $\frac{1}{8}$ in. coiled borosilicate glass column packed with

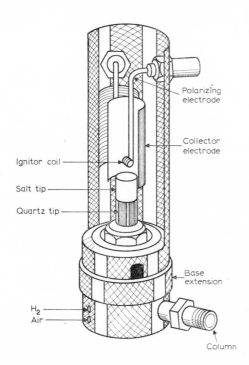

Fig. 32. Varian Aerograph phosphorus detector.

5% SE-30 on 60–80 mesh Chromosorb W was used with nitrogen flow rate of 36 ml/min, hydrogen 30 ml/min and compressed air 175 ml/min. Electrometer settings were 6.4×10^{-9} afs or 1.28×10^{-8} afs. The inlet and column temperatures were 210 and 185°C. A second column packed with a mixture of 3 parts of 2% QF-1 and 2 parts of 2% SE-30 on 90–100 mesh Anakrom ABS was used to separate many of the pesticides which could not be separated on the SE-30 column. Gas flows and temperatures were the same as given for the first column.

The determination of DDVP in DDVP–trichlorfon formulations by electron-capture gas chromatography after acetonitrile–hexane partitioning was described by Ellis and Bates[117]. Combinations of DDVP and trichlorfon which are frequently encountered in commercial fly-bait formulations cannot be determined by current spectrophotometric methods[118] because of interferences between 9.9 and 10.4 μ. A Jarrell-Ash Model 26-700 gas chromatograph was equipped with a 100 mc tritium electron-capture detector and a 4 ft. × $\frac{1}{4}$ in. stainless steel column containing 0.075% neopentyl glycol succinate and 0.675% SE-30 on Chromosorb W. The operating temperatures for the injector, detector, and column were 230, 199 and 177°C, respectively. Nitrogen was the carrier gas at 158 ml/min.

A variety of TLC procedures exists for the separation and identification of DDVP and related organophosphorus insecticides. Ackermann[119] described the TLC–enzymatic identification of 20 insecticide esterase inhibitors (including DDVP). Table 17 lists the R_F values of the insecticides on Silica Gel G developed with benzene and three benzene–acetone solvent mixtures (19:1, 9:1 and 2:1). Table 18

TABLE 17

R_F VALUES OF 20 INSECTICIDE ESTERASE INHIBITORS ON SILICA GEL G

Compound	R_F values with solvents			
	Benzene	Benzene–acetone		
		19:1	9:1	2:1
Bromophos	0.88	0.92	0.95	0.95
Bromoxon	0.06	0.49	0.63	0.78
Dibrom	0.19	0.42	0.70	0.82
DDVP	0.12	0.35	0.56	0.76
Dimethoat	0.00	0.13	0.35	0.60
PO-Dimethoat	0.00	0.00	0.05	0.18
Methylparathion	0.57	0.63	0.72	0.92
Methylparaoxon	0.00	0.28	0.45	0.71
Parathion	0.82	0.83	0.86	0.93
Paraoxon	0.04	0.53	0.66	0.78
Systox	0.00	0.27	0.53	0.82
Isosystox	0.00	0.26	0.52	0.82
Isosystoxsulfoxide	0.00	0.27	0.53	0.82
Isosystoxsulfone	0.00	0.00	0.17	0.60
Tinox	0.02	0.16	0.38	0.79
Isotinox	0.00	0.00	0.37	0.42
Isotinoxsulfoxide	0.00	0.00	0.00	0.04
Isotinoxsulfone	0.00	0.18	0.07	0.27
Trichlorphone	0.00	0.05	0.08	0.29
Carbaryl	0.06	0.43	0.56	0.75

depicts the limits of detection of these compounds with and without activation. The employment of different activation techniques considerably improved the detection seasitivity of the initially not, or only weakly inhibiting thiono- or dithiophosphoric ncid esters as well as of the phosphoric acid ester trichlorphon (the latter on treatment with ammonium hydroxide was converted into DDVP).

TABLE 18

LIMITS OF DETECTION (ng) OF ORGANOPHOSPHATES WITH AND WITHOUT ACTIVATION

	Without activation	$Br_2{}^a$ vapor (30 sec)	Aq. bromine[b] solution (15 min)	UV[c] (20 min)	Aq. ammonia[d] (15 min)
Bromophos	(−)	1.0	0.010	1.0	(−)
Bromoxon	0.010	0.5	0.010	0.010	0.010
Dibrom	1.0	(−)	1.0	1–5	1.0
DDVP	0.2	(−)	0.5	1–5	0.2
Dimethoat	(−)	500	500	10	(−)
PO-Dimethoat	5–10	10–20	5–10	5–10	10
Methylparathion	(−)	0.5–1.0	0.010	0.050	(−)
Methylparaoxon	0.002	0.5–1.0	0.002	0.002	0.005
Parathion	(−)	0.2	0.005	0.010	(−)
Paraoxon	0.001	0.01	0.001	0.001	0.002
Systox	10	10	1.0	1.0	10
Isosystox	5	5	1.0	1.0	10
Isosystoxsulfoxide	10	10	1.0	10	10
Isosystoxsulfone	1.0	1.0	1.0	1.0	1.0
Tinox	50	50	30	20	50
Isotinox	10	10	10	5	10
Isotinoxsulfoxide	10	10	10	5	10
Isotinoxsulfone	10	10	10	5	10
Trichlorphone	50–100	50–100	50–100	50–100	0.2–0.5
Carbaryl	0.5	0.5	0.5	0.5	0.5

(−) = Not detected.

[a] Saturated solution of bromine in carbon tetrachloride.
[b] Saturated aqueous bromine solution.
[c] UV analytical lamp "S 375" without filter.
[d] Water–conc. ammonium hydroxide (4:1).

The combined TLC–enzymatic detection method of Ackermann[119] was used to detect DDVP in air[120]. The sample was obtained by passing 50 l of air through 50 ml of water at the rate of 100 l/h, extracting the aqueous solution twice with 40 ml portions of chloroform. The chloroform extract was dried, filtered, diluted, and chromatographed on Silica Gel G plates with 90% benzene in acetone as developing solvent. The enzyme detection solution was prepared by first homogenizing 1 part of beef liver with 9 parts of water and centrifuging; 1 part of the supernatant was diluted with 4 parts of water and to this substrate solution was mixed prior to use 5 mg of naphthol-free β-naphthyl acetate in 2 ml ethanol with 15 mg of Fast Blue Salt B (zinc-stabilized bis-diazotized di-o-anisidine) in 8 ml of water. The developed chromatogram, after incubating 30–60 min at 38 °C, was sprayed with the above enzymic substrate solution yielding bright spots on a rose-red background. The detection limit was 0.5ng DDVP and the R_F value was 0.40. If the original chloroform extract is concentrated from 10 to 1 ml *in vacuo* at 30–35°C (after addition of a drop of silicone oil), ≥ 0.0005 mg of DDVP/m^3 air can be detected.

Mendoza *et al.*[121] evaluated the esterase from livers of beef, pig, sheep, monkey, and chicken for detection of organo esterase insecticides. DDVP, ethion, and dimethion with or without bromine exposure were detected with the 2000 or 3000 g supernatants of the water or tris buffer extracts of the livers. Table 19 illustrates the characteristic inhibition of esterases from different livers by pesticides.

A modified enzymatic detection method (utilizing a bee-enzyme solution) for the TLC of organophosphorus pesticides was described by Ernst and Schuring[122]. The pesticides were first separated on Silica Gel G with chloroform–ether (96:4) then oxidized on the plate by exposure to bromine vapor then sprayed with a honey-bee enzyme solution. After incubation for 30 min at 37°C, the plates were sprayed with 2-naphthyl acetate–Fast Blue B reagent, allowed to stand at 37°C for an additional 15 min whereupon the pesticides appeared as white spots on a magenta-colored background. The lower limit of detection for DDVP was 0.2 ng as well as for trichlorphon which was initially hydrolyzed into DDVP prior to detection[117].

The identification and determination of a number of organophosphorus and carbamate insecticides by TLC was described by Ramasamy[123]. Silica Gel G and Aluminum Oxide G were used as adsorbents with developing solvents (a) hexane–acetone (3:1), (b) chloroform–acetone (9:1), and (c) hexane – acetone (5:1); detecting reagents were: (1) 0.5% 2,6-dibromo-p-benzoquinone-4-chlorimine in cyclohexane, (2) p-nitrobenzene diazonium fluoroborate, and (3) bromine vapor followed by ammoniacal 4-methyl umbelliferone.

Polyamide-layer chromatography has also been utilized for the analysis of pesticides[124]. Polyamide layers 15 × 15 cm were developed with (a) acetone–ethanol–water (2:2:4) and (b) ethanol–ammonia–water (5:2:4) and were detected using (1) UV exposure (15-W germicidal lamp at a distance of 10 cm for 15 min), (2) 5% (v/v) bromine water in carbon tetrachloride, (3) 0.25% fluorescein sodium solution in dimethylformamide, and (4) 1% o-toluidine in 95% ethanol. Table 20 depicts the chromatographic data for seven organophosphorus, 5 chlorinated hydrocarbons and two carbamate pesticides. The minimum amounts of the compounds detectable were 0.1 μg of organophosphorus compounds, 2μg of chlorinated hydrocarbon, and 0.5μg of carbamate. The application of gel chromatography for the clean-up and separation of organophosphorus pesticides was described by Ruzicka *et al.*[125]. The utility of Sephadex LH-20, a modified dextran gel of lipophilic character, was investigated using acetone and ethanol as eluents.

TABLE 19

CHARACTERISTIC INHIBITION OF ESTERASES FROM DIFFERENT LIVERS BY PESTICIDES

Figures after pesticide names were the weights (ng) spotted on the plates (g = centrifugal force, H = water, T = tris buffer, Br = with bromine).

Chemicals		Beef 2000 g H	T	Beef 3000 g H	T	Sheep 2000 g H	T	Sheep 3000 g H	T	Pig 2000 g H	T	Pig 3000 g H	T	Monkey 2000 g H	T	Monkey 3000 g H	T	Chicken 2000 g H	T	Chicken 3000 g H	T
(1) Carbaryl (1 ng)	No Br	0	+	0	+	0	+	0	+	+	+	+	+	+	+	0	+	0	0	0	0
	Br	+	+	+	+	+	+	+	+	+	+	+	+	+	+	+	+	+	+	+	+
(2) Dichlorovos (8 ng)	No Br	+	+	+	+	+	+	+	+	+	+	+	+	+	+	+	+	+	+	+	+
	Br	+	+	+	+	+	+	+	+	+	+	+	+	+	+	+	+	+	+	+	+
(3) Ethion (10 ng)	No Br	+	+	+	+	+	+	+	+	+	+	+	+	+	+	+	+	+	+	+	+
	Br	+	+	0	+	+	+	+	+	+	+	+	+	+	+	+	+	+	+	+	+
(4) Oxydemeton-methyl (50 ng)	No Br	0	+	0	+	0	+	0	+	0	+	0	+	0	+	0	+	0	0	0	0
	Br	+	0	+	0	+	0	+	0	+	0	+	0	+	+	+	+	0	0	0	0
(5) Demeton (50 ng)	No Br	0	+	0	+	0	+	0	+	+	+	+	+	+	+	+	+	0	0	0	0
	Br	+	0	+	0	+	0	+	0	+	0	+	0	+	0	+	0	0	0	0	0
(6) Demeton sulfone thiol isomer (50 ng)	No Br	0	+	0	+	0	+	0	+	0	+	0	+	0	+	0	+	0	0	0	0
	Br	+	0	+	0	+	0	+	0	+	0	+	0	+	0	+	0	0	0	0	0
(7) Dimethoate (10,000 ng)	No Br	+	+	+	+	+	+	+	+	+	+	+	+	+	+	+	+	0	0	0	0
	Br	+	+	+	+	+	+	+	+	+	+	+	+	+	+	+	+	+	+	+	+
(8) Dimethoxon (10,000 ng)	No Br	+	+	+	+	+	+	+	+	+	+	+	+	+	+	+	+	+	+	+	+
	Br	+	+	+	+	+	+	+	+	+	+	+	+	+	+	+	+	+	+	+	+

TABLE 20

CHROMATOGRAPHIC DATA OF PURE COMPOUNDS

Pesticides	R_F value[a]		Color of spot in each stage[b]			
	I	II	Spraying reagent			UV exposure
			Bromine	Fluorescein	o-Toluidine	
DDVP	0.81	0.92	−[c]	−	−	B
Malathion	0.78	0.85	B	Y	D	DG
Diszinon	0.78	0.67	B	Y	D	DG
Ethylparathion	0.65	0.46	B	Y	D	DG
Methylparathion	0.64	0.48	B	Y	D	DG
Sumithion	0.64	0.56	B	Y	D	DG
Ethion	0.58	0.32	B	Y	D	DG
Lindane	0.45	0.26	−	−	−	DB
Endrin	0.40	0.21	−	−	−	B
Aldrin	0.36	0.20	−	−	−	B
DDT	0.31	0.07	−	−	−	B
Kelthane	0.21	0.31	−	−	−	B
Sevin	0.68	0.55	B	Y	D	Gr
Zineb	0.00	0.78	B	Y	D	DG

[a] I, Acetone–ethanol–water (2:2:4). II, Ethanol–ammonia–water (5:2:4).
[b] Y, yellow; B, brown; DG, dark green; DB, dark brown; R, red; D, dark; Gr, gray.
[c] −, Negative reaction to detection reagent.

6. TRICHLORFON

Trichlorfon (O,O-dimethyl-2,2,2-trichloro-1-hydroxyethyl phosphonate, Dipterex, Dylox, Neguvon) is made by the condensation of dimethyl hydrogen phosphite and chloral and is a non-systemic contact and stomach insecticide used against flies and household pests for crop protection and for the control of ectoparasites of domestic animals. Trichlorfon readily rearranges with loss of HCl to yield DDVP. Indeed its activity is attributed to its metabolic conversion to DDVP under physiological pH, since trichlorfon itself is devoid of anticholinesterase activity whereas DDVP is a very strong inhibitor. Some evidence[126, 127] suggests that direct P–C cleavage of the parent compound occurs, for the glucuronide of trichloroethanol was excreted by treated animals thus suggesting the reaction

$$(CH_3O)_2P(O)CH-CCl_3 \xrightarrow{H_2O} (CH_3O)_2P(O)OH + CH_2OHCCl_3$$
$$|$$
$$OH$$

Trichlorfon

Trichlorfon is a weak carcinogen causing local sarcomas of treated rats following subcutaneous injections[128, 129]. Dedek and Lohs[130] reported that trichlorfon methylates certain nucleophilic centers in human blood serum.

Early efforts to develop a gas-chromatographic method for the determination of trichlorfon residues in plant and animal tissues centered on the use of a Dohrmann microcoulometric gas chromatograph equipped with a halide-sensitive titration cell[131-133]. Anderson et al.[134] described an electron-capture gas chromatographic method for the determination of trichlorfon and the possible metabolites chloral hydrate and trichloroethanol in plant and animal tissues. An F&M Model 700 gas chromatograph equipped with an ^3H electron-capture detector was used with a 6 ft. × 4 mm i.d. borosilicate glass column with 60–80 mesh acid-washed Chromosorb W containing 20% XF-1150 (an A.G.E. acrylonitrile substituted silicone fluid) and 95% argon, 5% methane was the carrier gas. The temperature settings (°C) for the various compounds analyzed were as follows.

	Trichlorfon– chloral hydrate	Trichloro- ethanol	Trichloro- ethyl acetate
Inlet	270	250	210
Detector	200	200	200
Column oven	100	140	130

The vapor-phase chromatography of trichlorfon was dependent upon its thermal breakdown to chloral. The column temperature which gave optimum chromatographic results for chloral was 100°C. At this temperature, chloral was eluted in about 3 min and trichloroethanol in about 45 min. Trichloroethanol was optimally eluted in approximately 7.5 min at a column temperature of 140°C.

A microcoulometric gas chromatographic method sensitive to 0.1 p.p.m. of trichlorfon with clean-up from animal tissues hydrolysis was described by Barry et al.[135]. El-Rafai and Giuffrida[136] reported a GLC method for determining trichlorfon residues in the presence of DDVP residues, with a sodium thermionic detector. A general method for the determination of organophosphorus pesticide residues (including trichlorfon and DDVP) in river waters and effluents by gas, thin-layer, and gel chromatography was described by Askew et al.[137]. Table 21 lists the gas, thin-layer, and gel chromatographic results in terms of retention times, R_F values. and elution volumes. The use of a modified ammonium molybdate spray permitted sole detection of organophosphorus pesticides as blue spots on a buff background on the chromatoplates. Forty-two organophosphorus pesticides including trichlorfon and DDVP were chromatographed on binder-free thin-layer plates prepared from silica gel, aluminum oxide, and magnesium silicate, developed with five ternary solvent systems[138]. The three selective chromogenic sprays used and their range of sensitivity were (a) silver nitrate–bromerosol green for detecting the thiophosphoryl

TABLE 21

USAGE AND CHROMATOGRAPHIC RESULTS OF ORGANOPHOSPHORUS PESTICIDES

	Usage[a]	Gas-chromatographic results[b]			Thin-layer chromatographic results[c]			Gel-chromatographic results[d]
		Apiezon L	S.E. 30	X.E. 60	Solvent (i)	Solvent (ii)	Solvent (iii)	
Azinphos-ethyl	A, H	995*	970*	870*	0.33	0.90	—	113
Azinphos-methyl	A, H	840*	870*	870*	0.19	0.88	—	133
Bromophos	V	135	102	63	0.85	0.93	—	100
Carbophenothion	V	385*	270*	187*	0.83	0.96	—	98
Chlorfenvinphos	A, H, V	129	140	102	0.24	0.79	—	77
Coumaphos	V	N.D.	N.D.	N.D.	0.33	0.90	—	99
Crufomate	V	N.D.	164	141	0.06	0.43	0.86	73
Demeton-S	A, H	19	21	13	0.33	0.93	—	79
Demeton-S-methyl	A, H	22	33	27	0.17	0.73	—	86
Diazinon	A, H, F, V	38	41	18	0.61	0.95	—	77
Dibrom	A, H	26	6	25	0–0.22†	0–0.89†	—	86
Dichlofenthion	V	66	58	29	0.77	0.96	—	89
Dichlorovos	A, H, F, V	3#	4#	3.5#	0.22–0.27	0.73	—	83
Dimefox	A, H	1#	2#	2#	0.08	0.44	0.66	71
Dimethoate	A, H	43	65	95	0.05	0.37	0.59	95
Disulfoton	A, H	47	45	26	0.82	0.97	—	87
Ethion	A, H	224*	220*	156*	0.77	0.97	—	82
Ethoate-methyl	A, H	47	68	95	0.07	0.61	0.76	87
Fenchlorphos	A, H, F, V	86	70	40	0.84	0.93	—	101
Fenitrothion	F	81	88	95	0.49	0.91	—	114
Formothion	A, H	62	82	129	0.15	0.75	—	—
Haloxon	V	N.D.	N.D.	N.D.	0.04	0.71	0.86	112
Malathion	A, H, F, V	66	85	75	0.37	0.95	—	86
Mecarbam	A, H	117	127	116	0.42	0.95	—	85
Menazon	A, H	N.D.	N.D.	N.D.	0	0.02	0.38	—
Mevinphos	A, H	8, 10#	13, 16#	12, 15#	0.10	0.64	0.69, 0.82	79
Morphothion	A, H	212	285*	356*	0.06	0.49	0.77	106
Oxydemeton-methyl	A, H	N.D.	N.D.	N.D.	0	0.05	0.20	79

Compound	Usage							
Parathion	A, H	100	100	100	0.57	0.91	—	100
Phenkapton	A, H	640*	420*	290*	0.74	0.97	—	104
Phorate	A, H	31	29	18	0.80	0.97	—	89
Phosalone	A, H	730*	600*	685*	0.39	0.97	—	104
Phosphamidon	A, H	41, 55	67, 91	73, 110	0.04	0.34	0.60	72
Pyrimithate	A, H	78	79	42	0.62	0.96	—	81
Schradan	A, H	78	130	74	0	0.02	0.16	59
Sulfotep	A, H	19	26	18	0.75	0.92	—	75
Tepp	A, H	N.D.	N.D.	26	0	0–0.50†	0.03, 0.62	73
Thionazin	A, H	21	25	16	0.45	0.92	—	89
Trichlorphon	A, H, V	2‡	4‡	N.D.‡	0.03	0.18	0.61	80
Vamidothion	A, H	N.D.	N.D.	N.D.	0.01	0.16	0.30	73

[a] Usage: A denotes agricultural, V veterinary, F food storage, and H horticultural.

[b] Gas-chromatographic results: the values shown are retention times relative to that of parathion = 100. The columns used contained the following stationary phases: (i) Apiezon L, 2% and Epikote 1001, 0.2%; (ii) S.E. 30, 4% and Epikote 1001, 0.4%; and (iii) X.E. 60, 2% and Epikote 1001, 0.2%, coated on acid-washed, dimethyldichlorosilane-treated, 80–100 mesh Chromosorb G. All columns were 150 cm in length with 0.3 cm o.d. Varian Aerograph Model 205-B with a phosphorus specific detector was used.

Retention times were determined at 195°C, except where marked * (= 220°C) and ‡ (= 150°C). The retention times of parathion on the three columns were (i) 220°C 1.70 min, 195°C 4.0 min, and 150°C 19 min; (ii) 220°C 1.80 min, 195°C 4.50 min, and 150°C 22 min; and (iii) 220°C 1.60 min, 195°C 4.25 min, and 150°C 21 min. N.D. denotes not detected.

[c] Silica gel thin-layer chromatographic results: the values shown are the R_F values in the solvent (i) hexane–acetone (9:1), (ii) chloroform–acetone (9:1) and (iii) chloroform–acetic acid (9:1). † Denotes streaking.

[d] Gel-chromatographic results: the values shown are the elution volumes relative to that of parathion = 100 when eluting with ethanol from a Sephadex LH 20 column.

configuration (0.1–1 μg), (*b*) 4-(*p*-nitrobenzyl)-pyridine for general detection (0.5 to 5 μg), and (*c*) serum cholinesterase inhibition for detecting the P=0 configuration (5–100 μg).

Tables 22–25 depict the R_F values of the phosphate-type insecticides on Silic AR-4, Silic AR-7, acid alumina and Florisil plates developed with (*a*) cyclohexane–acetone–chloroform (70:25:5), (*b*) acetone–isopropyl ether–cyclohexane (40:40:20), (*c*) acetonitrile–2,2,4-trimethylpentane–ethyl acetate (50:20:15), (*d*) acetonitrile–cyclohexane–ethyl acetate (15:20:65), respectively.

The separation of a mixture of trichlorfon, DDVP, chloral and demethyl phosphorus acid by TLC has been described by Kolyakova[139]. This mixture dissolved

TABLE 22

R_F VALUES FOR SOLVENT SYSTEM (CYCLOHEXANE–ACETONE–CHLOROFORM, 70:25:5)

No.	R_F values			
	Silic AR-4	Silic AR-7	Acid alumina	Florisil
1D	0.26	0.19		0.32
2D	0.09	0.10	0.00	0.17
3D	0.23	0.23	0.21	0.30
4D	0.13	0.15	0.00	0.00
5D	0.03	0.04	0.10	0.13
6D	0.31	0.30		0.44
7D	0.42	0.38		0.00
8D	0.53	0.44		0.00
9D	0.23	0.21	0.24	0.38
10D	0.13	0.09	0.16	0.22

TABLE 23

R_F VALUES FOR SOLVENT SYSTEM
(ACETONE–ISOPROPYL ETHER–CYCLOHEXANE, 40:40:20)

No.	R_F values			
	Silic AR-4	Silic AR-7	Acid alumina	Florisil
1D	0.93	0.81	1.00	1.00
2D	0.72	0.48	1.00	0.62
3D	0.91	0.76	1.00	0.89
4D	0.83	0.61	1.00	0.00
5D	0.22	0.15	1.00	0.44
6D	1.00	0.85	1.00	1.00
7D	1.00	0.87	1.00	0.00
8D	1.00	0.90	1.00	0.00
9D	0.86	0.68	1.00	0.96
10D	0.67	0.40	1.00	0.74

TABLE 24

R_F VALUES FOR SOLVENT SYSTEM
(ACETONITRILE–2,2,4-TRIMETHYLPENTANE–ETHYL ACETATE, 50:20:15)

No.	R_F values		
	SilicAR 4	SilicAR7	Acid alumina
1D	0.00	0.37	
2D	0.26	0.25	0.15
3D	0.45	0.39	0.44
4D	0.36	0.29	0.00
5D	0.09	0.09	0.17
6D	0.41	0.34	0.47
7D	0.59	0.37	
8D	0.56	0.41	
9D	0.39	0.29	0.36
10D	0.25	0.20	0.33

TABLE 25

R_F VALUES FOR SOLVENT SYSTEM
(ACETONITRILE–CYCLOHEXANE–ETHYL ACETATE, 15:20:15)

No.	R_F values		
	SilicAR 4	SilicAR 7	Acid alumina
1D	0.54		
2D	0.77	0.40	0.23
3D	0.59		0.71
4D	0.14	0.50	0.00
5D	0.66	0.12	0.27
6D	0.67	0.65	0.79
7D	0.64	0.69	
8D	0.68	0.76	
9D	0.51	0.52	0.58
10D	0.37	0.37	0.52

D. Phosphates

1D Coumaphos O-analog (3-chloro-7-hydroxy-4-methylcoumarin diethyl phosphate)
2D Trichlorfon
3D Ruelene ® (4-*tert*.-butyl-2-chlorophenyl methyl methylphosphoramidate)
4D Bomyl ® (dimethyl-3-hydroxyglutaconate dimethyl phosphate)
5D Bidrin ® (3-hydroxy-N,N-dimethyl-*cis*-crotonamide dimethyl phosphate)
6D Ciodrin ® (α-methylbenzyl-3-hydroxycrotonate dimethyl phosphate)
7D Dichlorovos
8D Naled
9D Mevinphos
10D Phosphamidon

in acetone or ethanol was separated on silica gel developed with chloroform–methanol (9:1). The components were detected with Mitchell's reagent[97] under UV.

Ackermann et al.[140] described the TLC[141] separation of trichlorfon and DDVP residues in milk with a sensitivity of 1–5 µg/spot and 0.5–1 µg/spot, respectively. In milk, these compounds can be detected at residue concentrations of 0.01 and 0.002 mg/l, respectively.

7. 2,4,5-T AND DIOXINS

7.1 2,4,5-T

The herbicide 2,4,5-T (2,4,5-trichlorophenoxy acetic acid) is produced by the interaction of sodium monochloracetate with 2,4,5-trichlorophenol. The latter may be made by the action of alkali on 1,2,4,5-tetrachlorobenzene which is obtained as a by-product of the commercial synthesis of the pesticide lindane. A trace impurity produced in the manufacture of 2,4,5-T is the chloracnegen, 2,3,7,8-tetrachloro-dibenzo-p-dioxin (TCDD) which can arise under the following conditions.

2,4,5-T, its salts and esters, have considerable application in agriculture and forestry for the destruction of mixed growths of plants and herbaceous weeds and are effective against many woody plants not controlled adequately with 2,4-D. The amine forms of 2,4,5-T are isopropanol, triethanol, diethanol, dimethyl, and triethyl-amine. The esters (including isopropyl, butyl, and iso-octyl and propylene glycol butyl ether) are suited for dominant application when the material must penetrate the bark of woody plants.

The pharmacologic and toxicologic[142, 143], as well as the fetocidal and tera-togenic (in the mouse)[144, 145] properties of 2,4,5-T have been described.

A variety of chromatographic methods exist for the determination of 2,4,5-T and its derivatives in commercial formulations. A GLC method for the determination of residues of 2,4,5-T and its propylene glycol butyl ether esters in tissues and fluids was described by Clark[146]. Both compounds were converted to the methyl ester of 2,4,5-T with boron trifluoride–methanol reagent (Applied Science) and analyzed on a Micro-Tek 2500 R gas chromatograph fitted with a Dohrmann microcoulometer filled with 15% Dow 710 silicone on acid-washed Chromport XXX (60–80 mesh). The column effluent passed through a heated aluminum transfer line into a heated quartz tube, where the organic compounds in the column effluent were combusted in the presence of oxygen. The resulting hydrogen halides were subsequently titrated coulometrically in the Dohrmann T-300 halogen titration cell. The operating conditions for the analysis were column temp. 210 °C isothermal, injection port 225 °C, transfer line 212 °C, detector inlet 245 °C, combustion tube 845 °C, carrier gas prepurified nitrogen 50 ml/min 42 p.s.i., oxygen 65 ml/min 20 p.s.i., sweep gas prepurified nitrogen 20 ml/min 42 p.s.i., bias voltage 240 mV, and sensitivity range 450 ohms. Figure 33 illustrates chromatograms of control fat and omental fat

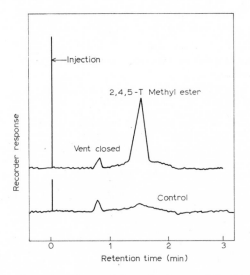

Fig. 33. Chromatograms of extracts of control fat and of omental fat to which 2,4,5-T was added.

to which 2,4,5-T was added. Average recoveries of 2,4,5-T added to fat, lean tissue, urine, and blood at levels from 0.05 p.p.m. to 20 p.p.m. were 89.3, 93.0 and 93.6%, respectively.

Stanley[147] described the gas chromatographic separation of the methyl and ethyl derivatives of 2,4-D, 2,4,5-T, 2-(2,4,5-trichlorophenoxy)propionic acid, α-(2,4-dichlorophenoxy)butyric acid, pentachlorophenol, o-phenylphenol, and 4,6-dinitro-o-sec.-butylphenol from each other on one column and the 3-alkyl phosphoric acids from each other on another column. The organic acids were readily esterified with diazomethane or diazoethane[148] and analyzed on Micro-Tek Model 500 R, 2000 R and 2000 MF gas chromatographs all equipped with both flame-ionization and electron-affinity detectors and with all-glass inlet system and 0.5 m × 6 mm glass column containing 2% SE-30 on 100–110 mesh Anakrom ABS and 1 m × 6 mm glass column containing 0.2% Versamid on 100–110 mesh Anakrom.

Tables 26 and 27 list the relative retention times for the esters of the chlorinated carboxylic acids and ethers of phenols, respectively.

TABLE 26

RELATIVE RETENTION TIMES FOR CHLORINATED CARBOXYLIC ACIDS

Acid	Relative retention times[a]			
	0.5-m column, 115°C		1-m column, 140°C	
	Methyl ester	Ethyl ester	Methyl ester	Ethyl ester
2,4-D	1.00	1.23	1.00	1.34
2-(2,4,5-T)P	1.65	2.08	1.69	2.09
2,4,5-T	2.23	2.88	2.03	2.63
α-(2,4-D)B	2.88	3.80	2.69	3.69

[a] Retention times relative to 2,4-D methyl ester, retention time 2.6 min on the 0.5-m column and 3.2 min on the 1-m column.

TABLE 27

RELATIVE RETENTION TIMES FOR PHENOLS

Phenol	Relative retention times[a]					
	0.5-m column, 115°C			1-m column, 140°C		
	Methyl-ether	Ethyl-ether	Free	Methyl-ether	Ethyl-ether	Free
o-Phenylphenol	0.596	0.654	0.924	0.406	0.500	0.969
Pentachlorophenol	1.31	1.69		1.34	1.78	
Dinitro-o-sec.-butylphenol	2.58	3.15		2.66	3.16	

[a] Retention times relative to 2,4-D methyl ester, retention time 2.6 min on the 0.5-m column and 3.2 min on the 1-m column.

Table 28 shows the relative retention times for the phosphate esters separated on a 1 m and 3 m × 6 mm o.d. copper column packed with 10% PEG 20M on 100 to 110 mesh Anakrom ABS.

Pursley and Schall[149] described the gas chromatographic analysis of 2,4-D and 2,4,5-T and their derivatives in commercial formulations by an examination of their corresponding methyl esters prepared *via* transesterification reaction of the esters of 2,4-D and 2,4,5-T with boron trichloride in methanol. An Aerograph Hy-Fi Model A-600 B instrument was used equipped interchangeably with electron-capture and hydrogen-flame-ionization detectors, 5 ft. × ⅛ in. stainless steel columns packed with 5% SE-30 silicone gum on 60–80 mesh acid-washed Chromosorb W. A 75 ml/min

TABLE 28

RELATIVE RETENTION TIMES FOR PHOSPHATE ESTERS

Phosphate ester	Relative retention times[a]	
	1-m column, 130°C	3-m column, 150°C
(MeO)$_3$PO	1.00	1.00
(MeO)$_2$(EtO)PO		1.147
(MeO)(EtO)$_2$PO		1.246
(EtO)$_3$PO	1.41	1.29
(MeO)$_2$(BuO)PO	1.43	
(MeO)(BuO)$_2$PO	2.94	
(EtO)$_2$(BuO)PO	2.84	
(EtO)(BuO)$_2$PO	5.08	

Retention times relative to (MeO)$_3$PO, retention time 3.7 min on the 1-m column and 5.5 min on the 3-m column.

nitrogen flow was used with the electron-capture detector and 25 ml/min with the hydrogen-flame-ionization detector. The column temperature was 165–175°C. Figure 34 illustrates a chromatogram of a methyl 2,4-D and 2,4,5-T mixture.

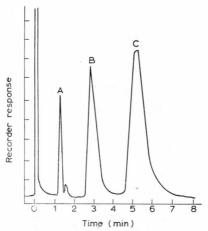

Fig. 34. Chromatogram of methyl 2,4-D and 2,4,5-T mixture. A, Internal standard (α-bromonaphthalene in n-hexane); B, methyl 2,4-D; C, methyl 2,4,5-T.

A GLC method has been described by Yip[150] for the determination of 7 herbicides in vegetable oils. The method involved extraction of 50 g of oil with sodium bicarbonate acidification and extraction of the herbicides (with chloroform). The herbicides separated were 2,3,6-TBA (2,3,6-trichlorobenzoic acid), MCPA (2-methyl-4-chlorophenoxy-acetic acid), PCP (pentachlorophenol), 2,4-D (2,4-dichlorophenoxy acetic acid), 2,4,5-T, 2,4,5-TP (2,4,5-trichlorophenoxy propionic acid), and 2,4-DB (4-(2,4-dichlorophenoxy)butyric acid). After esterification with diazomethane, the residue was analyzed by programmed-temperature gas chromatography using a Micro-Tek Model GC 2503 R instrument equipped with a microcoulometric halide detector and a 4 ft. × ¼ in. o.d. aluminum tube packed with 9% DC-200 silicone oil on Anakrom ABS 80–90 mesh (the column was conditioned at 225°C for two days). The initial column temperature was 133°C and was held at that temperature for 5 min before programming at 1°/min. The nitrogen flow rate was 100 ml/min, the sensitivity at 128 ohms and inlet block, outlet block and sample transfer line were at 250°C. Figure 35 illustrates a chromatogram of a 0.02 p.p.m. sample of a mixture of the above 7 herbicides and a control.

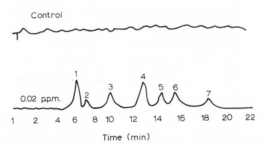

Fig.35. Gas chromatogram of a 0.02p.p.m. sample of a 7-herbicides mixture. 1, 2,3,6-TPA; 2, MCPA; 3, 2,4-D; 4, PCP; 5, 2,4,5-TP; 6, 2,4,5-T; 7, 2,4-DB. Both curves represent 25g of cotton seed oil.

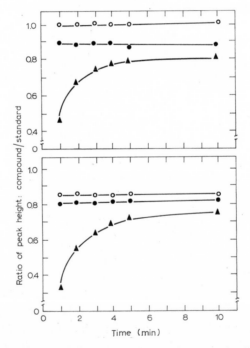

Fig.36. Methylation of 2,4-D and 2,4,5-T. ○, By diazomethane; ●, by dimethyl sulfate in methanol; ▼, by 5% perchloric acid in methanol.

Scoggins and Fitzgerald[151] compared the efficacy of dimethylsulfate in methanol with diazomethane and acid-catalyzed reactions for the methylation of chlorophenoxy acetic acid herbicides. Esterification is necessary to reduce the polarity of the carboxylic acid group to separate these compounds on many columns. The methyl esters are generally preferred because of ease of separation, short retention times and sharp symmetrical peaks that may be quantitated by peak height. Figure 36 compares the methylation of 2,4-D and 2,4,5-T by diazomethane, dimethylsulfate in methanol, and 5% perchloric acid in methanol. The dimethylsulfate reagent approaches the quantitative methylation of diazomethane and has the advantages of short preparation time, a long shelf life, considerably less toxicity and is non-explosive. Both diazomethane and dimethylsulfate react rapidly with a linear response from 1 to 10 min. The analyses were performed on a F & M Model 810 gas chromatograph equipped with an electron-capture detector and a 6 ft. $\times \frac{1}{8}$ in. o.d. stainless steel column packed with 10% SE-30 on 80–100 mesh Diatoport S. The parameters were injection port 230°C, oven 170°C, detector 210°C, attenuation 4, range 10, carrier gas helium, purge gas argon 5% in methane 95%, flow rates of 50 ml/min for both gases.

Devine and Zweig[152] described the GLC determination of 7 chlorophenoxy herbicides and their esters in water. A Micro-Tek Model MF-220 gas chromatograph was equipped with a ^{63}Ni electron-capture detector. The three columns used and their operating parameters were column 1, 5% SE-30 on 60–80 mesh Chromosorb W, 65 ml nitrogen/min, column temp. 175°C, column 2, 2% QF-1 on 90–100 mesh Anakrom ABS, 25 ml nitrogen/min, column temp. 175°C, and column 3, 20% Carbowax 20M on 60–80 mesh acid-washed Chromosorb W, 180 ml nitrogen/min, column temp. 220°C; the inlet and detector temperatures for all three columns were 230 and 275°C, respectively. The retention times relative to aldrin of the seven herbicides are listed in Table 29 and Fig. 37 illustrates a typical chromatogram of the various herbicide standards on column 1 (5% SE-30 on 60–80 mesh Chromosorb W).

TABLE 29

RETENTION TIME FACTORS (RELATIVE TO ALDRIN) OF SEVEN HERBICIDES

Herbicide	Column 1[a]	Column 2[a]	Column 3[a]
Dicamba[b]	0.190	0.439	0.375
MCPA[b]	0.219	0.504	0.421
2,4-D[b]	0.312	0.878	0.750
Isopropyl ester of 2,4-D	0.382	0.983	0.681
Silvex[b]	0.452	0.894	0.723
2,4,5-T[b]	0.531	1.31	1.21
n-Butyl ester of 2,4-D	0.706	1.67	1.19
Aldrin	1.00 (9 min)	1.00 (5 min)	1.00 (10 min)

[a] See text for column packings.
[b] Methyl ester.

References pp. 267–272

The estimated minimum detectable amounts of the herbicides and esters studied based on the relative instrument, the sample size and the final volume were 0.01 to 0.05 p.p.b. except 2 p.p.b. for MCPA.

Gutnick and Zweig[153] described two procedures for the quantitative determination of micro amounts of 2,4-D, 2,4,5-T, and 4-chlorophenoxy acetic acids. The

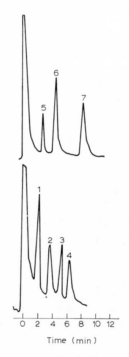

Time (min)

Fig. 37. Typical chromatograms for the methyl esters of some chlorophenoxy herbicides, using Column 1: 1, 2 ng dicamba; 2, 4 ng 2,4-D; 3, 0.5 ng silvex; 4, 1 ng 2,4,5-T; 5, 50 ng MCPA; 6, 3 ng isopropyl ester of 2,4-D; 7, 3 ng n-butyl ester of 2,4-D.

first procedure was based on the separation of the methyl esters by GLC and the colorimetric determination of collected fractions as the hydroxymate–Fe^{III} complex. The second procedure involved the formation of the ^{14}C-methyl esters, their separation by GLC, and the radioassay of collected fractions by liquid-scintillation spectrometry.

An Aerograph Model A-90 C instrument was used equipped with a thermal conductivity detector (hot-wire katharometer) and a 6 ft. $\times \frac{1}{4}$ in. o.d. copper tubing packed with 20 % Dow-11 high-vacuum silicone grease on 30–60 mesh Chromosorb P at a column temperature of 210 °C and a helium gas flow of 60 ml/min. Figure 38 illustrates a chromatogram of methyl esters of PCPA, 2,4-D and 2,4,5-T on 20 % Dow-11 at 210 °C. Table 30 shows the yield of esterification as determined by microcoulometric methods.

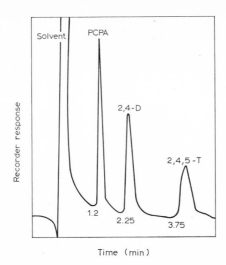

Fig. 38. Gas chromatography of methyl esters of PCPA, 2,4-D, and 2,4,5-T on a Dow 11 high-vacuum silicone column (20% on 30–60 mesh chromosorb) at 210°C.

TABLE 30

YIELD OF ESTERIFICATION AS DETERMINED BY MICROCOULOMETRIC METHODS[a, 93]

Compound	Peak area (sq. in.)	Wt. of ester recovered (μg)[b]	Wt. of acid (μg)	Recovery (%)
2,4,5-T	1.70	3.06	2.89	96.3
2,4-D	1.29	3.03	2.84	94.7
PCPA	0.81	3.20	2.96	98.7

[a] Average of two independent experiments in which 3 μg quantities were esterified.
[b] Micrograms chlorinated phenoxy acid

$$= \frac{\left(\dfrac{peak}{area}\right) \times \left(\dfrac{recorder}{sensitivity}\right) \times 35.5 \dfrac{g}{equiv.} \times 60 \dfrac{sec}{min} \times 10^6 \dfrac{\mu g}{g} \times 10^{-3} \dfrac{V}{mV} \times 10^2}{\left(\dfrac{recorder\ input}{resistance,\ \Omega}\right) \times (\%\ chlorine\ in\ pesticide) \times 96{,}500 \dfrac{C}{equiv.}}$$

Recorder sensitivity = 1.09 mV/in.
Recorder input resistance = 64 Ω.

Garbrecht[154] reported the rapid esterification of dicamba (3,6-dichloro-2-methoxybenzoic acid) and chlorophenoxy acids with N,O-bis(trimethylsilyl)acetamide for gas chromatographic analysis. The acids were extracted from a formulation, dried and then esterified with N,O-bis(trimethylsilyl)acetamide. Esterification was complete and quantitative within 15 min for dicamba, 2,4-D, 2,4,5-T, 2,4,5-TP, MCPA, MCPP, and 2,4-DB. A Perkin-Elmer Model 811 chromatograph was used with a thermal-conductivity detector with WX filaments and a 6 ft. × 1 mm i.d.

column packed with 5% DC-200 on 80–100 mesh Chromosorb WHP (Supelco). The helium flow was 40 ml/min, column oven, injector, and detector temperatures were 190, 220, and 240°C, bridge current 200 mA and attenuation was 4. Figure 39 shows a chromatogram of six common chlorophenoxy acids and dicamba as trimethylsilyl

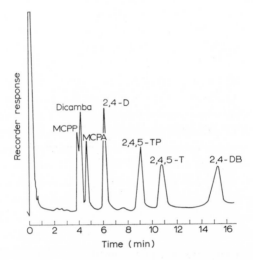

Fig. 39. Gas chromatogram of six common chlorophenoxy acids and dicamba as trimethylsilyl esters.

esters. A better separation of dicamba, MCPA, and MCPP could be obtained with a temperature-programmed analysis if all were found in the same formulation.

Radioisotopic and gas chromatographic methods for measuring absorption and translocation of 2,4,5-T and carboxyl-labeled and unlabeled butoxyethyl esters and ammonium salts of 2,4,5-T by mesquite were described by Morton et al.[155]. A Barber–Colman Model 5300 gas chromatograph was used equipped with an electron-capture detector ([226]Ra) and a 6-ft. spiral glass column packed with 80–100 mesh Chromosorb W coated with 2% SE-30. The injector, column, and detector temperatures were 290, 200, and 250°C, respectively. The flow rate of nitrogen carrier gas was approximately 75 ml/min for the methylated 2,4,5-T derivative samples and 100 ml/min for butoxyethyl ester samples.

The degradation of the n-butyl ester of 2,4,5-T in woody plants was studied by Fitzgerald et al.[156]. Leaf extracts were studied by GLC employing an F & M Model 810 gas chromatograph equipped with a pulsed electron-capture detector and a 6 ft. × ⅛ in. stainless steel column packed with 3.8% SE-30 on 80–100 mesh Diatoport S. The detector, injection port, and column temperatures were 200, 240, and 150°C, respectively. Helium was the carrier gas at 100 ml/min and a mixture of 5% methane and 95% argon with a flow rate of 50 ml/min was used as a purge gas. 2,4,5-Trichlorophenol was identified as the main product of degradation of the 2,4,5-T ester in leaf extracts (with no apparent formation of 2,4,5-trichloroanisole).

The evaluation of eight gas chromatographic columns for 50 chlorinated pesticides including the isopropyl ester of 2,4,5-T was described by Thompson et al.[157]. A column of 3% DEGS (at 195°C) and the mixed liquid-phase columns of 2% OV–1/3% QF-1 (at 180°C) and 1.5% OV–17/1.95% QF-1 (at 200°C) were the columns of choice by virtue of their capacity for high efficiency, response, practical retention times, and compound separation characteristics.

Leoni and Puccetti[158] described the utility of OV-17 stationary phase for the GLC of 2,4,5-T methyl ester and a variety of chlorinated and organophosphorus pesticides. The gas chromatograph used was a Carlo Erba Model C equipped with an electron-capture detector (tritium source) and a glass column 2 m × 4 mm o.d. packed with 3% OV-17 on 80–100 mesh Gas Chrom Q. The column was conditioned 3 days at 250°C with a moderate nitrogen flow rate operating. The operation conditions were column 198°C, injection block 220°C, detector 200 ± 1°C, carrier gas nitrogen at 70 ml/min, detector voltage 30 V d.c. (3.1 × 10⁻⁸ A; attenuation × 100 × 4) at which 1 ng of aldrin causes ½ a full-scale deflection. The retention time of 2,4,5-T methyl ester relative to aldrin was 0.64 min (retention time of aldrin 7.8 min). The results obtained at 198°C with the OV-17 stationary phase are comparable with those obtained at lower temperatures with the mixed phase QF-1 + DC-200; however, OV-17 was found to effect better separations.

Burke and Holswade[47] described the GLC of the isopropyl ester of 2,4,5-T on equal portions of 15% QF-1 and 10% DC-200 on 80–100 mesh Gas Chrom Q at 200 and 210°C with electron-capture and microcoulometric detection, respectively.

The fate of 2,4,5-T in the dairy cow was studied by St. John et al.[159]. The analysis of milk and urine for residues of 2,4,5-T[160] involved acetone phosphorous acid extraction, boron trifluoride methylation and electron-capture gas chromatography of the herbicide methyl esters in hexane. Animals fed 2,4,5-T eliminated the herbicide as soluble salts entirely in the urine. Figure 40 shows chromatograms of methylated 2,4,5-T in cow urine one day after the feeding of 5 p.p.m. of 2,4,5-T began, and control urine.

Courtney[161] studied the excretion pattern, serum levels, placental transport and metabolism of 2,4,5-T and 2,4-D in the rat. Both single and two-dimensional thin-layer chromatography were used to resolve 2,4,5-T and 2,4,5-trichlorophenol from biological material. The solvent for the first system (A) was benzene–dioxan–glacial acetic acid (45:10:2) and for the 90° development system (B) isopropanol–ethyl acetate–ammonia (35:45:2). For resolution of the compounds, the plates were sprayed with 0.05% brown caesol green (BCG) in methanol. The plates run in the basic system were first sprayed with a 5% solution of acetic acid, dried and then sprayed with BCG. The chlorophenoxy acids produced a yellow spot. The plates were then sprayed with ammoniacal silver nitrate and exposed to UV light as described by Nash et al.[162] or sprayed directly omitting the BCG step. The R_F values for 2,4-D, 2,4,5-T, 2,4-dichlorophenol, and 2,4,5-trichlorophenol are shown in Table 31. 2,4,5-T was found to be comparatively little metabolized in the rat (approximately

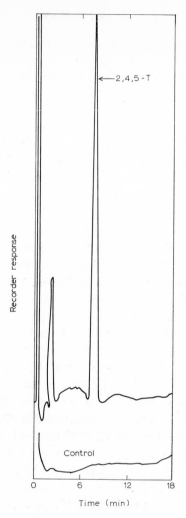

Fig. 40. Chromatograms of 2,4,5-T in cow urine one day after the feeding of 5 p.p.m. of 2,4,5-T began and control urine.

TABLE 31

R_F VALUES FOR 2,4-D, 2,4,5-T, 2,4-DICHLOROPHENOL AND 2,4,5-TRICHLOROPHENOL
FOR ONE- AND TWO-DIMENSIONAL SYSTEMS

Compounds	One-dimensional		Two-dimensional			
	System A[a]	System B[b]	System A then System B		System B then System A	
2,4-D	58	84	64	88	83	49
2,4,5-T	74	63	75	49	56	30
2,4-Dichlorophenol	81	79	79	75	85	58
2,4,5-Trichlorophenol	89	73	99	63	70	74

[a] System A contains benzene–dioxan–glacial acetic acid (45:10:2).
[b] System B contains isopropanol–ethyl acetate–ammonia (35:45:2).

90% of a subcutaneous administered dose of 2,4,5-T was excreted unchanged in the urine after 72 h).

The distribution and elimination of 2,4-D and 2,4,5-T in rats, pigs, calves, and chickens was studied by Erne[163]. When administered orally as amine or alkali salts, the compounds were readily absorbed and distributed over the organism in all species studied. The highest tissue levels of 2,4-D and 2,4,5-T were found in liver, kidney, lung, and spleen, the levels sometimes exceeding the plasma level. Elimination of the compounds was rapid, the plasma half-life being about 3 h in rats, about 8 h in calves and chickens, and about 12 h in pigs. The major excretory route appeared to be *via* the kidneys in all species studied. The chlorophenoxy acetic acids were isolated from body fluids and tissues by solvent extraction, separated from extractives by TLC and quantitatively determined photometrically[164].

A new chromatographic technique for the separation of and identification of halogenated aromatic pesticides and herbicides was described by Ceresia and Sanderson[165]. The method designated as tri-gradient tri-chromatographic (TGT) and di-gradient dichromatographic (DGD) chromatography has effected the separation of 2,4-D and 2,4,5-T in several 5-membered mixtures of pesticides as well as the separation of polyhalogenated pesticides from the di-, tri-, and tetra-halogenated ones, *e.g.* perthane, methoxychlor and rhothane. TGT is the exposure of a chromatogram to three different liquids in three different fields with the chromatogram air-dried before being developed in the next solvent. The third field or migration solvent surface may be at right angles to the first two. TCG or DCG involves the development of the spot color after each gradient or after the last two gradients, respectively, thus two to three R_F values may be obtained on one chromatogram to verify the identity of a particular chlorinated aromatic.

7.2 Dioxins

2,3,7,8-Tetrachlorodibenzo-*p*-dioxin (TCDD, dioxin) as pointed out earlier, is a contaminant in the preparation of 2,4,5-T. (When manufacture of the herbicide is carefully controlled, the TCDD content is less than 1 p.p.m.; higher concentrations have been noted, however.) As early as 1940, an unknown toxic compound associated with the production of 2,4,5-trichlorophenol was observed to cause chloroacne among factory workers. TCDD is known to be extraordinarily toxic (the acute oral LD_{50} in male guinea pigs is about 10^{-6} g/kg) and has been implicated in a variety of pathologic phenomena including neurological disturbances and birth defects in the mouse. Dioxins may be liberated from trichlorophenols or pentachlorophenols subject to excessive heat. The polychlorophenols find diverse applications as disinfectants, fungicides and anti-slime agents in the paint and paper industry. A related dioxin, 1,2,3,7,8,9-hexachlorodibenzo-*p*-dioxin, has been identified as one of the components of the chick edema factor which caused extensive damage and death of uncounted chicks in 1957. Chicks afflicted with edema or hydropericardium, suffer an accumulation of fluid in the heart sac and gross kidney and liver damage (5 μg of this dioxin is lethal in the chick). The herbicide 2,4,5-T or derivatives of it had been taken into plants and had ultimately appeared in vegetable oils. They were processed at high temperatures to liberate fatty acids but some dioxin which has extreme thermal stability was formed. Although originally associated with low-grade animal fats, chick edema factor was later found in certain batches of commercial oleic acids. An earlier bioassay[166] for the chick edema factor has largely been supplanted by gas chromatographic techniques. Firestone et al.[167] initially described the application of micro-coulometric gas chromatography following adsorption chromatography of extractable unsaponifiables on alumina to the detection of chick edema factor in fats or fatty acids. Toxic fats yield chromatographic peaks with retention times relative to aldrin of 5 or more (all samples which failed to reveal these peaks were shown to be non-toxic in the chick bioassay). A Dohrmann microcoulometric gas chromatograph was used with 3 ft. × ¼ in. o.d. aluminum tube containing 20% Dow Corning High Vacuum Grease or Dow Corning DC-200 silicone fluid on 80% acid-washed Chromosorb W. The column was conditioned at 275°C for 48–72 h passing nitrogen through at 20 ml/min. The column was operated at 248°C using a nitrogen flow rate of 60 ml/min so that aldrin eluted in 2.3–3 minutes. The injection block temperature was 270°C. Table 32 lists the relative retention times of chlorinated pesticides and materials isolated from toxic fats.

TABLE 32

RELATIVE RETENTION TIMES OF CHLORINATED PESTICIDES AND MATERIALS ISOLATED FROM TOXIC FATS

3 ft. × ¼ in. diam. column, 20% silicone grease, 80% Chromosorb W; carrier gas flow rate, about 60 ml/min; column temperature, 248°C; injection block temperature, 270°C

Sample	Retention time vs. aldrin (R_A)
Chlordane	1.0
Heptachlor	0.9
Kepone	2.2
Mirex	3.6
Strobane	0.5–3.5
Tedion	3.4
Toxaphene	0.6–3.8
Toxic factor from triolein	5.0
Inactive analogue from triolein	9.0
Concentrate from a toxic fat	2.3, 3.6, 5.4

The chick edema-producing factors isolated by Wooton et al.[168] had retention times relative to methyl arachidate of 1.17, 3.02, and 3.17 when chromatographed at 280°C on a 20% silicone column.

Higginbotham *et al.*[169] described the detection of chick edema factor in fats and fatty acids by electron-capture gas chromatography at 200°C. Extracted unsaponifiable was first chromatographed on an alumina column and a polar fraction (eluted with 25% ethyl ether in petroleum ether) was subjected to sulfuric acid clean-up and then chromatographed. Samples contaminated with the chick edema factor showed peaks with long retention times (10–25 that of aldrin).

A Barber–Colman Model 5360 chromatograph was used with a concentric-type Packard tritium source detector. The column was a 7 ft. $\times \frac{1}{4}$ in. i.d. glass tube packed with 2.5–3% SE-52 silicone gum rubber on 80–90 mesh Anakrom ABS, 60–80 mesh Gas Chrom Q or 60–80 mesh acid-washed Chromosorb W. Table 33 lists the analytical results obtained by electron GLC in the detection of the chick edema factor.

TABLE 33

DETECTION OF CHICK EDEMA FACTOR BY ELECTRON CAPTURE GAS CHROMATOGRAPHY[a]

Sample	Equiv. starting sample (mg)	R_A peaks (R_A 10–25)
Blank	1/200 residue	None
Low positive reference[b]	50.0	10, 12, 17, 20
Toxic fat used for preparation of low positive reference	5.0	10, 12, 17, 20
5% toxic fat in USP cottonseed oil[c]	50.0	10, 12, 17, 20
Toxic oleic acid	10.0	10, 12, 17, 20
Toxic glyceryl mono-oleate	10.0	10, 12, 17, 20
USP cottonseed oil	50.0	None
USP corn oil	50.0	None

[a] Barber–Colman Model 5360, with a concentric-type Packard tritium source detector. GLC conditions: electrometer setting 1×10^{-9} A full scale for 1 mV recorder (sensitivity 100, attenuator 1), column 7 ft. $\times \frac{1}{4}$ in. i.d. packed with 2.5% SE-52 on Gas Chrom Q, column temperature 200°C, nitrogen flow rate 100 ml/min, detector voltage 40 V d.c.
[b] 1.56% reference toxic fat in USP cottonseed oil.
[c] Submitted to collaborators in duplicate.

Figures 41–43 illustrate chromatograms of alumina fractions from toxic fats and fatty acids.

Refinements in the rapid screening method for the identification of chick edema factor in fats and fatty acids have been further described by Higginbotham *et al.*[170, 171]. The improved rapid screening method involves (*a*) preliminary treatment of a 2.5 g sample with sulfuric acid, (*b*) alumina column chromatography of the petroleum ether extract from sulfuric acid treatment, (*c*) sulfuric acid clean-up of the third alumina fraction (25% ethyl ether in petroleum ether), and (*d*) electron-capture gas chromatography at 200°C. Gas chromatographic peaks with retention times *vs.* aldrin (R_A values) between 10 and 25 are indicative of chick edema factor.

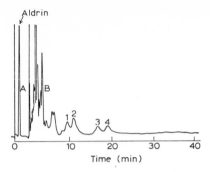

Fig.41. Electron-capture gas chromatograms of A, aldrin standard (0.1 ng); B, alumina fraction 3 from USP cottonseed oil containing 1.5% toxic fat. (1) R_A 10; (2) R_A 12; (3) R_A 18; (4) R_A 20.

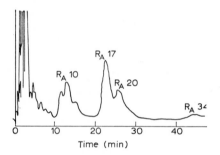

Fig.42. Electron-capture gas chromatogram of alumina fraction 3 from a toxic commercial oleic acid.

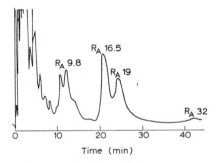

Fig.43. Electron-capture gas chromatogram of alumina fraction 3 from a toxic commercial glyceryl mono-oleate.

Chemical and toxicological evaluations of isolated and synthetic chloro derivatives of dibenzo-*p*-dioxin were described by Higginbotham *et al.*[172]. A number of commercial chlorophenols were pyrolyzed according to the procedure of Konita *et al.*[173], benzene extracts of the reaction mixtures were fractionated over an alumina column and the benzene effluents extracted with petroleum ether at room temperature, and the petroleum ether finally removed. The resulting products were sufficiently pure in most cases for examination by electron-capture GLC and biological testing by the chick embryo assay[174]. Table 34 shows retention times obtained at 200°C using a 7 ft. × ⅛ in. column containing 2.5% SE-52. A concentrate from an unsaponifiable fraction of a certain commercial toxic fatty acid material known to contain traces of the chick edema factor was used as a GLC reference (the toxic fatty acid material or

TABLE 34

EXAMINATION OF CHLOROPHENOL PYROLYSATES

Sample pyrolyzed[a]	R_A at 200°C of components[b] in pyrolysate	Chicken embryo bioassay	
		$\mu g/egg$[c]	% mortality[d]
2,4-Dichlorophenol (R)	1.0	500.0	70
2,4,5-Trichlorophenol (T)	3.5, 6.5	0.25	100
2,4,6-Trichlorophenol (R)	2.6	5.0	50
2,3,4,6-Tetrachlorophenol (T)	8.9, 10.0, 11.2, 17.4, 20.0, 35.0	1.0	100
Pentachlorophenol (R)	35.0	5.0	27
Chlorinated dibenzo-p-dioxin	1.0, 1.8, 3.5, 6.5	0.05	100
Reference toxic fat components	1.0, 1.8, 3.5, 6.5, 8.9, 10.0, 11.2, 17.4, 20.0, 35.0	3.0	100

[a] (R) = reagent grade; (T) = technical grade.
[b] R_A = retention times of peaks relative to the retention time of aldrin.
[c] 1 μg ≅ 20 p.p.b. in the egg. Sample injected into the air cell of fresh fertile eggs before incubation. Solvents used: ethanol, acetone and chloroform.
[d] Embryonic mortality at 21 days. The mortality of non-injected and solvent-injected controls was 10–15%.

by-product obtained from the manufacture of oleic and stearic acids has been previously used as a reference standard[175]). The pyrolysate from technical grade 2,3,4,6-tetrachlorophenol was the only product that showed a peak pattern (R_A values of 10 or more) indicating the presence of a chick edema factor (hydropericardium factor).

The technical grade 2,3,4,6-tetrachlorophenol pyrolysate, chlorinated dibenzo-p-dioxin and components from the reference toxic fat were each fractionated by preparative GLC (thermal conductivity detection), using a 12 ft. × 4 mm column at 250°C packed with 10% DC-200 on 60–80 mesh Gas Chrom Q. Isolated components from the 2,3,4,6-tetrachlorophenol pyrolysate and from the chlorinated dibenzo-p-dioxin were further examined by electron-capture GLC, IR and mass spectroscopy with the results shown in Table 35.

The two principal components isolated from the chlorinated dibenzo-p-dioxin were 2,3,7-trichlorodibenzo-p-dioxin and 2,3,7,8-tetrachlorodibenzo-p-dioxin (chicken embryo assay showed that the tetrachloro compound was more toxic than the trichloro compound).

TABLE 35

COMPONENTS ISOLATED FROM CHLORINATED DIBENZO-*p*-DIOXIN AND 2,3,4,6-TETRACHLOROPHENOL
PYROLYSATE

Mixture	GLC peak no.	R_A at 200°C	Molecular weight	No. of Cl atoms
Chlorinated dibenzo-*p*-dioxin	1	1.8	286	3
	2	3.5	320	4
2,3,4,6-Tetrachlorophenol pyrolysate	1	8.9	388	6
	2	10.0	388	6
	3	11.2	388	6
	4	11.6	388	6
	5	17.4	422	7
	6	20.0	422	7

The components isolated by preparative GLC from the reference toxic fat, the chlorinated dibenzo-*p*-dioxin and the 2,3,4,6-tetrachlorophenol pyrolysate were analyzed by electron-capture GLC at 200°C and by the chick embryo assay and the results are shown in Table 36. The peaks with high R_A values (those equal to or greater than 8.9) correspond to those of the hexachloro and heptachloro positional isomers isolated from the pyrolysate of 2,3,4,6-tetrachlorophenol. The overall results of this study

TABLE 36

GLC ANALYSIS AND EGG EMBRYO ASSAY OF ISOLATED COMPONENTS

Reference toxic fat components		Components from chlorinated dibenzo-p-dioxin and 2,3,4,6-tetrachlorophenol pyrolysate		
R_A at 200°C	Chicken embryo assay[a], % mortality	R_A at 200°C	Chicken embryo assay[b]	
			µg/egg	% Mortality
1.0	20			
1.8	20	1.8[c]	0.5	15
3.3	100	3.5[c]	0.2	100
5.1	40			
6.5	100			
8.9	80	8.9	2.0	30
10.0	80	10.0	1.0	100
		11.2	1.0	100
11.6	100	11.6	0.5	100
17.4	80	17.4	2.5	15
20.0	100	20.0	1.25	100
35.0	40			

[a] It was estimated that about 20 µg of material was injected into each egg.
[b] See footnotes *c* and *d*, Table 34.
[c] Components from chlorinated dibenzo-*p*-dioxin.

strongly suggest that chlorophenols are indeed precursors of the chick edema factor. This aspect has been given further credence in the recent studies of Higginbotham *et al.*[176] regarding the extraction and GLC detection of pentachlorophenol and 2,3,4,6-tetrachlorophenol in fat oils and fatty acids. The analysis involved treating a 5.0 g sample with conc. sulfuric acid and Celite, followed by a series of liquid–liquid extractions with petroleum ether, aqueous alkali, and chloroform. The residue from the final extract, after further treatment with conc. sulfuric acid, was analyzed by electron-capture GLC. A Barber–Colman Model 5360 instrument was used equipped with concentric-type Packard tritium source detector and a 7 ft. × ¼ in. i.d. glass column packed with 10% DEGS and 2% phosphoric acid on 60–80 mesh Gas Chrom P. The column temperature was 170°C, nitrogen flow rate 170 ml/min and detector voltage, 43 V d.c. The addition of phosphoric acid to the liquid phase suppressed tailing and permitted the determination of free chlorophenols without the necessity of preparing derivatives. The extraction and GLC method yielded recoveries in the range of 35–75% with a sensitivity of 0.5 p.p.m. on samples of corn oil and oleic acid fortified with each chlorophenol, *e.g.* 35–75% for pentachlorophenol and 35–50%

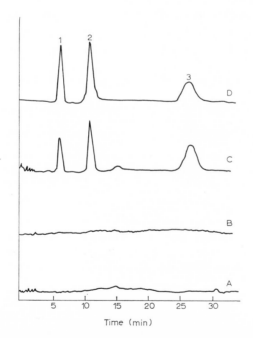

Time (min)

Fig. 44. Gas chromatograms of A, blank (injection equivalent to 5.0 mg sample); B, corn oil control equivalent to 5.0 mg sample; C, corn oil control containing 0.5 p.p.m. each standard chlorophenol (injection equivalent to 5.0 mg sample); D, 1 ng each standard. Standards: 1, 2,4,5-Trichlorophenol; 2, 2,3,4,6-tetrachlorophenol; 3, pentachlorophenol. GLC conditions: Barber-Colman Model 5360 with concentric type Packard tritium source detector; sensitivity 10^{-9} afs for a 1 mV recorder; column, 7 ft. × 1/4 in. i.d. packed with 10% DEGS + 2% H_3PO_4 on 60–80 mesh Gas Chrom P; column temperature 170°C; nitrogen flow rate 170 ml/min; detector voltage, 43 V d.c.

for 2,3,4,6-trichlorophenol. Several toxic oleic acids were examined by the method
and gave results which indicated the presence of pentachlorophenol at estimated levels
ranging from about 2 to 7 p.p.m. The presence of pentachlorophenol in the toxic oleic
acid was confirmed by TLC using Gelman Instant I.T.L.C. media, Type SG sheets
developed with heptane–petroleum ether (1:1) and detection with Rhodamine B
reagent followed by long-wave UV (350 nm). Polychlorophenols were detected as pink
spots on a purple background. Figure 44 illustrates gas chromatograms of (A) blank
(injection equivalent to 5.0 mg of an original 5.0 g sample), (B) corn oil control equivalent to 5.0 mg sample, (C) corn oil control containing 0.5 p.p.m. each standard
chlorophenol (injection equivalent to 5.0 mg sample), and (D) 1 ng of each standard.
The results of the analysis of commercial samples for chlorophenols and chick edema
factor (CEF) are shown in Table 37. Figure 45 illustrates chromatograms from the
analysis of toxic oleic acid and indicates that the toxic oleic acid sample contained
about 6.0 p.p.m. of pentachlorophenol.

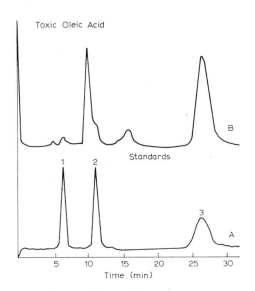

Fig. 45. Gas chromatograms of A, standard chlorophenols of 1 ng each of 1, 2,4,5-trichlorophenol;
2, 2,3,4,6-TCP; 3, PCP; B, toxic oleic acid, chromatogram obtained by injection aliquot (1 μl/10 ml)
equivalent to 0.5 mg original sample.

The identification of pentachlorophenol in fats, oils, and fatty acids is of
importance since pentachlorophenol and its salts have extensive and varied agricultural and industrial applications, more than any other chlorophenol currently
employed. Technical grades of pentachlorophenol generally contain as much as 10%
2,3,4,6-tetrachlorophenol. Further, the use of polychlorophenol as a defoliant for
cotton and soybean plants presents a potential source of contamination of vegetable
oils used for food and of oil mill products used for feed. In addition, contamination

TABLE 37

ANALYSIS OF COMMERCIAL SAMPLES FOR CHLOROPHENOLS AND CEF

Sample	CEF	2,3,4,6-TCP and PCP[a]
Commercial oleic acid 1	Positive	Positive
Commercial oleic acid 2	Positive	Positive
Commercial oleic acid 3	Negative	Trace
Commercial stearic acid	Negative	Negative
Corn oil	Negative	Negative

[a] Results represent GLC analyses only.

of animal fat (one of the major raw materials used in the commercial production of food-grade fatty acids) could result from the misuse of pentachlorophenol on or near animals.

8. 3,4-DICHLOROPROPIONANILIDE, 3,4-DICHLOROANILINE

AND CHLOROAZOBENZENES

3,4-Dichloropropionanilide (Propanil, DCPA) is made by the reaction of 3,4-dichloroaniline with propionic acid in the presence of thionyl chloride. It is a contact herbicide used to protect such economically important crops as rice, soybeans, cotton and tomatoes. DCPA is stable in emulsion concentrates but is hydrolyzed in acid and alkaline media to 3,4-dichloroaniline and propionic acid. It has been demonstrated[177,178] that the bio-degradation of DCPA inhibited soil respiration and produced residues identified as 3,4-dichloroaniline (DCA) and 3,3',4,4'-tetrachloroazobenzene (TCAB) (produced by the condensation of two molecules of DCA). The following scheme of transformation of DCPA in soil was proposed by Bartha and Pramer[178] based on coupling of the product of biological oxidation with the amine.

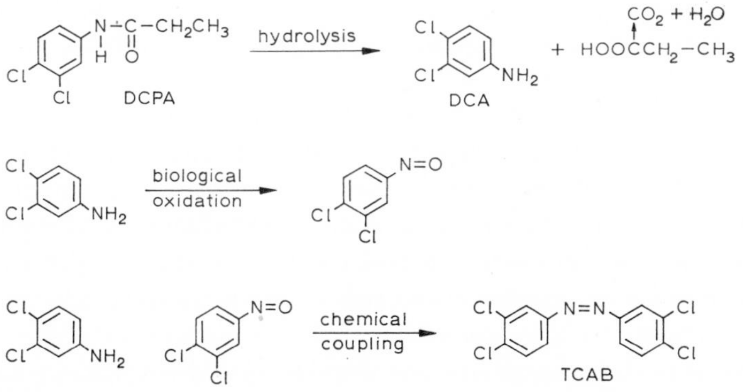

DCPA is also cleaved enzymatically to form DCA in plants[179-181], and mammalian liver[182]. 3,4-Dichloroaniline is also formed *via* the soil microbial degradation of the herbicides dicryl [N-(3,4-dichlorophenyl)-methacrylamide][183], karsil [N-(3,4-dichlorophenyl)-2-methylpentanamide][183], and diuron [3-(3,4-dichlorophenyl)-1,1-dimethylurea][184]. The cytogenic effects of DCPA and DCA on *Allium cepa* L.[185] as well as their mutagenic effects and that of TCAB in *Aspergillus nidulans*[186] have been described.

Bartha and Pramer[178] utilized GLC and TLC for the isolation and identification of the biotransformation products of DCPA. Soil extracts after clean-up were chromatographed on Eastman Chromagram type K301R plates developed with benzene. The R_F values found were DCPA 0.20, DCA 0.67, and TCAB 0.94. GLC analyses were performed on an Aerograph Model 660 chromatograph equipped with a flame-ionization detector and a 1.5 m × 0.3 cm o.d. stainless steel column packed with 5% SE-30 on Chromosorb W. The carrier gas was nitrogen at 60 ml/min. The retention times at 155°C were DCPA 5 min and DCA 1 min; retention times at 200°C were DCPA 1 min and TCAB 6 min. The identity of synthetic TCAB and of the pesticide transformation product isolated from soil was established by comparing their movements on thin-layer plates, retention times by GLC, IR spectra, and melting points.

The metabolism of unlabeled and ^{14}C-ring-labeled propanil was studied in 5 different soil types at two concentrations[187]. Ring-^{14}C-3,4-dichloroaniline was prepared from ^{14}C-ring-labeled diuron by basic hydrolysis in 2-methoxy ethanol. The radioactive compounds were purified by TLC on silica gel using a solvent system of *n*-hexane–benzene–acetone (7:3:1) and detected on no-screen X-ray film. TCAB was prepared by the method of Corbett and Holt[188]. Its purity as well as that of propanil was established by TLC, GLC, melting point, and mass spectral analysis. The GLC analyses of reaction products were performed by flame-ionization gas chromatography with a 6-ft. column packed with 10% methylvinyl silicone gum rubber on 80–100 mesh Diatoport S. The carrier gas was nitrogen with a flow rate of 40 ml/min. The injection port and detector temperatures were 270 and 310°C, respectively. The column temperatures were 200, 200, and 245°C for propanil, DCA and TCAB, respectively, yielding the respective retention times of 4 min 35 sec, 1 min 20 sec, and 5 min 57 sec. The limits of quantitative analysis were 0.5, 0.2 and 0.3 p.p.m. for propanil, DCA and TCAB, respectively. Soil type was found to influence propanil disappearance and TCAB formation. ^{14}CO$_2$ evaluation from carbonyl-^{14}C-propanil was rapid and varied between 60 and 80% of the original ^{14}C applied depending on soil and concentration. Ring ^{14}CO$_2$ was slower and amounted to less than 3% of the original ^{14}C. Figure 46 illustrates distribution of radioactive spots on thin-layer chromatograms from ring ^{14}C-propanil in non-sterile and sterile soil and reveals the formation of 6 unknown metabolites in non-sterile soil and 5 in sterile soil. Figure 47 illustrates the proposed pathways for propanil metabolism in soils. A major unknown ($m/e = 333$) appears to be related to TCAB by the addition of an –NH group to the azo compound.

A further reaction product, 4-(3,4-dichloro-anilino)-3,3′,4′-trichloroazobenzene ($m/e = 443$) has been described by Rosen et al.[189] and may arise by further reaction of TCAB with DCA as indicated by reaction G.

The biochemical transformation of a number of anilide herbicides (propanil, dicryl, karsil and Ramrod) in soil was described by Bartha[183]. TLC and GLC were utilized for the determination of purity of the herbicides and their metabolites and the isolation of reaction products. Analyses were performed with an F & M Model 700

Fig. 46. Distribution of radioactive spots on thin-layer chromatograms from ring-^{14}C propanil in soil E (non-sterile) and soil C (sterile) developed first with benzene and second with hexane–benzene–acetone (7:3:1 v/v).

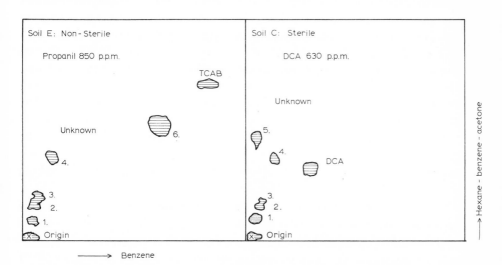

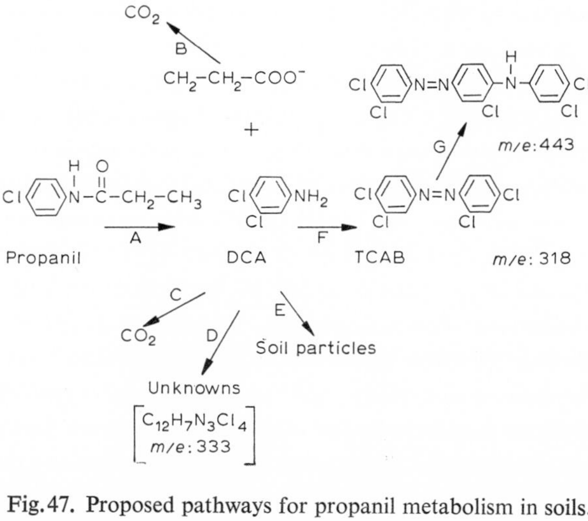

Fig. 47. Proposed pathways for propanil metabolism in soils.

References pp. 267–272

gas chromatograph having dual flame-ionization detectors and stainless steel columns 1.8 m × 3 mm o.d. packed with 5% UC-W98 on Chromosorb W. The carrier gas was helium at 30 ml/min. The temperatures were injection port 270°C, detector 300°C, oven 150°C, isothermal for aniline, 200°C for DCA and Ramrod (N-isopropyl-2-chloroacetanilide) and 250°C for all other compounds. Thin-layer chromatographic separations were performed on Eastman Chromagram sheets (silica gel with fluorescent indicator) developed with benzene (spots were located under UV light). Table 38 illustrates the melting points, R_F values and retention times of some anilide herbicides and degradation products. The aliphatic moieties of propanil, dicryl and Karsil were oxidized in soil and in each case a substantial portion of the liberated 3,4-dichloroaniline was condensed to 3,3′,4,4′-tetrachloroazobenzene (TCAB). Alkyl substitution of the acetanilide nitrogen in N-isopropyl-2-chloroacetanilide (Ramrod) increased persistence in soil and prevented the transformation of this herbicide to aniline or azobenzene residues.

TABLE 38

MELTING POINTS, R_F VALUES, AND RETENTION TIMES OF SOME ANILIDE HERBICIDES
AND DEGRADATION PRODUCTS

Compound	M.p. (°C)	R_F	Retention time (sec)	
			200°C	250°C
Propanil	92	0.10		33
Dicryl	124	0.25		38
Karsil	105	0.30		56
Ramrod	75	0.15	70	
DCA	71	0.65	38	
TCAB	158	0.95		140

The metabolism of ring-labeled ^{14}C-propanil in rice plants (*Oryza sation* L. var Bluebonnet 50) was studied by Yih *et al.*[190]. 3,4-Dichloroaniline was initially formed which in turn conjugated with carbohydrates. The soluble aniline–carbohydrate complexes account for only a small fraction of the hydrolyzed propanil. The major portion of the 3,4-dichloroaniline moiety was found complexed with polymeric cell constituents, mainly lignen. Figure 48 shows a diagram of radioautograph of 4 metabolites isolated from the treatment of rice plants with labeled propanil. Metabolite M-1 was identified as N-(3,4-dichlorophenyl-1-glucosylamine). Metabolite M-2 was identified as a 3,4-dichloroaniline–saccharide conjugate which contained glucose, xylose and fructose but was not further identified. Metabolite M-3 was unstable and readily decomposed to metabolite M-1 and metabolite M-4 was a 3,4-dichloroaniline–sugar derivative. Treatment of rice plants with ^{14}C-labeled 3,4-dichloroaniline produced the same 4 metabolites.

Still[181] had previously found three metabolites of 3,4-dichloroaniline when rice plants were treated with propanil. In both rice studies[181, 190], no 3,3',4,4'-tetrachloro-azobenzene was recovered.

Belasco and Pease[191] described the investigation of diuron, linuron and propanil treated soils. No detectable 3,3',4,4'-tetrachloroazobenzene (TCAB) was found in the

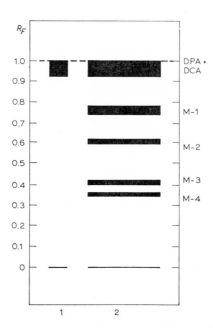

Fig.48. Diagram of a radioautograph of silica gel chromatogram developed with *n*-butanol–pyridine–water (6:4:3). 1, Standard mixture of ^{14}C-propanol and ^{14}C-3,4-dichloroaniline; 2, methanol extract from ^{14}C-ring labeled propanil-treated rice leaves (3-day treatment).

soil following treatment with diuron and linuron under practical *field* conditions. The incubation of 500 p.p.m. of diuron and linuron in soil (equivalent to a field treatment of 500 lb./acre to a depth of 3 in.) yielded minimal levels of 3,4-dichloroaniline (approximately 1 p.p.m. or less) and no detectable amounts of TCAB under the same laboratory conditions. Additional studies with 3,4-dichloraniline in soil up to concentrations of 500 p.p.m. showed the soil content of TCAB to be considerably less than anticipated, suggesting that 3,4-dichloraniline is *not* the prime precursor of TCAB formed in the soil.

GLC was used to study the soil extracts. A Micro-Tek Model MT-220 gas chromatograph was equipped with a Dohrmann microcoulometer, T-300S titration cell and a S-100 sample inlet/combustion unit. The glass column was a 4 ft. × ¼ in. o.d. and contained 5% XE-60 plus 0.2% Epon Resin 1001 on 60–80 mesh Gas Chrom Q. The gas chromatograph was equilibrated as follows: vaporizer 230°C, transfer 250°C, furnace 850°C, column 200°C, carrier flow helium at 100 ml/min, purge flow helium

50 ml/min and oxygen flow 50 ml/min. The column was conditioned by maintaining
its temperature at 225°C for at least 48 h. 3,4-Dichloroaniline in soil extracts was
analyzed at a temperature range of 90–190°C (the temperature programmed at
7.5°C/min). Figures 49 and 50 illustrate chromatograms of TCAB standard solution

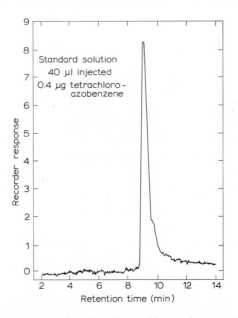

F g. 49. Gas chromatogram of a 3,3′,4,4′-tetrachloroazobenzene standard solution.

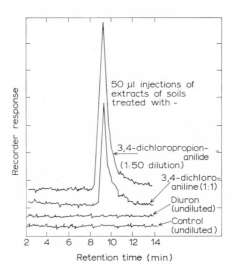

Fig. 50. Gas chromatograms of extracts of soils incubated with 3,4-dichloropropionanilide, 3,4-
dichloroaniline, and diuron.

and extracts of soil incubated with propanil, 3,4-dichloroaniline and diuron. TCAB was analyzed at an initial column temperature of 140°C, thence programmed at 5°C/min to an upper limit of 190°C. The retention time for TCAB from the start of programming was about 9 min.

Plimmer et al.[192] recently described the isolation of a major metabolite 1,3-bis(3,4-dichlorophenyl) triazene from ^{14}C-ring-labeled propanil in a Chikugo light-clay soil. The triazene was believed to arise by formation of an intermediate diazonium cation which subsequently couples with 3,4-dichloroaniline. It was proposed that soil nitrite initially reacts with 3,4-dichloroaniline to form the intermediate diazonium cation. The isolation of the triazene from the soil extract was accomplished using TLC on Silica Gel GF (Merck) with benzene as the first developing solvent and solvent A, hexane–benzene–acetone (7:3:1). Chromatograms were exposed to no-screen X-ray film for 10 days. A number of yellow spots visible on the silica plate corresponded to spots visible on the developed X-ray film. The most intensely radioactive spot (R_F in benzene 0.68, in solvent A 0.30) was subsequently scraped from the plate, extracted with ethanol and further analyzed by mass, NMR, UV, and IR spectroscopy. Figure 51

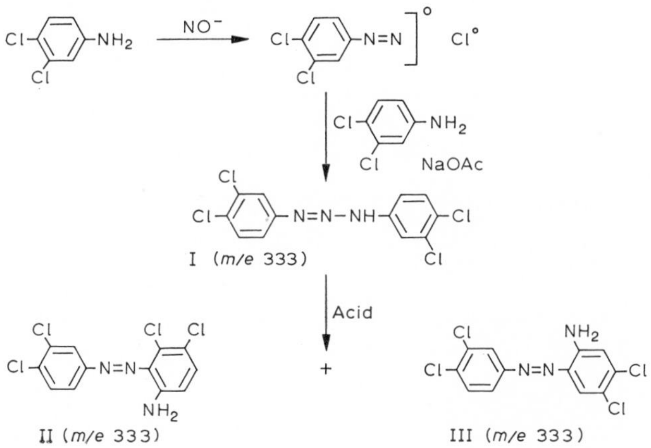

Fig. 51. Formation and rearrangement of 1,3-bis(3,4-dichlorophenyl)triazene.

illustrates the formation and rearrangement of 1,3bis-(3,4-dichlorophenyl) triazene from 3,4-dichloroaniline and Fig. 52 shows the fragmentation pattern of the triazene.

Bartha et al.[193] described the transformation in soil and by peroxidase of 11 chloroanilines. Reaction extracts were concentrated and subjected to gas chromatographic analysis with a flame-ionization detector; the stainless steel column was 1.8 m × 3 mm o.d. and packed with 5% UC-W98 on Chromosorb W. Figure 53 illustrates the similarity in specificity for aniline substrates exhibited by the soil and peroxidase. No azo compounds were formed from aniline, but all monochloro and some dichloroanilines were transformed to their corresponding dichloro- and tetra-

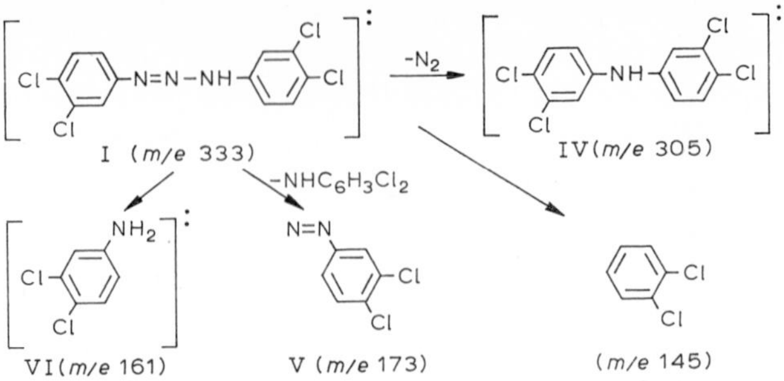

Fig. 52. Fragmentation of 1,3-bis(3,4-dichlorophenyl)triazene.

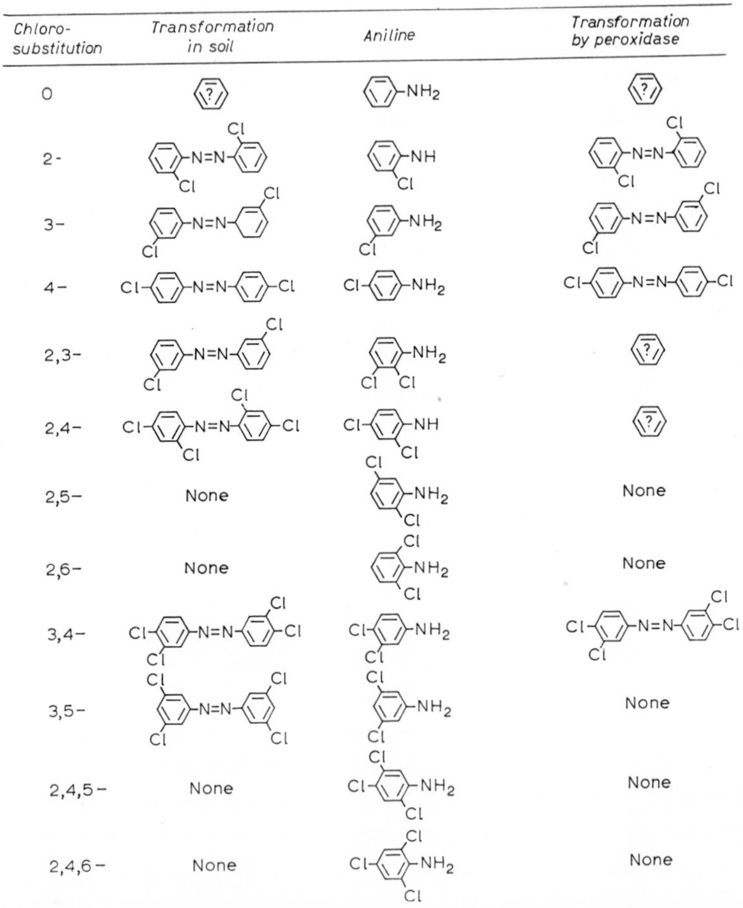

Chloro-substitution	Transformation in soil	Aniline	Transformation by peroxidase
0	(?)	NH₂	(?)
2-	N=N	NH	N=N
3-	N=N	NH₂	N=N
4-	Cl—N=N—Cl	Cl—NH₂	Cl—N=N—Cl
2,3-	N=N	NH₂	(?)
2,4-	Cl—N=N—Cl	Cl—NH	(?)
2,5-	None	NH₂	None
2,6-	None	NH₂	None
3,4-	Cl—N=N—Cl	Cl—NH₂	Cl—N=N—Cl
3,5-	N=N	NH₂	None
2,4,5-	None	Cl—NH₂	None
2,4,6-	None	Cl—NH₂	None

Fig. 53. Formation of azo compounds from anilines in soil and by peroxidase. The question mark indicates unidentified aromatic products.

chloroazobenzenes while other dichloroanilines and trichloroanilines were stable in soil. The correspondence in the range of substrates used and the products formed in both soil and peroxidase systems suggested a peroxidasic mechanism for the synthesis of azo compounds from anilines in soil. It is important to note that some azo compounds are carcinogenic[194].

Kearney *et al.*[195] described th formation of mixed chloroazobenzenes in soil. Condensation of 3-chloroaniline and 3,4-dichloroaniline to form 3,3,4'-trichloroazobenzene in addition to 3,3'-dichloroazobenzene (DCAB) and 3,3',4,4'-tetrachloroazobenzene (TCAB) has been found in Nixon sandy loam soil. The mixed chloroazobenzene was also found during the synthesis of TCAB *via* reduction of 1,2-dichloro-3-nitrobenzene. The purity of the reaction products from synthesis as well as an examination of soil extracts was determined by GLC and GLC–mass spectroscopy. The chloroazobenzenes were examined by GLC using a Research Specialties Model 600 with a flame detector and a 6 ft. × $\frac{1}{4}$ in. column packed with 5% SE-30 on 60–80 mesh Chromosorb W isothermally at 238 °C. Mass spectral analysis was performed with a Perkin-Elmer Model 270 combination gas chromatograph–mass spectrometer. The GLC column used was a 50 ft. surface coated open tubular column of 0.02 in. internal diameter coated with SE-30 on Chromosorb W. Retention order was the same as that from the packed column. The detection of 3,3',4-trichloroazobenzene suggests that mixed azobenzenes may be formed in soils receiving one or more aniline herbicide treatments. Further condensation products of aniline in soil probably occur as TLC of soil extracts revealed at least 10 colored compounds not present in the control. TLC was performed on silica gel plates developed with benzene–hexane–ethyl acetate (3:7:1). Figure 54 illustrates the proposed pathways for the formation of TCAB, DCAB, and 3,3',4-trichloroazobenzene in soil.

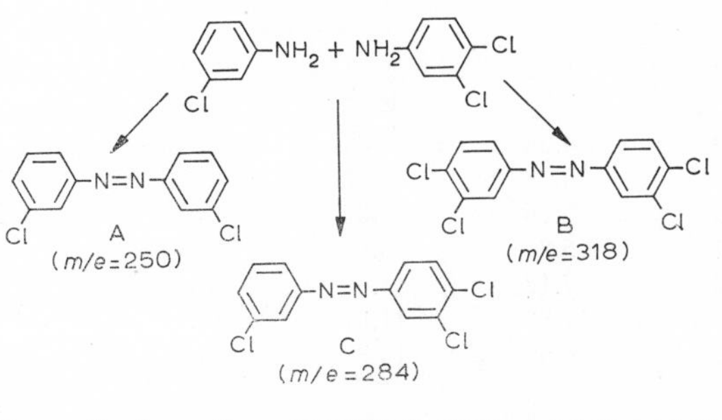

Fig. 54. Proposed pathways for formation of TCAB, DCAB, and 3,3',4-trichloroazobenzene in soil.

Mono- and dichloroanilines are important intermediates used in the synthesis of a variety of organic chemicals. They are generally prepared by reduction of the appropriate chloronitrobenzene and hence any chloroaniline isomer so prepared may be expected to contain other isomers unless the material has received extensive purification. Gas chromatography had been used to separate some isomers in mixtures of chloro- and dichloroaniline. James[196] reported retention data for 2,3-, 2,4-, and 3,4-dichloroaniline on one substrate. The separation of mono- and dichloroanilines by GLC has been described by Bombaugh[197]. Using temperature-programmed columns containing dodecylbenzene sodium sulfonate and silicone oil, complete resolution of all mono- and dichloro isomers (except 2,3 and 2,4) was obtained in 38 min (Fig. 55). The 2,3- and 2,4-dichloro isomers were separated using either silicone oil or polyethyleneglycol 20M. Table 39 lists the relative retention data by several substrates.

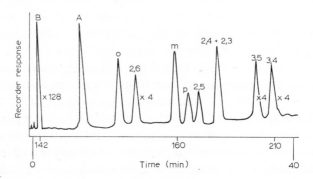

Fig. 55. Separation of mono- and dichloroaniline mixture using programmed-temperature operation. Column: 1.3 m Siponate DS10 at 15 wt. % plus 0.7 m of silicone oil 550 at 20 wt. % on acid-washed white Chromosorb precoated with 2% NaOH. Sample size, 10 µl; temperature, programmed from 142 to 210°C at 4°C/min.

TABLE 39

RELATIVE RETENTION DATA BY SEVERAL SUBSTRATES

Chloroaniline isomers	B.p. (°C)	PEG 20M	Silicone	Tide	Siponate	SipSo
o-	208	1.94	2.41	2.19	1.79	1.97
2,6-		2.21	4.23	3.52	1.93	2.32
m-	229	4.28	3.84	4.96	4.20	4.34
p-	231	4.18	3.62	5.78	6.10	5.66
2,5-	245	6.16	7.04	8.22	6.39	6.20
2,4-	251	5.89	6.82	10.07	9.94	8.37
2,3-	252	7.12	8.48	10.83	8.73	8.83
3,5-	260	13.85	11.61	18.74	24.85	16.74
3,4-	272	16.49	13.97	24.88	28.15	22.20

Aniline = 1

9. 3-AMINO-1H-1,2,4-TRIAZOLE

The non-selective, systemic herbicide 3-amino-1H-1,2,4-triazole (aminotriazole, amitrole) is produced by the condensation of formic acid and aminoguanidine. It is an effective phytocide and defoliant used alone or in combination with ammonium thiocyanate or soil sterilant type herbicides for general highway and industrial weed control and non-cropland. Aminotriazine is also used for certain crop uses (quack-grass control prior to planting corn and weed control in dormant grape vineyards). No tolerance level greater than zero has been permitted for residues of aminotriazole.

Aminotriazole has been shown to have antithyroid properties producing thyroid adenomas and liver hepatomas on chronic feeding in rats[198]. The chemistry of s-triazoles[199] and the degradation and mode of action[200-202] of aminotriazole have been recently reviewed. Fang et al.[203-204] studied the oral metabolism of 3-amino-1,2,4-triazole-5-^{14}C in rats. Analysis for radioactivity showed only traces of ^{14}C in expired air. Seventy to ninety-five percent of the activity appeared in the urine within 24 h as unchanged aminotriazole and two unidentified metabolites[203]. After absorption, aminotriazole was distributed throughout most of the body tissues.

The average half-time for aminotriazole clearance in various organs was 4.2 h. To define the chemical nature of the radioactivity in the urine and alcohol extracts of feces and tissues, small aliquots of these samples were chromatographed one-dimensionally on Whatman No. 1 paper using isopropanol–ammonium hydroxide–water (10:1:1) as developing solvent. Both direct scanning and radioautographic techniques were used following paper chromatographic separation. Analysis of rat liver extracts revealed two major radioactive spots. One spot (R_F 0.40) was unchanged aminotriazole while the other was metabolite-1 (R_F 0.00). Fang et al.[204] further studied the effect of a mixture of tritiated aminotriazole and ^{14}C-aminotriazole in the rat. The

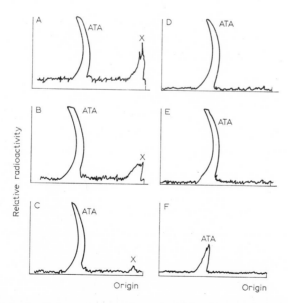

Fig. 56. Scanning records of chromatograms from the 24-h urine sample of rats fed with varying doses of ATA-5-^{14}C; X = rat ATA metabolite. A, 1 mg; B, 5 mg; C, 10 mg; D, 50 mg; E, 100 mg; F, 200 mg.

change of $^3H/^{14}C$ ratio in rat-aminotriazole metabolite indicated that 5 hydrogen atoms in the ring of aminotriazole had been substituted. Similar studies were carried out in rats using ^{14}C-labeled or ^3H- and ^{14}C-labeled aminotriazole metabolites isolated from bean plants and it was found that the pattern of elimination for metabolites 1 and 3 differed greatly in adult rats.

In this latter study[204] urine samples were analyzed for aminotriazole and its metabolites by paper chromatography using isopropanol–ammonium hydroxide–water (8:1:1). Figure 56 illustrates scanning records of chromatograms from a 24-h urine sample of rats fed varying doses of ^{14}C-aminotriazole. The rat appears to have a limited ability to metabolize the aminotriazole molecule. It is of note that no trace of an acetylated product, *e.g.* 3-acetamino-1,2,4-triazole, was found. Table 40 lists the R_F values of aminotriazole (ATA) and metabolites 1 and 3 from bean plants and the labeled compounds isolated from the urine of rats fed aminotriazole metabolites.

TABLE 40

R_F VALUES OF ATA, METABOLITE-1, AND METABOLITE-3 FROM BEAN PLANTS, AND LABELED COMPOUNDS FROM THE URINE OF RATS FED ATA BEAN METABOLITES

Compounds	R_F in solvent systems	
	Butanol–acetic acid–H_2O (4:1:1.5)	2-Propanol–NH_4OH–H_2O (3:1:1)
ATA	0.48	0.45
Bean metabolite-1	0.16	0.07
Bean metabolite-3	0.23	0.10
Urine of rat fed metabolite-1		
Spot 1	0.13	0.06
Spot 2	0.27	0.11
Spot 3	0.21	0.15
Urine of rat fed metabolite-3		
Spot 1	0.23	0.10
Spot 2	0.48	0.45

The metabolism of aminotriazole in a variety of plants has been reported. Massini[205] described the formation of β-(3-amino-1,2,4-triazolyl-1)-α-alanine (ATX) as the major metabolic product of aminotriazole in bean and tomato plants. In order to facilitate the comparisons of ATX with metabolites of aminotriazole found by other investigators, aminotriazole and ATX were chromatographed on strips of Whatman No. 1 paper by the ascending method at 25 °C in various solvents (Table 41). ATX was detected by the H-acid reagent of Racusen[206] which forms a red dye with aminotriazole and its derivatives, and by ninhydrin (0.2% in 96% ethanol). Table 42

TABLE 41

R_{AT} VALUES OF ATX AND OF SEVERAL UNKNOWN METABOLITES OF AMINOTRIAZOLE
DESCRIBED EARLIER[a]

Solvent no.	Solvent system	ATX	Unknown	Ref.
1	Methanol–formic acid–water 80:15:5	0.39	0.30	Racusen "X"[206]
2	n-Butanol–acetic acid–water 27:7:17	0.55	0.29	
3	Isopropanol–NH₄OH–water 6:2:2	0.75	0.47	
4	Pyridine–water 8:2	0.20	0.33	
5	Phenol–water 8:2	0.76	0.73	
6	n-Butanol–propionic acid–water 2:1:1.4	0.42	0.35	Carter and Naylor "I"[207]
7	Phenol–water 72:28	0.76	0.68	
3	Isopropanol–NH₄OH–water 6:2:2	0.75	0.75	Herrett and Linck "II"[208]
8	n-Butanol–ethanol–water 52.5:32:15.5	0.09	0.12	
9	n-Butanol–ethanol–water 1:4:1	0.14	0.26	Miller and Hall "Y"[209]
10	n-Propanol–ethyl acetate–water 6:1:3	0.31	0.68	
11	n-Butanol–acetic acid–water 4:1:5 (top layer)	0.33	0.42	
1	Methanol–formic acid–water 80:15:5	0.39	0.84	
12	Isopropanol–NH₄OH–water 80:5:15	0.21	0.34	
11	n-Butanol–acetic acid–water 4:1:5 (top layer)	0.33	0.24	Miller and Hall "X"[209]
1	Methanol–formic acid–water 80:5:15	0.39	0.51	
12	Isopropanol–NH₄OH–water 80:5:15	0.21	0.23	
9	n-Butanol–ethanol–water 1:4:1	0.14	0.12	
10	n-Propanol–ethylacetate–water 6:1:13	0.31	0.42	
13	tert.-Butanol–formic acid–water 70:15:15	0.27		Massini[210]
14	α-Picoline–water 6:4	0.57		

[a] $R_{AT} = \dfrac{\text{distance travelled by the metabolite}}{\text{distance travelled by AT (aminotriazole)}}$

depicts the color reactions of aminotriazole (AT), ATX and of several metabolites of aminotriazole reported earlier.

The metabolism of 3-amino-1,2,4-triazole-5-^{14}C in the fruit of bean *Phaseolus vulgaris* has been described by Shimabukuro and Linck[211]. Plant extracts were passed

TABLE 42

COLOR REACTIONS OF AT, ATX AND OF SEVERAL METABOLITES OF AT DESCRIBED EARLIER

Test	AT	ATX	Racusen X	Carter and Naylor I	Herrett and Linck II
Ninhydrin		Purple	Purple	Blue-green	Blue-green
H-acid	Red	Red	Orange-red	Red	Orange-pink
Nitroprusside	Blue			Green	
p-Dimethylamino-benzaldehyde	Yellow	Yellow	Yellow	Yellow	

through columns of Amberlite 1R-120 cation exchange resin (H$^+$ form) washed with water, eluted with 1.0 N NH$_4$OH, the aqueous and basic fractions concentrated under vacuo and chromatographed on Whatman No. 1 paper for sugars, amino acids, and aminotriazoles and its derivatives. The sugar chromatograms were developed in n-butanol–acetic acid–water (52.5:32:15.5) for 48 h. Amino acids were separated by two-dimensional chromatography using phenol saturated with water in one direction and collidine–lutidine in the second direction. The chromatograms for aminotriazole and its derivatives were developed in isopropanol–ammonia–water (6:2:2). Table 43

TABLE 43

SEPARATION AND DETECTION OF COMPOUNDS IN THE AQUEOUS AND BASIC FRACTIONS

Fraction	Solvent	No. of positive spots				Radio-active spots
		p-Ani-sidine	Nin-hydrin	AgNO$_3$	"H-acid"	
NH$_4$OH	Isopropanol–ammonia–water	0	7	0	0	1
NH$_4$OH	Phenol and collidine–lutidine	0	8	0	0	0
NH$_4$OH	n-Butanol–acetic acid–water	0	a		0	1
HOH	Isopropanol–ammonia–water	1	0	2	0	4
HOH	n-Butanol–acetic acid–water	3				4
HOH	Phenol and collidine–lutidine	1				1

a The detection agent was not used.

depicts the paper chromatographic separation and detection of compounds in the aqueous and basic fractions. The young fruits, buds, and flowers of the bean rapidly metabolized the labeled ^{14}C-aminotriazole which had been absorbed by a leaf and translocated to the reproductive structures and between 80–95% of the aminotriazole was converted into sucrose, glucose, fructose, and other compounds.

Herrett and Bagley[212] described the metabolism and translocation of amino-triazole by Canada thistle and postulated that aminotriazole must undergo meta-bolism to an active form prior to translocation within the phloem. Three chromatographically distinct metabolites were found. Two compounds (unknown I and II) were similar to metabolites previously described and were relatively inactive. A third metabolite (unknown III) was herbicidally more active than aminotriazole. Table 44 shows the R_F values of unknown I from thistle compound with the R_F values of a similar compound obtained from other plants.

TABLE 44

R_F VALUES OF UNKNOWN I FROM THISTLE AS COMPARED WITH THE R_F VALUES OF A SIMILAR COMPOUND OBTAINED FROM OTHER PLANTS IN VARIOUS SOLVENT SYSTEMS

Plant	Metabolite Y	R_F value	Ref.
Isopropanol–NH_4OH–H_2O (80:5:15)			
Cotton	Metabolite Y	0.18	209
Thistle	Unknown I	0.14	
Isopropanol–NH_4OH–H_2O (6:2:2)			
Bean	Compound Y	0.63	206
Thistle	Unknown I	0.63	
Ethanol–n-Butanol–H_2O (4:1:1)			
Cotton	Metabolite Y	0.20	209
Thistle	Unknown I	0.24	
Methanol–HCOOH–H_2O (80:15:5)			
Cotton	Metabolite Y	0.61	209
Thistle	Unknown I	0.66	

Figure 57 illustrates a scheme relating aminotriazole metabolism to its systemic herbicidal properties. Once aminotriazole has penetrated the leaf, the compound may be either detoxified to unknowns I and II or activated to unknown III. In the activation portion, aminotriazole is metabolized to an active transport form. Translocation of aminotriazole requires both time and light whereas unknown (III) occurs very rapidly and in the absence of light.

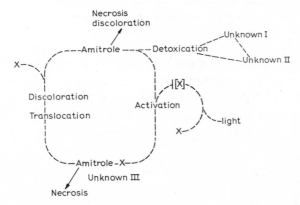

Fig. 57. Proposed reaction sequences of amitrole metabolism in plants[211].

Carter and Naylor[213, 214] described studies on an unknown metabolic product of aminotriazole in bean plants (*Phaseolus vulgaris* var. Black Valentine). Figure 58 illustrates a chromatographic map of derivatives of aminotriazole-^{14}C which occur in plants. Table 45 lists the color reactions of unknown I and aminotriazole.

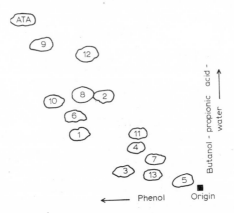

Fig. 58. Chromatographic map of derivatives of ATA-5-^{14}C which occur in plants. Numbers have been arbitrarily assigned to the unidentified compounds.

TABLE 45

COLOR REACTIONS OF COMPOUND "I" AND AMINOTRIAZOLE

Color reagent	Compound "I"	Aminotriazole
Ninhydrin	Blue-green	No visible reaction
Ehrlich's	Yellow	Faint yellow
p-Anisidine	No visible reaction	No visible reaction
Phenol–HCl	Yellow	Yellow
"H-acid"	Yellow or red[a]	Yellow or red
Nitroprusside–ferrocyanide	Green	Green

[a] The "H-acid" reagent produced a yellow color with "I" on chromatograms but a red color on paper before chromatography.

The principal metabolites of aminotriazole exhibit azo dye reactions, ninhydrin sensitivity, and zwitterion behavior. It has been suggested[200] that in all probability "X"[206], ATX[205], "I"[214], and similar compounds reported by other workers[208, 209] are the same compound which Massini[205] identified as 3-(3-amino-1,2,4-triazole-1-yl)-2-aminopropionic acid (3-ATAL) formed as follows.

The formation of 3-ATAL apparently represents detoxification. The majority of the studies of aminotriazole degradation in plants indicate that the extractable ^{14}C from plants treated with aminotriazole-5-^{14}C remain in the intact ring as free aminotriazole or conjugates. Considerable amounts of aminotriazole are attached to protein[215] or bound in an insoluble form[206].

The degradation of aminotriazole in soils has been studied by Ercegovich and Frear[216] and Kaufman et al.[217, 218] and it was shown that aminotriazole degradation obeys first-order kinetics, suggesting a chemical reaction. Ashton[219] found aminotriazole-5-^{14}C to be rapidly degraded in unsterile soil. The principal metabolite was CO_2 but at least 13 additional compounds were detected by single and two-dimensional chromatography and radioautography. Two-dimensional chromatography was performed on Whatman No. 3 paper using phenol–water (5:1) for the first development and n-butanol–propionic acid–water (45:22:33) for the 90° second development.

Plimmer et al.[220] examined the degradation of aminotriazole by free-radical-generating systems including Fenton's reagent, UV light, and UV light plus riboflavin. All three systems produced essentially the same result, aminotriazole was principally converted to carbon dioxides, urea and cyanamide by ring cleavage and to polymers proposed to be formed through an amine-free radical.

However, the reagent of Castelfranco et al.[221] consisting of ascorbic acid, cupric sulfate, and molecular oxygen, liberated no CO_2 although aminotriazole was degraded[220]. The ascorbate–copper reagent was suggested to reduce aminotriazole to a free radical which then polymerizes. The activation and degradation of aminotriazole was suggested to proceed as shown in Fig. 59.

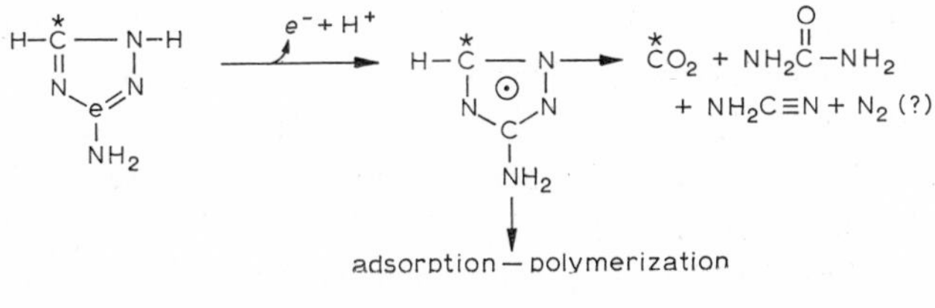

Fig. 59. Activation and degradation of aminotriazole[220, 221].

Castelfranco and Brown[222] suggested that aminotriazole, in the presence of a free-radical-generating system accepts electrons and becomes a free radical which is capable of alkylating proteins and possibly amino acids indiscriminately.

Hilton[202] recently reviewed the inhibition of growth and metabolism of aminotriazole and reported that the toxicity of aminotriazole could be circumvented by riboflavin acting through a photochemical mechanism in the light or through a biological interaction in the dark. The chemical interaction of riboflavin and aminotriazole is visualized in the TLC radiochromatograms shown in Fig. 60. Solutions of light and dark exposed riboflavin-2-^{14}C, with and without aminotriazole-^{12}C, and of aminotriazole-5 ^{14}C, with and without riboflavin-^{12}C, were compared. Riboflavin

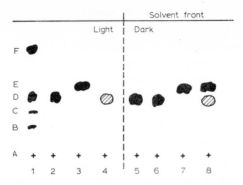

Fig. 60. Thin-layer radiochromatograms showing riboflavin–amitrole interactions after 3-h exposure to light (250 ft.-candles, columns 1–4) or darkness (columns 5–8) in growth medium (Ames *et al.*[223]) supplemented with 2×10^{-4} M histidine. Solutions 1 and 5, riboflavin-2-^{14}C 3×10^{-6} M; 2 and 6, riboflavin-2-^{14}C plus amitrole 2×10^{-2} M; 3 and 7, amitrole-5-^{14}C plus 1.3×10^{-5} M; 4 and 8, amitrole-5-^{14}C plus riboflavin 2×10^{-4} M. Vertical designations: A, Origin; B, C, F, unidentified photodegradation products of riboflavin; D, riboflavin (hatched spots show non-radioactive riboflavin); E, amitrole. Solvent system: 1-butanol–propionic acid–water (15:7:10).

photodecomposed to three products; aminotriazole protected riboflavin from photo-decomposition (Fig. 60, second column) but was itself degraded in the process (Fig. 60, column 4). In darkness, no decomposition of either riboflavin or aminotriazole could be detected.

The quantitative analysis of aminotriazole in biological systems was described by Herrett and Linck[224]. The method consists of the use of H-acid reagent (8-amino-1-naphthol-3,6-disulfonic acid) in a diazotization and coupling procedure for the determination of aminotriazole in extracts from biological systems freed of contaminating metabolite of aminotriazole by passage over Dowex 50W-X8 or Amberlite IR-120 and elution with 1.0 N ammonium hydroxide, thence re-chromatography of the concentrated eluate over a weakly acidic cation exchange resin IRC-50 (H$^+$ form) and elution with 0.6 N HCl. The cationic-exchange resins for the absorption of aminotriazole and its subsequent desorption have been used in clean-up procedures in residue analysis[225, 226], analysis of commercial aminotriazole[227], and for the aqueous formulations of aminotriazole with ammonium thiocyanate[227] and to the amino-triazole salt of 2,2-dichloropropionic acid[227]. Paper chromatographic analysis of aminotriazole has been described. Mitchell[228] used aqueous dioxan as developer and ammoniacal silver nitrate with heat for the detection of 0.5γ amounts of amino-triazole. Detection of aminotriazole on paper chromatograms has also been accomplished using diazotization followed by coupling with H-acid[206].

Coulson[229] described the selective detection of nitrogen compounds (including aminotriazole) in electrolytic conductivity gas chromatography. A Coulson Gas Chromatograph, Model 1 was used equipped with a linear temperature programmer coupled with a solid-state proportional temperature controller for the column oven. The effluent stream from the gas chromatograph column is delivered into a high-temperature micro-tube furnace fitted with a quartz tube in which the components eluted from the column are hydrogenated. The products of hydrogenation are contacted with an acid absorption packing which removes any H_2S and hydrogen halides. The resulting gas stream passed through an electrolytic conductivity detector cell. The gas chromatographic column was a coiled Pyrex tube 4 ft. × 2 mm i.d. packed with 80–100 mesh Gas Chrom Q coated with 10% DC 200 oil (12,500 centistokes). Helium was the carrier gas at 35 ml/min. The quartz reduction tube was 8 mm o.d., 6 mm i.d. and 11 in. in length and was packed with a nickel catalyst and a strontium hydroxide acid absorber as shown in Fig. 61. Figure 62 illustrates the separation of aminotriazole, prometryne, and nitrobenzene.

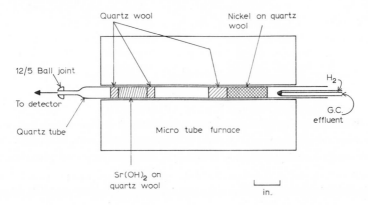

Fig. 61. Reduction apparatus for the GLC detection of nitrogen in organic compounds.

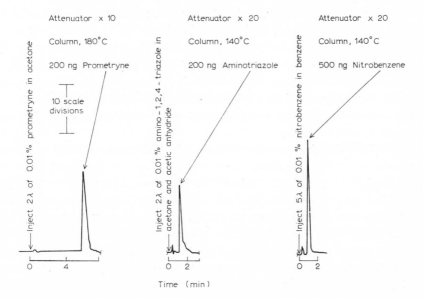

Fig. 62. Nitrogen-specific detection of prometryne, aminotriazole, and nitrobenzene.

References pp. 267–272

Aminotriazole has also been separated on Silica Gel G chromatoplates[230] using (a) benzene–acetone (1:1), (b) methanol, and (c) acetone as developing solvents. The R_F values of aminotriazole with the above developers were 0.05, 0.60, and 0.14, respectively. Detection was accomplished with an aqueous methanolic (1:2) solution of the sodium salt of 3,5-dihydroxy pyren-8,10-disulfonic acid or the sodium salt of 3-hydropyren-5,8,10-trisulfonic acid.

10. CHLOROBENZILATE

Chlorobenzilate (ethyl 4,4'-dichlorobenzilate) is prepared by the reaction of dichlorobenzilic acid with ethyl sulfate. The technical product has approximately 90% purity and is generally used in emulsions or wettable powders. It is a non-systemic acaricide with little insecticidal action and is primarily used against phytophagous mites.

The toxicity[231, 232] and carcinogenicity[233] of chlorobenzilate have been described. Chlorobenzilate is readily hydrolyzed by alkali and strong acid to 4,4'-dichlorobenzilic acid and ethanol. The ready hydrolysis of chlorobenzilate is utilized for clean-up purposes in analytical methods by selective extraction procedures[234, 235].

The microcoulometric gas chromatographic analysis of grapes and cottonseed for chlorobenzilate residues with a sensitivity of 0.05 p.p.m. has been described by Beckman and Bevenue[236]. Two gas chromatographs were used: (1) an F & M instrument equipped with a thermal conductivity detector and a 2-ft. stainless steel column containing 2.5% SE-30 on 60–70 mesh Chromosorb W. The temperature was programmed at 15°/min with a starting temperature of 100°C. Helium was the carrier gas at 60 ml/min; chlorobenzilate was eluted at 210°C. (2) A Dohrmann gas chromatograph was used equipped with a microcoulometric detector and a 6-ft. stainless steel column containing 20% Dow-11 silicone grease on acid-washed Chromosorb P. The injection port and column temperature were 270 and 260°C, respectively, and the nitrogen and oxygen flow rates were 100 ml/min. Benzene was used as the extraction solvent for the removal of the acaricide and clean-up was by partition into nitromethane followed by Florisil for cottonseed extracts while clean-up with Nuchar Carbon was used for grapes.

The utility of a mixed column of QF-1 and OV-17 for the separation of a variety of pesticides including chlorobenzilate has been described by Windham[237]. A 6 ft. × 4 mm i.d. glass column coiled to 4-in. diameter was packed with four parts by weight of 10% QF-1 on 80–100 mesh Gas Chrom Q plus one part of 10% OV-17 on 80–100 mesh Gas Chrom Q. The operating parameters were injection port 225°C, column oven 200°C, tritium electron-capture detector 200°C, nitrogen carrier gas flow 120 ml/min, voltage to detector 50 V, and sensitivity setting 3×10^{-9} afs. The approximate relative retention time of chlorobenzilate compared to aldrin (2.5 min) was 3.18 min.

The microbial degradation of ^{14}C-chlorobenzilate and ^{14}C-chloropropylate (isopropyl-4,4'-dichlorobenzilate) has been studied by Miyazaki *et al.*[238, 239]. Metabolites resulting from the degradation of either acaricide by a yeast *Rhodotorula gracilis* were identified as 4,4'-dichlorobenzilic acid and 4,4'-dichlorobenzophenone. The probable steps of the degradation pathway are illustrated in Fig. 63.

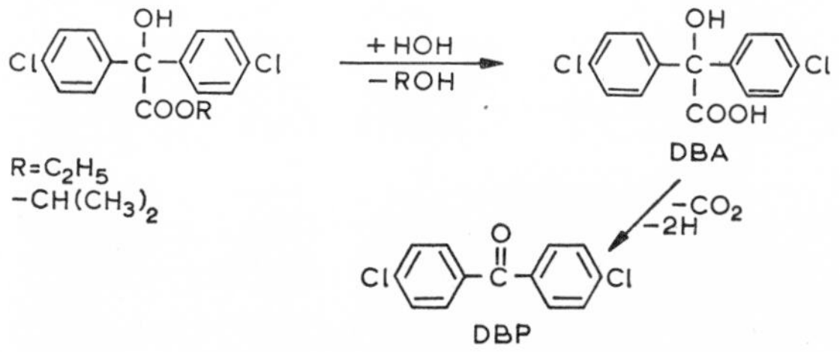

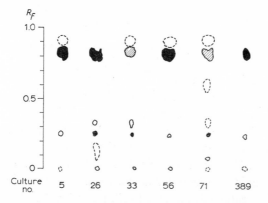

Fig. 63. Proposed pathway of chlorobenzilate and chloropropylate degradation by *Rhodotorula gracilis*. DBA = 4,4'-dichlorobenzilic acid; DBP = 4,4'-dichlorobenzophenone.

Thin-layer chromatography was used to separate the acaricides and their metabolites, *e.g.* (*1*) Eastman No. 6060 chromatographic sheets were developed with hexane–ethanol–acetic acid (17:2:1) and the resulting chromatograms autoradiogrammed on Kodak medical X-ray film of high contrast for 4 weeks; (*2*) preparative TLC using Silica Gel GF plates were used with cyclohexane–ether (3:1) and (14:1) for further purification of the acaricide metabolites. The resulting chromatograms were subjected to autoradiography with the previously described X-ray film for 3 days. Figure 64 shows the results of the metabolism of chloropropylate by promising microorganisms after an 8-week incubation at 30 °C. The R_F value of the parent compound was 0.80 and those of 4,4'-dichlorobenzilic acid and 4,4'-dichlorobenzophenone

Fig. 64. Autoradiographic presentation of thin-layer chromatograms of ^{14}C-chloropropylate and its metabolites produced by cultures of certain microorganisms. Spots showing strongest radioactivity are represented by black; medium radioactivity, shaded; weak radioactivity, open circles with solid line; and the weakest, open circles with dotted line.

were 0.27 and 0.90, respectively. Figure 65 shows the results of metabolism of ^{14}C-labeled chlorobenzilate and chloropropylate by *Rhodotorula gracilis* in basal medium fortified with 1 % sucrose. Comparisons of R_F values (as well as IR spectra and melting points) of authentic reference compounds established the identity of 4,4'-dichloro-

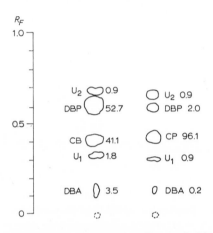

Fig. 65. Metabolism of chlorobenzilate, and chlorobenzilate and chloropropylate by *Rhodotorula gracilis* incubated in basal medium supplemented by 1 % sucrose as carbon source. CB, chlorobenzilate; CP, chloropropylate; DBA, 4,4'-dichlorobenzilic acid; DBP, 4,4'-dichlorobenzophenone; U$_1$ and U$_2$, unknown metabolites. Numerical figures at the right side of each compound represent percentages of radioactivity in that particular spot.

benzilic acid (DBA) and 4,4'-dichlorobenzophenone (DBP) as metabolites. The gas–liquid radiochromatography of residues of the related acaricide ^{14}C-isopropyl-4,4'-dibromobenzilate in field-weathered soil has been described by Cannizzaro et al.[240]. Chloroform extracts of soil samples without clean-up were injected directly into a Barber–Colman Series 5000 gas chromatograph equipped with a radioactive monitoring (ram) system. The response for standard ^{14}C-isopropyl-4,4'-dibromobenzilate was directly compared to fortified check and treated, field-weathered sample extracts. Only the parent compound peak was observed in the field-treated and fortified extracts. Greater than 95 % of the residual radioactivity was identified as the unchanged parent compound. The operating parameters for the analyses were: column 2 ft. $\times \frac{1}{4}$ in. o.d. Pyrex tubing packed with 3 % FFAP on 60–80 mesh Gas Chrom Q, injection port 225 °C, column 210 °C, transfer line 260 °C, combustion furnace 630 °C, carrier gas argon 60 ml/min, quench gas propane 6 ml/min, detector voltage 1500–1700 V, sensitivity 1–3 $\times 10^3$. Total radioactivity in sample extracts was determined using a Beckman LS-133 liquid scintillation system. Figure 66 shows gas–liquid radiochromatograms of field-weathered, untreated soil, treated soil and fortification recovery extracts.

The ram system (radioactive monitoring gas chromatograph) provided excellent means of organic ^{14}C-determinations with the above labeled acaricide, without inter-

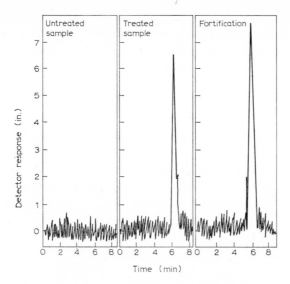

Fig. 66. Typical ramgrams of soil extracts. Left, untreated soil, equivalent to 1.05 g, 1.68×10^3 dpm ^{14}C isopropyl-4,4′-dibromobenzilate found. Right, fortification, equivalent to 1.15 g soil, 2.94×10^3 dpm ^{14}C isopropyl-4,4′-dibromobenzilate found (102 % recovery). Sensitivity: 1×10^3.

ference from extractable extraneous substances in the residue, hence allowing relatively simple clean-up procedures. This is in contrast to flame-ionization, electron-capture and the microcoulometric detection systems, for which more rigorous clean-up is required.

11. PENTACHLORONITROBENZENE

Pentachloronitrobenzene (PCNB, terrachlor, quintozene) is made by the chlorination of nitrobenzene at 60–70°C with iodine as catalyst. PCNB is a highly effective and widely used soil fungicide of specific use for seed and soil treatment, effective against bunt of wheat, botrytis, *Rhizoctonia* and *Sclerotinia* spp. and is generally formulated in dusts, wettable powders or in emulsions. The carcinogenicity[233] and teratogenicity[241] of PCNB in the mouse have been reported.

A variety of chromatographic procedures has been described for the analyses of PCNB in admixture with other fungicides, in residues in crops and soil and in mammalian metabolism. Kilgore and White[242] described gas chromatographic separations of a number of mixed chlorinated fungicides. The instruments employed were an Aerograph Hy-Fi, Model 600-B fitted with an electron-capture detector for isothermal studies and an F&M Model 810, dual column modular gas chromatograph equipped with differential flame for programmed temperature studies. A 1 mV Honeywell "Electronik" 16 recorder was used with the electron-capture instrument and a Sargent Model SR recorder was operated in conjunction with the hydrogen flame unit.

References pp. 267–272

The chromatographic columns used in the Hy-Fi instrument included a 5 ft. × ⅛ in. spiral borosilicate glass column packed with 5% Dow-11 on acid-washed 60–80 mesh Chromosorb W and a 5 ft. × ⅛ in. stainless steel column containing 5% QF-1 on acid-washed 60–80 mesh Chromosorb P. The dual columns in the F&M instrument were made of stainless steel (spiral 6 ft. × ⅛ in.) packed with 5% QF-1 on acid-washed 70–80 mesh dimethyl-dichlorosilane (DMCS) treated Chromosorb G. Figure 67 illustrates the isothermal separation of the chlorinated fungicides PCNB, TCNB (2,3,5,6-tetrachloronitrobenzene), Botran (2,6-dichloro-4-nitroaniline), Phygon (2,3-dichloro-1,4-naphthoquinone), HCB (1,2,3,4,5,6-hexachlorobenzene), and an isomer of trichloronitrobenzene. Figure 68 is a programmed separation of 9 fungicides (PCNB, TCNB, Botran, Phygon, HCB, Spergon (tetrachloro-*p*-benzoquinone), Dyrene (2,4-dichloro-6-(2-chloroaniline)-1,3,5-triazine) and Lanstan (1-chloro-2-nitropropane)) on 5% QF-1 programmed from 100°C to 200°C at 4°C/min. The chromatogram shown in Fig. 69 illustrates the effective separation of a mixture of 17 fungicides, including both isomers and derivatives.

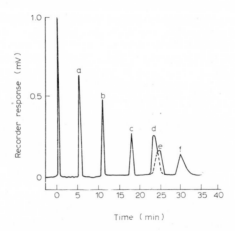

Fig. 67. Isothermal separation of chlorinated fungicides on QF-1 column. a, HCB; b, PCNB; c, TCNB; d, botran; e, an isomer of trichlorodinitrobenzene; f, phygon. Operating conditions: 5 ft. × 1/8 in. stainless steel column containing 5% QF-1 on acid washed Chromosorb P was used at 180°C; instrument: Aerograph Hy-Fi, Model 600 B; detector: electron-capture; carrier gas: nitrogen, 15 ml/min.

Gorbach and Wagner[243] utilized both TLC and GLC for the analysis of PCNB residues in potatoes. PCNB residues were found mostly in the peel and one of the two metabolites detected was shown to be pentachloroaniline. The active residues were extracted with a mixture of isopropanol and benzene and the determination was carried by GLC on extracts washed with sodium chloride solution. The equipment consisted of a self-assembled gas chromatograph whose special feature is the mounting of the injection block, the column and the combustion chamber on the inside of the door of a commercial thermostat from Heraeus (Hanau) Type FTU 340. The

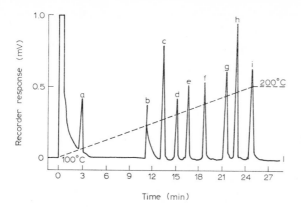

Fig. 68. Programmed separation of chlorinated fungicides on 5% QF-1 column. a, Lanstan; b, PCP; c, HCB; d, TCNB; e, spergon; f, PCNB; g, botran; h, phygon; i, dyrene. Operating conditions: Column, 6 ft. × 1/8 in. stainless steel column containing 5% QF-1 on acid-washed DMCS treated Chromosorb G; temperature: programmed from 100 to 200°C at 4°C/min; instrument: F&M Model 810; detector: flame-ionization; carrier gas: helium.

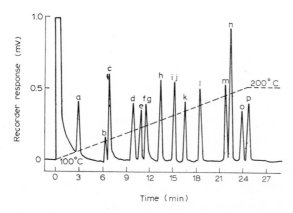

Fig. 69. Programmed separation of seventeen chlorinated fungicides, isomers, and derivatives on 5% QF-1 column. a, Lanstan; b, 2,4,6-trichlorphenyl, methyl ether; c, 2,4,5-trichlorophenol, 2,4,6-trichlorophenol; d, 2,4,5-trichlorophenyl, methyl ether; e, MeTCP; f, TCP; g, PCP; h, HCB; i, MePCP; j, TCNB; k, spergon; l, PCNB; m, Chemago 2635 (1,2,3-trichloro-3,5-dinitrobenzene); n, phygon; o, Chemago 2635 (1,2,3-trichloro-4,6-dinitrobenzene); p, dyrene. Operating conditions: Same as Fig. 68.

combustion chamber was thus included in the thermostatically controlled system. The system (injection, column, combustion tube, outlet) consisted entirely of quartz. The column was a 1 m × 4 mm i.d. quartz tube filled with silicone grease (according to Cassil)[244] on Chromosorb W. The detector consisted of a Dohrmann Coulometer C200 and a chlorine cell T300. The operating parameters were column 170°C, carrier gas nitrogen at 1.6 atm., oxygen 4.3 l/h, and temperature of combustion chamber 800°C. TLC on Silica Gel GF_{254} layers was also used to isolate the metabolites of PCNB from potato homogenates (Table 46) and confirm the separation of penta-chloroaniline.

Until recently, the presence of PCNB in crops and soil was detected and deter-mined using the spectrophotometric procedures of Ackermann *et. al.*[245, 246], Zweig[225], and Klein and Gajan[247] and polarography[247, 248].

TABLE 46

R_F VALUES OF PCNB AND METABOLITES I AND II ON THIN-LAYER CHROMATOPLATES

Solvent system	PCNB	R_F[a]		Pentachloro-aniline
		Metabolite I	Metabolite II	
Chloroform + benzene 1:1 (v/v)	0.73	0.67	0.77	0.67
CCl₄	0.52	0.36	0.69	0.36

[a] The R_F values are for Silica Gel GF₂₅₄ layers, 0.3 mm at 25 °C in a saturated chamber.

Methratta *et al.*[249] described the determination of PCNB in crops and soil by electron-capture gas chromatography. A Jarrell-Ash 26-700 gas chromatograph was equipped with an electron-capture detector and a 4 ft. × ¼ in. o.d., U-shaped stain-less steel tube with 90–100 mesh Anakrom ABS containing a 2% loading of SE-30. The operating conditions were injection port 220°C, column oven 170°C, detector 210°C, detector standing current 5×10^{-9} A at 19 V, and nitrogen carrier gas at 160 ml/min. The retention time for PCNB was 94 sec and amounts as low as 0.01 p.p.m. of PCNB were recovered from soil and crops (celery, flax seed, lettuce, peanuts, peanut hay, peanut shells, potatoes, radishes, and strawberries) following initial extraction with hexane and clean-up with a silicic acid column.

St. John *et al.*[250] described the fate of PCNB as well as the herbicides, simazine (2-chloro-4,6-bis(ethylamino)-*s*-triazene) and 4,6-dinitro-2-isobutyl phenol (DNOSBP), in the dairy cow. No residues of these compounds were found in the milk. About 3.5% of intact DNOSBP and 1% of simazine were found in the urine and PCNB was largely (45%) eliminated as pentachloroaniline (PCA) in the urine as shown by electron-capture gas chromatography. Figure 70 shows chromatograms of urine one day after feeding began showing the pentachloroaniline peak and pre-feeding control urine. The retention times for PCA and PCNB were 6.9 and 5.4 min, respectively. Betts *et al.*[251] have previously identified pentachloroaniline as a metabolite of PCNB in rabbits.

Kuchar *et al.*[252] described analytical studies of the metabolism of PCNB in beagle dogs, rats, and plants. Tissues from dogs fed PCNB in their daily diet for 2 years were analyzed by TLC and GLC for residual PCNB, impurities and meta-bolites. PCNB was not found in fat or tissue and the metabolic products were penta-chloroaniline and methylpentachlorophenylsulfide (further established by infrared and mass spectroscopy). Chromatographic analyses of extracts of rat tissues from rats fed PCNB and from plants grown in PCNB treated soil are indicative of an identical metabolism. Two instruments were used, (*1*) a Varian Aerograph Model 1521 with thermal-conductivity detector and a 2 m × $\frac{1}{4}$ in. o.d. aluminum tubing containing 15% SE-30 on 80–100 mesh Chromosorb W and (*2*) a Jarrell-Ash Model 26-700 gas chromatograph was used with an electron-affinity detector and a 4 ft. × $\frac{1}{8}$ in. i.d.

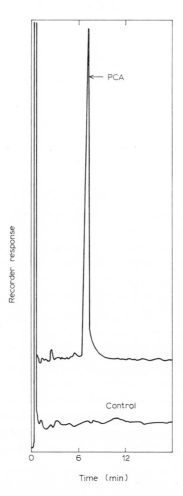

Fig. 70. Chromatograms of pentachloroaniline (PCA) metabolite in cow urine one day after feeding 5 p.p.m. of PCNB, and control urine.

stainless steel U-tube containing 2% SE-30 on 80–100 mesh acid-washed DMCS treated Chromosorb G. The column was conditioned at 210°C with nitrogen flow of 100 ml/min for about two days and the exit was disconnected from the detector. The instrument conditions used were injection port 210°C, detector 210°C, column

170°C, nitrogen 100–150 ml/min, detector standing current 5×10^{-9} A at 15–20 V, chart speed $\frac{1}{2}$ in./min. The relative retention data *vs*. PCNB were pentachlorobenzene (PCB) 0.42, 2,3,4,5-tetrachloronitrobenzene 0.67, hexachlorobenzene (HCB) 0.85, pentachloroaniline (PCA) 1.4, and methyl pentachlorophenylsulfide 2.0. Hexane extracts of tissues, excretion products and plant material were chromatographed and each component calculated by the procedure of Methratta *et al*.[249]. Figures 71 and 72 illustrate the separation of compounds in a steam distillate from dog feces analyzed

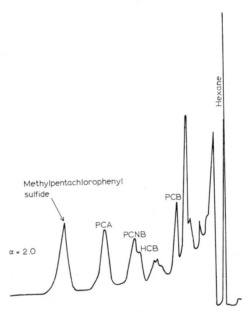

Fig. 71. Separation of compounds in steam distillate from dog feces. Varian Aerograph 1521. Thermal conductivity column, 2 m × 1/4 in. o.d. aluminum tubing. 15 % SE-30 silicone gum on 80–100 mesh Chromosorb W at 237 °C. Injection temperature, 205 °C. Detector temperature, 225 °C. Helium at 60 ml/min.

by thermal conductivity and electron gas chromatography, respectively. The presence of pentachloroaniline suggests a reduction mechanism and methylpentachlorophenyl sulfide a possible alkaline hydrolysis of PCNB to form a salt of a thiophenol which is subsequently methylated.

The microbial degradation of PCNB and DDT has been described by Chacko *et al*.[253]. Nine actiomyces and eight fungi were tested and all except *Streptomyces albus* degraded PCNB to an unknown metabolic product (Fig. 73) (*Streptomyces aureofaciens* degraded PCNB to pentachloroaniline). An Aerograph Model 600 D gas chromatograph with an electron-capture detector was used to detect the pesticides and their degradation products. For routine determinations a 1.8 m × 0.32 cm column containing 5 % DC-11 on Gas Chrom Q was used. Injection port and detection temperatures were 260 and 180 °C, respectively, and the flow rate of nitrogen was 60 ml/min. The column temperature for PCNB was 155°C. (The retention time for PCNB was 2.2 min and that for its degradation product was 3 min.)

Caseley[254] described the loss of three chloronitrobenzene fungicides (PCNB, TCNB and trichloronitrobenzene) from soil. The fungicides were determined with a Varian Aerograph Model 204 gas chromatograph fitted with a tritium foil electron-capture detector and for the analysis of PCNB and TCNB a 5 ft. × $\frac{1}{8}$ in. o.d. stainless steel column packed with 5% SE-30 on 60–80 mesh, acid-washed DMCS treated Chromosorb W. For the determination of trichloronitrobenzene (an isomeric mixture of 80% 1,2,4-trichloro-4,6-dinitrobenzene and 20% 1,2,3-trichloro-4,6-dinitroben-

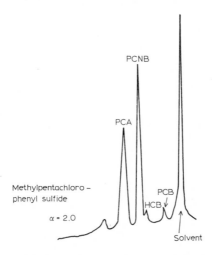

Fig. 72. Chromatogram of hexane extract of dog feces with an electron-capture detector.

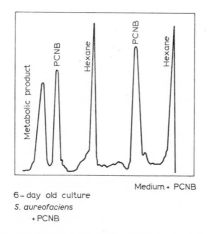

Fig. 73. Gas chromatogram of PCNB after incubation for 6 days in a medium with and without *Streptomyces aurofaciens*. PCNB was partially degraded to pentachloroaniline by *S. aureofaciens*.

zene), a 5 ft. × $\frac{1}{8}$ in. o.d. stainless steel column was packed with 4% XE-60 on 70–80 mesh acid-washed, DMCS treated Chromosorb G. The operating parameters were (a) for PCNB and TCNB: inlet 225°C, column 175°C, detector 200°C, carrier gas nitrogen at 55 ml/min, (b) for trichloronitrobenzene: inlet 210°C, column 160°C, detector 200°C, nitrogen carrier gas at 70 ml/min. The elution times were PCNB 3.0 min, TCNB 1.5 min, 1,2,4-trichloro-2,5-dinitrobenzene isomer 5 min, and 1,2,3-trichloro-4,6-dinitrobenzene isomer 6 min. The analyses of the fungicidal degradation products after sterilized and unsterilized soil treatment were performed on two additional instruments, a Varian Aerograph Model 202 gas chromatograph fitted with a thermal conductivity detector and a 5 ft. × $\frac{1}{4}$ in. o.d. stainless steel column packed with 4% Dow 11 on 60–80 mesh acid-washed Chromosorb G. The carrier gas was helium at a flow rate of 50 ml/min and the inlet temperature was 150°C, detector 275°C, and the column programmed 200–250°C at 4°/min. The compounds corresponding to the peaks on the chart were each collected in capillary tubes from the outlet of the detector. The compounds were then dissolved in a small volume of benzene and chlorine determinations were made by microcoulometric analysis using a Dohrmann gas chromatograph equipped with a 6 ft. × $\frac{1}{4}$ in. o.d. glass tube packed with 5% XE-60 on Chromosorb W. Helium was the carrier gas at 100 ml/min and the operating temperatures were inlet 240°C, column 200°C, and furnace 825°C. Figure 74 illustrates gas chromatograms of the three fungicides and their soil microbiological and chemical degradation products. The microcoulometric determinations revealed that the compounds producing the peaks contain chlorine which indicates that they are degradation products of the starting materials. The three fungicides bear a close structural resemblance to one another and the breakdown products of each compound have closely corresponding retention times.

Jain et al.[255] have described a rapid method for the extraction and identification of 23 pesticides (including PCNB) from blood. The pesticides were extracted from blood with a mixture of acetone and ethyl ether in equal volumes, the extracts evaporated to dryness and the residue dissolved in a known quantity of hexane and the aliquot injected directly into an Aerograph Hy-Fi Model 600 instrument equipped with an electron-capture detector and a 4.5 ft. × $\frac{1}{4}$ in. o.d. (0.070 i.d.) spiral glass tube packed with 5% SE-52 on 60–80 mesh acid-washed hexamethyldisilizane-treated Chromosorb W. The operating conditions were as follows column 190°C, injector 230°C, flow rate of nitrogen carrier 70 ml/min, input impedance 10^7 ohm and output sensitivity 1×, corresponding to 3.3×10^{-9} A for full-scale recorder deflection and detector voltage was –90 V.

In a later study, Jain and Kirk[256] utilized a 3-ft. glass column 1% Hi-Eff-8B on 100–120 mesh silanized Gas Chrom Q at 190 and 220°C to effect the separation and identification of 24 chlorinated and organophosphorus pesticides for toxicological analysis.

The thin-layer chromatographic separation on Silica Gel G of PCNB from a variety of pesticides has also been reported by Salo and Salminen[230]. The developing solvents used for PCNB were (*a*) hexane, (*b*) hexane–chloroform (7:3), (*c*) hexane–chloroform (1:1), (*d*) hexane–acetone (19:1), (*e*) hexane–acetone (9:1), (*f*) benzene–acetone (1:1), and (*g*) chloroform. The R_F values for PCNB in the above solvents were 0.28, 0.74, 0.81, 0.63, 0.64, 0.84, and 0.95, respectively.

Fehringer and Ogger[257] reported the TLC identification of 27 chlorine-containing pesticides on aluminum oxide (Camag D-5) layers containing hydroquinone, 2′,7′-dichlorofluorescein or silver nitrate and developed with 2% acetone–*n*-hexane and *n*-heptane solvents.

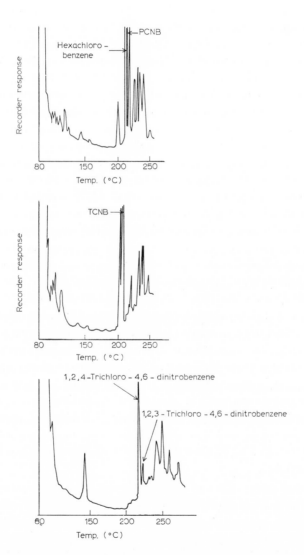

Fig. 74. Chromatography of PCNB, TCNB, and trichloronitrobenzene and their soil microbial and chemical degradation products.

References pp. 267–272

12. MIREX

Mirex (dodecachlorooctahydro-1,3,4-metheno-2H-cyclobuta[cd]pentalene) is prepared by the dimerization of hexachlorocyclopentadiene in the presence of aluminum chloride. It is a stomach insecticide with little contact activity and has found its widest utility against ants (*e.g.* fire and harvester ants). The toxicity[258–261] and carcinogenicity (in mice)[233] of Mirex have been reported.

Mirex has been analyzed by a variety of GLC procedures previously described for other pesticides and related derivatives in this chapter (*e.g.* Burke and Holswade[47], Bowman and Beroza[95], Thompson *et al.*[157], Firestone *et al.*[167], and Windam[237]). Thompson[262] described a device for the rapid peak identification of 25 chlorinated pesticides (including Mirex) and related compounds. A hand computer was described which permitted the dialing of relative retention values obtained from chromatograms thus making possible the tentative identification of compounds from data obtained using six different column liquid phases at two sets of operating parameters, *e.g.* column temperatures and carrier gas flow rates. Conversely, the device is useful in the selection of columns and operating conditions for maximum peak resolution in a mixture of known compounds. The retention data were obtained on a Micro-Tek Model MT 220 gas chromatograph using an electron-capture detector and operated at column inlet temperature of 230°C and a detector temperature of 200°C. The liquid-phase and column temperatures used were DC-200 at 180 and 200°C, OV-1 at 180 and 200°C, SE-30/QF-1 at 180 and 190°C, DC-200/QF-1 at 190°C, QF-1 at 160 and 180°C, and DEGS at 180 and 200°C. All column packings were prepared with Gas-Chrom Q, 80–100 mesh.

Walker and Beroza[263] described a TLC procedure for the analysis of 62 pesticides (including Mirex) using 19 solvent systems based on (*a*) chloroform, (*b*) chloroform–ethyl ether (9:1), (*c*) chloroform–ethyl acetate (9:1), (*d*) chloroform–acetone (9:1), and (*f*) chloroform–acetic acid (9:1). The foregoing systems were repeated with benzene and *n*-hexane in place of chloroform. To the *n*-hexane series, a seventh system of *n*-hexane–acetone (8:2) was added. The chromogenic reagents were (*1*) a solution of 5% of bromine in carbon tetrachloride used as a source of bromine vapor, (*2*) the fluorescein solution of Mitchell[264] modified by diluting 1 ml of the conc. fluorescein solution (0.25% fluorescein in N,N-dimethylformamide) to 50 ml with ethanol instead of to 200 ml and (*3*) the silver nitrate reagent of Mitchell[97] (1.7 g silver nitrate in 5 ml of water added to 10 ml 2-phenoxyethanol and diluted to 200 ml with acetone).

13. β-HYDROXYETHYLHYDRAZINE

β-Hydroxyethylhydrazine (Omaflora, HEH) is prepared by the reaction of hydrazine with ethanol and is used primarily as a growth regulator, for inducing flowering in pineapples[265], hence controlling the date of harvest (possibly by reduc-

TABLE 47

GAS CHROMATOGRAPHY OF HYDROXYETHYL DERIVATIVES

$$R\text{–}CH_2CH_2OH$$

Compound no.	R	Mol. wt.	Name	$t_r^{o\,a}$ 10%[b] C20M	$t_r^{o\,a}$ 4%[c] C20M
1	H_2N-	61	Ethanolamine	0.15	0.9, 2.75
2	CH_3NH-	75	2-(Methylamino)ethanol		2.10
3	H_2NNH-	76	β-Hydroxyethylhydrazine	0.30	2.15[d]
4	$HS-$	78	2-Mercaptoethanol		3.65
5	$Cl-$	80	2-Chloroethanol		1.45
6	$H\overset{O}{\overset{\|}{C}}-NH-$	88	N-(β-Hydroxyethyl)formamide	6.7	
7	$(CH_3)_2N-$	89	2-(Dimethylamino)ethanol		0.60
8	$(CH_3)_2CHNH-$	103	2-(Isopropylamino)ethanol		2.90
9	$CH_3\overset{O}{\overset{\|}{C}}-NH-$	103	N-(β-Hydroxyethyl)acetamide	5.6	
10	$CH_3\overset{O}{\overset{\|}{C}}-O-$	104	β-Hydroxyethyl acetate		5.7, 9.4
11	(pyrazol-1-yl)	112	N-(β-Hydroxyethyl)pyrazole		3.50[d]
12	(pyrrolidin-1-yl)	115	N-(β-Hydroxyethyl)pyrrolidine		5.6
13	(phenyl)	122	Phenethyl alcohol		2.8[d]
14	(pyridin-2-yl)	123	2-(β-Hydroxyethyl)pyridine	1.7	
15	(piperidin-1-yl)	129	N-(β-Hydroxyethyl)piperidine		4.1
16	$CH_2=\overset{H_3C}{\overset{\|}{C}}-\overset{O}{\overset{\|}{C}}-O-$	130	β-Hydroxyethyl methacrylate		1.70[d]
17	$HN\diagdown\diagup N-$ (piperazin-1-yl)	130	1-(β-Hydroxyethyl)piperazine	1.55	4.25[d]
18	(morpholin-4-yl)	131	N-(β-Hydroxyethyl)morpholine		1.95[d]
19	(cyclohexyl)$-NH-$	143	N-(β-Hydroxyethyl)cyclohexylamine	1.25	
20	$-\overset{H}{N}-\overset{H}{N}-$	120	N,N′-Bis(β-hydroxyethyl)hydrazine		2.20[d]

[a] Retention time in minutes from solvent front.

[b] 10% Carbowax 20M on 60–80 mesh Chromosorb W; 8 ft. × 0.125 in. o.d. stainless steel column; column temperature 220°C; 50 p.s.i.g. helium; 150 mA filament current; detector temperature 250°C; hot-wire detector.

[c] 4% Carbowax 20M terminated with terephthalic acid on 60–80 mesh HMDS-pretreated Chromosorb W; 6 ft. × 0.25 in. o.d. glass coil column; column temperature 73°C and nitrogen carrier flow 63 ml/min except for those designated footnote d; hydrogen 45 ml/min, air 300 ml/min; detector temperature 200°C; hydrogen-flame detector.

[d] As footnote c, except column temperature 110°C and nitrogen carrier 81 ml/min.

ing auxin concentration). The compound has also been found to be a potent inhibitor *in vivo* of animal diamine oxidase[266], significantly increases alkaloid formation in *Datura stramonium* plant[267] and is carcinogenic in the mouse[233].

HEH has been determined in pineapples[268] spectrophotometrically. Following blending and extracting the HEH from the fruit with water, the interfering pigments in the extract are removed with an ion-exchange resin (Dowex 1-X8 and 50W-X8, 50–100 mesh), thence reaction of HEH with cinnamaldehyde producing a yellow color which is measured at 420. HEH, as well as 20 additional hydroxyethyl derivatives, has been analyzed by GLC[269]. Analyses were carried out using two systems (*1*) an F & M Model 500 gas chromatograph with a hot-wire detector and a 10% Carbowax 20M column housed in an F & M Model 720 dual column oven and (*2*) an F & M Model 1609 gas chromatograph equipped with a hydrogen-flame detector and a 4% Carbowax 20M column. The specific analytical operating conditions and the chromatographic results for the hydroxyethyl derivatives are shown in Table 47. The derivatives required the use of several columns. The low temperatures employed for most of the derivatives (73 and 110°C on 4% Carbowax 20M) were not sufficient for several of the more strongly retained compounds (*e.g.* compounds 6 and 9). The influence of hydrogen bonding *via* the hydroxyl proton on the chromatographic

TABLE 48

INFLUENCE OF SILYLATION ON CHROMATOGRAPHIC ELUTION OF HYDROXYETHYL DERIVATIVES

Compound	R	Name	$t_r^{o\ a}$		Elution enhance- ment (%)[b]
			RCH_2- CH_2OH	RCH_2- CH_2OSi- $(CH_3)_3$	
16	$CH_2=\overset{CH_3}{\underset{}{C}}-\overset{O}{\underset{}{C}}-O-$	2-Hydroxyethyl methacrylate	1.70	0.60	65
18	$O\!\!\diagdown N-$	N-(β-Hydroxyethyl)morpholine	1.95	0.90	54
3	H_2NNH-	β-Hydroxyethylhydrazine	2.15	0.60	72
20	$-\overset{H}{N}-\overset{H}{N}-$	N,N'-Bis(β-hydroxyethyl)-hydryzine	2.20	0.60, 1.90	73, 14
13	(phenyl)-	Phenethyl alcohol	2.80	0.80	71
11	pyrazolyl N-	N-(β-Hydroxyethyl)pyrazole	3.50	0.55, 0.90	84, 74
17	$HN\!\!\diagdown N-$	1-(β-Hydroxyethyl)piperazine	4.25	2.00	53

[a] For column and conditions see footnote *d*, Table 47.

[b] Calculated by $100 - \dfrac{[t_r^0(RCH_2CH_2OSi(CH_3)_3)]\ 100}{t_r^0(RCH_2CH_2OH)}$

behavior of hydroxyethyl derivatives can be seen on examination of the elution results on 4% Carbowax 20M at 110°C between several hydroxyethyl compounds and their trimethylsilyl derivatives given in Table 48. (The TMS derivatives were prepared by reaction with hexamethyldisilazane in pyridine.) Elution enhancement *via* the silyl ether derivatives was generally in the order of 50–80%.

REFERENCES

1   C. HUGHES AND S. P. SPRAGG, *Biochem. J.*, 70 (1958) 205.
2   I. HOFFMAN AND E. V. PARUPS, *J. Agr. Food Chem.*, 10 (1962) 453.
3   G. J. STONE, in J. W. ZUCKEL (Ed.), *A Literature Summary on Maleic Hydrazide, 1949–1957*, U.S. Rubber Co., Naugatuck Chem. Div., Naugatuck, Conn., 1957.
4   C. ANGLIN AND J. H. MAHON, *J. Assoc. Offic. Agr. Chemists*, 41 (1958) 177.
5   I. HOFFMAN AND E. V. PARUPS, *Residue Rev.*, 7 (1964) 96.
6   A. S. CRAFTS, *The Chemistry and Mode of Action of Herbicides*, Wiley-Intersciences, New York, 1961, Chap. 15.
7   S. S. EPSTEIN AND N. MANTEL, *Intern. J. Cancer*, 3 (1968) 325.
8   S. S. EPSTEIN, J. ANDREA, H. JAFFEE, H. FALK AND N. MANTEL, *Nature*, 215 (1967) 1388.
9   R. E. MCCARTHY AND S. S. EPSTEIN, *Life Sci.*, 7 (1968) 1.
10 C. E. NASRAT, *Nature*, 207 (1965) 439.
11 J. H. NORTHRUP, *J. Gen. Physiol.*, 46 (1963) 971.
12 S. SHIMOMURA, *J. Pharm. Soc. Japan*, 6 (1958) 589.
13 T. TAKEUCHI, N. YUKOUCHI AND K. ONODA, *Japan Analyst*, 5 (1956) 399.
14 D. M. MILLER, *Can. J. Chem.*, 34 (1956) 1510.
15 I. HOFFMAN, *J. Assoc. Offic. Agr. Chemists*, 44 (1961) 723.
16 M. MASHIMA, *J. Chem. Soc. Japan, Pure Chem. Sect.*, 83 (1962) 981.
17 O. OHASHI, M. MASHIMA AND M. KUBO, *Can. J. Chem.*, 42 (1964) 970.
18 J. R. LANE, D. K. GULLSTROM AND J. E. NEWELL, *J. Agr. Food Chem.*, 6 (1958) 671.
19 J. R. LANE, *J. Assoc. Offic. Anal. Chemists*, 46 (1963) 261.
20 W. A. ANDREAE, *Can. J. Biochem. Physiol.*, 36 (1958) 71.
21 P. K. BISWAS, O. HALL AND B. D. MAYBERRY, *Physiol. Plantarum*, 20 (1967) 819.
22 G. H. N. TOWERS, A. HUTCHINSON AND W. A. ANDREAE, *Nature*, 181 (1958) 1935.
23 D. L. MAYS, G. S. BORN, J. E. CHRISTIAN AND B. J. LISKA, *J. Agr. Food Chem.*, 16 (1968) 356.
24 A. STOESSL, *Chem. Ind. (London)*, (1964) 580.
25 A. STOESSL, *Can. J. Chem.*, 43 (1965) 2430.
26 L. D. NOODEN, *Plant Physiol.*, 45 (1970) 46.
27 L. FISHBEIN AND W. L. ZIELINSKI, JR., *J. Chromatog.*, 18 (1965) 581.
28 W. L. ZIELINSKI, JR. AND L. FISHBEIN, *J. Chromatog.*, 20 (1965) 140.
29 D. STEFANYL AND W. L. HOWARD, *J. Org. Chem.*, 19 (1954) 115.
30 R. BENTLEY, C. C. SWEELEY, M. MAKITA AND W. W. WELLS, *Biochem. Biophys. Res. Commun.*, 11 (1963) 14.
31 R. J. LUKENS AND H. D. SISLER, *Phytopathol.*, 48 (1958) 235.
32 E. SOMERS, D. V. RICHMOND AND J. A. PACKARD, *Nature*, 215 (1967) 214.
33 R. G. OWENS AND G. BLAAK, *Contrib. Boyce Thompson Inst.*, 20 (1960) 475.
34 D. V. RICHMOND AND E. SOMERS, *Ann. Appl. Biol.*, 62 (1968) 35.
35 M. K. LIU AND L. FISHBEIN, *Experientia*, 23 (1967) 81.
36 E. M. BOYD AND C. J. KRIJNEN, *J. Clin. Pharmacol.*, 8 (1968) 225.
37 J. MCLAUGHLIN, JR., E. F. REYNALDO, L. K. LAMAR AND J. P. MARLIAC, *Toxicol. Appl. Pharmacol.*, 14 (1969) 641.
38 M. LEGATOR AND J. VERRETT, in *Conf. on Biolog. Effects of Pesticides in Mammalian Systems*, N.Y. Acad. Sci., N.Y., 1967, Abstr. No. 6, p. 17.
39 K. JONES AND J. G. HEATHCOTE, *J. Chromatog.*, 24 (1966) 106.

40 H. M. WINEGARD, G. TOENNIES AND R. J. BLOCK, *Science*, 108 (1948) 506.
41 E. D. MOFFAT AND R. I. LYTLE, *Anal. Chem.*, 31 (1959) 926.
42 W. W. KILGORE, W. WINTERLIN AND R. WHITE, *J. Agr. Food Chem.*, 15 (1967) 1035.
43 J. N. OSPENSON, D. E. PACK, G. K. KOHN, H. P. BURCHFIELD AND E. E. STORRS, in G. ZWEIG (Ed.), *Analytical Methods for Pesticides, Plant Growth Regulators and Food Additives*, Academic Press, New York, 1964, Vol. III, p. 11.
44 A. BEVENUE AND J. N. OGATH, *J. Chromatog.*, 36 (1968) 531.
45 I. H. POMERANTZ, L. J. MILLER AND G. KAVA, *J. Assoc. Offic. Anal. Chemists*, 53 (1970) 154.
46 I. H. POMERANTZ AND R. ROSS, *J. Assoc. Offic. Anal. Chemists*, 51 (1968) 1058.
47 J. A. BURKE AND W. HOLSWADE, *J. Assoc. Offic. Anal. Chemists*, 49 (1966) 374.
48 E. S. GOODWIN, R. GOULDEN AND J. G. REYNOLDS, *Analyst*, 86 (1961) 697.
49 J. A. BURKE AND L. GIUFFRIDA, *J. Assoc. Offic. Anal. Chemists*, 47 (1964) 326.
50 T. E. ARCHER AND J. B. CORBIN, *Food Technol.*, 23 (1969) 101.
51 T. E. ARCHER AND J. B. CORBIN, *Bull. Environ. Contam. Toxicol.*, 4 (1969) 55.
52 R. ENGST AND W. SCHNAAK, *Nahrung*, 11 (1967) 95.
53 L. FISHBEIN, J. FAWKES AND P. JONES, *J. Chromatog.*, 23 (1966) 476.
54 F. FEIGL, *Spot Tests in Organic Analysis*, Elsevier, Amsterdam, 1960, p. 246.
55 S. J. PURDY AND E. J. TRUTER, *Chem. Ind. (London)*, (1962) 506.
56 W. W. KILGORE AND E. R. WHITE, *J. Agr. Food Chem.*, 15 (1967) 1118.
57 W. MOJE, D. E. MUNNECKE AND L. T. RICHARDSON, *Nature*, 202 (1964) 831.
58 S. C. CHANG, P. H. TERRY AND A. B. BORKOVEC, *Science*, 144 (1964) 57.
59 J. PALMQUIST AND L. E. LACHANCE, *Science*, 154 (1966) 915.
60 R. M. KIMBROUGH AND T. B. GAINES, *Nature*, 211 (1966) 146.
61 T. B. GAINES AND R. KIMBROUGH, *Bull. World Health Organ.*, 13 (1964) 737.
62 S. C. CHANG, P. H. TERRY, C. W. WOODS AND A. B. BORKOVEC, *J. Econ. Entomol.*, 60 (1967) 1623.
63 R. B. MARCH, R. L. METCALF AND T. R. FUKUTO, *J. Agr. Food Chem.*, 2 (1954) 732.
64 S. C. CHANG AND A. B. BORKOVEC, *J. Econ. Entomol.*, 62 (1969) 1417.
65 S. AKOV, J. E. OLIVER AND A. B. BORKOVEC, *Life Sci.*, 7 (1968) 1207.
66 A. R. JONES AND H. JACKSON, *Biochem. Pharmacol.*, 17 (1968) 2247.
67 P. H. TERRY AND A. B. BORKOVEC, *J. Med. Chem.*, 11 (1968) 958.
68 A. R. JONES, *Biochem. Pharmacol.*, 19 (1970) 603.
69 C. S. HAINES AND F. A. ISHERWOOD, *Nature*, 164 (1949) 1107.
70 A. C. TERRANOVA AND C. H. SCHMIDT, *J. Econ. Entomol.*, 60 (1967) 1659.
71 S. AKOV AND A. B. BORKOVEC, *Life Sci.*, 7 (1968) 1215.
72 R. L. JASPER, E. L. SILVERS AND H. O. WILLIAMSON, *Federation Proc.*, 24 (1965) 641.
73 F. S. PHILIPS AND J. B. THIERSCH, *J. Pharmacol. Exptl. Therap.*, 100 (1950) 398.
74 J. T. NORTH, *Mutation Res.*, 4 (1967) 225.
75 S. C. CHANG, A. B. DE MILO, C. W. WOODS AND A. B. BORKOVEC, *J. Econ. Entomol.*, 61 (1968) 1357.
76 S. C. CHANG, C. W. WOOD AND A. B. BORKOVEC, *J. Econ. Entomol.*, 63 (1970) 1510.
77 S. M. BUCKLEY, C. C. STOCK, M. L. CROSSLEY AND C. P. RHOADS, *Cancer*, 5 (1952) 144.
78 W. L. WILSON AND J. G. DE LA GARZA, *Cancer Chemotherapy Rept.*, 48 (1965) 49.
79 G. T. BRYAN AND A. L. GORSKE, *J. Chromatog.*, 34 (1968) 67.
80 J. M. PRICE, *J. Biol. Chem.*, 211 (1954) 117.
81 B. L. OSER AND M. OSER, *Toxicol. Appl. Pharmacol.*, 2 (1960) 441.
82 S. B. STERNBERG, H. POPPER, B. L. OSER AND M. OSER, *Cancer*, 13 (1960) 780.
83 S. STERNBERG, H. POPPER, B. L. OSER AND M. OSER, *Am. J. Pathol.*, 35 (1959) 691.
84 W. D. KAPLAN AND R. SEECOF, *Drosophila Inform. Serv.*, 41 (1966) 101.
85 O. G. FAHMY AND M. J. FAHMY, *Nature*, 177 (1956) 996.
86 A. L. WALPOLE, *Ann. N.Y. Acad. Sci.*, 68 (1958) 750.
87 F. A. GUNTHER, R. C. BLINN, M. J. KOLBEZEN AND J. H. BARKLEY, *Anal. Chem.*, 23 (1951) 1835.
88 M. E. BROKKE, U. KIIGEMAGI AND L. C. TERRIERE, *Anal. Chem.*, 6 (1958) 27.
89 T. E. ARCHER, *Bull. Environ. Contam. Toxicol.*, 3 (1968) 71.
90 S. O. GRAHAM, *Science*, 139 (1963) 835.
91 R. C. BLINN AND F. A. GUNTHER, *J. Assoc. Offic. Anal. Chemists*, 46 (1963) 204.

92 R.A.HENRY, *J. Org. Chem.*, 27 (1962) 2637.
93 D.M.COULSON, L.A.CAVANAGH, J.E.DE VRIES AND B.WALTHER, *J. Agr. Food Chem.*, 8 (1960) 399.
94* J.BURKE AND L.JOHNSON, *J. Assoc. Offic. Anal. Chemists*, 45 (1962) 348.
95 M.C.BOWMAN AND M.BEROZA, *J. Assoc. Offic. Agr. Chemists*, 48 (1965) 943.
96 W.A.BOSIN, *Anal. Chem.*, 35 (1963) 833.
97 L.C.MITCHELL, *J. Assoc. Offic. Agr. Chemists*, 41 (1958) 781.
98 Shell Chemicals, Vapona in slow release strips, *Safe Use Manual No.2/V*, Sept. 1965.
99 Shell International Chemical Co., Ltd., *Safety Evaluation of Vapona Resin Strips*, July, 1968.
100 G.LÖFROTH, C.KIM AND S.HUSSAIN, *Environ. Mutagen Soc. Newsletter*, 2 (1969) 21.
101 G.LÖFROTH, *Naturwissenschaften*, 57 (1970) 393.
102 K.SAX AND J.H.SAX, *Japan J. Genet.*, 43 (1968) 89.
103 M.C.IVEY AND H.V.CLABORN, *J. Assoc. Offic. Anal. Chemists*, 52 (1969) 1248.
104 G.DRAEGER, *Pflanzenschutznachr. Bayer*, 21 (1968) 373.
105 J.W.MILES, *Chemical Memorandum No.10, TDL 6-23-69*, U.S. Public Health Service, H.E.W., Savannah, Ga.
106 J.W.MILES, personal communication.
107 R.D.O'BRIEN, *Toxic Phosphorus Esters: Chemistry, Metabolism, and Biological Effects*, Academic Press, New York, 1960, p.74.
108 S.D.FAUST AND I.H.SUFFET, *Residue Rev.*, 15 (1966) 44.
109 R.MÜHLMANN AND G.SCHRADER, *Z. Naturforsch.*, 12B (1957) 196.
110 J.H.RUZICKA, J.THOMSON AND B.B.WHEALS, *J. Chromatog.*, 31 (1967) 37.
111 J.RUZICKA, J.THOMSON AND B.B.WHEALS, *J. Chromatog.*, 30 (1967) 92.
112 L.W.GETZIN AND I.ROSEFIELD, *J. Agr. Food Chem.*, 16 (1968) 598.
113 C.E.COOK, C.W.STANLEY AND J.E.BARNEY, II, *Anal. Chem.*, 36 (1964) 2354.
114 J.E.BARNEY, II, C.W.STANLEY AND C.E.COOK, *Anal. Chem.*, 35 (1963) 2206.
115 M.SALAME, *Ann. Biol. Clin.*, 26 (1968) 1011.
116 G.ZWEIG AND J.M.DEVINE, *Residue Rev.*, 26 (1969) 17.
117 J.ELLIS AND J.BATES, *J. Assoc. Offic. Agr. Chemists*, 48 (1965) 1115.
118 L.E.MITCHELL, *J. Assoc. Offic. Agr. Chemists*, 47 (1964) 268.
119 H.ACKERMANN, *J. Chromatog.*, 36 (1968) 309.
120 G.FECHNER, H.ACKERMANN AND H.TOEPFER, *Z. Anal. Chem.*, 246 (1969) 250.
121 C.E.MENDOZA, D.L.GRANT, B.BRACELAND AND K.A.MCCULLY, *Analyst*, 94 (1969) 805.
122 G.F.ERNST AND F.SCHURING, *J. Chromatog.*, 49 (1970) 325.
123 M.RAMASAMY, *Analyst*, 94 (1969) 1078.
124 R.T.WANG AND S.S.CHOU, *J. Chromatog.*, 42 (1969) 416.
125 J.H.RUZICKA, J.THOMSON, B.B.WHEALS AND N.F.WOOD, *J. Chromatog.*, 34 (1968) 14.
126 B.W.ARTHUR AND J.E.CASIDA, *Proc. Entomol. Soc. Am. 3rd Ann. Meeting, Cincinnati, 1955.*
127 B.W.ARTHUR AND J.E.CASIDA, *J. Agr. Food Chem.*, 5 (1957) 186.
128 R.PREUSSMAN, *Food Cosmet. Toxicol.*, 6 (1968) 576.
129 H.DRUCKREY, *Konferenz über aktuelle Probleme der Tabaksforschung, Freiburg, 1967*, p.42.
130 W.DEDEK AND K.LOHS, *Z. Naturforsch.*, 25B (1970) 94.
131 T.J.OLSON, *Chemagro Corp. Rept. No.8442*, Jan. 25, 1962, revised April 17, 1962.
132 T.J.OLSON, *Chemagro Corp. Rept. No.8839*, March 14, 1962.
133 T.J.OLSON, *Chemagro Corp. Rept. No.10,081*, Oct. 25, 1962, revised Nov. 27, 1962.
134 R.J.ANDERSON, C.A.ANDERSON AND T.J.OLSON, *J. Agr. Food Chem.*, 14 (1966) 508.
135 H.C.BARRY, J.G.HINDLEY AND L.Y.JOHNSON, *Pesticides Analytical Manual*, U.S. Food and Drug Administration, Washington, D.C., July, 1965, Vols. I and II.
136 A.R.A.EL-RAPAI AND L.GIUFFRIDA, *J. Assoc. Offic. Agr. Chemists*, 48 (1965) 374.
137 J.ASKEW, J.H.RUZICKA AND B.B.WHEALS, *Analyst*, 94 (1969) 275.
138 M.E.GETZ AND H.G.WHEELER, *J. Assoc. Offic. Anal. Chemists*, 51 (1968) 1101.
139 V.Y.KOLYAKOVA, *Tr. Vses. Nauchn.-Issled. Inst. Vet. Sanit.*, 29 (1967) 348.
140 H.ACKERMANN, R.ENGST AND G.FECHNER, *Z. Lebensm.-Untersuch.-Forsch.*, 137 (1968) 303.
141 H.ACKERMANN, *Nahrung*, 10 (1966) 273.
142 V.A.DRILL AND T.HIRATZKA, *Arch. Ind. Hyg. Occup. Med.*, 7 (1953) 61.
143 V.K.ROWE AND T.A.HYMAS, *Am. J. Vet. Res.*, 15 (1954) 622.

144 K.D.COURTNEY, D.W.GAYLOR, M.D.HOGAN, H.L.FALK, R.R.BATES AND I.MITCHELL, *Science*, 168 (1970) 168.
145 K.D.COURTNEY, personal communication.
146 D.E.CLARK, *J. Agr. Food Chem.*, 17 (1969) 1168.
147 C.W.STANLEY, *J. Agr. Food Chem.*, 14 (1966) 321.
148 A.F.MCKAY, W.L.OTT, G.W.TAYLOR, N.N.BUCHANAN AND J.F.CROOKER, *Can. J. Res.*, 28B (1950) 683.
149 P.L.PURSLEY AND E.D.SCHALL, *J. Assoc. Offic. Agr. Chemists*, 48 (1965) 327.
150 G.YIP, *J. Assoc. Offic. Agr. Chemists*, 47 (1964) 1116.
151 J.E.SCOGGINS AND C.H.FITZGERALD, *J. Agr. Food Chem.*, 17 (1969) 156.
152 J.M.DEVINE AND G.ZWEIG, *J. Assoc. Offic. Anal. Chemists*, 52 (1969) 187.
153 D.L.GUTNICK AND G.ZWEIG, *J. Chromatog.*, 13 (1964) 319.
154 T.P.GARBRECHT, *J. Assoc. Offic. Anal. Chemists*, 53 (1970) 70.
155 H.L.MORTON, F.S.DAVIS AND M.G.MERKLE, *Weed Sci.*, 16 (1968) 88.
156 C.H.FITZGERALD, C.L.BROWN AND E.G.BECK, *Plant Physiol.*, 42 (1967) 459.
157 J.F.THOMPSON, A.C.WALKER AND R.F.MOSEMAN, *J. Assoc. Offic. Anal. Chemists*, 52 (1969) 1263.
158 V.LEONI AND G.PUCCETI, *J. Chromatog.*, 43 (1969) 388.
159 L.E.ST.JOHN, JR., D.G.WAGNER AND D.J.LISK, *J. Dairy Sci.*, 67 (1964) 1267.
160 D.J.LISK, W.H.GUTENMANN, C.A.BACHE, R.G.WARNER AND D.G.WAGNER, *J. Dairy Sci.*, 46 (1963) 1435.
161 K.D.COURTNEY, *Pesticides Symposia of the 7th Inter-American Conference on Toxicology and Occupational Medicine, Miami Beach, Fla., (Aug., 1970)*, Halos, Miami Beach, Fla., 1970, p.277.
162 R.J.NASH, T.A.SMITH AND R.L.WAIN, *Ann. Appl. Biol.*, 51 (1968) 481.
163 K.ERNE, *Acta Vet. Scand.*, 7 (1966) 240.
164 K.ERNE, *Acta Vet. Scand.*, 7 (1966) 77.
165 G.B.CERESIA AND W.W.SANDERSON, *J. Water Pollution Control Federation*, 41 (1969) R34.
166 D.FIRESTONE, W.IBRAHIM AND W.HORWITZ, *J. Assoc. Offic. Agr. Chemists*, 46 (1963) 406.
167 D.FIRESTONE, W.IBRAHIM AND W.HORWITZ, *J. Assoc. Offic. Agr. Chemists*, 46 (1963) 384.
168 J.C.WOOTON, N.R.ARTMAN AND J.C.ALEXANDER, *J. Assoc. Offic. Agr. Chemists*, 45 (1962) 739.
169 G.R.HIGGINBOTHAM, D.FIRESTONE, L.CHAVEZ AND A.D.CAMPBELL, *J. Assoc. Offic. Agr. Chemists*, 50 (1967) 874.
170 G.R.HIGGINBOTHAM, J.RESS AND D.FIRESTONE, *J. Assoc. Offic. Agr. Chemists*, 50 (1967) 884.
171 G.R.HIGGINBOTHAM, J.RESS AND D.FIRESTONE, *J. Assoc. Offic. Agr. Chemists*, 51 (1968) 940.
172 G.R.HIGGINBOTHAM, A.HUANG, D.FIRESTONE, J.VERRET, J.RESS AND A.D.CAMPBELL, *Nature*, 220 (1968) 702.
173 M.KONITA, S.UEDA AND M.NARISADA, *Yakugaku Zasshi*, 79 (1959) 186.
174 M.J.VERRETT, J.MARLIAC AND J.MCLAUGHLIN, *J. Assoc. Offic. Agr. Chemists*, 47 (1964) 1003.
175 P.NEAL, *J. Assoc. Offic. Anal. Chemists*, 50 (1967) 1338.
176 G.R.HIGGINBOTHAM, J.RESS AND A.ROCKE, *J. Assoc. Offic. Anal. Chemists*, 53 (1970) 673.
177 R.BARTHA, R.P.LANZILOTTA AND D.PRAMER, *Appl. Microbiol.*, 15 (1967) 67.
178 R.BARTHA AND D.PRAMER, *Science*, 156 (1967) 1617.
179 M.ADACHI, K.TONEGAWA AND T.UESHIMA, *Noyaku Seisan Gijutsu*, 14 (1966) 19.
180 D.S.FREAR AND G.G.STILL, *Phytochem.*, 7 (1968) 913.
181 G.G.STILL, *Science*, 159 (1968) 992.
182 C.H.WILLIAMS AND K.H.JACOBSON, *Toxicol. Appl. Pharmacol.*, 9 (1966) 495.
183 R.BARTHA, *J. Agr. Food Chem.*, 16 (1968) 602.
184 R.L.DALTON, A.W.EVANS AND R.C.RHODES, *Weeds*, 14 (1966) 31.
185 I.PRASAD AND D.PRAMER, *Cytologia*, 34 (1969) 351.
186 I.PRASAD, *Can. J. Microbiol.*, 16 (1970) 369.
187 H.CHISAKA AND P.C.KEARNEY, *J. Agr. Food Chem.*, 18 (1970) 854.
188 J.F.CORBETT AND P.F.HOLT, *J. Am. Chem. Soc.*, 85 (1963) 2385.
189 J.D.ROSEN, M.SIEWIERSKI AND G.WINNETT, *158th Meeting ACS, New York, Sept. 1969*.

190 R.Y.Yih, D.H.McRae and H.F.Wilson, *Science*, 161 (1968) 376.
191 I.J.Belasco and H.C.Pease, *J. Agr. Food Chem.*, 17 (1969) 1414.
192 J.R.Plimmer, P.C.Kearney, H.Chisaka, J.B.Yount and U.I.Klingebiel, *J. Agr. Food Chem.*, 18 (1970) 859.
193 R.Bartha, H.A.B.Linke and D.Pramer, *Science*, 161 (1968) 582.
194 J.H.Weisburger and E.K.Weisburger, *Chem. Eng. News*, 44 (1966) 124.
195 P.C.Kearney, J.R.Plimmer and F.B.Guardia, *J. Agr. Food Chem.*, 17 (1969) 1418.
196 A.T.James, *Anal. Chem.*, 28 (1956) 1564.
197 K.J.Bombaugh, *Anal. Chem.*, 37 (1965) 72.
198 T.H.Jukes and C.B.Shaffer, *Science*, 132 (1960) 296.
199 K.T.Potts, *Chem. Rev.*, 61 (1961) 87.
200 M.C.Carter, in P.C.Kearney and D.D.Kaufmann (Eds.), *Degradation of Herbicides*, Marcel Dekker, New York, 1969, p.187.
201 C.M.Menzie, Metabolism of pesticides, *Bureau of Sport Fisheries and Wildlife Special Scientific Report, Wildlife No.127*, Washington, 1969, p. 40.
202 J.L.Hilton, *J. Agr. Food Chem.*, 17 (1969) 182.
203 S.C.Fang, M.George and T.C.Yu, *J. Agr. Food Chem.*, 12 (1964) 219.
204 S.C.Fang, S.Khanna and A.V.Rao, *J. Agr. Food Chem.*, 14 (1966) 262.
205 P.Massini, *Acta Botan. Neerl.*, 12 (1963) 64.
206 D.Racusen, *Arch. Biochem. Biophys.*, 74 (1958) 106.
207 M.C.Carter and A.W.Naylor, *Plant Physiol.*, 34 (1959) Suppl.VI.
208 R.A.Herrett and A.J.Linck, *Plant Physiol.*, 14 (1961) 767.
209 C.S.Miller and W.C.Hall, *J. Agr. Food Chem.*, 9 (1961) 210.
210 P.Massini, *2nd U.N. Conf. At. Energy*, 27 (1958) 58.
211 R.H.Shimabukuro and A.J.Linck, *Physiol. Plantarum*, 18 (1965) 532.
212 R.A.Herrett and W.P.Bagley, *J. Agr. Food Chem.*, 12 (1964) 17.
213 M.C.Carter and A.W.Naylor, *Botan. Gaz.*, 122 (1960) 138.
214 M.C.Carter and A.W.Naylor, *Physiol. Plantarum*, 14 (1961) 20.
215 J.C.Brown and M.C.Carter, *Weed Sci.*, 16 (1968) 222.
216 C.D.Ercegovich and D.E.H.Frear, *J. Agr. Food Chem.*, 12 (1964) 26.
217 D.D.Kaufman, *Weed Soc. Am. Abstr.*, (1967) 78.
218 D.D.Kaufman, J.R.Plimmer, P.C.Kearney, J.Blake and F.S.Guardia, *Weed Sci.*, 16 (1968) 226.
219 F.M.Ashton, *Weeds*, 11 (1963) 161.
220 J.R.Plimmer, P.C.Kearney, D.D.Kaufman and F.S.Guardia, *J. Agr. Food Chem.*, 15 (1967) 996.
221 P.Castelfranco, A.Oppenheim and S.Yamaguchi, *Weeds*, 11 (1963) 111.
222 P.Castelfranco and M.S.Brown, *Weeds*, 11 (1963) 116.
223 B.N.Ames, B.Garry and L.A.Herzenberg, *J. Gen. Microbiol.*, 22 (1961) 369.
224 R.A.Herrett and A.J.Linck, *J. Agr. Food Chem.*, 9 (1961) 466.
225 G.Zweig (Ed.), *Analytical Methods for Pesticides, Plant Growth Regulators and Food Additives*, Academic Press, New York, 1964, Vol.4, p.17.
226 R.W.Storherr and J.Burke, *J. Assoc. Offic. Agr. Chemists*, 44 (1961) 196.
227 B.D.Wills, *Analyst*, 91 (1966) 468.
228 L.C.Mitchell, *J. Assoc. Offic. Agr. Chemists*, 43 (1960) 87.
229 D.M.Coulson, *J. Gas Chromatog.*, 4 (1966) 285.
230 T.Salo and K.Salminen, *Z. Lebensm. Untersuch.-Forsch.*, 129 (1966) 149.
231 H.J.Hurn, R.B.Bruce and O.E.Paynter, *J. Agr. Food Chem.*, 3 (1955) 752.
232 R.Gasser, *Phytiat.-Phytopharm. No. Spec.*, (1954) 357.
233 S.R.M.Innes, B.M.Ulland, M.G.Valerio, L.Petrucelli, L.Fishbein, E.R.Hart, A.J.Pallotta, R.R.Bates, H.L.Falk, J.J.Gart, M.Klein, I.Mitchell and J.Peters, *J. Natl. Cancer Inst.*, 42 (1969) 1101.
234 R.C.Blinn, F.A.Gunther and M.J.Kolbezen, *J. Agr. Food Chem.*, 2 (1954) 1080.
235 H.J.Harris, *J. Agr. Food Chem.*, 3 (1955) 939.
236 H.Beckman and A.Bevenue, *J. Agr. Food Chem.*, 12 (1964) 183.
237 E.S.Windham, *J. Assoc. Offic. Anal. Chemists*, 52 (1969) 1237.

238 S.Miyazaki, G.M.Boush and F.Matsumura, *Appl. Microbiol.*, 18 (1969) 972.
239 S.Miyazaki, G.M.Boush and F.Matsumura, *J. Agr. Food Chem.*, 18 (1970) 87.
240 R.D.Cannizzaro, T.E.Cullen and R.T.Murphy, *J. Agr. Food Chem.*, 18 (1970) 728.
241 *Report of the Secretary's Commission on Pesticides and Their Relationship to Environmental Health, Parts I and II*, Dept. H.E.W., Washington, 1969, p.666.
242 W.W.Kilgore and E.R.White, *J. Chromatog. Sci.*, 8 (1970) 166.
243 S.Gorbach and U.Wagner, *J. Agr. Food Chem.*, 15 (1967) 654.
244 C.C.Cassil, *Residue Rev.*, 1 (1962) 35.
245 H.J.Ackermann, L.J.Carbone and E.J.Kuchar, *J. Agr. Food Chem.*, 11 (1963) 297.
246 H.J.Ackermann, H.A.Bactrush, H.H.Berges, D.E.Brookover and B.B.Brown, *J. Agr. Food Chem.*, 6 (1958) 747.
247 A.K.Klein and R.I.Gajan, *J. Assoc. Offic. Agr. Chemists*, 44 (1961) 712.
248 C.A.Bache and D.I.Lisk, *J. Agr. Food Chem.*, 8 (1960) 459.
249 T.P.Methratta, R.W.Montagna and W.P.Griffith, *J. Agr. Food Chem.*, 15 (1967) 648.
250 L.E.St.John, Jr., J.W.Ammering, D.G.Wagner, R.J.Wagner and D.J.Lisk, *J. Dairy Sci.*, 48 (1965) 502.
251 J.J.Betts, S.P.James and W.V.Thorpe, *Biochem. J.*, 61 (1955) 611.
252 E.J.Kuchar, F.O.Geenty, W.P.Griffith and R.J.Thomas, *J. Agr. Food Chem.*, 17 (1969) 1237.
253 C.I.Chacko, J.L.Lockwood and M.Zabik, *Science*, 154 (1966) 893.
254 J.C.Caseley, *Bull. Environ. Contam. Toxicol.*, 3 (1968) 180.
255 N.C.Jain, C.R.Fontan and P.L.Kirk, *J. Pharm. Pharmacol.*, 17 (1965) 362.
256 N.C.Jain and P.L.Kirk, *Microchem. J.*, 12 (1967) 265.
257 N.V.Fehringer and J.D.Ogger, *J. Chromatog.*, 25 (1966) 95.
258 H.Martin, *Guide to the Chemicals Used in Crop Protection, Publication 1093*, 4th edn., Queens Printer, Ottawa.
259 E.C.Naber and G.W.Ware, *Poultry Sci.*, 44 (1965) 875.
260 G.W.Ware and E.E.Good, *Toxicol. Appl. Pharmacol.*, 10 (1967) 54.
261 G.W.Ware and E.E.Good, *J. Econ. Entomol.*, 50 (1967) 530.
262 J.F.Thompson, *J. Gas Chromatog.*, 6 (1968) 560.
263 K.C.Walker and M.Beroza, *J. Assoc. Offic. Agr. Chemists*, 46 (1963) 250.
264 L.C.Mitchell, *J. Assoc. Offic. Agr. Chemists*, 43 (1960) 810.
265 D.P.Gowing and R.W.Leeper, *Science*, 122 (1955) 1267.
266 D.J.Reed, *Science*, 148 (1965) 1097.
267 L.A.Sciuchetti and G.K.Nielsen, *J. Pharm. Sci.*, 56 (1967) 244.
268 M.P.Thomas and H.J.Ackermann, *J. Agr. Food Chem.*, 12 (1964) 432.
269 L.Fishbein and W.L.Zielinski, Jr., *J. Chromatog.*, 28 (1967) 418.

Chapter 5

DRUGS

1. CHLORAL HYDRATE

Chloral hydrate (hydrated trichloroacetaldehyde) is prepared by chlorination of a mixture of ethanol and acetaldehyde followed by hydration of the intermediate trichloroacetaldehyde or by the action of hypochlorous acid on trichloroethylene. Medicinally, chloral hydrate is a common central nervous system depressant used primarily as a hypnotic and sedative. However, chloral hydrate has been used in large quantities in the production of DDT (*via* condensation with chlorobenzene), herbicidal formulations and in the crosslinking of nylon fibers.

The hypnotic action of chloral hydrate depends on its reduction to trichloroethanol in the liver (catalyzed by alcohol dehydrogenase) and other tissues including whole blood. Chloral hydrate is also a metabolite of the extensively employed solvent trichloroethylene (probably *via* an initial step leading to the formation of trichloroethanol, then conjugation and excretion of the latter as a glucuronide).

The toxicity[1], pharmacology[1, 2], and mutagenicity[3, 4] of chloral hydrate have been described. Analytical procedures for chloral hydrate include colorimetry[5-8], titrimetry[9], and polarography[10].

The gas chromatographic analysis of trichloroethanol, chloral hydrate, trichloroacetic acid, and trichloroethanol glucuronide (urochloralic acid) has been reported by Garrett and Lambert[11]. The procedure involved the initial extraction of trichloroethanol and chloral hydrate from alkanized water or urine with ether, and a portion of the ether extract assayed with chlorobutanol and chloroform as internal standards. (Trichloroacetic acid and trichloroethanol glucuronide are not extracted under these conditions.) The trichloroacetic acid was decarboxylated as the potassium salt and the resultant product (chloroform) extracted into ether and assayed. The trichloroethanol glucuronide was enzymatically hydrolyzed and the hydrolysate assayed for the trichloroethanol formed. The sensitivities obtained were 0.5 μg of trichloroethanol and chloral hydrate (3.3 and 3.02 mμ moles, 10^{-9} moles, respectively), 1.0 μg of trichloroacetic acid (6.1 mμ moles) in 2.0 ml of sample and 0.5 μg of trichloroethanol glucuronide (1.54 mμ moles) in 3.0 ml of sample. An F & M Model 700 gas chromatograph was used equipped with an electron-capture detector, Minneapolis Honeywell recorder and disc integrator. The column was 8 ft. × ¼ in. stainless steel containing 20% Carbowax 20M on 60–80 mesh Chromosorb W. The operating parameters were injection port 160°C, column 125°C, detector 190°C, helium carrier gas at 60 ml/min while the purge gas (90% argon–10% methane) was fixed at 140 ml/min. Figure 1 illustrates a chromatogram of ether-extracted chloral hydrate and trichloroethanol with chloroform and chlorobutanol as internal standards.

Better resolution of the chloral hydrate and chloroform peaks could be obtained by decreasing either the carrier gas flow rate or the column temperature, but only at the expense of prolonging the retention time and increasing the width of the trichloroethanol and chlorobutanol peaks.

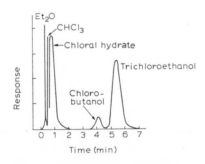

Fig. 1. A typical chromatogram for a 5 μl aliquot of 1:1 ether extract of a solution of chloral hydrate and trichloroethanol (1.40 and 1.36 mg/100 ml, respectively). Pulse = 15. Sensitivity: 0.001 of maximum.

The GLC determination of trichloroethanol in urine has also been described by Sedivec and Flek[12]. A Carlo Erba Model GD gas chromatograph was used equipped with a flame-ionization detector and a 2 m × 2 mm stainless steel column containing 5% polyethylene glycol 20M on 60–80 mesh Chromosorb W. The column temperature and injection port were 130 and 150°C, respectively, the carrier was nitrogen at 16 ml/min and the detector gases were hydrogen at 20 ml/min and air at 300 ml/min.

GLC methods for the determination of chloral hydrate in capsules and liquids have been developed by Snell[13] using electron-capture detection and Fehringer[14] using flame-ionization detection for the analysis of mono- and dichlorobenzene contaminants in chloral hydrate. Chloral hydrate was reacted with sodium hydroxide and the reaction mixture extracted with petroleum ether, concentrated and chromatographed on 18% UNCON 75-H-90,000-coated Anakron ABS 100/120. An electron-capture gas–liquid chromatographic method[15] for chloral hydrate as a possible metabolite of the organophosphorus pesticide trichlorfon in plant and animal tissues has been described in Chapter 4.

The use of ion-exchange chromatography for the determination of chloro derivatives of acetic acid has been described[16]. Trichloroacetic acid (or its sodium salt) is separated from chloral hydrate by passing the solution through an anionic column of AV-17 (Cl⁻ form) with subsequent elution of chloral hydrate and trichloroacetate anions with water and 1 N NaOH, respectively.

2. ANTINEOPLASTIC DRUGS

2.1 Cyclophosphamide

The alkylating antineoplastic agent cyclophosphamide (N,N-bis(2-chloro-ethyl)-N′, O-propylenephosphoric acid ester diamide, endoxan) is prepared by treating bis(2-chloroethyl)-phosphoramide dichloride with propanolamine. Its anti-tumor spectrum and chemotherapy in neoplastic disease have been well documented[17, 18]. The agent was prepared with an inert transport moiety and activation only *in vivo* (to nor · HN2) in an attempt to circumvent some of the toxic effects of the other alkylating agents. Anti-tumor agents such as cyclophosphamide exhibit marked alopecic properties[19]. Dolnick *et al.*[20] have recently described the utility of cyclophosphamide as a chemical defleecing agent in sheep.

The teratogenicity of cyclophosphamide in the rat[21, 22], mouse[23-25] as well as its mutagenicity in mice[26] and *Drosophila*[27] and induction of chromosome aberrations in marine and human cells *in vivo*[28] and human leukocytes *in vitro*[29] have been described.

The pharmacologic characterization[30] and the *in vivo* activation of cyclophosphamide to cytostatic and alkylating metabolites has been described by Brock and Hohorst[31] and suggested to occur as shown in Fig. 2. In rat liver microsomes,

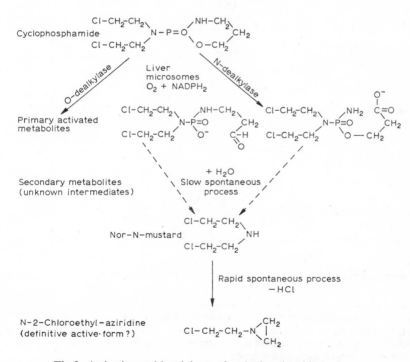

Fig. 2. Activation and breakdown of cyclophosphamide *in vivo*.

cyclophosphamide is transformed into the primary active metabolite *via* N- or O-dealkylation by means of a mixed function oxidase. The primary metabolite is unstable, spontaneously but slowly decomposed to form nor-N-mustard, which may be rapidly transformed into N-chloroethyl-aziridine by intramolecular alkylation. It is of importance to note that, in contrast to cyclophosphamide, nor-N-mustard, and N-2-chloroethyl aziridine, the activated metabolite, is highly hydrophilic and hence will permeate the cell membranes less readily and thus its distribution in the tissues may be reduced. (This may possibly lead to a higher specificity of the active metabolite for tumor cells.) The spontaneous secondary breakdown of the primary metabolite into nor-N-mustard and N-2-chloroethyl aziridine does not occur in the blood or body fluids. Table 1 illustrates the R_F values of alkylating metabolites and

TABLE 1

CHROMATOGRAPHY[a] OF RAT'S SERUM AFTER i.p. ADMINISTRATION
OF CYCLOPHOSPHAMIDE (1000 mg/kg)

	Alkylating activity		
Time after injection (min)	*R_F mean values (% rel. conc.)*		
30	0.87 (52)	0.71 (39)	0.60 (9)[b]
60	0.87 (50)	0.71 (35)	0.59 (15)[b]

Compound	*Comparative substances*	
	Formula	*R_F*
Nor-N-mustard	$Cl-CH_2-CH_2 \diagdown NH$ $Cl-CH_2-CH_2 \diagup$	0.60
Cyclophosphamide	$ClCH_2-CH_2 \diagdown \quad NH-CH_2 \diagdown$ $N-P=O \qquad CH_2$ $ClCH_2-CH_2 \diagup \quad O--CH_2 \diagup$	<0.95[c]
N,N-Bis(2-chloroethyl)-O-(3-amino-propyl)phosphoric acid amido-ester	$ClCH_2-CH_2 \diagdown \quad O^- \qquad {}^+NH_3 \diagdown$ $N-P=O \qquad\qquad CH_2$ $ClCH_2-CH_2 \diagup \quad O-CH_2-CH_2 \diagup$	0.67
N-2-Chloroethyl-aziridine	$ClCH_2-CH_2-N \diagup\!\!\overset{CH_2}{\big|}\!\!\diagdown CH_2$	0.45

[a] Ascending chromatography at 6 °C of neutralized perchloric acid extracts on S & S paper 2043b
Solvent, *n*-butanol–acetic acid–water (6:2:2). Corroborative detection of the alkylating fractions
on air-dried paper by the NBP reagent in the form of violet spots. Sensitivity, 0.015 μmole.
[b] Amount extractable by methylene chloride.
[c] Does not directly react with NBP; corroborative test after 10 min HCl hydrolysis at 100 °C.

comparative reference substances found in rat serum at 30 and 60 min after administration of cyclophosphamide and shows that only two cytostatic alkylating metabolites were found but practically no nor-N-mustard or N-2-chloroethyl aziridine. Thus,
the breakdown of the primary active metabolite is said to take place, if at all, only within
the cells and possibly only at the site of interaction with the cellular receptor.

The fate of cyclophosphamide labeled with carbon-14 or tritium acting on animal cells *in vitro* has been reported[32]. Cyclophosphamide labeled mainly with ^{14}C
was injected intraperitoneally into rats with 50–70% of the ^{14}C appearing in the urine
within 24 h. After the contact of ^{14}C-labeled cyclophosphamide with animal cells,
paper chromatography in methanol–pyridine–water (4:2:1) gave a fraction with
$R_F = 0.9$ (presumably unaltered ^{14}C-cyclophosphamide) and two fractions with R_F
values of 0.67 and 0.09, respectively.

Paper chromatography was also used to elaborate the incubation of slices of
rat liver, kidney slices and Yoshida ascites cells or homogenates of solid Yoshida
tumors with ^{14}C-cyclophosphamide. Results suggested that metabolic changes of
cyclophosphamide were greater for kidney than for liver, with a minimum metabolism of cyclophosphamide in the tumor preparations tested.

Rauen and Norpoth[33] studied the metabolism of cyclophosphamide by an
analysis of deproteinized blood filtrates from rats which had been "activated" by
phenylethylbarbituric acid prior to drug administration. High-voltage electrophoresis,
column and paper chromatographic techniques were used in this elaboration. The
latter techniques utilized Schleicher and Schüll No. 2043b paper with *n*-butanol–
acetic acid–water (6:6:2) in a descending development at 6°C for the analysis of
cyclophosphamide metabolites. The following systems were used for the analysis of
aliphatic acids: (*a*) *n*-propanol–eucalyptol–formic acid (50:50:20), ascending development at room temperature, (*b*) *n*-propanol–water (saturated with SO_2), ascending
development at room temperature, (*c*) methyl cellosolve–eucalyptol–formic acid
(75:75:30), ascending development at room temperature, and (*d*) *n*-butanol–acetic
acid–water (6:2:2), descending development at 6°C. Detection was accomplished
with an indicator solution composed of Bromocresol Green and Bromophenol Blue.

This study, together with earlier observations[31, 34] led to the establishment of
a metabolic scheme for cyclophosphamide *in vivo*, *e.g.*

In addition, 3-aminopropyl phosphoric acid and hydracrylic acid (β-hydroxy-propionic acid) have been identified as metabolic products.

The metabolism of cyclophosphamide *in vitro* has been studied by Dede and Farabollini[35]. Livers originating from pregnant or tumor-bearing rats metabolized cyclophosphamide as shown by paper chromatography using methanol–pyridine–water (8:4:2) as developer. (Cyclophosphamide, R_F 0.82, was converted into an unidentified compound with R_F value 0.70.) Cyclophosphamide was not metabolized by normal livers or by normal livers incubated with placenta or fetus homogenate.

The degradation of cyclophosphamide in aqueous media has been studied by Hirata *et al.*[36] and Friedman *et al.*[37, 38]. Figure 3 illustrates the overall reaction

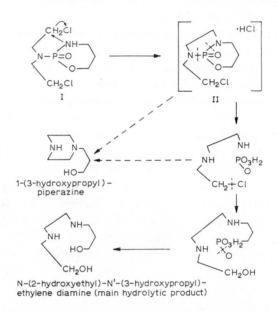

Fig. 3. Hydrolysis of cyclophosphamide in boiling water.

sequence of the hydrolysis of cyclophosphamide (heated to reflux in distilled water) according to Friedman[37]. The main pathway apparently involves an initial intra-molecular alkylation (I → II) takes place, followed by a sequence of simple hydro-lytic cleavages of the amide (P–N) and ester (O–P) bonds.

Hirata *et al.*[36] found the aqueous hydrolysis of cyclophosphamide to be inde-pendent of pH except at pH < 1 and > 11. The primary decomposition product as isolated by TLC was suggested to be 2-[N-(2-chloroethyl)-N-(2-hydroxyethyl)amino]-perhydro-2H-1,3,2-oxazophosphorine-2-oxide.

Two major metabolites (I and II) were recently found in the urine of dogs which were given cyclophosphamide by intravenous injection[39]. The spectroscopic properties of II suggest that it is a 4-keto cyclophosphamide, while the other component of the urinary metabolites is likely to be the ring-opened product (III), *viz.*,

$$\text{III}$$

This finding appears to confirm the previous hypothesis that activation of cyclophosphamide involves oxidation at C_4 and raises the question of the tautomeric behavior of the ketocyclophosphamide which would influence the nature of its hydrolysis and therapeutic reactivity. It thus appears probable that 4-ketocyclophosphamide is either the active form of cyclophosphamide or a precursor of it since II has been found to be 50 times more effective than cyclophosphamide in restricting the development of HE p-2 cells (a cultured strain of cells from a human neck cancer).

2.2 BCNU

BCNU [1,3-bis(2-chloroethyl)-1-nitrosourea], prepared by the reaction of 2-chloroethylamine in base with 3-(2-chloroethyl)-1-methyl-1-nitrosourea, is a potent antitumor agent that is highly effective in intraperitoneal and intracerebral L-1210 leukemia, a distinct feature seldom found in most conventional agents. Although considered to be an alkylating agent, BCNU differs from typical derivatives of 2-chloroethylamine in having several reaction sites in addition to the carbon–chlorine bond which are potentially liable to attack by a variety of reagents under normal physiological conditions. The resultant transient chemical species may undergo further extensive biotransformations.

The chemical properties of the haloalkyl nitrosoureas[40], the physiological disposition of BCNU in man and animals[41], its clinical trials[42], inhibition of DNA synthesis[43], *in vivo* antiviral activity[44], hepatotoxicity in rats[45], and production of chromosomal defects in human lymphocytes[46] have all been reported. Loo *et al.*[47, 48] elaborated the chemical and pharmacological properties of BCNU. It was found to be most stable at pH 4 (in more acidic, and in solutions of pH above 7, it decomposed rapidly). In plasma, BCNU has a half-life of 20 min *in vitro* and less than 15 min *in vivo*. BCNU's alkylating action is not caused by the slow hydrolysis of chlorine nor does it appear to belong to the group of alkylating agents such as bis(2-chloroethyl)-methylamine. As suggested by the authors, if it is an alkylating agent at all, it must owe the alkylating action to one of its degradation products. BCNU was 80 % bound to human plasma protein and when administered intravenously to the dog, it entered the cerebrospinal fluid (CSF) readily and disappeared rapidly from the plasma and CSF. The total amount of unchanged drug excreted in the urine in 4 h was less than

0.1% of the dose. Both colorimetry and paper chromatography were utilized for the analysis of BCNU and its decomposition and metabolic products. Whatman No. 3MM paper was used in ascending development with either 0.1 M sodium acetate buffer of pH 4.4 or petroleum ether (b.p. 110–115°C) as solvents. Depending on BCNU concentration, detection was achieved by any of the following procedures. (1) Direct illumination with UV light at 254 nm, (2) sprayi g first with sulfanilamide in 0.2 N HCl, followed by N-(1-naphthyl)ethylenediamine dihydrochloride in 95% butanol, (3) diphenylamine–palladium chloride reagent[49] (1.5% diphenylamine–0.1% palladium chloride (5:1) and 2% sodium chloride in water) followed immediately by irradiation with UV light at 254 nm.

DeVita et al.[41] studied the stability and physiological disposition of ^{14}C-BCNU (uniformly labeled on all four carbons of the chloroethyl groups) in man, monkeys and mice. Radioactivity was excreted slowly in man and monkeys and rapidly in mice. Urinary excretion accounted for the major portion of the isotope although as much as 10% was excreted as CO_2. The compound was rapidly degraded (no intact drug was found promptly after administration), although plasma levels of the isotope were prolonged by protein binding of a portion of the drug.

Thin-layer chromatography on Silica Gel G with chloroform as developer was used to elaborate the homogeneity of ^{14}C-BCNU as well as to effect the separation and identification of urinary and tissue metabolites and degradation products. Detection was accomplished by spraying the plates with Griess reagent and 4 N hydrochloric acid and heated for 10 min in an oven at 90°C to develop color. (BCNU formed a purple spot against a light-pink background.) Radioautograms were prepared by covering the chromatogram with thin polyethylene plastic film and placing in an 8 × 10 in. cassette for exposure to Kodak single-coated blue-sensitive medical X-ray film for three days. The mechanism of action of BCNU is unknown. It has been suggested by Johnston[50] that BCNU may cleave at the bond between the nitrosated nitrogen and the carbonyl leading to the intermediate formation of chloroethyl isocyanate.

In aqueous solution the decomposition products of BCNU were identified as 2-chloroethylamine hydrochloride, acetaldehyde, nitrogen, and carbon dioxide[51]. Presumably, 2-chloroethylamine hydrochloride is derived from the non-nitrosated half of the BCNU molecule through the intermediate formation of the isocyanate. It was suggested that the nitrosated half of the BCNU molecule under proper chemical or physiological conditions, is converted into the unstable diazoalkane which in the absence of a substrate to be alkylated decomposes into acetaldehyde, one of the degradation products of BCNU in aqueous solution.

The distribution of radioactivity from ^{14}C-labeled BCNU in tissues of mice and hamsters after intraperitoneal administration of the agent has been reported by Wheeler et al.[52]. The purity of the labeled agent as well as isolation of metabolites was determined by a combination of TLC on silica gel with chloroform as developer and radioautography of chromatograms using no-screen blue-sensitive X-ray film.

Comparable quantities of ^{14}C from BCNU labeled in all 4 carbon atoms of the two chloroethyl groups and from BCNU labeled only in the carbonyl carbon were found in various tissues of mice including the brain and equal quantities of ^{14}C were present in the sensitive and resistant plasma cytomas.

Montgomery *et al.*[53] have used GLC procedures for the elucidation of the stability and modes of decomposition of BCNU in various solutions. F & M Models 1720 and 5657A gas chromatographs were used with 10% Versamid 900 on Ultraport and 10% silicone gum rubber W-98 on 80–100 mesh Diatoport S-S columns, respectively. The sequence of reactions of BCNU with two equivalents of 0.1 N NaOH was as follows.

$$ClCH_2CH_2N - \underset{\underset{ON}{|}}{C} - \underset{\underset{H}{|}}{N} - CH_2CH_2Cl \ + \ OH^- \longrightarrow$$

$$Cl^- + CH_3CHO + N_2 + H_2O + O=C=NCH_2CH_2Cl \longrightarrow$$

$$[\ ^-O_2CNHCH_2CH_2Cl\ ] \longrightarrow \text{(ring)}=O + Cl^- \longrightarrow H_2NCH_2CH_2OH$$

2.3 Myleran

The alkylating agent Myleran (1,4-di(methanesulfonyloxy)butane, busulfan) is prepared by the reaction of 1,4-butanediol and methanesulfonyl chloride in the presence of a base such as pyridine or dimethylaniline. Myleran is a clinically useful chemotherapeutic agent active against chronic myeloid leukemia[54, 55]. Its acute and chronic toxicity[56, 57], the synthesis of tritiated[58, 59] ^{35}S- and ^{14}C-Myleran[59, 60], and the metabolism of labeled and unlabeled Myleran in the rat[60–64], mouse[62], rabbit[64], and human[65, 66] have been described. The comparative effects of Myleran on spermatogenesis[67], its implication in human teratogenesis[68], mutagenicity in rats[69], mice[69, 70], and *Drosophila*[71, 72], and alkylation of DNA[73] have also been reported.

Roberts and Warwick[74, 75] identified the major metabolites of Myleran and proposed the route of metabolism as shown in Fig. 4. Myleran (I) reacts with cysteine or a cysteinyl moiety (II) to form a cyclic sulfonium ion (III) which undergoes decomposition to tetrahydrothiophene (IV), which is converted to tetrahydrothiophene-1,1-dioxide (V) and then to 3-hydroxy-tetrahydrothiophene-1,1-dioxide (VI) (which was found in the urine of rats, mice and rabbits). While III, IV and V were not isolated, it was shown that S-ß-L-alanyl tetrahydrothiophenium mesylate (VII) and IV and V were metabolized almost entirely giving (VI). The urinary metabolites of the rat, rabbit and mouse were isolated by paper chromatography following the intraperitoneal administration of 2,3-^{14}C-Myleran and ^{35}S-IV, V, VI, and ^{14}C-VII. Whatman No. 1 paper was used with the following solvent systems: (*a*) *n*-butanol–ethanol–propionic acid–water (20:10:10:4), (*b*) *n*-butanol–acetone–dicyclohexylamine–water (20:20:10:4), and methylethyl ketone–acetic acid–water (3:1:1). An autoradiograph of paper chromatogram prepared from urine excreted during the first 24 h after injection is illustrated in Fig. 5.

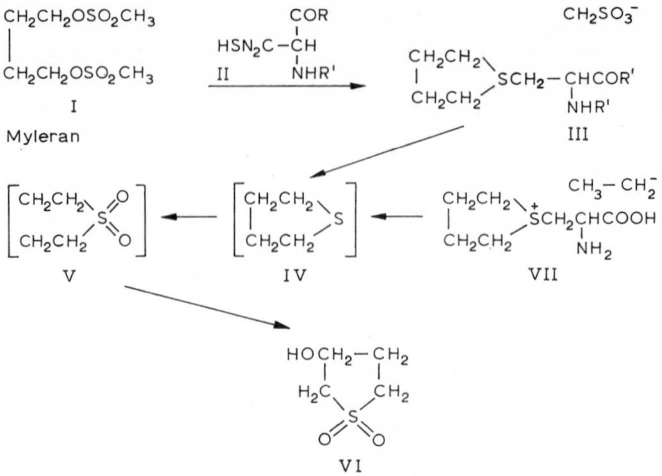

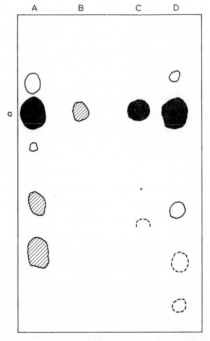

Fig.4. Metabolism of Myleran. R = OH, amino acid, peptide or protein residue; R′ = H, amino acid, peptide or protein residue.

Fig.5. Autoradiograph of a unidimensional paper chromatogram of urine excreted from the rat after injection of A, 2,3-^{14}C-Myleran; B, 2,3-^{14}C-S-β-alanyl-tetrahydrothiophenium mesylate; C, ^{35}S-tetrahydrothiophene; D, ^{35}S-tetrahydrothiophene-1,1-dioxide. Spot a is radioactive 3-hydroxy-tetrahydrothiophene-1,1-dioxide. Chromatogram developed with n-butanol–ethanol–propionic acid–water (20:10:10:4). Spots showing strongest radioactivity are represented by black, medium radioactivity shaded, weak radioactivity open circles with solid line, and the weakest open circles with dotted line.

Peng[76] studied the distribution and metabolic fate of ^{35}S-Myleran in normal and tumor-bearing rats to obtain information for the optimum therapeutic use of the drug, especially during prolonged therapy. ^{35}S-labeled Myleran was prepared by the reaction of methane-^{35}S-sulfonylchloride[274] with 1,4-butanediol in anhydrous ether. Paper chromatography and radioautography were utilized both for the elaboration of radio-purity of ^{35}S-Myleran as well as urinary metabolites following the intraperitoneal administration of the labeled drug to normal and lymphosarcoma-bearing rats of the Slonaker strain. Characterization of the urinary transformation products was effected by ascending paper chromatography on Whatman No. 1 paper using 95% ethanol as solvent. The radioautograms were prepared from the paper chromatograms using DuPont Xtra Fast, Type 508, X-ray film. For quantitation, the paper chromatograms were cut in sections and counted in a windowless gas-flow counter.

The main biotransformation product in the urine was methanesulfonic acid, existing mostly in the free form and to some extent in a combined state. In addition, some unchanged Myleran and a small amount of labeled sulfate were also recovered from the urine. The spleen and bone marrow were found to exhibit selective accumulation of the drug. The high activity observed in the spleen, especially in tumor-bearing rats, is thought to be due to the selective affinity this organ possesses for the chemical; this is in accordance with clinical observations of the regression of the enlarged spleen in leukemic patients undergoing Myleran therapy[77]. Paper electrophoretic studies of the plasma of the treated animals failed to reveal any stable binding between the labeled Myleran or its metabolites and the various plasma protein fractions.

The distribution and fate of ^{35}S- and ^{14}C-labeled Myleran (I and II) as well as ^{32}P-Tepa and ^{14}C-TEM following oral and i.v. administration to cancer patients was reported by Nadkarni et al.[65]. The chemical and radio-purity of each drug and the elaboration of urinary metabolites was studied by paper chromatography. Ion-exchange chromatography and inverse isotope dilution were also utilized for urinary metabolic studies.

$$
\begin{array}{cc}
\overset{*}{\diagup}\text{CH}_2\text{OSO}_2\text{CH}_3 & \overset{*}{\diagup}\text{CH}_2\text{OSO}_2\text{CH}_3 \\
\text{CH}_2 & \text{CH}_2 \\
| & | \\
\text{CH}_2 & \text{CH}_2 \\
\diagdown\overset{*}{\text{CH}_2\text{OSO}_2\text{CH}_3} & \diagdown\text{CH}_2\text{OSO}_2\text{CH}_3 \\
 & \overset{*}{} \\
\text{I} & \text{II}
\end{array}
$$

Whatman No. 1 paper was used with the solvent systems n-butanol–acetic acid–water (4:1:1), methanol–water (19:1), and acetone–acetic acid–water (8:1:1).

With ^{14}C-Myleran, the average urinary excretion was of the order of 25–30% of the injected dose. Radioactivity from ^{35}S-Myleran, however, was excreted to the extent of 45–60%. The quantitative difference between the excretions of sulfur-labeled and carbon-labeled Myleran indicated a more rapid elimination of the methanesulfonyl moieties as compared with the four-carbon chain. The main urinary metabolite of ^{35}S-Myleran was identified as an alkali salt of methanesulfonic acid by co-chromatography with several reference compounds and the hydrolytic product of the drug itself. Quantitatively methanesulfonic acid accounted for over 95% of the excreted radioactivity. Table 2 depicts the R_F values of Myleran, Myleran hydrolysate, and urinary metabolite of Myleran.

Ion-exchange elution of urine collected after ^{14}C-Myleran treatment resulted in the separation of twelve radioactive components. No unchanged drug was found to be present in the urine.

TABLE 2

R_F VALUES OF MYLERAN, MYLERAN HYDROLYSATE, AND URINARY METABOLITE OF MYLERAN

Test material	R_F values		
	A^a	B^a	C^a
Myleran	0.83	0.68	0.96
Myleran hydrolysate	0.09	0.40	0.17
Urinary metabolite	0.10	0.38	0.17
NH$_4$-methanesulfonate	0.10	0.42	0.18

^{a} Solvent systems: A, n-butanol–acetic acid–water (4:1:1); B, methanol–water (19:1); C, acetone–acetic acid–water (8:1:1).

In comparison of all the drugs tested in the above study, e.g. ^{35}S- and ^{14}C labeled Myleran, ^{32}P-Tepa and ^{14}C-TEM, a common pattern of metabolism of alkylating drugs in the human was indicated. A study of urinary metabolites of each drug suggested detachment of the alkylating moiety from the "carrier" portion of the drug molecule. For example, whenever the "carrier" moiety was labeled, e.g. the methanesulfonyl unit in Myleran or the P atom in Tepa, there was primarily a single metabolite which accounted for almost complete excretion. However, when the alkylating moiety was labeled, e.g., the butylene unit in Myleran or the ethylene-imino unit in TEM, the drug yielded a large number of metabolites in the urine. It is also pertinent to note the physiological disposition of Myleran, TEM, and Tepa in laboratory animals and in the human; the drugs were shown to be excreted in the urine mostly as inert catabolic products which possess no known anti-tumor action.

The synthesis, uptake, distribution and metabolic fate of ^{14}C- and ^{35}S-labeled Myleran in the rat have been reported by Trams et al.[60]. Ion-exchange and paper chromatography were used for the analysis of urinary metabolites following intravenous and oral administration of the tagged drugs. Tables 3 and 4 depict the R_F values of ^{14}C- and ^{35}S-labeled Myleran and urinary metabolites.

TABLE 3

R_F VALUES OF VARIOUS GLYCOLS, RELATED ALIPHATIC ACIDS AND R_F VALUES OF THE MAJOR URINARY METABOLITE AND OXIDATION PRODUCTS OF ^{14}C-LABELED MYLERAN

Compound	R_F			
	A^a	B^a	C^a	D^a
[14]C-Myleran	0.82	0.95	0.95	
Neutral volatile urinary product	0.56	0.80	0.38	
1,4-Butanediol	0.67	streak	0.38	
1,3-Butanediol	0.73		0.39	
2,3-Butanediol	0.77		0.47	
1,2-Propanediol	0.69		0.15	
1,3-Propanediol	0.71		0.28	
1,2-Ethanediol	0.66		0.07	
Oxidized urinary metabolic product	0.60	0.80		0.78
Oxalic acid	0.55	0.23		streak
Succinic acid	0.73	0.80		0.70
Malonic acid	0.73	0.70		0.67
Glycolic acid	0.68	0.57		0.59
Lactic acid		0.70		0.72
Malic acid	0.66	0.52		0.51
Tartaric acid	0.55	0.38		0.39

[a] Solvent systems: A, 95% methanol; B, methylethylketone–propionic acid–water (75:25:30); C, diethyl ether–chloroform (1:1) saturated with water; and D, n-butanol–ethanol (1:1).

The overall metabolic pathways of Myleran were only partially established From the ^{35}S-Myleran injected rats the urinary radiosulfur was recovered almost quantitatively in the forms of methanesulfonate and unchanged drug. The small yield of inorganic ^{35}S indicated that only minute amounts of methanesulfonate were split further. The relatively high yield of [14]CO$_2$ from ^{14}C-Myleran established that a breakdown of the butane chain occurred following the initial metabolic cleavage of the drug. The major metabolic product of ^{14}C-Myleran was found to be neutral

TABLE 4

R_F VALUES OF ^{35}S-LABELED MYLERAN, URINARY ^{35}S-LABELED METABOLITE AND METHANESULFONIC ACID

Compound	R_F		
	A^a	B^a	C^a
[35]S-Myleran	0.82	0.84	0.96
[35]S-Labeled rat urine	0.80	0.10	0.17
Methanesulfonic acid	0.80	0.11	0.17

[a] Solvent systems: A, 95% methanol; B, n-butanol–acetic acid–water (4:1:1); and C, acetone–acetic acid–water (8:1:1).

References pp. 337–344

and volatile and exhibited the characteristics of a glycol. It was shown that the methanesulfonyl moiety was readily removed from the butane chain as indicated by the high yield of methanesulfonate in the urine of the rat. Both with sulfur- and with carbon-labeled Myleran, excretion of some unchanged drug in the urine was observed in rats.

The comparative metabolism of ^{35}S-Myleran in the rat, mouse and rabbit has been studied by Fox *et al.*[64, 78]. The radioactive urinary metabolites following intraperitoneal administration of the tagged drug in the above species were separated by single-dimensional paper chromatography using two solvent systems, *viz.* *n*-butanol–dioxan–2 *N* ammonia (4:1:5) and *n*-butanol–2 *N* acetic acid (1:1) and further identified by radioautography.

Methanesulfonic acid was the only material excreted in the urine of a rabbit given ^{35}S-Myleran. Rat and mouse urine contained, in addition to methanesulfonic acid and a small percentage of Myleran, two other unidentified components (U-1 and U-2) which appear to be produced by reaction of labeled material with a substance not present in the rabbit. Table 5 depicts the R_F values and distribution of radioactivity in urine after administration of ^{35}S-labeled Myleran.

TABLE 5

R_F VALUES AND DISTRIBUTION OF RADIOACTIVITY IN URINE AFTER ADMINISTRATION
OF ^{35}S-LABELED MYLERAN

Metabolite	R_F		Distribution of radioactivity in urine (%)		
	A^a	B^a	Rat	Mouse	Rabbit
Sulfate	0.00	0.00			
Methanesulfonic acid	0.16	0.12	90	91	100
U-1	0.11	0.18			
Myleran	0.85	0.86	5	4	
U-2	0.98	0.93	2	3	

^{a} Solvent systems: A, *n*-butanol–dioxan–ammonia (4:1:5); B, *n*-butanol–2 *N* acetic acid (1:1).

The uilization of Myleran labeled with tritium in positions 2 and 3 of the butanediol moiety (20 mC/mmole) for the determination of the size of DNA molecules in cells has been studied by Verly and Petitpas-Dewandie[73]. The reliability of the labeled "Myleran" method for the estimation of the weight of an *in situ* DNA molecule biologically characterized depends on (*a*) the stability of the alkyl groups bound to the DNA, and (*b*) the knowledge of the percentage of bound alkyl groups forming bridges between the complementary strands of the macromolecule. In the first set of experiments, calf thymus DNA was left for 3 h in contact with tritiated Myleran before precipitation (addition of solid sodium chloride to bring the concentration to 1 *M*, then of an equal volume of 95% ethanol). The alkylated DNA of stable specific activity was hydrolyzed for 30 min in 98% formic acid at 175°C, passed through an ion-exchange column (Amberlite R-120 H$^+$) eluted with 1 N ammonium hydroxide. (This revealed that most of the radioactivity firmly bound to the DNA is in alkyl groups attached to the bases of the macromolecule.) This base fraction was submitted to two-dimensional paper chromatography, e.g. development in the first direction with methanol–ethanol–conc. HCl–water (50:25:6:19), and in the second direction with n-butanol–conc. ammonium hydroxide–water (86:1:13). Ultraviolet detection revealed fluorescent spots of the four bases in addition to two faint extra ones at the sites given by Brookes and Lawley[79] for the 7-alkyl derivative of guanine (R_F 0.48 and 0.15) and the corresponding diguanyl compound (R_F 0.26 and 0.00). The stable radioactivity fixed on DNA after reaction with tritiated Myleran was found to be mostly in alkylated guanines with about 8% forming bridges between two neighboring guanine residues.

3. MONOALKYL METHANESULFONATES

3.1 Ethyl methanesulfonate ("half-Myleran", EMS)

The mono-functional alkylating agents generally manifest their biological effects (their ability to inhibit tumor growth at concentrations 50–100 times those of the bifunctional compounds). However, it was considered that any in vivo alkylation reactions would be comparable to those undergone by their di- or polyfunctional analogs and permit a more facile determination of the reaction products.

The metabolic fate in the rat of ^{14}C-ethyl methanesulfonate ("half-Myleran") (EMS) ($CH_3CH_2OSO_2CH_3$) was studied by Roberts and Warwick[62].

Urinary metabolites were examined by paper chromatography following intraperitoneal administration of the drug to normal and Walker-tumor-bearing rats. Single- and two-dimensional paper chromatography was performed using Whatman No. 4 paper with the following solvent systems: n-butanol–ethanol–propionic acid–water (20:10:10:4), n-butanol–acetone–water–dicyclohexylamine (20:20:10:4), n-butanol saturated with 3% ammonia, n-butanol–acetic acid–water (4:1:1), and phenol–water (containing sodium cyanide + 0.3% ammonia). Paper chromatograms were exposed to X-ray films for periods of up to four weeks for the preparation of radioautograms and then exposed with various spot reagents (prepared and used by the methods summarized by Block et al.[80]). ^{14}C-Ethyl methanesulfonate was shown to be metabolized by two major pathways, one involving hydrolysis to ethanol, which is rapidly metabolized and excreted. The other route involves the reaction with the thiol group of cysteine or cysteine-containing compounds such as glutathione or protein. Thus, while a number of radioactive metabolites are present in the urine, almost all the radioactivity is incorporated in N-acetyl-S-ethylcysteine and other conjugates or derivatives of S-ethylcysteine. (This was demonstrated by comparing autoradiograms of chromatograms of urine obtained after injection of ^{14}C-ethyl methanesulfonate and S-^{14}C-ethylcysteine.) Methyl methanesulfonate and propyl methanesulfonate have also been shown in this study to be excreted as derivatives of the corresponding alkylated cysteines.

The carcinogenicity of EMS in neonatal mice[81] and its mutagenicity in a wide variety of organisms including *Drosophila*[82], phage[83], *E. coli*[84], hamster cells *in vitro*[85], and induction of dominant lethal mutations in mice[86] and rats[69] have been noted.

3.2 Methyl methanesulfonate

Methyl methanesulfonate (MMS) is prepared from sulfur trioxide and methane. The oncogenic[87] and leukomogenic effects of MMS (resembling X-rays and nitrogen mustard)[88], its utility in cancer chemotherapy[89] and in the sterilization of houseflies *(Musca domestica)*[90] as well as its antifertility effects in mice and male rats[91] have been described. The mutagenicity of MMS in mice[92, 93], bacteriophage[94], and *S. pombe*[95] as well as its methylation proficiency of DNA[96] have also been reported.

The metabolism of MMS in the rat has been studied by Barnsley[97]. Urine collected from rats during the first 16 h after i.v. administration of ^{14}C-MMS was chromatographed on Amberlite CG-400 resin with sodium formate buffer (pH 4.1) yielding seven groups of radioactive fractions which were then examined by paper chromatography. The following components were identified: methylmercapturic acid sulfoxide (acid hydrolysis of the reduction product yielded S-methyl-2-cysteine), 2-hydroxy-3-methylsulfinylpropionic acid (after reduction yielded 2-hydroxy-3-methylthiopropionic acid), methylsulfinylacetic acid (yielded methylthioacetic acid on reduction), and a mixture of methylmercapturic acid and N-(methylthioacetyl)-glycine. The sulfoxide-containing fractions together represented about 60% and the thioether-containing fractions about 20% of the radioactivity of the urine. (The ratio of these values was found to be similar to that found among the metabolites of S-methyl-2-cysteine.) Chromatography of urine on Whatman No. 1 paper with *n*-butanol–acetic acid–water (12:3:5) showed four radioactive peaks at R_F 0.28, 0.47, 0.76, and 0.83. The material at R_F 0.47 represented about 58% of the radioactivity and at this R_F value the sulfoxide metabolites appeared and were not separated when present in urine. The compounds at R_F 0.76 and 0.83 were chromatographically identical with N-(methylthioacetyl)-glycine and methylmercapturic acid, respectively, and together represented about 20% of the radioactivity of the urine. It was concluded that about 80% of the radioactivity excreted was accounted for by metabolites resulting from an initial methylation of cysteine residues by methyl methanesulfonate.

The metabolism and mode of action of simple alkanesulfonic esters has been studied in rodents using ^{14}C-MMS and whole-animal autoradiography, liquid scintillation and single-dimension ascending paper chromatography[98]. The two solvent systems used were *n*-butanol–dioxan–ammonia (4:1:5) and *n*-butanol–acetic acid–water (12:3:5). The detecting reagents employed were ninhydrin, iodoplatinate,

Ehrlich's reagent, pentacyanoaquoferriate, and potassium dichromate–silver nitrate. Less than 30% of the injected radioactivity was excreted or exhaled by the rat after 24 h. (The majority of the radioactivity was fixed in the tissues.) The reaction with glutathione in the liver was found to be the main metabolic process and the S-methyl glutathione formed was excreted in bile. N-Acetyl-S-methylcysteine, S-methylthio-acetic acid and S-methylcysteine were several of the urinary metabolites although the main metabolite remained unidentified. Although it was possible to correlate the pattern of metabolites with the observed biological effects, the significant reaction was believed to be methylation of cellular components.

Swann[99] described the methylation *in vivo* of guanine of DNA and RNA in rat testes following a single i.v. administration of ^{14}C-MMS. Within 24 h, 25% of the injected radioactivity was recovered, 25% as ^{14}CO$_2$ and 20% in the urine. Only 3% was recovered in the urine in the second 24 h. The rate of excretion indicated that reaction occurred within a few hours. Both columns (Dowex 50) and two-dimensional paper chromatography on Whatman No. 1 paper confirmed the isolation of radioactive 7-methylguanine in both DNA and RNA of rat testes. The solvent systems for the paper chromatographic separations were methanol–conc. HCl–water (7:2:1) for the first development and *n*-butanol–aq. ammonia (sp. gr. 0.880)–water (85:2:12) for the 90° development.

4. ANTIBIOTICS

4.1 Mitomycin C

Mitomycin C (I) is an antibiotic isolated from the broth of *Streptomyces caespitosus* and is distinguished from other mitomycin fractions by its thermal stability, high melting point, ultraviolet absorption peak and solubility in organic solvents. Although the primary utility of mitomycin C is as an antineoplastic agent in the treatment of Hodgkin's disease, it has also produced responses in experimental animals in a wide range of solid tumors, such as lung, breast, colon, stomach, pancreas, and osteogenic sarcoma[100] and also possesses strong activity against a variety of bacteria and viruses. Mitomycin C can be considered as a derivative of urethan and of ethylenimine. It is biologically inactive in its natural state, but it becomes a mono- and bifunctional alkylating agent upon chemical or enzymic reduction.

I

The isolation, chemistry and elaboration of structure[101, 102] of mitomycin C, its pharmacology and toxicology[103, 104], clinical trials[100], mode of action[105–107], cellular effects[108-110], crosslinking of DNA[108], mutagenicity in *Drosophila*[111, 112], and induction of chromosome aberrations in cultured human leukocytes[113, 114] have been described.

The isolation and characterization of new fractions of antitumor mitomycins from *S. caespilotus* by column and paper chromatography and paper electrophoresis was described by Wakaki *et al.*[102]. The R_F values for the paper chromatography of mitomycin C in various solvent systems were as follows: aq. butanol 0.70, 80% phenol 0.80, 50% acetone 0.80, *n*-butanol–methanol–water–methyl orange (4:1:2:1.5) 0.70, *n*-butanol–methanol–water (4:1:2) 0.70, and benzene–methanol (4:1) 0.70, and in various concentrations of aqueous ammonium chloride ranging from 0.5% to saturated, 0.70.

Stevens *et al.*[101] elaborated the chemistry and structure of mitomycin C and utilized column, paper, and thin-layer chromatography to effect separation and detection of various hydrolytic products and elaborate product homogeneity. Whatman No. 1 paper was used with the following systems: (A) *n*-propanol–1% ammonium hydroxide (2:1), (B) isopropanol–1% ammonium hydroxide (2:1), (C) *n*-butanol–acetic acid–water (4:2:1), (D) *n*-butanol–morpholine–water (3:1:3), (E) acetone–water (3:1), and (F) methanol–water (1:1). TLC was done using Silica Gel G (Merck) with ethyl acetate–isopropanol (1:2) as solvent. Crude mitomycin C in methanol was purified over alumina (Wohlm, neutral, Brockman activity grade 1), and the methanol eluate chromatographed on paper with solvent systems A, D, and E, above, yielding mitomycin C as a single spot with R_F values of 0.78, 0.83, and 0.85, respectively.

The renal excretion in dogs and metabolism of mitomycin C by tissue homogenates was described by Schwartz and Phillips[104]. Qualitative identification of mitomycin was made by descending chromatography on Whatman No. 1 or 3 paper developed with 0.001 M sodium phosphate buffer (pH 7.0). After development, the spots were identified by absorption under ultraviolet illumination. Spots were also identified by auxanograms using plates of *B. subtilis* spores. For this purpose, either 0.5-cm strips were cut from front to origin through the center of each chromatogram channel (Whatman No. 3) and left on the plates during incubation or entire chromatograms (Whatman No. 1) were placed on plates for 5 min and removed prior to incubation. The metabolism of mitomycin was usually measured by changes in absorption at 363 nm. Mitomycin C disappeared rapidly from the plasma of dogs and rats. (One-third or less was recovered unchanged in the urine.) *In vitro* mitomycin C was metabolized mostly by liver tissue homogenates. The metabolism is characterized by a loss of antibiotic potency due to reductive transformation and changes in the UV spectrum.

4.2 Streptozotocin

Streptozotocin is a broad-spectrum antibiotic isolated from *Streptomyces achromogenes*. Its isolation, structure and chemistry[115, 116], as well as its assay, stability, and antibacterial properties[117, 118] have been described. The phage-inducing capacity[119, 120] of streptozotocin, its antitumor activity (against Ehrlich carcinoma, Sarcoma 51784, and Walker 256 in the mouse and rat)[121], utility in the treatment of human malignant tumors of the pancreatic inlet[122, 123], diabetogenic action[121, 124], carcinogenicity in the Chinese hamster[125], induction of renal tumors in rats[126] and mutagenicity[127–129] have been described.

At room temperature, fully active samples are present after 30 days, while at 4°C, streptozotocin is stable for periods of up to 6 months. Streptozotocin exhibits maximum stability at pH 4, with stability decreasing rapidly at either higher or lower pH[115]. The kinetics of solution degradation of streptozotocin have been studied polarographically, spectrophotometrically and by bioassay[130]. The possibility that streptozotocin acts as an alkylating agent *via* the formation of diazomethane is strongly supported by the presence of N-nitrosomethylamide portion of its structure. (Treatment of the antibiotic with 2 *N* aq. NaOH at 0°C yields diazomethane.) Also, the action spectrum of streptozotocin in terms of mutant classes affected and the degree of this response are similar to that of nitrosoguanidine.

The differentiation of streptozotocin from other known antibiotics by descending paper chromatography has been described by Vavra *et al.*[131]. The activity was located by placing developed strips on trays of agar seeded with *P. vulgaris*, *S. aureus* or *E. coli*.

Figure 6 illustrates a papergram pattern of streptomycin obtained after development with six solvent systems using descending development on Whatman No. 1 paper.

Herr *et al.*[115] described the isolation and purification of streptozotocin. The crude product was purified by partition chromatography on a Dicalite column using the solvent system of *n*-butanol–cyclohexane–McIlwaine's pH 4.0 buffer (20:4:4). The product from the peak fraction was again chromatographed on Dicalite using methylethyl ketone–cyclohexane–McIlwaine's pH 4.0 buffer (9:1:1.43) and the bioactive material from this column countercurrently distributed between methylethyl ketone and water and streptozotocin finally obtained from the anhydrous methylethyl ketone solution following removal of water by azeotropic distillation.

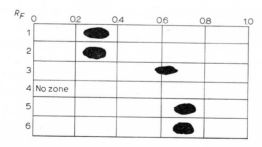

Fig. 6. Papergram pattern of streptozotocin. 1, *n*-butanol–water (84:16); 2, *n*-butanol–water (84:16) plus 0.25% *p*-toluenesulfonic acid; 3, *n*-butanol–acetic acid–water (2:1:1); 4, *n*-butanol–water (84:16), 2 ml piperidine added to 98 ml of *n*-butanol–water mixture; 5, water–*n*-butanol (96:4); 6, water–*n*-butanol (96:4) plus 0.25% *p*-toluenesulfonic acid.

4.3 Patulin

The α,β-unsaturated lactone antibiotic patulin (4-hydroxy-4H-furo[3,2-c]-pyran-2(6H)-one) is derived from the metabolism of several species of *Aspergillus* and *Penicillium* (*e.g. A. clavatus, A. claviforme, P. patulum, P. expansum, P. meliniis,* and *P. leucopus*). Some of these fungal species are likely contaminants of foods. For example, *P. expansum*, the common storage rot of fruits, *A. clavatus, A. terreus, P. cyclopsium,* and *P. urticae*, isolated from flour[132] and *Byssochlomys nivea* and the heat-resistant fruit juice contaminant identified by Keuhn[133] as the *Gymnoascus* species of Karow and Foster[134] have all been shown to produce patulin[135, 136]. Patulin is also produced by fungi in apples[137] and by field crops[138] and has been isolated from a *Penicillium* which infected a malt feed responsible for the death of cows[139]. The potential health hazard of patulin in foods or animal feeds has been stressed by Kraybill and Shimkin[140] and Mayer and Legator[141].

The isolation[138, 142, 143], synthesis[144], and bio-synthesis[145] of patulin as well as its toxicity[135], carcinogenicity[146, 147], inhibition of cell division, nuclear division, or both in bacteria[148], plants[149], and in cell culture[150], its mutagenicity in *Saccharomyces cerevisiae*[141] and induction of chromosome aberrations in human leukocyte cell culture[151], and avian eggs in mitosis[152] have all been reported.

The stability of patulin and penicillic acid in fruit juices and flour has been studied by Scott and Somers[153]. A semiquantitative assay for these compounds based on TLC was developed and applied to the determination of their stability in apple, grape, and orange juices and in whole wheat and bleached flours. Thin layers (0.25 mm) of Adsorbosil 5 (Applied Science) were activated at 80 °C for 2 h, spotted with the food extract and standards of patulin and penicillic acid (1 mg/5 ml of chloroform), developed with anhydrous ether in an equilibrated tank and lightly sprayed with 0.3% aqueous ammonia, then with 4% aqueous phenyl hydrazine hydrochloride[154] and heated at 100 °C for 2 or 3 min. Amounts of patulin were estimated visually by comparing intensities of the yellow spots with standards at the same R_F. Penicillic acid was estimated similarly by yellow fluorescence under long-wave ultraviolet light. (R_F values for patulin and penicillic acid were 0.77 and 0.50, respectively, and limits of their detection in foods were of the order 0.1–0.3 p.p.m.) Patulin was found to be stable for several weeks over the pH range 3.3–6.3 and was slowly inactivated at pH 6.8[155] and stable in grape and apple juice but not in orange juice or the flours (partially due to reaction with thiols)[156]. The reaction of patulin with thiols at a pH as low as 4.5 has been reported previously[157]. Patulin was decomposed by glutathione at pH's as low as 2.3 and 3.0. It has been shown that if high concentrations of patulin are initially present in fruit juices of low thiol content, appreciable concentrations may remain in the processed juice.

Pohland *et al.*[158−160] investigated the chemical stability of patulin in a variety of foods and grain feeds in various solvents and described spectrophotometric[159], thin-layer[158], and gas–liquid chromatographic[160] techniques for its analysis. Patulin was found to be completely stable in apple juice and dry corn. However, in wet corn, Durham wheat, and sorghum and in the presence of SO_2 in water, patulin was observed to be unstable. Sulfur dioxide is often used as a preservative for apple juice and other fruit juices and as a fumigant in the storage of grains and other dry commodities. Hydration of SO_2 and reaction with aqueous patulin is believed to involve the addition of bisulfite in ($HSO_3{}^-$) to the incipient aldehyde function of the hemiacetal moiety of patulin[159]. The quantitative estimation of patulin in corn, wheat, rye, oats and sorghum involved an initial extraction of the grain with acetonitrile–hexane (4:1) followed by further clean-up through preparative TLC. The patulin concentration was estimated using quantitative TLC on Silica Gel 7GF (Mallinckrodt) with development using benzene–methanol–acetic acid (90:5:5) and comparison with a pure patulin standard. Detection was achieved by first spraying the plates with 3% ammonium hydroxide and then with 4% phenyl hydrazine in water[153, 154] and heating the plate 3 min at 110 °C. The lower limit of detection of patulin in corn by this method is 40 μg/kg (the amount observed on the plate at this level was about 0.12 μg while a spot of approximately 0.04 μg pure patulin could be detected).

The quantitative analysis of patulin in apple juice[160] involved an initial ethyl acetate extraction followed by derivatization of patulin (silyl ether, acetate or chloroacetate) and subsequent GLC analysis. A Packard Model 7821 instrument was used with electron-capture and flame-ionization detectors. Table 6 describes the GLC conditions used for the detection of patulin derivatives.

TABLE 6

GLC CONDITIONS USED FOR DETECTION OF PATULIN DERIVATIVES

Column	6 ft. × 4 mm i.d., glass coil	6 ft. × 4 mm i.d., glass coil	10 ft. × 4 mm i.d., glass coil
Column packing	3% JXR on Gas Chrom W	3% JXR on Gas Chrom W	1% SE-30 on Gas Chrom Q
Temperature (°C)	110	130	110
Electrometer sensitivity (A)	3×10^{-11}	3×10^{-11}	3×10^{-11}
Hydrocarbon number	15.2	17.2	
Detector	Hydrogen flame	Hydrogen flame	Hydrogen flame
Detection limit (ng patulin)	ca. 60	ca. 40[a]	ca. 100

[a] Preliminary work with the electron-capture detector showed that the lower limit for this detector was 12 ng.

The use of BSA (bis-trimethylsilyl acetamide) as the silylating reagent permitted the silylation of patulin in 90–95% yields. However, TLC analysis revealed that 5–10% of original patulin remained after silylation. The silyl derivative was unstable even under refrigeration and its GLC analysis resulted in coating of the hydrogen-flame detector requiring frequent cleaning. The acetate and chloroacetate derivatives of patulin proved more amenable for GLC analysis. Using the acetate derivative with hydrogen-flame detection, recoveries of patulin in spiked apple juice were 90% or better with the limit of detection being about 0.7 µg/ml. Figure 7 illustrates gas chromatograms of the acetate and chloroacetate derivatives of patulin from spiked

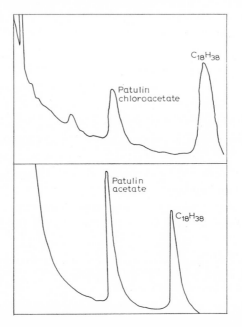

Fig. 7. Gas chromatograms of patulin derivatives prepared from spiked apple juice extract: chloroacetate (top) and acetate (bottom) and internal standard ($C_{18}H_{38}$).

apple juice extract. The chloroacetate derivative of patulin was more stable than the acetate (refrigerated solutions showed no decomposition for periods of up to one week). A lower detection limit of detection for the chloroacetate derivative using flame-ionization and electron-capture detection was 40 ng and 12 ng, respectively.

The paper chromatographic separation of patulin from a number of antibiotics has been described by Betina[161].

4.4 Griseofulvin

The antibiotic griseofulvin (7-chloro-4,6,2'-trimethoxy-6'-methyl-gris-2'-en-3:4'-dione) is a fermentation product of several species of *Penicillium, e.g. P. patulum, P. griseofulvin,* and *P. janczewski.* Since 1959 it has been widely used as an effective oral antifungal agent in the treatment of human and veterinary dermatophytoses.

Griseofulvin is demethylated by sensitive fungi[162, 163] and microsomal liver enzymes[164-166] and induces synthesis of δ-aminolevulinic acid synthetase, a rate-limiting enzyme in porphyrin metabolism in primary tissue cultures of liver cells[167].

Griseofulvin in oral doses at the 1 % level in rats has been reported to produce growth inhibition, testicular atrophy[168], hypercholesterolemia, hepatic porphyrin, hepatomegaly[165], co-carcinogenicity[169], and hepatotoxicity and hepatocarcinogenicity[170-172]. The hepatocarcinogenicity of griseofulvin following parenteral administration of mg quantities to infant mice[173, 174], the enhancement of its acute toxicity by combination with pesticidal synergists (piperonyl butoxide)[175], its binding to nucleic acids and proteins of sensitive fungi[162], potentiation of colchicine toxicity in mice[176], and production of metaphase delay and multipolar mitoses in *Vicia faba* root tips *in vitro* and mammalian cells *in vivo*[177] and teratogenicity in rats[178] have also been described. The isolation[179], structure[180], total synthesis[181], and biosynthesis[182] of griseofulvin have been described.

The metabolism of ^{36}Cl-griseofulvin in mammals has been studied by Barnes and Boothroyd[166]. The major metabolite present in the urine of rats, rabbits and man was characterized as 6-demethyl griseofulvin following oral administration of the tagged antifungal agent (less than 1 % of the administered dose of griseofulvin was detected in all the urine samples). Paper chromatography was used to elaborate the homogeneity of the tagged starting material and for the separation and identification of the urinary metabolites. The mobile solvent system used was the organic phase of benzene–cyclohexane–methanol–water (5:5:6:4) to which 0.5 % of acetic acid had been added after separation. When run on Whatman No. 1 paper by descending chromatography (6 h at 24 °C), griseofulvin appeared as a bright fluorescent spot under UV (R_F 0.9) and 6-demethyl griseofulvin appeared as a darker blue fluorescent spot (R_F 0.15).

The metabolism of ^{14}C-griseofulvin *in vitro* was studied by Symchowicz and Wong[183]. Labeled griseofulvin was incorporated into liver slices from the incubation medium and extensively metabolized by this tissue. At least two metabolites, 6- and 4-demethyl griseofulvin (present in free and conjugated forms) were identified by paper chromatography in the incubation medium as well as in liver slices. There was no noticeable metabolism of griseofulvin by the heart, kidney, lung, and skin slices, although the uptake by the organs from the incubation medium was quite high. The purity of ^{14}C-griseofulvin was determined by ascending paper chromatography on Whatman No. 1 paper using the solvent system of Barnes and Boothroyd[166] as well as *n*-butanol–ammonia (20:1) as the mobile phase and chloroform as the stationary phase for additional chromatographic identification of griseofulvin and its metabolites. (Their positions were located by UV light and the radioactivity pattern was obtained by scanning the strips with a Vanguard 4π automatic strip counter.) A Hewlett-Packard integrator (Model 5202L) attached to the strip counter was used for quantitative evaluation of the radioactive peaks. Figure 8 illustrates a chromatographic pattern of the metabolic fate of ^{14}C-griseofulvin after incubation with rat liver slices. Treatment with Glusulase resulted in an appreciable increase in the relative amount of 4-demethyl griseofulvin, particularly in samples of the incubation medium. The number of observed metabolites was the same in the incubation medium and tissue slices, although the relative ratio differed. In the incubation medium, after Glusulase treatment, the 6-demethyl griseofulvin was the major metabolite (about 42%), followed by appreciable amounts of 4-demethyl griseofulvin (about 28%) and an unidentified material at the origin (about 18%). In liver slices, however, 4-demethyl griseofulvin was predominant (about 63%) and the other metabolites were present in small amounts. The Glusulase-treated samples contained about 12% and 18% of free griseofulvin in the incubation medium and tissue slices, respectively.

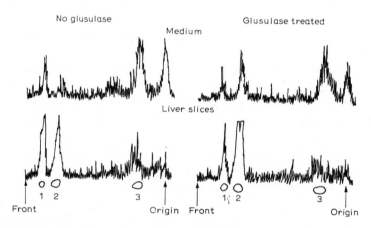

Fig.8. Metabolic fate of griseofulvin-^{14}C after incubation with rat liver slices. Zones marked 1, 2, and 3 correspond to the position on the paper strip of authentic griseofulvin, 4-demethylgriseofulvin, and 6-demethylgriseofulvin, respectively.

Symchowicz et al.[184] described a comparative study of ^{14}C-griseofulvin metabolism in the rat and rabbit. In the rat, griseofulvin metabolism is characterized by extensive biliary excretion represented by a minor pathway and the enterohepatic circulation could not be demonstrated. In the biliary cannulated rat during a 24-h period, about 77% of an i.v. dose was found in the bile and 12% in the urine, whereas in the rabbit only 11% was observed in the bile and 78% in urine. The major metabolite in rat bile was 4-demethyl griseofulvin with small amounts of the 6-demethyl derivative present. In rabbit bile, however, the major metabolite was the 6-demethyl griseofulvin with very small amounts of the 4-demethyl derivative present. In addition, small percentages of free griseofulvin and unidentified metabolite(s) were observed in bile of both species. In urine of the intact and cannulated rats, two major metabolites were present (the 4- and the 6-demethyl derivatives of griseofulvin). Under similar conditions rabbit urine contained 6-demethyl griseofulvin as the predominant metabolite. A comparison of the chromatographic pattern of ether-extractable radioactivity in bile samples (0–2 h) from the rat and rabbit is shown in Fig.9. Spots marked 1, 2, and 3 at the bottom correspond to the position on the chromatographic paper strip of authentic griseofulvin (R_F 0.89), 4-demethyl griseofulvin (R_F 0.78) and 6-demethyl griseofulvin (R_F 0.21), respectively, when developed with the solvent system of Barnes and Boothroyd[166]. The upper portion of the figure (no Glusulase) shows the metabolic pattern of bile samples prior to hydrolysis and represents about 30% of the radioactivity found during 0–2 h and illustrates that these samples contain appreciable amounts of non-conjugated metabolites of griseofulvin. The lower portion (Fig.9) represents the metabolic pattern found in ether extracts of Glusulase-treated samples and represents about 95% of the radioactivity present in the bile at this time period and indicates the metabolites of griseofulvin found from rat and rabbit with the percentage of total radioactivity as indicated.

It is of interest to note that Tomomatsu and Kitamura[185] have shown a breakdown product of griseofulvin, 3-chloro-4,6-demethoxy salicylic acid also to be present in rabbit urine.

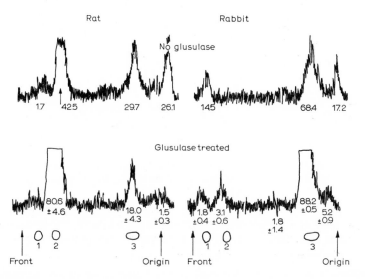

Fig. 9. Metabolic fate of griseofulvin-^{14}C in bile (0–2 h). Zones marked 1, 2, and 3 on the paper strip correspond to the position of authentic griseofulvin, 4-demethylgriseofulvin, and 6-demethylgriseofulvin, respectively. The numbers (±S.E.) show the percentage of total radioactivity on the paper strip associated with each peak.

The quantitative determination of griseofulvin (I) and griseofulvin alcohol (II) in plasma by fluorimetry on thin-layer chromatograms has been described by Fischer and Riegelman[186].

The chromatoplates consisted of a mixture of 2% Baymal (technical colloidal boehmite alumina, DuPont) with silica gel (Merck). The plates (250 μm) were

activated by heating in an oven at 110°C for 1 h. The inclusion of colloidal boehmite
alumina in the silica gel is said to provide an excellent binding with the glass surface
producing layers which are less subject to cracking, powdering and are more stable
to rough handling. Ether extracts of plasma following oral administration of griseofulvin were developed (45 min) on the above plates using a solvent mixture of anhydrous ether–acetone (3:2). The R_F values for griseofulvin and that of the 4'-alcohol
were 0.62 and 0.50, respectively. The chromatograms were scanned for fluorescence
with a modified Photovolt TLC Densitometer Model 530, 12–15 h after development.
(It was found that fluorescence of the spots increased to a small extent upon allowing
the plates to stand at room temperature overnight.) Figure 10 illustrates typical curves

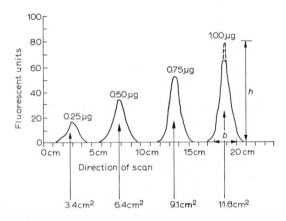

Fig. 10. Typical curves produced by scanning the fluorescent spots of known amounts of griseofulvin
on thin-layer chromatograms. The amount of griseofulvin applied to the plate in 50 λ of chloroform
appears above each peak. The respective areas shown under each curve were calculated using the
triangulation procedure shown on the 1.0 μg curve.

produced by scanning the fluorescent spots of known amounts of griseofulvin on thin-
layer chromatograms prepared and developed as described above. The area under the
symmetrical peaks was calculated using the triangulation technique shown on the
1.0 μg peak in this figure with area (cm²) = $\frac{1}{2} bh$. Linearity between the area under
the peaks produced by the scan and the amount of material applied to the spot existed
between 0.25 and 1.5 μg for both griseofulvin and griseofulvin-4'-alcohol.

Figure 11 depicts the TLC fluorimetric recordings of chromatograms containing
griseofulvin (G) and griseofulvin-4'-alcohol (A) extracted from plasma. Figure 12
shows the extraction curves of griseofulvin and griseofulvin-4'-alcohol from rabbit
plasma obtained from the TLC fluorimetric assay. The results of the TLC fluorimetric
assay and the spectrophotofluorimetric[187] assay on plasma samples following oral
doses of griseofulvin to rabbits were found to be in good agreement.

Cole et al.[188] described the TLC and GLC of griseofulvin and dechlorogriseofulvin extracted from Penicillium urticae. Chloroform extracts of a griseofulvin-

producing isolate were spotted on MN-Silica Gel G-HR (Brinkman) plates and developed with chloroform–acetone (93:7) to a height of 10 cm, and then examined under long-wave UV light and then in normal light after being sprayed with 50% sulfuric acid and heated at 110°C for 30 min. Griseofulvin and dechlorogriseofulvin

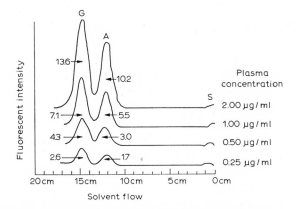

Fig. 11. TLC fluorimetric recordings of chromatograms containing griseofulvin (G) and griseofulvin-4'-alcohol (A) extracted from plasma. An ether extract of 1 ml of plasma to which had been added the indicated amounts of each compound was applied to the plate at point S and developed in an ascending manner in an ether–acetone (3:2) solvent system.

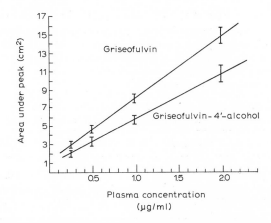

Fig. 12. Extraction curves of griseofulvin and griseofulvin-4'-alcohol from rabbit plasma obtained from the TLC fluorimetric assay. The 95% confidence intervals for each set of experimental points are shown.

appeared together as a bright blue spot at R_F 0.65. (The limit of detection of griseofulvin by TLC was 0.05 μg.) GLC analyses were made with a Barber-Colman Series 5000 gas chromatograph equipped with a hydrogen-flame-ionization detector and disc integrator. The liquid phases used were 1% QF-1 and 1–2% SE-30 coated on Anakrom

ABS 80 mesh and packed into silanized glass columns. The injection port, column and detector temperatures were 270, 235 and 270°C, respectively. Figure 13 shows a gas chromatogram of dechlorogriseofulvin and griseofulvin from *Penicillium urticae* cultures chromatographed on 2% SE-30. Table 7 lists the relative retention times (relative to cholestane) of griseofulvin and dechlorogriseofulvin on SE-30 and QF-1 columns. The structure and identity of the isolated dechlorogriseofulvin was established by comparison of its UV, IR, NMR and mass spectra with those of griseofulvin.

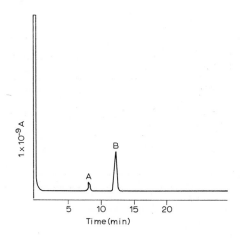

Fig.13. Gas–liquid chromatogram of A, dechlorogriseofulvin and B, griseofulvin from *Penicillium urticae* cultures. Chromatographic conditions, 2% SE-30; temperatures: injector 270°C, detector 270°C, column 235°C.

TABLE 7

RELATIVE RETENTION TIMES[a] OF GRISEOFULVIN AND DECHLOROGRISEOFULVIN FROM *Penicillium urticae* CULTURES COMPARED WITH AUTHENTIC GRISEOFULVIN

Compound	SE–30	QF–1
Authentic griseofulvin	6.31	0.81
Penicillium urticae griseofulvin	6.31	0.82
P. urticae dechlorogriseofulvin	4.11	0.56

[a] Relative to cholestane.

5. N-HYDROXYUREA

N-Hydroxyurea (hydroxyurea) is prepared by the reaction of (*a*) ethyl carbamate with an excess of hydroxylamine and (*b*) potassium cyanate. The latter reaction has been shown to yield both isomers, *i.e.* N-hydroxyurea (m.p. 140°C) and "isohydroxyurea" (m.p. 72°C).

H₂N—C—NHOH   H₂N—C—ONH₂
      ‖   ‖
      O   O

N-Hydroxyurea   Isohydroxyurea

Hydroxyurea is of current utility in cancer chemotherapy as an antineoplastic agent demonstrating activity in chronic myelogenous leukemia, acute lymphoblastic leukemia, and possibly in certain solid tumors. Its antitumor activity, clinical utility, and pharmacology have been well documented[189-191]. The antiviral[192], chemosterilant properties[193], teratogenicity in hamsters[194, 195], rats[196, 197], and chick embryos[197, 198], its induction of chromosome breaks in cultured, normal human leukocytes[199] and chromosome aberrations in Chinese hamster cells and mouse embryo cells in culture[200] and specific inhibition of DNA synthesis and mitosis[201-203] have also been described.

The distribution, excretion and metabolism of hydroxyurea-^{14}C in mice and rats following single parenteral and oral administration was studied by Adamson et al.[204]. It was found that 30–50% of an administered dose was recovered in the urine as urea-^{14}C. In vitro experiments demonstrated that liver and kidney mince could convert hydroxyurea to urea and it was suggested that this was an enzymatic reduction of the hydroxylamine group.

The radiopurity of the hydroxyurea-^{14}C was examined by paper chromatography in n-butanol–ethanol–water (4:1:1) and all of the ^{14}C was recovered at the R_F, 0.34, characteristic of hydroxyurea. Upon high-voltage paper electrophoresis at pH 12, 99.8% of the ^{14}C coincided with the hydroxyurea spot and less than 0.2% with the area corresponding to urea. Paper chromatographic separation of hydroxyurea from its metabolite urea was attempted in various solvent systems. Descending chromatography on Whatman No. 1 paper in isobutanol–water (4:1) failed to resolve hydroxyurea (R_F 0.23) and urea (R_F 0.27). Similarly, there was no separation of hydroxyurea (R_F 0.54) from urea (R_F 0.56) by similar chromatography using 95% ethanol–10 N NH₄OH (4:1). Because of the difference in pK_a of hydroxyurea (10.6) as compared with urea (13.8) high-voltage paper electrophoresis at pH 12 (Whatman No. 3 MM paper, 0.05 M borate buffer) proved excellent for separating the two compounds. Radioactive spots on electrophoresis papers were located by autoradiography on Kodak single-emulsion, blue-sensitive X-ray film. The identities of hydroxyurea-^{14}C and urea-^{14}C spots on electrophoresis papers were further confirmed by their characteristic color reactions with ferric chloride[205] and Ehrlich's p-dimethylaminobenzaldehyde reagent[206].

Colvin and Bono[207] showed that hydroxyurea was enzymatically reduced to urea by mouse liver tissue and that this reaction occurred to a large extent in the hepatic mitochondria. The determination of urea-^{14}C was carried out using the paper electrophoretic procedure of Adamson et al.[204] above.

The mechanism of action of hydroxyurea in cancer chemotherapy has been studied by Fishbein *et al.*[189, 208]. The presence of acetohydroxamic acid in the blood of patients with chronic myelogenous leukemia suggested that the drug is hydrolyzed yielding hydroxylamine, which then cleaves thioesters, particularly acetyl coenzyme A according to the scheme

$$H_2N\!-\!\underset{\underset{O}{\|}}{C}NHOH + 2H_2O \rightarrow NH_2OH + NH_4^+ + HCO_3^-$$

$$NH_2OH + CH_3\!-\!\underset{\underset{O}{\|}}{C}\!-\!S\!-\!CoA \rightarrow CH_3\!-\!\underset{\underset{O}{\|}}{C}NHOH + HS\!-\!CoA$$

Table 8 depicts the R_F values of acetohydroxamic acid (obtained from plasma extracts) compared to hydroxyurea on descending paper chromograms developed with 5 solvent systems. In each case, a major chromatographic spot was obtained which stained reddish-purple with ferric chloride (hydroxyurea stained green) and showed no difference in mobility when chromatographed against or in combination with pure acetohydroxamic acid prepared by a modification of the method of Jeanrenaud[209].

TABLE 8

R_F VALUES OF ACETOHYDROXAMIC ACID, PLASMA EXTRACT AND HYDROXYUREA OBTAINED BY DESCENDING
PAPER CHROMATOGRAPHY

Solvent system	Acetohydroxamic acid	Plasma extract	Hydroxyurea
n-Butanol, water saturated	0.52	0.52	0.19
Phenol, water saturated	0.75	0.75	0.50
Isobutyric acid, water saturated	0.55	0.55	0.52
n-Butanol–isopentanol–formic acid–water (2:2:1:3)	0.40	0.40	0.21
n-Butanol–formic acid–water (3:1:3)	0.50	0.50	0.32

Kofod[205] utilized paper chromatography for the analytical separation and identification of the isomeric hydroxyureas. The preferred solvent system was ethanol–10 N ammonia (20:80) which effected good resolution of the isomers in a 2–4 h development on Whatman No. 1 paper. The isomers were differentiated using a 1% ethanolic solution of picryl chloride. Isohydroxyurea yielded a bright red color, which gradually developed over 1–2 min, and which was further intensified by subsequent exposure of the dry chromatogram to ammonia vapor. Hydroxyurea yielded no original color when sprayed with picryl chloride alone. However, when the dry chromatogram was subsequently exposed to ammonia vapor, a bright orange color was produced spontaneously. The lower limit of sensitivity for both isomers was approximately 1 μg in a spot 1 cm in diameter.

Fishbein and Cavanaugh[210] separated the isomeric hydroxyureas on What-man No.1 paper using n-butanol–acetic acid–water (5:1:4) and affected their differentiation utilizing the multiple spray ninhydrin–pyridine followed by ferric chloride (0.2% solution of ninhydrin in acetone with 5% pyridine followed after air-drying with a 1% ferric chloride solution in methanol).

The detection of a reactive intermediate in the reaction between DNA and hydroxyurea has been described by Jacobs and Rosenkranz[211]. Exposure of DNA resulted in degradation of the polydeoxy nucleotide. This reaction, being dependent on time, temperature and pH, has been associated with the transformation of hydroxyurea to a reactive derivative N-carbamoyloxyurea (I), which was suggested to be responsible for DNA degradation.

$$H_2N-\underset{\underset{O}{\|}}{C}-\underset{\underset{OH}{|}}{N}-\underset{\underset{O}{\|}}{C}-NH_2$$

I

Carbamoyloxyurea was identified by paper and thin-layer chromatography and its chromatographic properties compared with hydroxyurea (Table 9). Spots on paper chromatograms were detected by examination with an ultraviolet lamp (carbamoyloxyurea absorbs in the UV) and by spraying the chromatograms either with 1 N FeCl$_3$ (for hydroxyurea) or with 1% picryl chloride in ethanol followed after drying by exposure to ammonia fumes. Thin-layer chromatography was carried

TABLE 9

R_F VALUES OF HYDROXYUREA AND N-CARBAMOYLOXYUREA

System	Solvent[a]	Hydroxyurea	Carbamoyl-oxyurea
Paper	A	0.41	0.21
Paper	B	0.24	0.12
Paper	C	0.48	0.30
Paper	D	0.14	0.07
Paper	E	0.44	0.27
Thin-layer	F	0.37	0.22
Thin-layer	G	0.78	0.03
Thin-layer	H	0.10	0.00
Thin-layer	I	0.25	0.13
Thin-layer	J	0.36	0.00

[a] For paper chromatography the following solvent systems were used: A, n-butyl alcohol–acetic acid–water (4:1:2); B, n-butyl alcohol–ethanol–water (4:1:1); C, chloroform–methanol–water (7:5:1); D, isobutyl alcohol saturated with water; E, ethanol–NH$_4$OH–water (80:13:7). For thin-layer chromatography the following solvent systems were used: F, ethanol–propyl alcohol (3:7); G, methanol–benzene (7:13); H, chloroform–acetic acid (99:1); I, butyl alcohol–acetic acid (19:1); J, chloroform–methanol (9:1).

out with Eastman Type K301R (silica gel) precoated chromatogram sheets. The spots could be seen after exposure of the chromatograms to iodine vapor.

In addition, they were sprayed either with 1% $FeCl_3$ in 50% ethanol or with 1% sodium aminoprusside in 0.006 M $MgCl_2$[212]. Areas containing hydroxyurea became colored when sprayed with either of these solutions while carbamoyloxyurea did not yield colored complexes, but could, however, be detected with picryl chloride and iodine vapors.

Carbamoyloxyurea has been shown to exist in aged hydroxyurea preparations[213] along with isohydroxyurea and an unidentified nitrosourea[214, 215]. Carbamoyloxyurea has been found to be bactericidal for *E. coli*[213] and the drug-induced killing was independent of cellular metabolism. Ribonucleic acid and protein syntheses were the processes most affected and the lethal action was accompanied by degradation of cellular DNA. In all of these effects the drug differs from hydroxyurea, a primarily bacteriostatic agent that inhibits DNA synthesis and whose lethal action ultimately depends on cellular activity.

6. ISONICOTINYLHYDRAZIDE

Isonicotinylhydrazide (isoniazid, INH) is generally prepared by condensing isonicotinic acid ethyl ester with hydrazine or from 4-cyanopyridine and hydrazine. INH is widely used in man and is the most effective tuberculostatic drug yet to be introduced.

Its utility[216], chemistry and pharmacology[216], mode of action[216–218], and metabolic fate[219–221] have been delineated. Its toxicity[222, 223], tumorigenicity[224, 225], and carcinogenic behavior[226, 227] have also been described. INH (as well as antidepressant mono- and disubstituted hydrazines) has also been found to produce hydrogen peroxide by interaction with oxygen and inactivate transforming DNA[228].

INH can be detoxicated by two primary routes of metabolism, *viz.* hydrolysis and direct conjugation. In man, the major mode of detoxication is direct conjugation. However, the extremely labile C–N bond between the aromatic carbonyl and hydrazine moieties results in a highly important metabolic degradation route by hydrolysis, *viz.*

$$R-CONHNH_2 + H_2O \rightarrow R-COOH + H_2NNH_2$$

In animals, such as dog, which apparently is deficient in the ability to acetylate free amines, the hydrolysis to free acid and hydrazine or its corresponding substituted derivative becomes a major pathway in the metabolism of isoniazid as well as N-acetyl-isoniazid[221]. Figure 14 summarizes the mammalian metabolism of INH. Quantitatively acetylation of INH to N'-acetyl-N²-isonicotinyl-hydrazine is one of the most important metabolic alterations of the drug in primates and results in complete loss of antimycobacterial activity. Acetyl-INH represents 91% of the total metabolite in man[229]. Acetylation of INH has been shown by Hughes[230] to be a significant reaction both toxicologically and chemotherapeutically.

Baretto *et al.*[231-235] have described a series of chromatographic studies on INH and its metabolic products from samples of biological material and have included centrifugal ultrafiltration combined with descending chromatography[231], paper electrophoresis[232,233], direct chromatographic separation of the metabolites

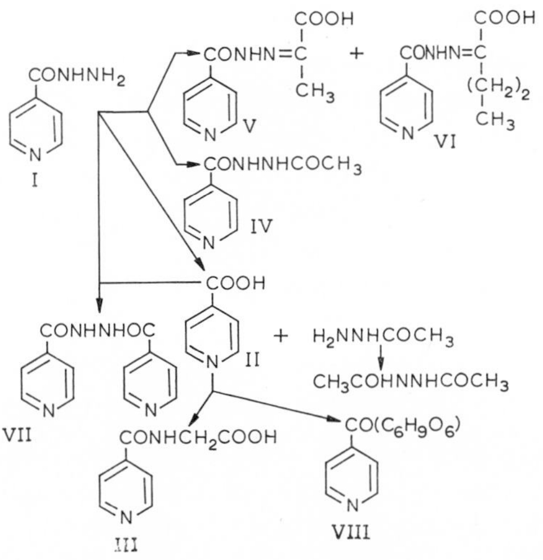

Fig. 14. Summary of the mammalian metabolism of isoniazid (INH). I, INH; II, isonicotinic acid; III, isonicotinyl-glycine; IV, acetyl-INH; V, isonicotinyl hydrazone; VI, α-keto-glutaric acid isonicotinyl hydrazone; VII, di-INH; VIII, isonicotinyl glucuronide.

TABLE 10

R_F VALUES OF SOME DERIVATIVES OF ISONICOTINIC ACID HYDRAZIDE IN VARIOUS SOLVENTS OBTAINED ON MACHEREY-NAGEL NO. 261 PAPER[231]

Compound	R_F values[a]							
	1	2	3	4	5	6	7	8
INH	0.85	0.61	0.52	0.83	0.78	0.61	0.90	0.71
Acetyl-INH	0.87	0.57	0.35	0.92	0.43	0.61	0.81	0.71
Acetaldehyde isonicotinyl-hydrazone	0.88	0.38	0.24	0.89	0.42	0.68	0.82	0.71
Na pyruvate isonicotinyl-hydrazone	0.28	0.40	0.20	0.58	tail	0.32	0.70	0.49
Pyridoxal phosphate isonicotinyl-hydrazone	0.06	0.00	0.19	0.03	0.02	0.08	0.04	0.03

[a] Solvent: 1, water–saturated *n*-butanol; 2, isopropanol–1% ammonium hydroxide (20:3); 3, *n*-butanol satd. with 1% ammonium hydroxide; 4, isopropanol–water (85:15); 5, isoamyl alcohol satd. with 0.5 N acetic acid; 6, *n*-butanol satd. with 0.5 N acetic acid; 7, propanol–ammonium hydroxide (70:30); 8, *n*-butanol–ethanol–water (40:10:50).

from the wet sample[233], and the isolation of INH derivatives by column chromatography followed by paper chromatographic identification[234].

Tables 10 and 11 list the R_F values of some derivatives of INH in various solvents following ascending chromatography on Macherey-Nagel No.261 paper.

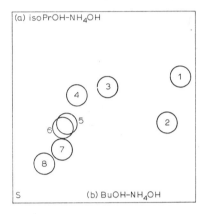

Fig.15. Map of a two-dimensional paper chromatogram of INH metabolites (Macherey-Nagel
No.261 filter paper, unbuffered) run with a, isopropanol–NH₄OH and b, butanol–NH₄OH, by the
ascending technique. 1, INAmide; 2, di-INH; 3, INH; 4, acetyl-INH; 5, Py-INHzone (56.2% of
the ammonium salt); 6, Ac-INH zone; 7, INAcid (ammonium salt); 8, Py-INHzone (47.7% of the
ammonium salt).

Table 12 summarizes the results obtained with a sequential procedure for the localization of INH and some of its metabolites on paper chromatograms. Figure 15
illustrates a map of a two-dimensional paper chromatogram of INH metabolites[232].

The utility of impregnated papers and paper electrophoresis for the separation
of INH and its metabolites was elaborated[232]. Tables 13 and 14 show the R_F values

TABLE 11

R_F VALUES OF SOME INH-METABOLITES ON MACHEREY-NAGEL NO. 261 FILTER PAPER[232]

Sample	R_F value[a]		
	1	2	3
Py-INHzone[b]	0.41–0.22	0.30–0.17	0.56
Ac-INHzone	0.40	0.29	0.70
INAcid	0.29	0.27	0.61
INAmide	0.66	0.93	0.72
Di-INH	0.37	0.85	0.67

[a] Solvent: 1, isopropanol–1% NH₄OH (20:3) (ascending); 2, butanol saturated with 1% NH₄OH
(ascending); 3, propanol–NH₄OH (70:30) (descending).
[b] As ammonium salts.

TABLE 12

RESULTS OBTAINED WITH THE SEQUENTIAL PROCEDURE FOR THE LOCALIZATION OF INH AND SOME OF ITS METABOLITES ON PAPER CHROMATOGRAMS[232]

The values given are the amounts in μg

Step	Description	INH	Acetyl-INH	Py-INHzone	Ac-INHzone	Di-INH	INAcid	INAmide
1	uv	Brownish 5	Brownish 15	Brownish 2	Slate 1	Slate 2	Brownish 15	Brownish 10
2	BrCN/visible		Brownish 15	Brownish 4	Brownish 1	Brownish 2		
3	BrCN/uv	Brownish 5	Yellow 1	Brownish 2	Brownish 0.6	Brownish 0.6	Brownish 10	Brownish 10
4	NH₄OH/visible	Yellow 0.4	Brownish 1	Brownish 4	Brownish 0.8	Brownish 0.6	Yellow 0.5	Yellow 0.5
5	NH₄OH/uv	Yellow 1	Yellow 0.5	Brownish 2	Brownish 0.2	Brownish 0.2	Yellow 0.5	Yellow 0.5
6	HCl/BrCN/uv		Yellow 1	Bluish 2	Brownish 0.6	Bluish 2		
7	Greulach–Haesloop[236] reagent (1% aq. ferric chloride–1% potassium ferricyanide (1:1))	Deep blue 1	Deep blue 0.2	Deep blue 0.6	Deep blue 0.2	Deep blue 0.2		

TABLE 13

R_F VALUES OF INH AND SOME OF ITS METABOLITES ON MACHEREY-NAGEL NO. 261
FILTER PAPER IMPREGNATED WITH 0.1 M TRIS BUFFER, pH 7.0[232]

Sample	R_F value[a]	
	1	2
INH	0.40	0.62
Acetyl-INH	0.39	0.57
Py-INHzone[b]	0.05	0.06–0.33
Ac-INHzone	0.47	0.63
INAcid[b]	0.08	0.23
INAmide	0.51	0.62
Di-INH	0.45	0.63

[a] Solvent: 1, isoamyl alcohol saturated with Tris buffer (ascending); 2, butanol saturated with Tris buffer (ascending).
[b] As ammonium salts.

TABLE 14

R_F VALUES FOR SOME INH-DERIVATIVES IN EDTA-IMPREGNATED MACHEREY-NAGEL NO. 261
PAPER[233]

Compound	R_F values[a]				
	1	2	3	4	5
INH	0.66	0.79	0.84	0.76	0.71
Acetyl-INH	0.79	0.87	0.81	0.77	0.58
Acetaldehyde INHzone	0.83	0.84	0.85	0.82	0.58
NH$_4$ pyruvate INHzone	0.09	0.06	0.56	0.57	0.05
	0.42	0.37	0.76		
NH$_4$ isonicotinate	0.48	0.34	0.57	0.50	0.23
Isonicotinamide	0.71	0.83	0.80	0.76	0.62
Diisonicotinyl hydrazide	0.80	0.85	0.82	0.85	0.58

[a] Solvent: 1, pyridine–amyl alcohol–0.2 M EDTA (40:35:satd.); 2, n-butanol satd. with 0.2 M EDTA; 3, n-butanol–ethanol–0.2 M EDTA (40:10:satd.); 4, isoamyl alcohol–pyridine–0.2 M EDTA (20:25:satd.); 5, isoamyl alcohol satd. with 0.2 M EDTA.

of INH and some of its metabolites on Macherey-Nagel No. 261 paper impregnated with 0.1 M Tris buffer, pH 7.0 and on EDTA-impregnated paper, respectively.

The electrophoretic mobilities found for the various metabolites of INH on Macherey-Nagel filter paper impregnated with 0.1 M Tris buffer, pH 7.0 after application of a potential of 300 V for 3 h are shown in Table 15.

Paper electrophoresis was found to be especially useful for the separation of the acid metabolites such as isonicotinic acid (1 N acid) and the isonicotinyl-hydra-

TABLE 15

ELECTROPHORETIC MOBILITY OF INH AND ITS METABOLITES

Macherey-Nagel No. 261 filter paper impregnated
with 0.1 M Tris buffer, pH 7.0; potential applied 300 V for 3 h.

Sample	Displacement	
	Pole	cm
INH	−	1.4
Acetyl-INH	−	1.4
Py-INHzone	+	5.8
Ac-INHzone	+	1.6
INAmide	−	1.4
INAcid	+	8.5
Di-INH	+	1.6

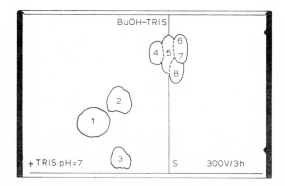

Fig. 16. Two-dimensional paper electrophoresis (0.1 M TRIS buffer, pH 7.0, 300 V, 3 h) and ascending paper chromatography (n-butanol satd. with TRIS buffer) of INH and some of its metabolic products. 1, INAcid; 2 and 3, Py-INHzone; 4, di-INH; 5, Ac-INHzone; 6, INH; 7, INAmide; 8, acetyl-INH.

zone of pyruvic acid, especially when combined with ascending paper chromatography[232] as shown in Fig. 16.

Because pyridine was found by Barretto and Sabino[233] to be the only solvent capable of dissolving all of the INH derivatives, its utility in chromatographic solvent mixtures was elaborated (Table 16). Figure 17 illustrates a map of a two-dimensional chromatogram of INH derivatives obtained following development with n-butanol saturated with 1% ammonium hydroxide in one direction followed by a 90° development with pyridine–amyl alcohol–water (40:35:30). The localization of the spots was achieved using the Greulach–Haesloop reagent[236].

The isolation of INH derivatives by column chromatography as well as the utility of a number of new reagents for their localization and identification on paper

TABLE 16

R_F VALUES FOR SOME INH-DERIVATIVES IN SOLVENT MIXTURES CONTAINING PYRIDINE[233]

Compound	R_F values[a]			
	1	2	3	4
INH	0.60	0.89	0.74	0.73
Acetyl-INH	0.66	0.92	0.71	0.54
Acetaldehyde INHzone	0.69	0.92	0.76	0.58
NH_4 pyruvate INHzone	0.42/0.59	0.89	0.44/0.59	0.68
NH_4 isonicotinate	0.37	0.87	0.42	0.55
Isonicotinamide	0.67	0.88	0.69	0.72
Diisonicotinyl hydrazide	0.71	0.92	0.74	0.55

[a] Solvent: 1, pyridine–amyl alcohol–water (40:35:30); 2, pyridine–water (65:35); 3, isoamyl alcohol–pyridine–water (20:25:20); 4, pyridine–isoamyl alcohol–1.6 N NH_4OH (20:14:20).

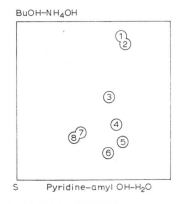

Fig. 17. Map of a two-dimensional paper chromatogram of INH-derivatives. 1, isonicotinamide; 2, di-INH; 3, INH; 4, acetyl-INH; 5, acetaldehyde INHzone; 6 and 7, NH_4 pyruvate INHzone; 8, NH_4-isonicotinate.

chromatograms has been reported[234]. INH and its derivatives were eluted from a 20 × 1 cm column packed with 4 g of anhydrous sodium sulfate using a mixture of chloroform–diethylamine (90:10) at a flow rate of 2 ml/min. Elution was checked by paper chromatography and it was found that 50 ml of chloroform–diethylamine (90:10) were sufficient to completely elute mixtures of INH, acetyl-INH and the INH hydrazones of pyruvic acid and acetaldehyde dissolved in blood serum (1 % w/v concentrations of each component) as shown in Table 17.

The utility of a number of new diethylamine solvent mixtures for the resolution of INH and its derivatives on paper chromatograms (Macherey-Nagel No. 261 paper) is illustrated in Table 18.

TABLE 17

ELUTION RATE OF VARIOUS INH-DERIVATIVES PURIFIED BY MEANS OF COLUMN CHROMATOGRAPHY[234]

Fraction (10 ml)	INH-derivatives present			
	INH	Acetyl-INH	INHzone acet.	INHzone pyr.
I	+	+	−	−
II	+	+	+	+
III	+	+	+	+
IV	−	−	+	+
V	−	−	+	+
VI	−	−	−	−

TABLE 18

R_F VALUES OF INH-DERIVATIVES IN SOLVENT MIXTURES CONTAINING DIETHYLAMINE[234]

Compound	Solvent mixture				
	I	II	III	IV	V
INH	0.53	1	0.53	0.38	0.22
Acetyl-INH	0.47	1	0.47	0.02	0.13
Pyruvic acid INHzone	0.42–0.56	1	0.41–0.53	0.09	0.08–0.19
Acetaldehyde INHzone	0.55	1	0.61	0.03	0.20
Isonicotinamide	0.76	1	0.67	0.58	0.52
Isonicotinic acid	0.54	1	0.49	0.14	0.18
Di-INH	0.55	1	0.60	0.03	0.20

Solvent systems: I, n-butanol–diethylamine–water (40:10:satd.); II, isopropyl alcohol–diethyl-amine–water (60:20:10); III, benzyl alcohol–diethylamine–water (40:10:satd.); IV, chloroform-diethylamine–water (40:20:satd.); V, n-amyl alcohol–diethylamine–water (40:10:satd.).

The results for the sensitivity of six reagents (sodium nitroprusside, Wachsmuth reagent[237], ninhydrin, Percheron reagent[238], isatin, and pyridine-acetaldehyde[239]) toward the detection of INH and its derivatives are shown in Table 19.

The determination of small amounts of a mixture of INH and isonicotinic acid by means of cationite paper has been reported[240]. The mixture was first separated into its components by chromatography on Whatman No. 2 paper with n-butanol saturated with water as the mobile phase. INH and isonicotinic acid were excised from the appropriate parts of the chromatogram on to cationite paper (a sulfonated phenolic ion-exchanger in the H^+ form) and developed with dioxan for 18 h. The spots were revealed by exposure of the paper to gaseous ammonia, followed by dipping into 1.5% aq. picryl chloride. The above technique was used to determine > 10 μg of both INH and isonicotinic acid in biological material.

312   DRUGS

TABLE 19

SENSITIVITY OF VARIOUS REAGENTS TOWARDS INH AND ITS METABOLIC DERIVATIVES[234]

INH-derivative	Reagent[a]					
	I	II	III	IV	V	VI
	Sensitivity					
INH	2	2	2		4	30
Acetyl-INH	2	5	20		4	
Pyruvic acid INHzone	2	2	2	0.5	8	20
Acetaldehyde INHzone	2	2	2	20	4	10
Isonicotinamide	2	10	20			
Isonicotinic acid	2	2	1			
Di-INH	2	2	2		4	50
	Color					
INH	Orange	Brownish	Yellow		Brown	Yellow
Acetyl-INH	Brownish	Brownish	Yellow		Yellow	
Pyruvic acid INHzone	Bordeaux	Bordeaux	Brown	Brown	Brown	Blue
Acetaldehyde INHzone	Brownish	Brownish	Orange	Brown	Yellow	Yellow
Isonicotinamide	Brownish	Bordeaux	Yellow			
Isonicotinic acid	Yellow	Bordeaux	Yellow			
Di-INH	Brownish	Yellow	Orange			

[a] Reagents: I, Sodium nitroprusside. 5 g of the salt are dissolved in a 10% solution (v/v) of acetaldehyde in water; before using, an equal volume of 2% (w/v) sodium carbonate is added; after spraying, the sample is heated at 120°C for 10 min; II, Wachsmuth reagent. 2 g of quinhydrone are dissolved in 95 ml of ethanol plus 5 ml of pyridine; after spraying, the sample is heated at 100°C for 2 min; III, Ninhydrin. 0.2% (w/v) solution in acetone; after spraying, the sample is heated at 120°C for 15 min; IV, Percheron reagent. 0.5 g barbituric acid dissolved in 100 ml ethanol containing 2 ml of 85% phosphoric acid; after spraying and heating at 120°C for 5 min, the sample is detected by observing under a UV lamp; V, Isatin. 1 g of the reagent is dissolved in 100 ml isopropanol containing 1 ml pyridine and 1.5 g zinc acetate; the sample is dried in an oven at 110°C and examined under a UV lamp; VI, Pyridine–acetaldehyde. The sample is sprayed with a mixture containing equal parts of the reagents, dried at 110°C and observed under a UV lamp.

LaRue[241] described the detection of 23 hydrazides (as the respective picryl hydrazides), utilizing a 0.25% aq. trinitrobenzene sulfonic spray reagent. The R_F values of the hydrazides in three solvents: (a) isopropanol–water (17:3), (b) upper phase n-butanol–acetic acid–water (4:1:5) and (c) 1.4 M potassium phosphate buffer pH 7.0 are shown in Table 20.

The thin-layer chromatographic separation of a number of antitubercular and anti-depressive drugs has been described by Alessondro et al.[242]. INH, iproniazid, nialamide, iproclozide, phenelzine, and isocarboxazid were separated on a 0.25 mm layer of Silica Gel G (activated at 110°C for 30 min) by a 30 min development with chloroform–methanol (1:1) or with a mixture of the chloroform phase of $CHCl_3$– conc. aq. NH_3 (2:1) and 1 part of methanol. The spray reagents used for location

TABLE 20

R_F VALUES OF ACYL HYDRAZIDES ON WHATMAN NO. 1 PAPER[241]

	$R_F \times 100$		
	Solvent A^a	Solvent B^a	Solvent C^a
Acetic hydrazide	62	str[b]	84
Adipic dihydrazide	25	str	76
p-Aminobenzoic hydrazide	53	str	43
β-Aspartic hydrazide	5	15	88
Benzoic hydrazide	78	93	39
Benzilic hydrazide	86	96	52
o-Cresotic hydrazide	91	95	27
Cyclopropanecarboxylic hydrazide	76	83	77
γ-Glutamic hydrazide	5	20	96
Glutaric hydrazide	25	str	81
Indole-3-acetic hydrazide	71	85	37
Indole-3-propionic hydrazide	70	84	36
Isonicotinic acid hydrazide	54	75	64
Lactic hydrazide	66	71	85
Malonic dihydrazide	17	str	86
Mandelic hydrazide	78	96	75
m-Nitrobenzoic hydrazide	81	90	54
p-Nitrobenzoic hydrazide	90	96	48
Oxalic dihydrazide	16	str	76
Phenylacetic hydrazide	79	99	64
Salicylic hydrazide	78	95	40
Suberic dihydrazide	55	99	73
Succinic dihydrazide	15	str	84

^{a} Solvents: A, isopropanol–water (17:3); B, upper phase of n-butanol–acetic acid–water (4:1:5);
C, 1.4 M potassium phosphate buffer pH 7.0. Spray reagent: 0.25% aq. trinitrobenzene sulfonic acid.
^{b} str indicates streaking or tailing of the hydrazide on the chromatogram.

and determination were (a) dil. (1:3) Folin-Ciocalteu reagent, followed by 10%
Na_2CO_3 soln., (b) a mixture (50:3) of 0.2% methanolic ninhydrin–anhyd. acetic
acid–2,4,6-trimethylpyridine (25:5:1) and 1% aq. $Cu(NO_3)_2$, followed by heating
the sprayed plate at 130–140°C, (c) 10% $FeCl_3$ soln. followed by 1% aq. fluorescein
and then by aq. NH_3, and (d) satd. aq. $(NH_4)_6Mo_7O_{24} \cdot 4H_2O$ followed by satd.
aq. oxalic acid.

The gas chromatographic separation of the antitubercular drugs INH, iproniazid and ethambutol was described by Calo et al.[243,244]. A Perkin-Elmer 801 gas
chromatograph was used with a differential flame-ionization detector and equipped
with a glass injector and a 1.80 m × 2 mm glass column containing 6% QF-1 on 80 to
100 mesh silanized Chromosorb G. The operating temperatures were column programmed 120 to 250°C at 8.33°/min, injector 250°C, and detector 250°C. The carrier gas was nitrogen at a flow rate of 50 ml/min. The sample mixture analyzed was
1 μl of a 1% benzene solution of TMS ethambutol, 0.5 μl of a 1% alcoholic solution

of INH, and 1 µl of a 1% ethereal solution of iproniazid. Figure 18 illustrates a chromatogram of the separation of the three drugs.

INH and *p*-aminosalicylic acid (PAS) are often administered together with organic or inorganic antacids in a single tablet. The use of an ion-exchange resin (Dowex 2-X 8 spheres in Cl⁻ form, 50–80 mesh) and elution with water and 1 N HCl

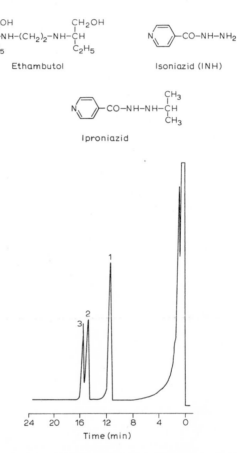

Fig. 18. Separation of 1, ethambutol; 2, isoniazid; 3, iproniazid[240].

for their separation and subsequent spectrophotometric assay was described by Fan and Wald[245].

The separation and determination of INH and its degradation products was studied by Inoue *et al.*[246]. INH was determined colorimetrically using sodium-1,2-naphthoquinone-4-sulfonate. Separation of isonicotinic acid (INA) was effected by passing the sample through a column of a weak cation-exchange resin (Amberlite CG-50, H⁺ form) by which INA alone was not absorbed and INA in the effluent was determined colorimetrically using cyanogen bromide. After the same solution was oxidized by ferricyanide in alkaline medium, the reaction mixture was passed through a column of a strong anion-exchange resin (Dowex 1-X8, Cl⁻ form) by which isonicotinamide (INAA) and isonicotinaldehyde remain unchanged and not absorbed by the resin. Isonicotinaldehyde in the effluent was determined colorimetrically using 2,4-dinitrophenylhydrazine and INAA was determined by measurement of UV absorbance of the effluent at 263 nm. 1,2-Diisonicotinoylhydrazine and isonicotinaldehyde and isonicotinoylhydrazine were determined by measurement of absorbance at 329 nm and 315 nm of sample solutions brought to pH 8.9 with 0.05 M borate and pH 1.0 with Clark-Labs buffers, respectively.

The interaction of INH with components of its formulations (magnesium oxide and lactose) was investigated by Wu *et al.*[247] utilizing TLC and spectrophotometric techniques. TLC of the reaction products using silica gel plate, development using methanol–chloroform (60:125) and detection by UV light and exposure of plates to iodine vapor confirmed the presence of isonicotinoyl hydrazones of lactose and hydroxymethyl furfural.

7. THALIDOMIDE

Thalidomide (α-phthalimidoglutarimide) (I) is prepared by heating N-phthalylglutamic acid or anhydride with ammonia at 140–160°C or with urea or thiourea in boiling xylene. It was first synthesized by King *et al.*[248] who also described its main pharmacological properties. The early studies indicated it to be a very effective nontoxic sedative hypnotic drug. It is unique in that it does not contain the grouping $R_1R_2C \cdot CO \cdot N\!<$ (R_1 and R_2 = alkyl or aryl radicals) which are present in many other sedatives and that it is a derivative of a naturally occurring amino acid, glutamic acid.

I

Despite its very low order of acute oral toxicity, its effect on human embryogenesis is perhaps one of the most specific biologic processes involving a chemical agent. Lenz and Knapp[249] established that the ingestion of thalidomide between the 27th and 40th days of pregnancy causes human phocomelia. It has since been reported that this agent also effects embryogenesis in the laboratory animals[250–254] and the embryotoxic and teratogenic effectiveness of thalidomide depends on the animal species and the method of administration. For example, after p.o. administration, thalidomide is a teratogen in monkeys[255, 256] and in rabbits[257], but causes few, if any, malformations in rats[258]. Thalidomide has also been shown to induce chromosome aberrations[259] in human cells *in vitro* and is mutagenic in the mosquito *Culex*

TABLE 21

DESCENDING CHROMATOGRAPHY OF THALIDOMIDE AND RELATED COMPOUNDS ON WHATMAN NO. 1 PAPER

No.	Compound (20 μg each)	R_F values × 100[a]				Appearance of spot		
		One-dimensional system		Two-dimensional system		In ultraviolet light	After ninhydrin spray[b]	In ultraviolet light after hydrazine spray[c]
		A	B	A	B			
I	Thalidomide	86–91	Streaks	86–91	65–70	Dark, becoming green after 4–5 min exposure		Greenish-blue
II	4-Phthalimidoglutaramic acid	45–49	61–64	45–49	53–58			Greenish-blue
III	2-Phthalimidoglutaramic acid	35–39	57–60	35–39	45–51	Dark		Greenish-blue
IV	α-(o-Carboxybenzamido)-glutarimide	36–39	61–65	36–39	39–42	Purplish-black	Red developing slowly	Weak greenish-blue
V	2-Phthalimidoglutaric acid	24–29	80–86	24–29	62–66	Dark		Greenish-blue
VI	4-(o-Carboxybenzamido)-glutaric acid	14–18	45–49	14–18	24–27	Dark		Weak greenish-blue
VII	2-(o-Carboxybenzamido)-glutaramic acid	15–19	47–52	15–19	36–40	Dark		Weak greenish-blue
VIII	2-(o-Carboxybenzamido)-glutaric acid	16–20	64–68	16–20	43–47	Dark		Weak greenish-blue
IX	Phthalic acid	56–59	72–76	56–59	58–61	Dark		Very weak blue
X	α-Aminoglutarimide	39–43	10–14	39–43	9–12	Greenish	Red	None
XI	Isoglutamine	11–14	11–15	11–14	8–11	Dark	Violet	None
XII	Glutamine	9–11	3–7	9–11	7–7	Dark	Violet	None
XIII	Glutamic acid	4–7	6–11	4–7	0–10	Dark	Violet	None

[a] Solvents: A, pyridine–n-amyl alcohol–water (7:7:6); B, n-butanol–acetic acid (10:1) saturated with water.
[b] 0.5% ethanolic ninhydrin followed by heating the chromatogram at about 80°C for 4 min.
[c] 5% ethanolic solution of hydrazine then heating the sprayed chromatogram in an oven at 100°C for 10 min.

pipiens molestus[260]. The evidence in favor of the concept of the direct binding effect of thalidomide through an acylation mechanism with DNA, RNA, histones, and phospholipids[261,262] suggested that it is perhaps an acylating radiomimetic mutagenic agent[260].

The important features of the thalidomide molecule in addition to its two halves, *viz.* the phalimide and glutarimide rings, are (*a*) the asymmetric carbon atom at position 3' of the glutarimide ring and (*b*) the four substituted amide bonds which are located at 1-2 and 2-3 in the phthalimide ring and 1'-2' and 1'-6' in the glutarimide ring. The drug as used is the optically inactive ($\pm$) form. Williams[263] has suggested the possibility of the teratogenic effect of thalidomide being associated with the instability of the amide bonds of the phthalimide portions of the molecule. This instability makes thalidomide a phthalylating agent which could phthalate the polyamines in the embryo and thus indirectly interfere with messenger RNA concerned with the enzymes involved in the initiation of the growth of certain structures during morphogenesis.

Fabro *et al.*[264] have found the three optical forms of thalidomide to be teratogenic in the New Zealand white rabbit and are CNS depressants in mice (but are also much more toxic when given orally than the ($\pm$) form).

Studies on the nature of the urinary excretory products of thalidomide in man, dog, rat, and rabbit showed that these products were those that would be formed if the drug were hydrolyzed at its amide bonds. These products could be formed enzymatically or spontaneously and it was also found[264,265] that thalidomide was unstable in aqueous solution and after storage of an aqueous solution of the drug, it gave rise to all 12 possible hydrolysis products. It was found possible to separate all these products by descending two-dimensional paper chromatography, the first solvent being pyridine–*n*-amyl alcohol–water (7:7:6) and the second solvent, *n*-butanol–glacial acetic acid (10:1) and to identify them by R_F values and color reactions as described by Schumacher *et al.*[266]. Table 21 lists these hydrolysis products as well as their R_F values determined by descending chromatography on Whatman No. 1 paper and detected with 0.5% ethanolic ninhydrin and 5% ethanolic hydrazine spray reagents. Figure 19 illustrates the pathways of hydrolysis of thalidomide.

Table 22 depicts the composition of a solution of thalidomide at pH 7.4 after 24 h at 37°C[266] and shows that the major breakdown products at pH 7.4 are those in which both the phthalimide and glutarimide rings have been opened. However, the glutarimide ring appeared to be the more stable since just over 50% of the material in solution at pH 7.4 after 24 h was α-(*o*-carboxybenzamido)glutarimide (IV) in which the glutarimide ring is intact (Fig. 19). Williams *et al.*[265,268] reported the isolation in crystalline form from the urine of rabbits dosed with thalidomide, of all the compounds listed in Table 21.

Table 23 shows the urinary metabolites of ^{14}C-thalidomide in various species (man, rat, and dog) and reveals the presence of metabolites identical to the hydrolysis products of thalidomide. Many of these, however, have been shown to be break-

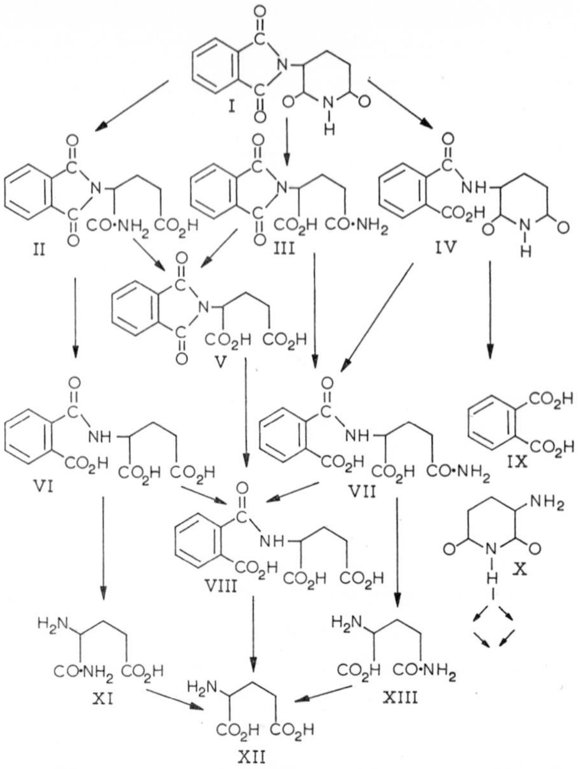

Fig. 19. Hydrolytic reactions of thalidomide. Splitting of the 1'-2' bond of the glutarimide ring gives glutamine derivatives whereas splitting of the 1'-6' bond gives isoglutamine derivatives[266, 267].

TABLE 22

COMPOSITION OF A SOLUTION OF THALIDOMIDE AFTER 24 h AT 37°C[a]

Product	% Present	No. of hydrolytic steps from thalidomide
Thalidomide	0.15	0
α-(o-Carboxybenzamido)glutarimide	52.5	
4-Phthalimidoglutaramic acid	2.0	1
2-Phthalimidoglutaramic acid	1.8	
2-Phthalimidoglutaric acid	0.1	
4-(o-Carboxybenzamido)glutaramic acid	23.0	
2-(o-Carboxybenzamido)glutaramic acid	20.0	2
Phthalic acid	0.1	
2-(o-Carboxybenzamido)glutaric acid	0.1	3

[a] Saturated solution of ^{14}C-thalidomide in phosphate buffer pH 7.4 kept 37°C for 24 h then chromatographed and the radioactivity of each spot determined.

TABLE 23

URINARY METABOLITES OF ^{14}C-THALIDOMIDE IN VARIOUS SPECIES

Compound found	As % of urinary ^{14}C in		
	Man[a]	Rat[a]	Dog[b]
Thalidomide	1.0	2.2	1.8
α-(o-Carboxybenzamido)glutarimide	29.5	35.54	22.9
4-Phthalimidoglutaramic acid	53.2	7.14	2.2
2-Phthalimidoglutaramic acid	5.0	8.16	13.6
2-Phthalimidoglutaric acid	3.5	6.15	9.0
4-(o-Carboxybenzamido)glutaramic acid ⎫ 2-(o-Carboxybenzamido)glutaramic acid ⎭		5.13	
2-(o-Carboxybenzamido)glutaric acid	2.3	2.4	4.7
Phthalic acid	2.7	2.9	6.1

[a] After Beckman[269].
[b] After Faigle et al.[270].

down products of the drug rather than true metabolites formed enzymatically[264]. The fate of ^{14}C-thalidomide orally administered to pregnant rabbits at the beginning of the sensitive phase of pregnancy has been studied by Fabro et al.[271]. At 12, 24, and 58 h after dosing, radioactivity was present in the embryo and the maternal tissues examined. The ^{14}C concentration in the embryo was at nearly all times higher than that in the plasma, brain, skeletal muscle, and fat but lower than that in the liver and kidney. Thalidomide was found in the embryo together with 7 of its products for more than 24 h after dosing and the accumulation of radioactivity in the embryo was found to be due to the retention of the polar hydrolysis products. Radioactive compounds in the embryo, maternal tissues, and excreta were identified by a reverse isotope-dilution technique after a preliminary two-dimensional chromatographic separation with the solvent systems of Schumacher et al.[272]. The fate of the hydrolysis products of thalidomide, e.g. α-(o-carboxybenzamido)glutarimide, 2-phthalimidoglutaramic acid, 2-phthalimidoglutaric acid, and 2-(o-carboxybenzamido)glutaramic acid in the pregnant rabbit was also evaluated by Fabro et al.[273]. None of these products was found to be teratogenic suggesting that thalidomide itself is the teratogenic agent.

Nicholls[274] described the absorption and excretion of ^{14}C-thalidomide in pregnant mice. After oral administration of the labeled drug, similar concentrations of the drug are found in several maternal tissues and in the placenta and foetus. Thalidomide and α-(o-carboxybenzamido)glutarimide have been identified in the foetus. Chromatograms of extracts of foetal tissue were run on Whatman No. 1 paper in isopropanol–water (4:1). The radioactive compounds were located by autoradiography and the non-radioactive marker substances by ultraviolet light and treatment with hydrazine[266]. The thin-layer chromatography of thalidomide and its 12 hydrolysis products has been described by Pischek *et al.*[275]. Silica Gel G and GF$_{254}$ plates were developed with 5 mixed solvents and the spots detected fluorometrically or with 5% iodine–chloroform, bromocresol green, ninhydrin or Fe-hydroxylamine reagents. Table 24 lists the R_F values of thalidomide and its hydrolysis products on Silica Gel G and Silica Gel GF$_{254}$ using 5 solvent systems.

TABLE 24

$R_F \times 100$ VALUES OF THALIDOMIDE AND ITS HYDROLYSIS PRODUCTS ON SILICA GEL G AND GF$_{254}$

Compound	Solvent system[a]					Detector[b]				
	1	2	3	4	5	a	b	c	d	e
Thalidomide	81	63	35	71	63	+	−	−	+	+
N-Phthaloylglutaramic acid imide	35	13	8		33	−	−	+	+	+
N-Phthalylglutamine	45	22	10	54	41	+	+	−	+	+
N-Phthalylisoglutamine	50	25	12	50	44	+	+	−	+	+
N-Phthaloylisoglutamine	25	7	5		14	−	+	−	+	+
N-Phthaloylglutamine	18	4	0		12	+	+	+	+	+
N-Phthalylglutaramic acid	74	46	25	42	29	+	+	−	+	+
N-Phthaloylglutaramic acid	42	11	3		16	+	+	+	+	+
Glutamine	0	0	0	33	25	−	−	+	−	−
Isoglutamine	0	0	0	29	23	−	−	+	−	−
Glutaramic acid imide (HCl)	0	0	0	51	40	−	−	+	−	+
L-Glutaramic acid	0	0	0	19	10	−	−	+	−	−
Phthalic acid	64	37	27	60	38	−	−	−	−	+

[a] Solvent systems: 1, chloroform–isopropanol–formic acid (75:20:5); 2, chloroform–isopropanol–formic acid (85:10:5); 3, benzene–dioxan–formic acid (75:20:5); 4, collidine–water (75:25); 5, pyridine–amyl alcohol–water (35:35:30).
[b] Detectors: a, 5% iodine–chloroform; b, bromocresol green; c, ninhydrin; d, ferric hydroxylamine[276]; e, uv fluorescence at 254 nm.

Fiedler and Heine[277] described the utility of paper chromatography for the isolation and detection of the hydrolytic products of thalidomide. FN4/L Paper (VEB) and Schleicher & Schüll No. 2043B paper chromatograms were developed with (*1*) *n*-butanol–acetic acid–water (4:1:1), (*2*) pyridine–amyl alcohol–water (7:7:6), and (*3*) dimethylformamide–methanol–water (25:70:5). The spray reagents were (*1*) 0.2% ethanolic ninhydrin, (*2*) a saturated ethanolic solution of hydroxyl-amine-HCl and ethanol potassium hydroxide (following removal of potassium chloride) and (*3*) ferric nitrate–nitric acid reagent[266]. The hydrolysis of thalidomide by potassium hydroxide was compared with the cleavage products of thalidomide *in vivo*. Six ninhydrin positive cleavage products were detected on paper chromatograms. N-Phthaloyl-DL-glutamic acid imide was isolated as an intermediate of the hydrolytic decomposition and phthalic acid and glutamic acid were identified as the final products of thalidomide decomposition.

The gas–liquid chromatography of thalidomide has been reported by Sandberg *et al.*[278]. An EIR gas chromatograph was used equipped with an argon ionization cell with a 20 mc ^{90}Sr foil and 4–6 ft. × 4–5 mm glass U-columns containing three stationary phases: (*a*) 2% XE-60, (*b*) 13% QF-1, and (*c*) 16% DEGS, all on 100–120 mesh Gas Chrom S, acid-washed and siliconized with dimethyl dichlorosilane. The operating parameters were column, cell, and inlet temperatures, 230, 240, and 290°C, respectively, flow rate 50 ml/min, pressure 30 p.s.i., and high voltage 1000 V.

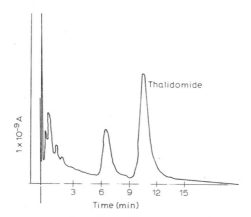

Fig. 20. Chromatogram of extract of thalidomide from serum on 2% XE-60.

Figure 20 represents a chromatogram of a chloroform extract of thalidomide from serum when chromatographed on a 2% XE-60 column. A thalidomide standard and the serum extract gave identical retention times when chromatographed on each of the 3 stationary phases described above. The column temperatures with the QF-1 and DEGS phases were 185°C.

8. VINCA ALKALOIDS

The *Vinca* alkaloids constitute a large number of derivatives isolated from the periwinkle plant *Vinca rosea* Linn.[279] *(Catharanthus roseus)*. A number of these derivatives which include vincristine and vinblastine form a unique class of leukopenic and antineoplastic agents on the basis of chemical structure and biologic actions. The dimeric indole alkaloids vinblastine (vincaleukoblastine) and vincristine (leurocristine) were originally isolated during the investigations[280, 281] of the reported hypoglycemic activity of *Vinca rosea* Linn. and were found to be potent oncolytic agents in some experimental systems[282]. Clinically, vinblastine has been useful in the treatment of lymphoma, including Hodgkin's disease, carcinomata of the breast, and bronchus. Vincristine has been used in acute lymphocytic and myelogenous leukemia[283, 284], carcinoma of the cervix, malignant lymphomata, neurobla toma, Wilson's tumor as well as in certain cranial gliomata. Vincristine produces profound embryocidal and teratogenic effects in a large variety of tissues and organs in several mammalian species, *e.g.* hamster[285], rats[286, 287], monkey *Macacca mulatta*[288], and mice[289, 290].

The *Vinca* drugs vincristine and vinblastine both *in vitro* and *in vivo* are powerful mitotic poisons that arrest cells at or near metaphase[291–294]. Vincristine and vinblastine both inhibit preferentially the synthesis of transfer RNA (t. RNA) in Ehrlich ascites cells *in vitro*[295, 296], and r RNA (and to a lesser extent t. RNA) in human cells (HE p-2) *in vitro*[297]. The inhibition of DNA synthesis has also been reported[298, 299].

The metabolism of tritiated vinblastine in the rat has been described by Beer *et al.*[300, 301] who found vinblastine to be eliminated (almost entirely as metabolites) in the bile.

The binding of tritiated vinblastine by platelets in the rat following i.p. injection has been reported by Hebden *et al.*[302]. Isotopic dilution analysis showed that unchanged vinblastine accounted for essentially all the platelet radioactivity but for only 50% of that in the plasma. Alkaloids in blood and blood fractions, including

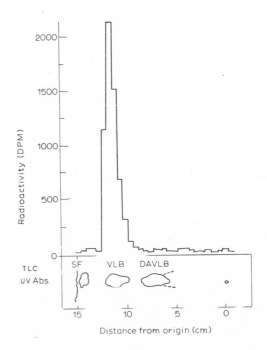

Fig. 21. Thin-layer chromatography on silica gel of benzene extract of platelets isolated from rat receiving vinblastine-³H. Carrier vinblastine (VLB) and deacetylvinblastine (DAVLB) were added to platelet pellet before extraction. Chromatogram developed with acetone. SF, solvent front. Alkaloids were visualized in short-wavelength ultraviolet (UV) light. Thin-layer chromatographic (TLC) sheet (plastic base) was cut into 0.5-cm strips for radioassay.

those to which carrier compounds had been added, were extracted into benzene at pH 7.5, then reextracted from the benzene into 0.1 N HCl, then purified for radio-assay and spectrophotometric analysis either on columns of cellulose phosphate or by TLC using chromatogram sheets No. 6060 (silica gel containing a fluorescent indicator, Eastman-Kodak) and developed with acetone. Figures 21 and 22 illustrate the

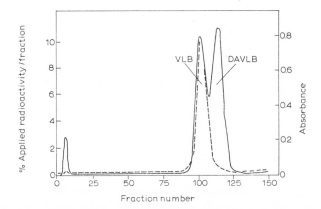

Fig. 22. Ion-exchange chromatography on cellulose phosphate of benzene extract of platelets isolated from rat receiving vinblastine-^{3}H. Carrier vinblastine (VLB) and deacetylvinblastine (DAVLB) were added to platelet pellet before extraction. ———, effluent continuously monitored for absorption in range 250–320 mμ. ---, radioactivity determined on individual fractions. Eluent was phosphate buffer (0.05 M, pH 3.4) containing 1.25 g NaCl/l.

results obtained by isotopic dilution and TLC and ion-exchange chromatography, respectively, of a benzene extract of platelets isolated from rat receiving vinblastine-^{3}H.

The single- and two-dimensional thin-layer chromatographic separation of four closely related dimeric alkaloids (vincristine, vinblastine, leurosine, and leurosidine)

TABLE 25

THIN-LAYER CHROMATOGRAPHIC SEPARATION OF VINCALEUKOBLASTINE, LEUROCRISTINE,
LEUROSINE AND LEUROSIDINE[a, 303]

Alkaloid	R_F value[b]	Chromogenesis following CAS reagent		
		Immediate	After 15 min	After 1 h
Leurosidine	0.06 ± 0.01	Orange-brown	Fades to tan	Tan
Leurocristine	0.16 ± 0.03	Blue	Light blue	Light blue
Vincaleukoblastine	0.24 ± 0.02	Orange-brown	Lavender	Light lavender
Leurosine	0.45 ± 0.03	Orange-brown	Yellow	Yellow
Ajmalicine (reference)	0.64 ± 0.03	Yellow-green	Yellow	Yellow

[a] On Silica Gel G plates (250 μ) using a chloroform–methanol (95:5) eluent. Development time was 10–12 min for 100 mm solvent front.
[b] Data were derived only from plates on which the reference alkaloid R_F value fell within a range of 0.60–0.68 and the solvent front (100 mm) was attained within 9–13 min.

has been described by Farnsworth and Hilinski[303]. Table 25 lists the R_F values and spot colors of the alkaloids obtained on Silica Gel G plates using chloroform–methanol (95:5) as eluent and ceric ammonium sulfate reagent[304, 305] for detection. Figure 23 illustrates a thin-layer chromatographic separation of the above 4 alkaloids and a reference alkaloid (ajmalicine) singly and in mixture under the conditions described in Table 25. Figure 24 depicts a two-dimensional chromatographic separation of the 4 *Vinca* alkaloids using chloroform–methanol (95:5) for the first development and methanol for the second.

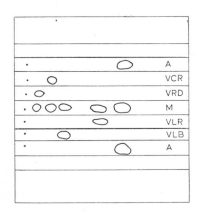

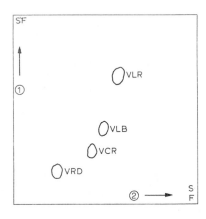

Fig. 23   Fig. 24

Fig. 23. A typical thin-layer chromatogram of vincaleukoblastine (VLB), leurocristine (VCR), leurosine (VLR), leurosidine (VRD). Matrix: Silica Gel G. Eluant in first direction: chloroform–methanol (95:5) to a solvent front (SF) of 120 mm. Average time of development 23–31 min. Eluent in second direction: methanol to a solvent front of 120 mm. Average development time 25–30 min.

Fig. 24. A typical two-dimensional thin-layer chromatogram of vincaleukoblastine (VLB), leurocristine (VCR), leurosine (VLR), and leurosidine (VRD). Matrix: Silica Gel G. Eluant in first direction: chloroform–methanol (95:5) to a solvent front (SF) of 120 mm. Average time of development 23–31 min. Eluent in secon direction: methanol to a solvent front of 120 mm. Average development time 25–30 min.

Cone *et al.*[305] described the TLC of 26 *Vinca* alkaloids using both alumina and silica gel plates and 7 eluent systems. Table 26 lists the R_F values and spot colors obtained for the alkaloids obtained in different solvent systems. Table 27 illustrates the systems of choice for the separation of closely related dimeric alkaloids.

Jakovljevic *et al.*[304] described the separation of leurosidine (R_F 0.23) and leurocristine (R_F 0.51) using alumina plates prepared with 0.5 N lithium hydroxide in conjunction with 5% absolute ethanol in acetonitrile as eluent. The same absorbent matrix together with a 30% acetonitrile in benzene eluent was required to separate leurosine (R_F 0.27) and vinblastine (R_F 0.36). The utility of 3 chromogenic reagents for the detection of 13 *Vinca* alkaloids was also described (Table 28). Approxi-

TABLE 26

R_F VALUES OF VINCA ALKALOIDS IN DIFFERENT SOLVENT SYSTEMS[305]

	Solvent system									Color: ceric ammonium sulfate
	1	2	3	4	5	6	7	8	9	
Ajmalicine	0.57	0.68	0.72	0.03	0.51	0.02	0.09	0.68	0.54	Yellow
Carosidine		0.58						0.59	0.10	Yellow
Carosine		0.71						0.65	0.24	Purple-grey
Catharanthine	0.77	0.59	0.74	0.12	0.77	0.03	0.37	0.58	0.38	Green (fades quickly)
Catharine	0.18	0.58	0.76					0.56	0.10	Yellow
Catharosine		0.56						0.58	0.08	Purple
Isoleurosine	0.35	0.22								Grey
Leurosine	0.27	0.35								Grey
Lochnericine	0.15	0.25	0.77	0.00	0.23	0.00	0.00	0.46	0.03	Blue
Lochneridine	0.00	0.00	0.21	0.00	0.20			0.04	0.00	Blue-green
Lochnerine	0.04	0.35	0.70					0.42	0.06	Pale grey
Neoleurocristine		0.27						0.43	0.03	Blue
Neoleurosidine		0.06						0.17	0.00	Yellow-brown
Perivine	0.05	0.30	0.48					0.39	0.11	Light brown
Pleurosine		0.51	0.42					0.07	0.03	Yellow
Serpentine	0.03	0.00	0.11					0.00	0.00	
Tetrahydroalstonine	0.76	0.60	0.73	0.05	0.66	0.04	0.29	0.76	0.69	Yellow-green
Vinblastine	0.25	0.24	0.66	0.00	0.17	0.00	0.00	0.33	0.04	Purple
Vincamicine	0.03	0.09	0.42					0.20	0.00	Bluish orange
Vincarodine		0.50						0.50	0.10	Blue (fades quickly)
Vindolicine	0.24	0.46	0.73					0.29	0.00	Blue
Vindolidine		0.15						0.57	0.20	Blue
Vindoline	0.44	0.37	0.68	0.00	0.53	0.03	0.06	0.44	0.13	Crimson
Vindolinine	0.55	0.54	0.70	0.00	0.51	0.00	0.09	0.48	0.09	Orange
Virosine	0.09	0.31	0.63					0.45	0.09	Colorless
Sitzirikine										Yellow-green

Solvent systems: 1, alumina:chloroform–ethyl acetate (1:1); 2, silica:chloroform–ethyl acetate (1:1); 3, alumina:ethyl acetate–absolute ethanol (3:1); 4, alumina:benzene (100%); 5, alumina:chloroform (100%); 6, silica:chloroform (100%); 7, alumina:benzene–chloroform (3:1); 8, silica:ethyl acetate–absolute ethanol (1:1); 9, silica:100% ethyl acetate; 10, alumina:100% ethyl acetate; 11, silica, prepared using 0.5 N KOH instead of H₂O: ethyl acetate–absolute ethanol (1:1).

TABLE 27

SEPARATION OF CLOSELY RELATED DIMERIC ALKALOIDS

Compounds	Suitable systems for separating
Leurosine VLB Isoleurosine	No. 2 or No. 8
Leurosidine Leurocristine VLB or leurosine	Develop first in No. 10; air dry 10 min; then develop in No. 3
Leurosidine sulfate Leurocristine sulfate VLB sulfate or Leurosine sulfate	No. 11

mately 200–300 meq. of the alkaloid dissolved in about 1 ml of the corresponding reagent was required to yield the colors shown in Table 28. Table 29 depicts the chromatographic and chromogenic behavior (using ceric ammonium sulfate in phosphoric acid) of 8 *Vinca* alkaloids.

TABLE 28

COLOR REACTIONS OF SOME *Vinca Rosea* Linn. ALKALOIDS

Name	1% Ceric ammonium sulfate in 85% H_3PO_4, room temp.	1% Ferric ammonium sulfate in 85% H_3PO_4 after 10 min in water bath	1% Ferric ammonium sulfate in 75% H_2SO_4, room temp.
Vincaleukoblastine	Purple	Violet	Blue
Leurocristine	Blue violet	Pink	Blue → grey blue
Leurosine	Pink	Pale yellow	Yellow → green
Leurosidine	Yellow → copper	Violet (strong)	Dark blue
Vindoline	Purple (strong)	Blue grey	Purple (strong)
Catharanthine	Indigo → green	Yellow	Blue violet
Virosine	Sky blue → yellow	Colorless	Pink → green
Perivine	Green → brown	Pale yellow	Umber
Reserpine	Sky blue → yellow	Lemon yellow	Blue → green
Vincarodine	Blue → green	Colorless	Blue → green → umber
Neoleurosidine	Violet → blue-violet	Magenta	Green → blue
Tetrahydroserpentine	Yellow → umber	Colorless	Violet → blue → green
Isoleurosine	Copper	Yellow → orange	Green → blue
(Indole)	Dark brown	Yellow	Yellow → green

Inagaki *et al.*[306] described the column and thin-layer chromatography of constituents of *Vinca rosea*. The fraction effective against AH-13 cancer was prepared from an extract of *V. rosea* and identified by TLC as leurosine. *Vinca rosea* extracts were first chromatographed over alumina and eluted with benzene–chloroform (1:1)

TABLE 29

SOME CHARACTERISTICS OF SEVERAL VINCA ALKALOIDS[29]

Name	Ultraviolet fluorescence	Color[a]	R_F values	
			5% EtOH in aceto-nitrile	30% Aceto-nitrile in benzene
Neo-leurosidine	Purple	Brown → lavender	0.04	
Leurosidine	Purple	Lavender	0.23	
Vincathicine	Bluish-white	Pink → yellow	0.41	
Leurocristine	Quenches	Lavender → pink	0.51	
Leurosine	Blue	Bluish lavender		0.27
Vincaleukoblastine	Purple	Lavender		0.36
Vindoline	Quenches	Violet → pink		0.56
Catharanthine	Quenches	Yellow		0.85

[a] Colors obtained using a 1% ceric ammonium sulphate in 85% phosphoric acid spray at room temperature.

and chloroform yielding 7 fractions, each of which were chromatographed at 120°C for 2 h on Silica Gel G plates using ethyl acetate–abs. ethanol (3:1) as developing solvent. Leurosine was identified using Dragendorff's reagent.

9. D-LYSERGIC ACID DIETHYLAMIDE

There are perhaps few drugs today as steeped in controversy as lysergic acid diethylamide (Delysid, lysergide, LSD) (I). The lysergic acid molecule consists of three groups, the A + B rings constitute an indole structure, the A + C rings naphthalene, and the C + D, N-methylquinoline. Because the C-5 and C-8 atoms are asymmetric, LSD has four isomers (d-, d-iso-, l-, and l-iso); only N,N-diethyl-d-lysergamide is reported to have significant hallucinogenic effect[307].

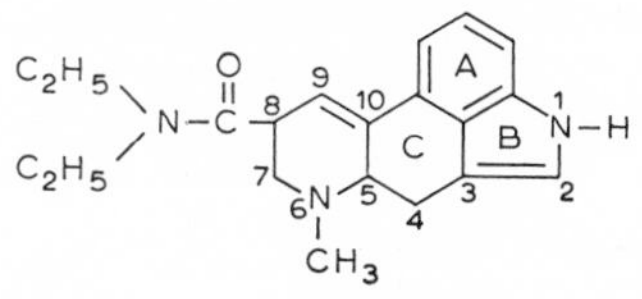

I

LSD was first prepared by Hoffman[308] in 1938, who accidentally noticed its hallucinogenic effect in 1943. It is the most potent of the presently known hallucinogens (ingestion of 100 μg or even less is sufficient to cause symptoms of the same intensity as elicited by 500 mg of mescaline). Three of the hallucinogens named in the Federal Register as "abuse drugs" are ergot alkaloids, namely, LSD, lysergic acid, and ergine (lysergic acid amide). Ergine is present in very small quantities in ergot, and is the hallucinogenic substance in morning-glory seeds. Lysergic acid, which is non-hallucinogenic, is obtained by the alkaline hydrolysis of crude ergot; however, its abuse potential exists because it is the necessary precursor in the preparation of LSD (*via* interaction with diethylamine) and other related derivatives.

The genetic effect of LSD in animals has been described in the teratogenic studies in mice[309-311], rats[312], hamsters[313], rabbits[314], and monkeys[315]. However, the lack of teratogenicity action by LSD in rats, mice and hamsters has also been reported[316].

An autoradiographic study on the placental transfer and tissue distribution of ^{14}C-LSD in mice[317] showed that in the early stage of pregnancy 2.5% and in the latter stage 0.5% of the radioactive dose passed the placental barrier into the fetus in 5 min (over 70% of this fetal radioactivity was unchanged ^{14}C-LSD). Information concerning possible teratogenic effects in humans is less available[318] and less convincing[319]. However, the possible teratogenic effect on two human embryos has been suggested[319,320]. The chromosome-breaking potential of LSD has been hotly contested. For example, chromosome breakage has been reported both *in vitro* and *in vivo* in human somatic cells[315,321-323] and germinal cells in mice[324,325]. Other reports, however, have indicated that the number of chromosome abnormalities in leukocytes from persons exposed to LSD was not significantly higher[326,327], nor were there any significant chromosomal aberrations in root tips of *Vicia faba*, cell cultures of Chinese hamster or *in vitro* cultures of human leukocytes treated with LSD[328].

The question of mutagenicity of LSD appears to be equally conflicting. Although mutagenic effects in *Drosophila* were noted by Browning[329] and Vann[330], Grace *et al.*[331] were unable to detect any mutagenic or chromosomal effects of LSD in *Drosophila* and Zetterberg[332] reported its non-mutagenicity in *ophiostoma multianulatam*.

The chemistry of LSD[307,333,334], its pharmacology[335-337], clinical and therapeutical aspects[336,337], abuse[338], possible relation to the induction of leukemia[339], and interaction with DNA[340] have been described.

The absorption, distribution and elimination of LSD has been extensively studied[341-344]. LSD *in vitro* is transformed in the liver to 2-hydroxy LSD. *In vivo* most of the LSD is metabolized to 4 or more different compounds (and excreted primarily in the bile) none of which is identical to 2-hydroxy LSD nor are any of the metabolites hallucinogenic.

Drug-abuse controls on hallucinogens has necessitated the requirement for rapid, sensitive micro identification procedures. A variety of chromatographic procedures have been developed for the identification of LSD and related compounds.

The thin-layer chromatographic identification and determination of LSD in the presence of heroin, other narcotics, and controlled drugs was described by Genest and Farmilo[345]. Three TLC systems were used.

(*A*) Silica Gel G (containing 0.1 *N* NaOH) was developed with chloroform–methanol (9:1) (10 cm, 20 min). This system was applied to basic compounds and their salts before and after irradiation.

(*B*) Silica Gel G with chloroform–methanol–ammonium hydroxide (28%) (40:40:20) (10 cm, 40 min development). This system was used to examine the acid-hydrolysis products of seizure material containing LSD and ergot alkaloids.

(*C*) Silica Gel G containing Eosin Y developed with morpholine–isopropyl ether–chloroform (10:80:10) (10 cm, 35 min development for the resolution of barbiturates). The chromogenic reagents used were (*a*) 0.5% *p*-dimethylaminobenzaldehyde (DMBA) in 95% ethanol containing hydrochloric acid[346] is used with system *A* above (LSD and ergot alkaloids yield blue-violet spots), (*b*) 0.2% ninhydrin in *n*-butanol containing 2 *N* acetic acid[347] (after spraying, the plates were heated for 5 min at 110°C to develop color; the reagent is applied to acid hydrolysis products chromatographed in system *B*), (*c*) potassium iodoplatinate[347] is applied to spots containing heroin and controlled drugs. Separation of LSD from heroin, other narcotics and controlled drugs was achieved by TLC in system *A* (Table 30). LSD (0.05 μg) was identified by light-blue fluorescence and by the DMBA spray. Table 31 illustrates

TABLE 30

R_F VALUES[a] OF LYSERGIC ACID DERIVATIVES, NARCOTICS AND CONTROLLED DRUGS

Compound	R_F value	Compound	R_F value
LSD	0.60	Dihydroergocristine	0.57
Ergonovine maleate	0.18	Ergotoxine	0.45–0.69–0.81
Ergometrinine	0.43	Heroin	0.48
Ergotamine tartrate	0.52	Morphine	0.10
Ergotaminine	0.72	Codeine	0.30
Ergocristine	0.71	Thebaine	0.53
Ergocristinine	0.77	Narcotine	0.77
Lysergic acid	0.05	Papaverine	0.77
Dihydroergotamine	0.35	Amphetamine	0.16
		Methamphetamine	0.17

[a] R_F values obtained on Silica Gel G (30 g and 0.1 *N* NaOH 60 ml) layers developed with chloroform–methanol (9:1) (10 cm, 20 min).

the R_F values of lysergic acid derivatives after irradiation with UV light. Compounds which migrate close to LSD before irradiation develop additional spots after treatment with UV light and the DMBA spray. (LSD, ergocristine and ergotamine gave one, dihydrocristine none, and ergotamine three DMBA positive spots.) Both methods, above, for the analysis of LSD in the presence of other lysergic acid derivatives were based on R_F measurements after DMBA treatment. Use was also made of the fact that, after acid hydrolysis of ergot alkaloids, the lysergic acid moiety is decomposed

TABLE 31

R_F VALUES[a] OF LYSERGIC ACID DERIVATIVES AFTER IRRADIATION WITH ULTRAVIOLET LIGHT

Compound	R_F value	
	DMBA-spray	Ultraviolet fluorescence[b]
LSD	0.61	0.80 GB
	0.38	0.73 Y
		0.61 B
		0.47 YB
Ergotamine tartrate	0.54	0.54 B
	0.51	0.35 Y
	0.31	
Ergocristine	0.72	0.72 B
	0.45	0.43 Y
Dihydroergocristine	0.57	0.66 YG
		0.35 YG
Ergotaminine	0.72	0.72 Gy
	0.52	0.52 Or
		0.20 Gy
		0.17 Ol
Heroin	0.48	0.48 B
		0.13 B

[a] TLC system as described in Table 30.
[b] G, green; Y, yellow; B, blue; Gy, grey; Or, orange; Ol, olive.

completely, whereas the peptide part at C(8) is hydrolyzed to yield amino acids[348]. If a strongly alkaline solvent is used for the chromatography of the hydrolysis products, the amino acids when sprayed with ninhydrin gave reddish spots in amongst the ergot alkaloids whereas LSD gave no reaction (Table 32).

TABLE 32

R_F VALUES[a] OF AMINO ACIDS AND ACID HYDROLYSIS PRODUCTS OF LYSERGIC ACID DERIVATIVES

Compound	R_F value	Compound	R_F value
LSD	No spot	Ergotoxine	0.49–0.60–0.68
Dihydroergotamine	0.49–0.78		(faint)–0.78
Ergotamine	0.48–0.78	Proline	0.49
Ergotaminine	0.49–0.78	Valine	0.61
Dihydroergocristine	0.49–0.79	Leucine	0.69
Ergocristine	0.49–0.77	Phenylalanine	0.79

[a] R_F values obtained on Silica Gel G (30 g and water 60 ml) developed with chloroform–methanol–ammonium hydroxide (28%) (40:40:20).

The presence of barbiturates is usually anticipated in LSD-containing seizure material. When preliminary color and crystal tests indicate barbiturates, TLC using system C (Table 33) was then employed. In this system using eosin-impregnated plates[349], barbiturates are revealed as dark spots on a light yellowish fluorescent background, whereas LSD yields a blue fluorescence. (Quinine, occasionally found in heroin seizures, also gives a light blue fluorescence with slightly different maxima but is separable from LSD in system A above (R_F quinine, 0.31).)

TABLE 33

R_F VALUES[a] OF BARBITURATES, HEROIN AND LSD

Compound	R_F value	Compound	R_F value
Pentobarbitone	0.40	Barbitone	0.39
Amylobarbitone	0.44	Hexobarbitone	0.46
Quinalbarbitone	0.49	LSD	0.34
Phenobarbitone	0.23	Heroin	0.35[b]
Butabarbital	0.44		

[a] R_F values obtained on Silica Gel G – Eosin Y (30 g + 30 mg + 30 ml water) plates developed with morpholine–isopropyl ether–chloroform (10:80:10).
[b] Elongated.

Spectrophotofluorometric analysis was also described in the above study for the analysis of LSD in the presence of heroin, other narcotics and controlled drugs with a standard error of $\pm 2\%$. When ergot alkaloids were present, TLC separation, elution and subsequent fluorometric analysis (using a Bowman-Aminco spectrophotofluorometer) are recommended.

Dal Cortivo et al.[350] utilized both Alumina G plates and silica gel Chromogram sheets (Types K-301R and K-301R2, Eastman-Kodak) with and without fluorescent indicator for the TLC of LSD. The most satisfactory solvents for the silica gel films was 1,1,1-trichloroethane–methanol (90:10) and for the Alumina G plates, 1,1,1-trichloroethane–methanol (98:2). Spots were examined under UV and a modified van Urk reagent[351], prepared by dissolving 0.8 g of p-dimethylaminobenzaldehyde in 100 ml of ethanol containing 10% conc. sulfuric acid, was used as a spray reagent. Table 34 lists the R_F values of LSD and related substances obtained on silica gel developed with 1,1,1-trichloroethane–methanol (90:10) and Fig. 25 illustrates the separation of LSD from related substances on Alumina G with 1,1,1-trichloroethane–methanol (98:2) as developer.

The analysis of preparations containing LSD was described by Martin and Alexander[352, 353]. LSD has been sold in the illicit drug market in various forms, i.e. powder, capsules, sugar cubes, liquids (both blue and light green to brown), bulk powder (both purple and light yellow), on candy, and in tablet form. The type of analysis thus depends upon the form of the sample. LSD can be detected by its

TABLE 34

R_F VALUES OF LSD AND RELATED SUBSTANCES

Silica Gel (100 μm), 1,1,1-trichloroethane–methanol (90 : 10)

Compound	R_F value	Compound	R_F value
Quinine	0.35	Ergotamine	0.47
Quinidine	0.30	Ergonovine	0.21
Tryptamine	0.06	Dihydroergotamine	0.45
Indole	1.00	Methylergonovine	0.25
Etryptamine	0.11	Methyl-methylergonovine	0.42
LSD	0.55		

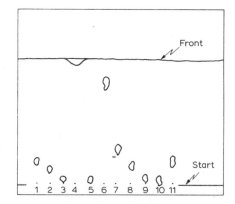

Fig. 25. Thin-layer chromatogram of LSD and related substances on Alumina G (250 microns) with TCE–methanol (98:2) as developing solvent. Compounds are: 1, quinine; 2, quinidine; 3, tryptamine; 4, indole; 5, etryptamine; 6, LSD; 7, ergotamine; 8, methyl-methylergonovine; 9, ergonovine; 10, methylergonovine; 11, dihydroergotamine.

fluorescence under UV light and by the production of a deep blue color when it is treated with a solution of p-dimethylaminobenzaldehyde and ferric chloride in sulfuric acid. LSD can be detected by TLC at a level as low as 0.2 μg and separated from iso-LSD if present by two different TLC systems, (1) Silica Gel G developed with chloroform–abs. methanol (1:4) and (2) silica gel–aluminum oxide plates developed with acetone. In the latter system, LSD migrates ca. 0.6–0.7 mm from origin and iso-LSD, when present, migrates 0.3–0.4 mm. The column chromatographed isolation of LSD has been accomplished using a mixture of Celite 545–citric acid with chloroform as eluent.

The identification of LSD and other indole alkaloids by the TLC analysis of their ultraviolet degradation products was reported by Anderson[354]. The method involved controlled degradation of the alkaloids in a chloroform solution irradiated for 13 min at 13 cm from a Chromato-Vue lamp Model XX-15C, then chromatographed on Silica Gel G developed with chloroform–acetone (1:4). (This system separated LSD from iso-LSD[355]; spots are then detected under short-wave UV light.) Table 35 is a computation of R_F values of 10 related indole-containing compounds and Fig. 26 shows the relative positions of each spot as well as their relative intensities. Irradiation in a solvent was found to produce more consistent spot patterns

TABLE 35

R_F VALUES FOR SOME INDOLE ALKALOIDS BEFORE AND AFTER IRRADIATION

Compound	R_F of each observable spot[a]							
LSD	0.27*							
LSD (irradiated)	0.0*	0.06*	0.27*	0.34*	0.39	0.55	0.63	0.70
Lysergic acid	0.0*							
Lysergic acid (irradiated)	0.0*	0.34	0.52	0.60	0.71			
Ergometrinine	0.19*							
Ergometrinine (irradiated)	0.0*	0.02*	0.17*	0.20*	0.24	0.44	0.51	
Ergotaminine	0.58*							
Ergotaminine (irradiated)	0.0	0.05	0.51*	0.57*	0.63*			
Ergocristine	0.50*							
Ergocristine (irradiated)	0.0*	0.15	0.45*	0.50*	0.57	0.62		
Ergotoxine	0.0	0.07	0.17	0.22	0.46*	0.49		
Ergotoxine (irradiated)	0.0*	0.14*	0.20	0.39*	0.45*	0.56	0.66	
Methysergide	0.09*							
Methysergide (irradiated)	0.0*	0.08*	0.10*	0.22				
Ergonovine	0.07*							
Ergonovine (irradiated)	0.0*	0.01*	0.06*	0.08*	0.19	0.50	0.59	
Methylergonovine	0.10*							
Methylergonovine (irradiated)	0.0*	0.02*	0.09*	0.12*	0.24	0.56	0.62	
Ergotamine	0.22*							
Ergotamine (irradiated)	0.0*	0.06*	0.17*	0.22*	0.36	0.55		

[a] Asterisk indicates spot is of equal or greater intensity than 1 μg unirradiated LSD.

than irradiation on the TLC plate after spotting. Clarke[356] described the thin-layer and paper chromatography of a number of psychedelic drugs. Two TLC systems were used: (1) Silica Gel G with methanol–ammonium hydroxide (100:1.5)[357] for development and (2) silica gel containing 0.1 N sodium hydroxide[345] with chloroform–methanol as developer. Detection (for TLC and PC) was accomplished using the following reagents: (a) iodoplatinate reagent, 10 ml of 5% platinum solution to 240 ml of 2% KI solution diluted with an equal volume of water, (b) 0.5% in bromocresol green (after spraying, the paper is exposed to ammonia vapor), (c) diazo reagent (p-nitroaniline in 2 N HCl–sodium nitrite) followed by an ethanolic sodium hydroxide spray, (d) 1% aq. potassium permanganate, (e) 0.5% ninhydrin in acetone (papers heated at 100°C for 5 min after spraying, and (f) 1% p-dimethylaminobenzaldehyde in ethanol containing 10 ml conc. hydrochloric acid.

    Paper chromatography was carried out by the method of Curry and Powell[357]; slightly modified Whatman No. 1 paper was impregnated with 5% sodium dihydro-

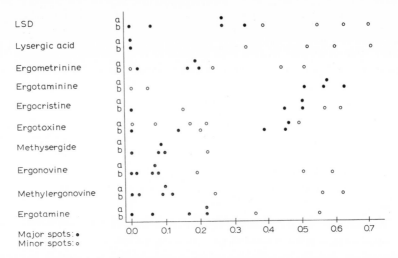

Fig. 26. R_F values of indole alkaloid. a, Unirradiated; b, irradiated. Solid circle is spot of equal or greater intensity than 1 μg unirradiated LSD; open circle is spot of less intensity than 1 μg unirradiated LSD.

gen citrate and the chromatograms developed with a solvent containing 2.4 g of citric acid in a mixture of 65 ml of water and 435 ml of n-butanol. Table 36 shows the results obtained on paper and thin-layer chromatograms for the lysergic acid hallucinogens compared with those for some common ergot alkaloids.

Look[358] described the identification and determination of LSD and its degradation products using paper chromatography. Whatman No. 1 paper was satu-

TABLE 36

PAPER AND THIN-LAYER CHROMATOGRAPHY OF LYSERGIC ACID HALLUCINOGENS
AND SOME COMMON ERGOT ALKALOIDS

	1	2	3	4	5	6	7	8
	PC R_F	UV	PtI$_4$	BCG	pDB	KMnO$_4$	TLC 1 R_F	TLC 2 R_F
"Ergotoxin"	0.82	Blue	−	−	Blue	+ +	0.73	0.78
"Dihydroergotoxin"	0.80	Green	−	−	Blue	+ +	0.71	0.68
Ergometrine	0.25	Blue	+	?	Blue	+ +	0.67	0.23
Methylergometrine	0.40	Blue	−	−	Blue	+ +	0.70	0.30
Lysergamide	0.11	Blue	+	+	Blue	+ +	0.60	0.18
Lysergide	0.47	Blue	+	?	Blue	+ +	0.66	0.63
Lysergic acid	0.18	Blue	−	?	Blue	+ +	0.58	0.01
Methysergide	0.45	Blue	−	?	Blue	+ +	0.66	0.49
Ergosine	0.65	Blue	+	−	Blue	+ +	0.72	0.58
Ergotamine	0.65	Blue	−	−	Blue	+ +	0.68	0.58
Dihydroergotamine	0.63	Green	−	−	Blue	+ +	0.63	0.56

+ + Strong Reaction   + Weak Reaction   ? Doubtful Reaction

rated with a methanolic solution of 25% formamide. The mobile solvent for development consisted of the upper layer of a saturated ether solution containing formamide. Paper chromatography of the products of partial racemization of LSD by base was also carried out. Spots appeared for both LSD and iso-LSD as well as for minor degradation products.

The separation of LSD from iso-LSD has also been accomplished by aluminum oxide column chromatography with chloroform containing 0.5% ethanol as the eluting agent[359]. Pioch[360], using the same column, eluted the LSD with benzene–chloroform (3:1) and the iso-LSD with benzene–chloroform. Martin and Alexander[353] separated LSD and iso-LSD on a pair of columns containing Celite–2% citric acid to trap the iso-LSD and Celite–8% citric acid to trap the LSD, with chloroform as the solvent.

Radecka and Nigam[361] described the detection of trace amounts of LSD in sugar cubes using a combination of GLC and TLC techniques. The free base was extracted from aq. sodium bicarbonate solution with methylene chloride, hydrogenated with Adam's catalyst, the product examined by GLC, and its identity confirmed by TLC of the effluents. An Aerograph Hy-Fi Model A-600C gas chromatograph was used equipped with a hydrogen-flame detector. A glass tube was inserted into the injection port to minimize degradation of the same at the metal surface of the preheater. The column, a stainless steel tube (5 ft. × $\frac{1}{8}$ in.) was packed with micro glass beads (60 mesh) coated with 0.2% SE-30.

By means of a stream splitter, effluents were collected in capillaries packed lightly with methanol-moistened glass wool. The column and injector temperatures were 270 and 290°C, respectively, and the carrier gas was nitrogen at 22 ml/min. Figure 27 illustrates gas chromatograms of hydrogenated LSD. A main peak (reten-

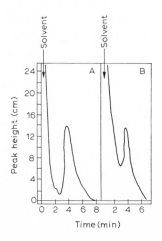

Fig.27. Gas chromatograms of hydrogenated LSD. A, Reference sample; B, sugar cube extract (exhibit J568). Column temperature: 270°C; injector temperature: 290°C; nitrogen flow: 22 ml/min; sample volume: 1 μl.

References pp. 337–344

tion time 3.6 min) and two slight shoulders (retention times 3.1 and 2.5 min, respectively) were observed. Thin-layer chromatographic analysis of the effluent emerging between 3 and 4 min served to confirm the gas chromatographic identification (Table 37). The chromatogram of hydrogenated LSD exhibited a blue fluorescent

TABLE 37

THIN-LAYER CHROMATOGRAPHY[a] OF HYDROGENATED LSD

Sample	Spot	Chromatographic data			
		Color observed		R_F value	Relative R_F value[b]
		Under UV light	After spraying		
Hydrogenated LSD	1	Blue fluorescence	Faint yellow	0.81	1.04
	2		Blue	0.78	1.00
Gas chromatographic eluate, 3–4 min	1	Blue fluorescence	Faint yellow	0.81	1.04
	2		Blue	0.78	1.00
Hydrogenated LSD soln. stored for 2 weeks	1	Blue fluorescence	Faint yellow	0.81	1.04
	2		Blue	0.78	1.00
	3		Blue	0.54	0.69

[a] Absorbent Silica Gel G, developing solvent chloroform–ethanol (96:4), detection UV light 366 nm then plates sprayed with p-dimethylaminobenzaldehyde (2 g in conc. HCl (20 ml) and ethanol (80 ml)).
[b] Reference: spot 2.

spot when viewed under UV light. After spraying with the p-dimethylaminobenzaldehyde reagent, this spot acquired a yellow coloration, while another intense blue spot appeared at a slightly lower R_F value. These same spots were observed when the gas chromatographic eluate was similarly examined. The direct microanalysis of LSD by gas chromatography was described by Katz et al.[362]. A Perkin-Elmer 881 instrument equipped with a flame-ionization detector was employed. A glass column (6 ft × ⅛ in.) with a built-in glass injector port was packed with 0.3% SE-30 on microglass beads. The operating temperatures for the injector column and detector were 315, 280, and 285°C, respectively, and the carrier gas was helium at a flow rate of 70 ml/min. LSD was analyzed following its extraction with methanol from a sugar cube, filter paper or bicarbonate capsule; the extract filtered, evaporated to dryness under vacuo, brought to a 0.2 ml volume with methanol and injected directly into the gas chromatograph. The retention time for LSD was 4.0 min.

REFERENCES

1   L.S.GOODMAN AND A.GILMAN, *The Pharmacological Basis of Therapeutics*, 3rd edn., MacMillan, New York, 1967, p.132.
2   F.J.MACKAY AND R.J.COOPER, *J. Pharmacol. Exptl. Therap.*, 135 (1962) 271.
3   A.BARTHELMESS, *Arzneimittel-Forsch.*, 6 (1956) 157.
4   A.GOLDSTEIN, in W.J.SCHULL (Ed.), *Mutations*, Univ. of Michigan Press, Ann Arbor, Mich., 1960, p.172.
5   P.J.FRIEDMAN AND J.R.COOPER, *Anal. Chem.*, 30 (1958) 1674.
6   B.E.CABANA AND P.K.GESSNER, *Anal. Chem.*, 39 (1967) 1449.
7   A.E.MAYER, *Arzneimittel-Forsch.*, 7 (1957) 194.
8   A.W.DAVIDSON, *J. Assoc. Offic. Anal. Chemists*, 51 (1968) 626.
9   N.STANCIV AND U.STOICESCU, *Farmacia (Bucharest)*, 4 (1956) 313; *Chem. Abstr.*, 52 (1958) 8459F.
10 J.BARLOT AND C.ALBISSON, *Clin. Anal. (Paris)*, 38 (1956) 313.
11 E.R.GARRETT AND H.J.LAMBERT, *J. Pharm. Sci.*, 55 (1966) 812.
12 V.SEDIVEC AND J.FLEK, *Collection Czech. Chem. Commun.*, 34 (1969) 1533.
13 R.P.SNELL, private communication in ref.8.
14 N.V.FEHRINGER, private communication in ref.8.
15 R.J.ANDERSON, C.A.ANDERSON AND T.J.OLSON, *J. Agr. Food Chem.*, 14 (1966) 508.
16 I.K.TSITOVICH AND E.A.KUZ'MENKO, *Zh. Analit. Khim.*, 22 (1967) 603.
17 H.ARNOLD, F.BOURSEAUX AND N.BROCK, *Naturwissenschaften*, 45 (1958) 64.
18 D.G.DECKER, E.MUSSEY, G.D.MALKASIAM AND C.E.JOHNSON, *Clin. Obstet. Gynecol.*, 11 (1968) 382.
19 E.R.HOMAN, R.P.ZENZIAN, W.M.BUSEY AND D.P.RALL, *Nature*, 221 (1969) 1059.
20 E.H.DOLNICK, I.L.LUNDAEHL, C.E.TERRILL AND P.J.REYNOLDS, *Nature*, 221 (1969) 467.
21 T. VON KREYBIG, *Arch. Exptl. Pathol. Pharmakol.*, 252 (1965) 173.
22 N.BROCK AND T. VON KREYBIG, *Arch. Exptl. Pathol. Pharmakol.*, 249 (1964) 117.
23 D.O.E.GEBHARDT, *Teratology*, 3 (1970) 273.
24 J.E.GIBSON AND B.A.BECKER, *Cancer Res.*, 28 (1968) 475.
25 H.NORDLINGER, *Experientia*, 25 (1969) 1296.
26 D.BRITTINGER, *Humangenetik*, 3 (1966) 156.
27 G.ROHRBORN, *Mol. Gen. Genet.*, 102 (1968) 50.
28 F.E.ARRIGHI, T.C.HSU AND D.E.BERGSAGEL, *Texas Rept. Biol. Med.*, 20 (1962) 545.
29 K.E.HAMPEL, M.FRITZSCHE AND D.STOPIK, *Humangenetik*, 7 (1969) 28.
30 N.BROCK, *Cancer Chemotherapy Rept.*, 51 (1967) 315.
31 N.BROCK AND H.J.HOHORST, *Cancer*, 20 (1967) 900.
32 E.H.GRAUL, H.HUNDESHAGEN AND H.WILLIAMS, *Proc. 3rd Intern. Congr. Chemotherapy, Stuttgart, 1963*, [No.2] 1964, p.1107.
33 H.M.RAUEN AND K.NORPOTH, *Arzneimittel-Forsch.*, 17 (1967) 599.
34 H.M.RAUEN AND K.NORPOTH, *Arzneimittel-Forsch.*, 16 (1966) 40.
35 A.DEDE AND F.FARABOLLINI, *Bull. Soc. Ital. Biol. Sper.*, 43 (1967) 1489.
36 M.HIRATA, H.KAGAWA AND M.BABA, *Shionogi Kenkyusho Nempo*, 17 (1967) 107; *Chem. Abstr.*, 69 (1968) 21955E.
37 O.M.FRIEDMAN, *Cancer Chemotherapy Rept.*, 51 (1967) 327.
38 O.M.FRIEDMAN, S.BIEN AND J.K.CHATRABARTI, *J. Am. Chem. Soc.*, 87 (1965) 4978.
39 Anon, *Nature*, 227 (1970) 16.
40 T.P.JOHNSTON, G.S.McCALEB, P.S.OPLINGER AND J.A.MONTGOMERY, *J. Med. Chem.*, 9 (1966) 802.
41 V.T.DE VITA, C.DENHAM, J.D.DAVIDSON AND V.T.OLIVERIO, *Clin. Pharm. Therap.*, 8 (1967) 566.
42 V.T.DE VITA, D.P.CARBONE, A.H.OWENS, JR., G.L.GOLD, M.J.KRANT AND J.EDMONDSON, *Cancer Res.*, 25 (1965) 1876.
43 G.P.WHEELER AND B.J.BOWDAN, *Cancer Res.*, 25 (1965) 1770.

44 R.W.Sidwell, G.J.Dixon, S.M.Sellers and F.M.Schabel, Jr., *Appl. Micriobiol.*, 13 (1965) 579.
45 G.R.Thompson and R.E.Larson, *J. Pharmacol. Exptl. Therap.*, 166 (1969) 104.
46 E.K.Harrod and J.A.Cortner, *J. Natl. Cancer Inst.*, 40 (1968) 269.
47 T.L.Loo, R.L.Dixon and D.P.Rall, *J. Pharm. Sci.*, 55 (1966) 492.
48 T.L.Loo and R.C.Dixon, *J. Pharm. Sci.*, 54 (1965) 809.
49 R.Preussman, D.Daiber and H.Hengy, *Nature*, 201 (1964) 502.
50 T.P.Johnston, private communication in ref.41.
51 T.P.Johnston, G.S.McCaleb, P.O.Pliger and J.A.Montgomery, *Abstr. Am. Chem. Soc.*, *147th Meeting, April, 1964*.
52 G.P.Wheeler, B.J.Bowdon and T.C.Herren, *Cancer Chemotherapy Rept.*, 42 (1964) 9.
53 J.A.Montgomery, R.James, G.S.McCaleb and T.P.Johnston, *J. Med. Chem.*, 10 (1967) 688.
54 A.Haddow and G.M.Timmis, *Lancet*, 1953-I, 207.
55 D.A.G.Galton, *Advan. Cancer Res.*, 4 (1956) 73.
56 S.Ishikawa, H.Yokotani, F.Watanabe, Y.Aramaki and K.Kajiwara, *Takeda Kenkyusho Nempo*, 16 (1957) 64.
57 B.E.Heard and R.A.Cooke, *Thorax*, 23 (1968) 187.
58 G.Koch, *Bull. Soc. Chim. Belges*, 68 (1959) 59.
59 E.G.Trams, R.A.Salvador, G.Maengwyn-Davies and V.De Quattro, *Proc. Am. Assoc. Cancer Res.*, 2 (1957) 256.
60 E.G.Trams, M.V.Nadkarni, V.De Quattro, G.D.Maengwyn-Davies and P.K.Smith, *Biochem. Pharmacol.*, 2 (1959) 7.
61 J.J.Roberts and G.P.Warwick, *Nature*, 183 (1959) 1509.
62 J.J.Roberts and G.P.Warwick, *Biochem. Pharmacol.*, 1 (1958) 60.
63 L.A.Elson, *Biochem. Pharmacol.*, 1 (1958) 39.
64 B.W.Fox, A.W.Craig and H.Jackson, *Biochem. Pharmacol.*, 5 (1960) 27.
65 M.V.Nadkarni, E.G.Trams and P.K.Smith, *Cancer Res.*, 19 (1959) 713.
66 M.V.Nadkarni, E.G.Trams and P.K.Smith, *Proc. Am. Assoc. Cancer Res.*, 2 (1957) 235.
67 H.Jackson, A.W.Craig and B.W.Fox, *Acta Unio Intern. Contra Cancrum*, 16 (1964) 611.
68 I.Diamond, M.A.Anderson and S.R.McCreadie, *Pediatrics*, 25 (1960) 85.
69 M.Partington and H.Jackson, *Genet. Res. Cambridge*, 4 (1963) 333.
70 J.Moutschen, *Genetics*, 46 (1961) 291.
71 O.G.Fahmy and M.J.Fahmy, *Genetics*, 46 (1961) 1111.
72 G.Rohrborn, *Z. Vererbungslehre*, 90 (1959) 457.
73 W.G.Verly and A.Petitpas-Dewandre, *Nature*, 203 (1964) 865.
74 J.J.Roberts and G.P.Warwick, *Nature*, 179 (1957) 1181.
75 J.J.Roberts and G.P.Warwick, *Biochem. Pharmacol.*, 6 (1961) 217.
76 C.T.Peng, *J. Pharmacol. Exptl. Therap.*, 120 (1957) 229.
77 D.A.G.Galton, *Lancet*, 1953-I, 208.
78 B.W.Fox, A.W.Craig and H.Jackson, *Brit. J. Pharmacol.*, 14 (1959) 149.
79 P.Brookes and P.D.Lawley, *Biochem. J.*, 77 (1960) 478.
80 R.J.Block, E.L.Durrum and G.Zweig, *A Manual of Paper Chromatography and Paper Electrophoresis*, Academic Press, New York, 1955.
81 M.A.Walters, F.J.C.Roe, B.C.U.Mitchley and A.Walsh, *Brit. J. Cancer*, 21 (1967) 367.
82 T.Alderson, *Nature*, 207 (1965) 164.
83 D.R.Krieg, *Genetics*, 48 (1963) 561.
84 N.M.Schwartz, *Genetics*, 48 (1963) 1357.
85 E.H.Y.Chu and H.V.Malling, *Proc. Natl. Acad. Sci. U.S.*, 61 (1968) 1306.
86 A.I.Tareeva and A.I.Yakovleva, *Khim. i Med.*, 13 (1960) 34; *Chem. Abstr.*, 55 (1961) 10679A.
87 N.K.Clapp, A.W.Craig and R.E.Toya, Sr., *Science*, 161 (1968) 913.
88 A.C.Upton, *Natl. Cancer Inst. Monograph*, 22 (1966) 329.
89 A.J.Bateman, R.L.Peters, J.G.Hazen and J.L.Steinfeld, *Cancer Chemotherapy Rept.*, 50 (1966) 675.
90 J.B.Kissam, J.A.Wilson and J.B.Hays, *J. Econ. Entomol.*, 60 (1967) 1130.

91   H. JACKSON, B. W. FOX AND A. W. CRAIG, *J. Reprod. Fertility*, 2 (1961) 447.
92   S. S. EPSTEIN AND H. SHAFNER, *Nature*, 219 (1968) 385.
93   U. H. EHLING, R. B. CUMMING AND H. V. MALLING, *Mutation Res.*, 5 (1968) 417.
94   E. BAUTZ AND E. FREESE, *Proc. Natl. Acad. Sci. U.S.*, 46 (1960) 1585.
95   N. LOPRIENO, *Mutation Res.*, 3 (1966) 486.
96   P. D. LAWLEY, *Nature*, 218 (1968) 580.
97   E. A. BARNSLEY, *Biochem. J.*, 106 (1968) 18P.
98   D. J. PILLINGER, B. W. FOX AND A. W. CRAIG, in L. J. ROTH (Ed.), *Isotopes in Experimental Pharmacology*, Univ. of Chicago Press, Chicago, Ill., 1965, p. 415.
99   P. F. SWANN, *Nature*, 214 (1967) 918.
100 S. K. CARTER, *Cancer Chemotherapy Rept.*, 1 [3] (1968) 99.
101 C. L. STEVENS, K. G. TAYLOR, M. E. MUNK, W. S. MARSHALL, K. NOLL, G. D. SHAH, L. G. SHAI AND K. UZU, *J. Med. Chem.*, 8 (1964) 1.
102 S. WAKAKI, H. MARUMO, K. TOMIOKA, G. SHIMIZU, E. KAO, H. KAMADA, S. KUDO AND Y. FUGI-MOTO, *Antibiot. Chemotherapy*, 8 (1958) 228.
103 F. S. PHILIPS, H. S. SCHWARTZ AND S. S. STERNBERG, *Cancer Res.*, 20 (1960) 1354.
104 H. S. SCHWARTZ AND F. S. PHILIPS, *J. Pharm. Exptl. Therap.*, 133 (1961) 335.
105 I. H. GOLDBERG, *Am. J. Med.*, 39 (1965) 722.
106 V. N. IYER AND W. SZYBALSKI, *Proc. Natl. Acad. Sci. U.S.*, 50 (1963) 355.
107 M. J. WARING, *Nature*, 219 (1968) 1320.
108 V. N. IYER AND W. SZYBALSKI, *Science*, 145 (1964) 55.
109 H. KERSTEN, *Biochim. Biophys. Acta*, 55 (1962) 558.
110 S. SHIKA, A. TIRAWAKI, T. TAGUCHI AND J. KAWAMATA, *Nature*, 183 (1959) 1056.
111 R. MUKHERJEE, *Genetics*, 51 (1965) 947.
112 D. T. SUZUKI, *Genetics*, 51 (1965) 635.
113 M. W. SHAW AND M. M. COHEN, *Genetics*, 51 (1965) 181.
114 P. C. NOWELL, *Exptl. Cell Res.*, 33 (1964) 445.
115 R. R. HERR, T. E. EBLE, M. E. BERGY AND H. K. JAHNKE, *Antibiot. Ann.*, (1960) 236.
116 R. R. HERR, H. JAHNKE AND A. ARGOUDELIS, *J. Am. Chem. Soc.*, 89 (1967) 4808.
117 L. LEWIS AND A. R. BARBIERS, *Antibiot. Ann.*, (1960) 247.
118 L. J. HANKA AND W. T. SOKOLSKI, *Antibiot. Ann.*, (1960) 255.
119 B. HEINMANN AND A. J. HOWARD, *Appl. Microbiol.*, 12 (1964) 234.
120 K. E. PRICE, R. L. BUCK AND J. LEIN, *Antimicrobial Agents Chemotherapy*, (1964) 505.
121 J. S. EVANS, G. C. GERRITSEN, K. M. MANN AND S. P. OWEN, *Cancer Chemotherapy Rept.*, 48 (1965) 1.
122 Y. ARNOULD, J. A. OOMS AND P. A. BASTENIE, *Lancet*, 1969-II, 1210.
123 I. M. MURRA-LION, AL. W. F. EDDLESTON, R. WILLIAMS, M. BROWN, B. M. HOGBIN, A. BENNET, J. C. EDWARDS AND K. W. TAYLOR, *Lancet*, 1968-II, 895.
124 N. RAKIETEN, M. L. RAKIETEN AND M. V. NADKARNI, *Cancer Chemotherapy Rept.*, 29 (1963) 91.
125 T. M. SIBAY AND J. A. HAYES, *Lancet*, 1969-II, 912.
126 R. N. ARISON AND E. L. FEUDALE, *Nature*, 214 (1967) 1254.
127 S. M. KOLBYE AND M. S. LEGATOR, *Mutation Res.*, 6 (1968) 387.
128 M. G. GABRIDGE, A. DENUNZIO AND M. S. LEGATOR, *Nature*, 221 (1968) 68.
129 M. G. GABRIDGE, E. J. OSWALD AND M. S. LEGATOR, *Mutation Res.*, 7 (1969) 117.
130 E. R. GARRETT, *J. Am. Pharm. Assoc.*, 49 (1960) 767.
131 J. J. VAVRA, C. DE BOER, A. DIETZ, L. J. HANKA AND W. T. SOKOLSKI, *Antibiot. Ann.*, (1959–1960) 230.
132 R. R. GRAVES AND C. W. HESSELTINE, *Mycopathol. Mycol. Appl.*, 29 (1966) 277.
133 H. H. KUEHN, *Mycologia*, 50 (1958) 417.
134 E. O. KAROW AND J. W. FOSTER, *Science*, 99 (1944) 265.
135 E. P. ABRAHAM AND H. W. FLOREY, in N. W. FLOREY *et al.* (Eds.), *Antibiotics*, Oxford Univ. Press, London, 1949, Vol. I, p. 273.
136 O. M. EFIMENKO AND P. A. YAKIMOV, *Tru. Leningr. Khim.-Farmatsevt. Inst.*, (1960) 88; *Chem. Abstr.*, 55 (1961) 21470.
137 W. P. BRIAN, G. W. ELSON AND D. LOWE, *Nature*, 178 (1956) 263.
138 F. A. NORSTADT AND T. M. MCCALLA, *Science*, 140 (1963) 410.

139 T. UKAI, Y. YAMAMOTO AND T. YAMAMOTO, *J. Pharm. Soc. Japan*, 74 (1954) 450.
140 H. F. KRAYBILL AND M. B. SHIMKIN, *Advan. Cancer Res.*, 8 (1964) 191.
141 V. W. MAYER AND M. S. LEGATOR, *J. Agr. Food Chem.*, 17 (1969) 454.
142 S. A. WAKSMAN, E. S. HORNING AND E. L. SPENCER, *Science*, 96 (1944) 202.
143 J. KENT AND N. G. HEATLEY, *Nature*, 156 (1945) 295.
144 R. B. WOODWARD AND G. SINGH, *J. Am. Chem. Soc.*, 72 (1950) 1428.
145 S. TANNENBAUM AND E. BASSETT, *J. Biol. Chem.*, 234 (1959) 1861.
146 F. DICKENS AND H. E. H. JONES, *Brit. J. Cancer*, 15 (1961) 85.
147 F. DICKENS, On cancer hormones essays, *Exptl. Biol.*, (1962) 107; *Chem. Abstr.*, 60 (1964) 9723.
148 B. BABUDIERI, *Rend. Ist. Super. Sanita*, 11 (1968) 577.
149 F. H. WANG, *Botan. Bull. Acad. Sinica*, 2 (1948) 265.
150 H. KEILOVA-RODOVA, *Experientia*, 5 (1949) 242.
151 R. F. J. WITHERS, *Symp. Mutational Process, Mech. Mutation Inducing Factors, Prague*, 1965, p. 359.
152 P. SENTEIN, *Compt. Rend. Soc. Biol.*, 149 (1955) 1621.
153 P. M. SCOTT AND E. SOMERS, *J. Agr. Food Chem.*, 16 (1968) 483.
154 T. YAMAMOTO, *J. Pharm. Soc. Japan*, 76 (1956) 1375.
155 E. G. JEFFERYS, *J. Gen. Microbiol.*, 7 (1952) 295.
156 F. DICKENS AND J. COOKE, *Brit. J. Cancer*, 15 (1961) 85.
157 W. B. GEIGER AND J. E. CONN, *J. Am. Chem. Soc.*, 67 (1945) 112.
158 A. E. POHLAND AND R. ALLEN, *J. Assoc. Offic. Anal. Chemists*, 53 (1970) 686.
159 A. E. POHLAND AND R. ALLEN, *J. Assoc. Offic. Anal. Chemists*, 53 (1970) 688.
160 A. E. POHLAND, K. SANDERS AND C. W. THORPE, *J. Assoc. Offic. Anal. Chemists*, 53 (1970) 692.
161 V. BETINA, *J. Chromatog.*, 15 (1964) 379.
162 M. A. EL-NAKEEB AND J. O. LAMPEN, *J. Gen. Microbiol.*, 39 (1965) 285.
163 B. BOOTHROYD, E. J. NAPIER AND G. A. SOMERFIELD, *Biochem. J.*, 80 (1961) 34.
164 D. BUSFIELD, K. J. CHILD AND E. G. TOMICH, *Brit. J. Pharmacol.*, 22 (1964) 137.
165 F. DE MATTEIS, *Biochem. J.*, 98 (1966) 23.
166 M. J. BARNES AND B. BOOTHROYD, *Biochem. J.*, 78 (1961) 41.
167 S. GRANICK, *J. Biol. Chem.*, 238 (1963) 2247.
168 J. SCHWARZ AND J. K. LOUTZENHISER, *Arch. Dermatol.*, 81 (1960) 694.
169 L. L. BARICH, J. SCHWARZ AND D. BARICH, *Cancer Res.*, 22 (1962) 53.
170 L. L. BARICH, J. SCHWARZ, D. J. BARICH AND M. G. HOROWITZ, *Chemotherapy*, 11 (1961) 566.
171 F. DE MATTEIS, A. J. DONELLY AND N. J. RUNGE, *Cancer Res.*, 26 (1966) 721.
172 E. W. HURST AND G. E. PAGET, *Brit. J. Dermatol.*, 75 (1963) 105.
173 S. S. EPSTEIN, J. ANDREA, S. JOSHI AND N. MANTEL, *Cancer Res.*, 27 (1967) 1900.
174 K. FUJII AND S. S. EPSTEIN, *Toxicol. Appl. Pharmacol.*, 14 (1969) 613.
175 S. S. EPSTEIN, J. ANDREA, P. CLAPP AND D. MACKINTOSH, *Toxicol. Appl. Pharmacol.*, 11 (1967) 442.
176 W. L. EPSTEIN AND M. A. LARSON, *J. Invest. Dermatol.*, 36 (1961) 5.
177 G. E. PAGET AND A. L. WALPOLE, *Arch. Dermatol.*, 81 (1960) 750.
178 N. N. SLONITSKAYA, *Antibiotiki*, 14 (1969) 44.
179 A. E. OXFORD, H. RAISTRICK AND F. SIMONART, *Biochem. J.*, 33 (1939) 240.
180 J. F. GROVE, D. ISMAY, J. MACMILLAN, T. P. C. MULHOLLAND AND M. A. D. ROGERS, *Chem. Ind. (London)*, (1951) 219.
181 H. BROSSI, *Helv. Chim. Acta*, 43 (1960) 1444.
182 A. J. BIRCH, R. A. MASSY-WESTROPP, R. W. RICHARDS AND H. SMITH, *J. Chem. Soc.*, (1958) 360.
183 S. SYMCHOWICZ AND K. K. WONG, *Biochem. Pharmacol.*, 15 (1966) 1601.
184 S. SYMCHOWICZ, M. S. STAUB AND K. K. WONG, *Biochem. Pharmacol.*, 16 (1967) 2405.
185 S. TOMOMATSU AND G. KITAMURA, *Chem. Pharm. Bull. (Tokyo)*, 8 (1960) 755.
186 L. J. FISCHER AND S. RIEGELMAN, *J. Chromatog.*, 21 (1966) 268.
187 L. J. FISCHER AND S. RIEGELMAN, *J. Pharm. Sci.*, 54 (1965) 1571.
188 R. J. COLE, J. W. KIRKSEY AND C. E. HOLADAY, *Appl. Microbiol.*, 19 (1970) 106.

189 W. N. Fishbein, P. P. Carbone, E. J. Friereich, J. Misra and E. Frei, *Clin. Pharmacol. Therap.*, (1964) 574.
190 R. H. Adamson, S. T. Yancey, M. Ben, T. L. Loo and D. P. Rall, *Arch. Intern. Pharmacodyn.*, 2 (1965) 153.
191 I. H. Krakoft, M. L. Murphy and H. Savel, *Proc. Am. Assoc. Cancer Res.*, 4 (1963) 35.
192 H. S. Rosenkranz, H. M. Rose, C. Morgan and K. C. Hsu, *Virology*, 28 (1966) 510.
193 J. B. Kissam, J. A. Wilson and S. B. Hays, *J. Econ. Entomol.*, 60 (1967) 1130.
194 V. H. Ferm, *Lancet*, 1965-I, 1338.
195 V. H. Ferm, *Arch. Pathol.*, 81 (1966) 174.
196 S. Chaube and M. L. Murphy, *Cancer Res.*, 26 (1966) 1448.
197 M. L. Murphy and S. Chaube, *Cancer Chemotherapy Rept.*, 40 (1964) 1.
198 S. Chaube, E. Simmel and C. Lacon, *Proc. Am. Assoc. Cancer Res.*, 4 (1963) 10.
199 J. J. Oppenheim and W. N. Fishbein, *Cancer Res.*, 25 (1965) 980.
200 W. K. Sinclair, *Science*, 24 (1965) 1729.
201 H. S. Rosenkranz and J. A. Levy, *Biochim. Biophys. Acta*, 95 (1965) 181.
202 C. W. Young, G. Schoehetman, S. Hodas and M. E. Balls, *Cancer Res.*, 27 (1967) 535.
203 B. A. Kihlman, T. Eriksson and G. Odmark, *Hereditas*, 55 (1966) 386.
204 R. H. Adamson, S. L. Ague, S. M. Hess and J. D. Davidson, *J. Pharm. Exptl. Therap.*, 150 (1965) 322.
205 H. Kofod, *Acta Chem. Scand.*, 9 (1955) 1575.
206 J. E. Milks and R. H. Janes, *Anal. Chem.*, 28 (1956) 846.
207 M. Colvin and V. H. Bono, Jr., *Cancer Res.*, 30 (1970) 1516.
208 W. N. Fishbein and P. O. Carbone, *Science*, 142 (1963) 1069.
209 A. Jeanrenaud, *Ber. Deut. Chem. Ges.*, 22 (1889) 1270.
210 L. Fishbein and M. A. Cavanaugh, *J. Chromatog.*, 20 (1965) 283.
211 S. J. Jacobs and H. S. Rosenkranz, *Cancer Res.*, 30 (1970) 1084.
212 E. Boyland and R. Nery, *J. Chem. Soc.*, (1966) 354.
213 H. S. Rosenkranz, *J. Bacteriol.*, 102 (1970) 20.
214 H. S. Rosenkranz and J. Rosenkranz, *Biochim. Biophys. Acta*, 195 (1969) 266.
215 H. S. Rosenkranz, R. D. Pollak and R. M. Schmidt, *Cancer Res.*, 29 (1969) 209.
216 L. S. Goodman and A. Gilman, *The Pharmacological Basis of Therapeutics*, 3rd edn., Macmillan, New York, 1967, p. 1322.
217 W. B. Schaefer, *Am. Rev. Tuber. Pulmonary Diseases*, 69 (1954) 125–127.
218 D. S. Feingold, *New Engl. J. Med.*, 269 (1963) 900.
219 L. B. Colvin, *J. Pharm. Sci.*, 58 (1969) 1433.
220 S. Hess, H. Weissbach, B. G. Redfield and S. Udenfriend, *J. Pharmacol. Exptl. Therap.*, 124 (1958) 189.
221 A. S. Yard, H. McKennis, Jr. and J. H. Weatherby, *J. Pharmacol. Exptl. Therap.*, 119 (1957) 195.
222 B. Toth, *Arch. Environ. Health*, 20 (1970) 343.
223 K. H. Harper and A. N. Worden, *Toxicol. Appl. Pharmacol.*, 8 (1966) 325.
224 J. Juhász, J. Baló and G. Kendrey, *Tuberkulozis*, 3–4 (1957) 49.
225 A. Peacock and P. R. Peacock, *Brit. J. Cancer*, 20 (1966) 307.
226 J. Juhász, J. Baló and B. Szende, *Z. Krebsforsch.*, 65 (1963) 434.
227 L. Severi and C. Biancifiori, *J. Natl. Cancer Inst.*, 41 (1968) 381.
228 E. Freese, S. Sklarow and E. B. Freese, *Mutation Rev.*, 5 (1968) 343.
229 J. N. Vivien and J. Grosset, *Advan. Tuberc. Res.*, 11 (1961) 45.
230 J. B. Hughes, *J. Pharmacol. Exptl. Therap.*, 109 (1953) 444.
231 R. C. R. Barreto, *J. Chromatog.*, 7 (1962) 82.
232 R. C. R. Barreto and S. O. Sabino, *J. Chromatog.*, 9 (1962) 180.
233 R. C. R. Barreto and S. O. Sabino, *J. Chromatog.*, 11 (1963) 344.
234 R. C. R. Barreto and S. O. Sabino, *J. Chromatog.*, 13 (1964) 435.
235 R. C. R. Barreto and D. B. Mano, *Biochem. Pharmacol.*, 8 (1961) 403.
236 V. A. Greulach and J. G. Haesloop, *Anal. Chem.*, 33 (1961) 1446.
237 H. Wachsmuth, *J. Pharm. Belg.*, 11 (1956) 86.
238 F. Percheron, *Bull. Soc. Chim. Biol.*, 44 (1962) 1162.

239 O. FURTH AND H. HERRMANN, *Biochem. Z.*, 280 (1935) 448.
240 A. LEWANDOWSKI AND H. SYBIRSKA, *Chem. Anal. (Warsaw)*, 13 (1968) 319.
241 T. A. LA RUE, *J. Chromatog.*, 32 (1968) 784.
242 A. ALESSANDRO, F. MARI AND S. SETTECASE, *Farmaco (Pavia)*, *Ed. Prat.*, 22 (1967) 437.
243 A. CALO, C. CARDINI AND V. QUERCIA, *J. Chromatog.*, 37 (1968) 194.
244 A. CALO, C. CARDINI AND V. QUERCIA, *Bull. Chim. Farm.*, 107 (1968) 296.
245 M. C. FAN AND W. G. WALD, *J. Assoc. Offic. Agr. Chemists*, 48 (1965) 1148.
246 S. INOUE, A. OGINO AND Y. ONO, *Yakuzaigaku*, 26 (1966) 302.
247 W. H. WU, T. F. CHIN AND J. L. LACH, *J. Pharm. Sci.*, 59 (1970) 1234.
248 W. KING, H. KELLER AND H. MUCKTER, *Arzneimittel-Forsch.*, 6 (1956) 426.
249 W. LENZ AND K. KNAPP, *Deut. Med. Wochschr.*, 87 (1962) 1232.
250 G. F. SOMERS, *Lancet*, 1962-I, 912.
251 C. S. DELAHUNT AND L. J. LASSEN, *Science*, 146 (1964) 1300.
252 S. FABRO, H. SCHUMACHER, R. L. SMITH AND R. T. WILLIAMS, *Life Sci.*, 3 (1964) 987.
253 H. P. DROBECK, F. COULSTON AND D. CORNELIUS, *Toxicol. Appl. Pharmacol.*, 7 (1965) 165.
254 I. D. FRATTA, E. B. SIGG AND K. MAIORANA, *Toxicol. Appl. Pharmacol.*, 7 (1965) 268.
255 J. G. WILSON AND J. A. GAVAN, *Anatom. Rec.*, 158 (1967) 99.
256 J. G. WILSON, R. FRADKIN AND A. A. HARDMAN, *Teratology*, 1 (1968) 223.
257 H. SCHUMACHER, D. A. BLAKE, J. M. GURIAN AND J. R. GILLETTE, *J. Pharmacol. Exptl. Therap.*, 160 (1968) 189.
258 R. CAHEN, *Clin. Pharmacol. Therap.*, 5 (1964) 480.
259 M. K. JENSEN, *Acta Med. Scand.*, 177 (1965) 1783.
260 J. D. AMIRKHANIAN, *Experientia*, 26 (1970) 796.
261 H. SCHUMACHER, D. A. BLAKE AND J. R. GILLETTE, *Federation Proc.*, 26 (1967) 730.
262 H. SCHUMACHER, D. A. BLAKE AND J. R. GILLETTE, *J. Pharmacol. Exptl. Therap.*, 160 (1968) 201.
263 R. T. WILLIAMS, *Arch. Environ. Health*, 16 (1968) 493.
264 S. FABRO, R. L. SMITH AND R. T. WILLIAMS, *Nature*, 215 (1967) 296.
265 R. T. WILLIAMS, *Lancet*, 1963-I, 723.
266 H. SCHUMACHER, R. L. SMITH, R. B. I. STAGG AND R. T. WILLIAMS, *Pharm. Acta Helv.*, 39 (1964) 394.
267 R. T. WILLIAMS AND D. V. PARKE, *Ann. Rev. Pharmacol.*, 4 (1964) 85.
268 H. SCHUMACHER, R. L. SMITH, R. B. L. STAGG AND R. T. WILLIAMS, *3rd Intern. Meeting Forensic Immunol. Med., Pathol. Toxicol. Plenary Session VIIA, London, 1963.*
269 R. BECKMAN, *Arzneimittel-Forsch.*, 13 (1963) 185.
270 J. W. FAIGLE et al., *Experientia*, 18 (1962) 389.
271 S. FABRO, R. L. SMITH AND R. T. WILLIAMS, *Biochem. J.*, 104 (1967) 565.
272 H. SCHUMACHER, R. L. SMITH AND R. T. WILLIAMS, *Brit. J. Pharmacol.*, 25 (1965) 324.
273 S. FABRO, R. L. SMITH AND R. T. WILLIAMS, *Biochem. J.*, 104 (1967) 570.
274 P. J. NICHOLLS, *J. Pharm. Pharmacol.*, 18 (1966) 46.
275 G. PISCHEK, E. KAISER AND H. KOCH, *Microchim. Acta*, (1970) 530.
276 M. FRAHM, A. GOTTESLEBEN AND K. SOEHRING, *Pharm. Acta Helv.*, 38 (1963) 785.
277 M. FIEDLER AND W. HEINE, *Acta Biol. Med. Ger.*, 13 (1964) 1.
278 D. H. SANDBERG, S. A. BOCK AND D. A. TURNER, *Anal. Biochem.*, 8 (1964) 129.
279 N. R. FARNSWORTH, R. N. BLOMSTER, D. DAMRATOSKI, W. A. MEER AND L. V. CAMMARATO, *Lloydia*, 27 (1964) 302.
280 C. T. BEER, *Brit. Empire Cancer Campaign*, 33 (1955) 487.
281 I. S. JOHNSON, H. F. WRIGHT AND G. H. SVOBODA, *J. Lab. Clin. Med.*, 54 (1959) 830.
282 J. H. CUTTS, *Cancer Res.*, 21 (1961) 168.
283 A. E. EVANS, S. FARBER, S. BRUNET AND P. J. MARIANO, *Cancer*, 16 (1963) 1032.
284 M. R. KARON, E. J. FREIREICH AND E. FREI, *Pediatrics*, 30 (1962) 791.
285 V. H. FERM, *Science*, 141 (1963) 426.
286 W. DE MYER, *Arch. Anat.*, 48 (1965) 181.
287 M. T. TAMAKI, T. SUGAWARA AND Y. KAMEYAMA, *Ann. Rept. Inst. Environ. Med. Nagoya Univ.*, 15 (1967) 61.
288 K. D. COURTNEY AND D. A. VALERIO, *Teratology*, 1 (1968) 163.

289 M. Joneja and S. Ungthavorn, *Teratology*, 2 (1969) 235.
290 S. Ungthavorn and M. Joneja, *Am. J. Anat.*, 126 (1969) 291.
291 G. Cardinaci, G. Cardinali and M. A. Enein, *Blood*, 21 (1963) 102.
292 U. Carpentieri and J. P. G. Williams, *Current Mod. Biol.*, 2 (1968) 4.
293 J. H. Cutts, *Cancer Res.*, 21 (1961) 168.
294 C. G. Palmer, D. Livengood, A. Warren, P. J. Simpson and I. S. Johnson, *Exptl. Cell Res.*, 20 (1960) 198.
295 W. A. Creasey and M. E. Markin, *Biochem. Pharmacol.*, 13 (1964) 135.
296 W. A. Creasey and M. E. Markin, *Biochem. Biophys. Acta*, 103 (1965) 635.
297 E. K. Wagner and B. Roizman, *Science*, 162 (1968) 569.
298 J. F. Richards, R. G. W. Jones and C. T. Beer, *Proc. Am. Assoc. Cancer Res.*, 4 (1963) 57.
299 W. A. Creasey, *Cancer Chemotherapy Rept.*, 52 (1968) 501.
300 C. T. Beer and J. F. Richards, *Lloydia*, 27 (1964) 352.
301 C. T. Beer, M. L. Wilson and J. A. Bell, *Can. J. Physiol. Pharmacol.*, 42 (1964) 368.
302 H. F. Hebden, J. R. Hadfield and C. T. Beer, *Cancer Res.*, 30 (1970) 1417.
303 N. R. Farnsworth and I. M. Hilinski, *J. Chromatog.*, 18 (1965) 184.
304 I. M. Jakovljevic, L. D. Seay and R. W. Shaffer, *J. Pharm. Sci.*, 53 (1964) 553.
305 N. J. Cone, R. Miller and N. Neuss, *J. Pharm. Sci.*, 52 (1963) 688.
306 I. Inagaki, S. Nishibe and T. Tokuhiro, *Shiritsu Daigaku Yokugokubl Kenkyu Nempo*, 12 (1964) 38.
307 A. Hoffmann, *Die Mutterkorn-Alkaloide*, Enke, Stuttgart, 1964, p. 190.
308 A. Stoll and A. Hoffmann, *Helv. Chim. Acta*, 26 (1943) 944.
309 R. Auerbach and J. A. Rugowski, *Science*, 157 (1967) 1325.
310 J. K. Hanaway, *Science*, 164 (1969) 574.
311 J. A. Dipaolo, H. M. Givelber and H. Erwin, *Nature*, 230 (1968) 490.
312 G. J. Alexander, B. E. Miles, G. M. Gold and R. B. Alexander, *Science*, 157 (1967) 459.
313 W. F. Geber, *Science*, 158 (1967) 265.
314 S. Fabro and S. M. Sieber, *Lancet*, 1968-I, 639.
315 S. Irwin and J. Egozcue, *Science*, 157 (1967) 313.
316 C. Roux, R. Dupuis and M. Aubey, *Science*, 169 (1970) 589.
317 J. E. Idanpaan-Heikkila and J. C. Schoolar, *Science*, 164 (1969) 1295.
318 G. Carakushansky, R. C. Neu and L. I. Gardner, *Lancet*, 1969-I, 1950.
319 H. Zellweger, J. S. McDonald and G. Abbo, *Lancet*, 1967-II, 1066.
320 F. Hecht, R. K. Beals, M. H. Lees, H. Jolly and P. Roberts, *Lancet*, 1968-II, 1087.
321 M. M. Cohen, M. J. Marinello and N. Beck, *Science*, 155 (1967) 1417.
322 M. M. Cohen, K. Hirschhorn and W. A. Forsch, *New Engl. J. Med.*, 277 (1967) 1043.
323 L. F. Oarvik and T. Lato, *Lancet*, 1968-I, 250.
324 N. E. Sicakkeback, J. Philip and O. J. Rafaelson, *Science*, 160 (1968) 1246.
325 M. M. Cohen and A. B. Mukherjee, *Nature*, 219 (1968) 1072.
326 L. Bender and S. Sankar, *Science*, 159 (1968) 749.
327 R. S. Sparkes, J. Melynk and L. P. Bozzetti, *Science*, 160 (1968) 1343.
328 S. Sturelid and B. A. Kihlman, *Hereditas*, 62 (1969) 259.
329 L. S. Browning, *Science*, 161 (1968) 1022.
330 E. Vann, *Nature*, 223 (1969) 95.
331 D. Grace, E. A. Carlson and P. Goodman, *Science*, 161 (1968) 694.
332 G. Zetterberg, *Hereditas*, 62 (1969) 262.
333 A. Stoll, A. Hoffmann and F. Trokler, *Helv. Chim. Acta*, 32 (1949) 506.
334 A. Stoll and A. Hoffmann, *Helv. Chim. Acta*, 38 (1955) 421.
335 A. Laitha, in H. H. Pennes (Ed.), *Psychopharmacology*, Harper (Hoeber), New York, 1958, p. 126.
336 E. Jacobsen, *Clin. Pharm. Therap.*, 4 (1963) 480.
337 T. M. Itil, A. Keskiner and J. M. C. Holden, *Gann Suppl.*, 30 (1969) 93.
338 R. C. De Bold and R. C. Leaf, *LSD, Man and Society, A Symposium*, Wesleyan Univ. Press, Middleton, Conn., 1967.
339 O. M. Garson and M. K. Robson, *Brit. Med. J.*, 1969-I, 800.
340 A. Macieira-Coelho, I. J. Hiu and E. Garcia-Giralt, *Nature*, 262 (1969) 1170.

341 U. A. LANZ, A. CERLETTI AND E. ROTHLIN, *Helv. Physiol. Pharmacol. Acta*, 13 (1955) 207.

342 E. S. BOYD, E. ROTHLIN, J. F. BONNER, I. H. SLATER AND H. C. HODGE, *J. Pharmacol. Exptl. Therap.*, 113 (1955) 6.

343 J. AXELROD, R. O. BRADY, B. WITKOP AND E. V. EVARTS, *Ann. N.Y. Acad. Sci.*, 66 (1967) 435.

344 A. STOLL, E. ROTHLIN, J. RUTSCHMANN AND W. R. SCHALCH, *Experientia*, 11 (1955) 396.

345 K. GENEST AND C. G. FARMILO, *J. Pharm. Pharmacol.*, 16 (1964) 250.

346 H. HELLBERG, *Acta Chem. Scand.*, 11 (1957) 219.

347 C. G. FARMILO AND K. GENEST, in C. P. STEWART AND A. STOLMAN (Eds.), *Toxicology, Mechanisms and Analytical Methods*, Academic Press, New York, p. 573.

348 A. STOLL AND A. HOFFMANN, *Helv. Chim. Acta*, 33 (1950) 1705.

349 H. EBERHARDT, K. J. FREUNDT AND J. W. LANGBEIN, *Arzneimittel-Forsch.*, 12 (1962) 1087.

350 L. A. DAL CORTIVO, D. R. BROICH, A. DIHRBERG AND B. NEWMAN, *Anal. Chem.*, 38 (1959) 1966.

351 Data Bulletin, *Delysid Substance*, Sandoz Pharmaceuticals, Hanover, New Jersey, April 30, 1965.

352 R. J. MARTIN AND T. G. ALEXANDER, *J. Assoc. Offic. Anal. Chemists*, 50 (1967) 1362.

353 R. J. MARTIN AND T. G. ALEXANDER, *J. Assoc. Offic. Anal. Chemists*, 51 (1968) 159.

354 D. L. ANDERSEN, *J. Chromatog.*, 41 (1969) 491.

355 A. ROMANO, personal communication in ref. 354.

356 E. G. C. CLARK, *J. Forensic Sci. Soc.*, 7 (1967) 46.

357 A. S. CURRY AND H. POWELL, *Nature*, 173 (1954) 1143.

358 J. LOOK, *J. Assoc. Offic. Anal. Chemists*, 51 (1968) 1318.

359 A. STOLL AND A. HOFFMANN, *U.S. Patent 2,438,259*, March 23, 1948.

360 R. P. PIOCH, *U.S. Patent 2,736,728*, Feb. 28, 1956.

361 C. RADECKA AND I. C. NIGAM, *J. Pharm. Sci.*, 55 (1966) 781.

362 M. A. KATZ, G. TADJER AND W. A. AUFRICHT, *J. Chromatog.*, 31 (1967) 545.

Chapter 6

FOOD AND FEED ADDITIVES AND CONTAMINANTS

1. CYCLAMATE AND CYCLOHEXYLAMINE

1.1 Cyclamate

Cyclamic acid (cyclohexanesulfamic acid) is prepared by the sulfuration of cyclohexylamine or the treatment of cyclohexylamine and triammonium nitrilo-sulfate. The sodium and calcium salts are prepared from cyclohexylammonium N-cyclohexylsulfamate by reaction with the respective hydroxides. Calcium and sodium cyclamates have been, till recently, extensively used alone or in combination with saccharin as non-caloric sweeteners in a wide variety of food products such as beverages, cereals and bakery products. Ten parts of cyclamates are usually mixed with one part of saccharin; the mixture is 60–100 times sweeter than sugar.

Perhaps no food additive has aroused more controversy and disquiet than cyclamate with no apparent sign of abating. Widespread restrictions have been placed on the use of cyclamate following the finding of bladder tumors in rats fed a cyclamate–saccharin mixture[1], as well as by cyclamate alone at a dietary level of 0.4%[2]. Both the carcinogenic and embryotoxic aspects of cyclamate are equally conflicting. Rudali et al.[3] described cyclamate as a weak carcinogen in mice increasing the incidence of malignant hepatomas and lung tumors on chronic feeding of sodium cyclamate. However, Roe et al.[4] failed to demonstrate any carcinogenic or tumor-promoting activity by feeding sodium cyclamate to mice for 18 months at a dietary level of 5%. Both the teratogenicity[5] and lack of teratogenic effects of cyclamate in the mouse[6] and rabbit[7] have been reported.

Cyclamate induced chromosome breakage in onion-root tips[8] and in both leukocyte and monolayer cultures from human skin and carcinoma of the larynx[9]. The synergistic radiomimetic effects of caffeine, alcohol and Sucaryl (which contains 8% sodium cyclamate and 0.8% saccharin) in onion-root tips[8] has also been described.

Other aspects of the biologic activity of cyclamate include (a) the induction of myocardial lesions accompanied by coronary sclerosis as well as soft tissue calcification in other organs following the oral administration of calcium cyclamate to hamsters[10] and (b) the induction of parenchymal damage to the liver and kidney epithelium following administration of cyclamate to guinea pigs in the drinking water at a concentration providing 2 g/kg/day[11]. The acute and chronic toxicity[12–15] of cyclamates have been also described. The metabolism of cyclamate has been studied in rats[16–19], dogs[16, 18], rabbits[16, 20], and man[21–26]. Cyclohexylamine has been found in the urine of man following oral administration[21–26]. Human excretion of cyclohexylamine after daily ingestion of soda containing 512 mg of cyclamate ranged from a trace to 28% of the dose on 3–5 days after ingestion. Other metabolites present in human urine included cyclohexanone, cyclohexinol and N-hydroxycyclohexylamine[27]. The conversion of cyclamate to cyclohexylamine in animals has also been reported following administration of cyclamate in saccharin mixtures[28].

Evidence of breakdown of cyclamate to cyclohexene in canned foods due to the presence of nitrite has been reported by Higuchi et al.[29]. In view of the formation of monochlorocyclohexane, cyclohexanone and cyclohexanol in addition to cyclohexene in the nitrous acid assay of cyclamate[30], and the formation of cyclohexylamine on hydrolysis of cyclamate by organic acids[31], there is a possibility that such breakdown products may also be formed during the processing of foods sweetened with cyclamate.

A variety of chromatographic techniques has been utilized both in the elaboration of cyclamate metabolism and its detection and analysis in food products. Urinary metabolites were isolated and detected by paper and thin-layer chromatography following oral administration of sodium cyclamate to both humans and dogs[21]. Ascending development was employed with Toyo Roshi No. 50 paper with the solvent systems (1) ethyl acetate–acetic acid–water (4:1:2), (2) n-butanol–benzyl alcohol–water–80% formic acid (45:45:9:1), (3) n-butanol–ethanol–water (4:2:1), (4) n-butanol–isopropanol–water (4:2:1), and (5) n-butanol–acetic acid–water (4:2:1). For TLC, silica gel (Wako Gel B-5) plates were developed with solvent systems 1, 2, and 5 above as well as with (6) ethyl acetate–80% formic acid–water (4:1:2). The detecting reagents for paper and thin-layer chromatography were (a) 1% ethanol quinhydrone (for the detection of cyclohexylamine), (b) equal mixture of 0.2% ethanolic naphthoresorcinol and 2% aq. trichloroacetic acid, and (c) aniline hydrogen pertholate reagent (1.66 g of phthalic acid and 0.93 g of aniline in 100 ml of n-butanol saturated with water) (reagents b and c were used for the detection of glucuronide). Urinary

TABLE 1

R_F VALUES OF THE METABOLITE FROM HUMAN AND DOG URINE

| | Solvent system[a] | | | | | | |
| | Paper | | | Thin-layer | | | |
	1	2	3	1	2	5	6
Cyclohexylamine	0.53	0.40	0.57	0.69	0.20	0.46	0.75
Metabolite of sodium cyclamate (human)	0.53	0.40	0.56		0.18	0.47	0.74
Metabolite of sodium cyclamate (dog)				0.68		0.46	0.76

[a] Solvent systems described in text.

extracts when chromatographed on paper and thin-layers revealed the presence of cyclohexylamine (Table 1).

The results indicated that sodium cyclamate was metabolized to cyclohexyl-amine in the human and dog, but not in the rabbit.

Kojima and Ichibagase[20] described the detection of metabolites of sodium cyclamate and cyclohexylamine in rabbit and rat urine and in rat liver homogenate by gas–liquid chromatography. Figure 1 illustrates the preparation procedure for the above sample solutions prior to GLC analysis.

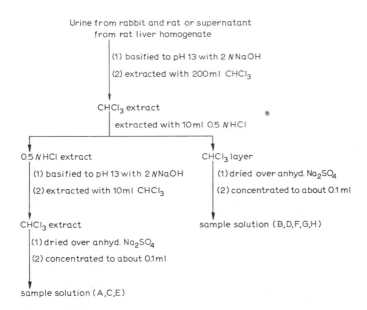

Fig. 1. Preparation procedure of sample solution. A, B, From urine of rabbit receiving CHS-Na; C, D, from urine of rat receiving CHS-Na; E, F, from rat liver homogenate; G, from urine of rabbit receiving cyclohexylamine; H, from urine of rat receiving cyclohexylamine.

A Shimadzu Model GC-1B dual gas chromatograph was used equipped with a Model HFD-1 dual hydrogen-flame ionization. The column was a 3 m × 4 mm i.d. stainless steel U tube containing 20% PEG 20M and 2.5% NaOH on 60–80 mesh Shimalite. The gas chromatogram of a sample of solution A (from urine of rabbit receiving oral sodium cyclamate) showed a peak (retention time 6.8 min) which corresponded with that of authentic cyclohexylamine (Fig. 2). However, the gas chromatogram of sample solution B (urine sample of rabbit receiving sodium cyclamate) showed two peaks at retention times of 5.8 and 7.2 min, which were analogous with those of authentic cyclohexanone and cyclohexanol (Fig. 3). Gas chromatograms of rat liver homogenate incubated with sodium cyclamate also indicated the presence

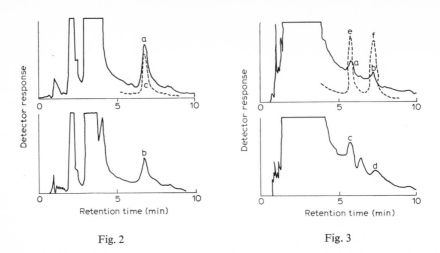

Fig. 2   Fig. 3

Fig.2. Gas chromatogram of cyclohexylamine from sample solutions A and C. Peak: a, cyclohexylamine from sample solution A; b, cyclohexylamine from sample solution C; c, authentic cyclohexylamine. Condition: column temp. 115°C; injector temp. 230°C; detector temp. 200°C. Gas flow rate, N_2 75 ml/min; H_2 55 ml/min; air 1000 ml/min; sensitivity 100; sample size 2 μl.

Fig.3. Gas chromatograms of cyclohexanone and cyclohexanol from sample solutions B and D. Peak: a, cyclohexanone from sample solution B; b, cyclohexanol from sample solution B; c, cyclohexanone from sample solution D; d, cyclohexanol from sample solution D; e, authentic cyclohexanone; f, authentic cyclohexanol. Condition: column temp. 150°C; injector temp. 230°C; detector temp. 200°C. Gas flow rate, N_2 65 ml/min; H_2 55 ml/min; air 1000 ml/min; sensitivity 100; sample size 2 μl.

of the metabolites cyclohexylamine, cyclohexanone, and cyclohexanol in small amounts (Fig. 4).

Following the oral administration of cyclohexylamine to both the rabbit and rat, cyclohexanone and cyclohexanol were detected by GLC as urinary metabolites in both species (Fig. 5). As a result of these experiments, it was postulated that in the rabbit and rat, sodium cyclamate was metabolized primarily by the formation of cyclohexylamine, which was further oxidized to cyclohexanone and cyclohexanol.

The metabolism of sodium cyclamate in the human following a single oral administration was described by Kojima and Ichibagase[26] who employed paper, thin-layer and gas–liquid chromatography. The ascending technique was employed with Toyo Roshi No. 50 paper with the following solvent systems: (1) decalin–dimethylformamide, (2) n-heptane–methanol, (3) n-butanol saturated with 1 N acetic acid, and (4) n-propanol–28% ammonium hydroxide (7:3). For TLC, Silica Gel G (Merck) activated at 105°C for 1 h was used with the solvent systems (5) benzene–ligroin (1:1), (6) benzene–ligroin (2:1), and (7) benzene–petroleum ether (1:1).

A Shimadzu Model GC-1B dual column gas chromatograph equipped with a hydrogen-flame-ionization detector and Model GC-3AF were used. The columns

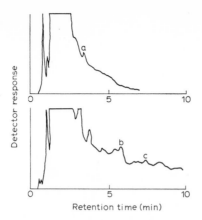

Fig. 4. Gas chromatograms of cyclohexylamine, cyclohexanone, and cyclohexanol from sample solutions E and F. Peak: a, cyclohexylamine from sample solution E; b, cyclohexanone from sample solution F; c, cyclohexanol from sample solution F. Condition: as described in Fig. 3.

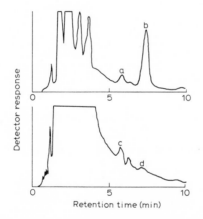

Fig. 5. Gas chromatograms of cyclohexanone and cyclohexanol from sample solutions G and H Peak: a, cyclohexanone from sample solution G; b, cyclohexanol from sample solution G; c, cyclohexanone from sample solution H; d, cyclohexanol from sample solution H. Condition: as described in Fig. 3.

were 300 cm × 4 mm i.d. stainless steel U tube (for Model GC-1B) and 300 cm × 3 mm i.d. stainless steel coil tube (for Model GC-3AF) containing a packing of 20% PEG 20M and 2.5% NaOH on 60–80 mesh Shimalite. The Model GC-1B instrument was used for the identification of some metabolites of sodium cyclamate and Model GC-3AF for the quantitative investigation of cyclohexylamine, cyclohexanol, cyclo-hexanone, and conjugated cyclohexanol as the metabolites of sodium cyclamate. Figure 6 illustrates the preparation of human urine for GLC analysis and Figs. 7 and 8 the gas chromatogram of urine extract sample solutions. As shown in Figs. 7 and 8,

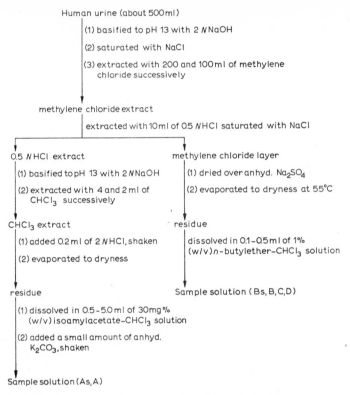

Human urine (about 500ml)

    (1) basified to pH 13 with 2 N NaOH

    (2) saturated with NaCl

    (3) extracted with 200 and 100 ml of methylene
        chloride successively

methylene chloride extract

    extracted with 10 ml of 0.5 N HCl saturated with NaCl

0.5 N HCl extract

    (1) basified to pH 13 with 2 N NaOH

    (2) extracted with 4 and 2 ml of
        CHCl$_3$ successively

methylene chloride layer

    (1) dried over anhyd. Na$_2$SO$_4$

    (2) evaporated to dryness at 55°C

CHCl$_3$ extract

    (1) added 0.2 ml of 2 N HCl, shaken

    (2) evaporated to dryness

residue

    dissolved in 0.1–0.5 ml of 1%
    (w/v) n-butylether-CHCl$_3$ solution

residue

    (1) dissolved in 0.5–5.0 ml of 30 mg%
    (w/v) isoamylacetate-CHCl$_3$ solution

    (2) added a small amount of anhyd.
    K$_2$CO$_3$, shaken

Sample solution (Bs, B, C, D)

Sample solution (As, A)

Fig. 6. Preparation procedures of sample solution As, Bs, A, B, C, and D.

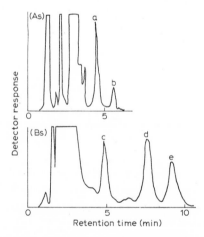

Fig. 7. Gas chromatogram of sample solution As and Bs. Apparatus: Shimadzu Model GC-3AF. Peak: a, isoamylacetate (internal standard); b, cyclohexylamine; c, n-butylether (internal standard); d, cyclohexanone; e, cyclohexanol. Condition: (As) column temp. 130°C; gas flow rate, N$_2$ 30 ml/min; H$_2$ 30 ml/min; air 1000 ml/min; sensitivity: 100; sample size, 2 μl; (Bs) column temp. 140°C; gas flow rate, N$_2$ 30 ml/min; air 1000 ml/min; sensitivity: 100; sample size, 2 μl.

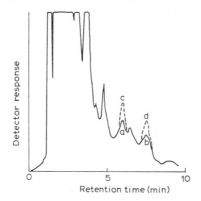

Fig. 8. Gas chromatogram of cyclohexanone and cyclohexanol from sample solution C. Apparatus: Shimadzu Model GC-1B. Peak: a, metabolite of CHS-Na; b, another metabolite of CHS-Na; c, standard cyclohexanone; d, standard cyclohexanol. Condition: column temp. 150 °C; injector temp. 230 °C; detector temp. 200 °C; gas flow rate, N_2 65 ml/min; H_2 55 ml/min; air 1000 ml/min; sensitivity: 1000; sample size, 2 µl.

two peaks were obtained which had the same retention times as those of standard cyclohexanone and cyclohexanol. Cyclohexanol was further characterized and confirmed by the paper chromatography of its 3,5-dinitrobenzoate (Table 2) and cyclohexanone by the TLC of the corresponding 2,4-dinitrophenylhydrazone (Table 3).

TABLE 2

PAPER CHROMATOGRAPHY[a] OF CYCLOHEXANOL AS A METABOLITE FROM HUMAN URINE
FOLLOWING ORAL ADMINISTRATION OF SODIUM CYCLAMATE

Compound	R_F in solvent system[b]		
	I	II	III
Cyclohexyl 3,5-dinitrobenzoate	0.74	0.84	0.93
Metabolite of sodium cyclamate (as 3,5-dinitrobenzoate derivative)	0.74	0.84	0.93

[a] Ascending technique using Toyo Roshi No. 50 paper. Cyclohexyl 3,5-dinitrobenzoate detected in ultraviolet light as a dark spot on a yellow fluorescent background after spraying with a methanolic solution of Rhodamine B (20 mg/l).
[b] Solvents: I, decalin–dimethylformamide; II, n-heptane–methanol; III, n-butanol saturated with 1 N acetic acid.

Identification of cyclohexylglucuronide as a urinary metabolite of sodium cyclamate was established by paper chromatography using n-propanol–28 % ammonium hydroxide (7:3) for development and detection of the glucuronide as a deep-blue spot (R_F 0.61) by spraying with a fresh solution of naphthoresorcinol (2 % in 33 % aq. trichloroacetic acid), drying at room temperature and then heating at 100 to 105 °C for about 30 min. The portion of the paper corresponding to R_F 0.61

TABLE 3

THIN-LAYER CHROMATOGRAPHY[a] OF CYCLOHEXANONE AS A METABOLITE FROM HUMAN URINE
FOLLOWING ORAL ADMINISTRATION OF SODIUM CYCLAMATE

Compound	R_F in solvent system[b]		
	V	VI	VII
Cyclohexyl 2,4-dinitrophenylhydrazone	0.32	0.54	0.43
Metabolite of sodium cyclamate (as 2,4-dinitrophenylhydrazone derivative)	0.32	0.54	0.43

[a] Ascending technique using Silica Gel G (Merck).
[b] Solvents: V, benzene–ligroin (1:1); VI, benzene–ligroin (2:1); and VII, benzene–petroleum ether (1:1).

above was excised, eluted with warm water, eluate concentrated to a small volume
in vacuo, incubated with about 10,000 units of β-glucuronidase at 38 °C for 3 h, saturated with sodium chloride, extracted with methylene chloride, the methylene chloride
extract concentrated to a small volume and finally gas chromatographed. Figure 9
shows a gas chromatogram of cyclohexanol obtained *via* the above enzymatic hydrolysis of cyclohexylglucuronide.

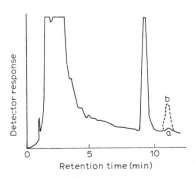

Fig. 9. Gas chromatogram of cyclohexanol obtained by enzymatic hydrolysis of cyclohexylglucuronide. Apparatus: Shimadzu Model GC-1B. Peak: a, hydrolysate of CHS-Na metabolite; b, standard cyclohexanol. Condition: column temp. 140 °C; injector temp. 230 °C; detector temp. 200 °C.
Gas flow rate, N_2 65 ml/min; H_2 55 ml/min; air 1000 ml/min; sensitivity, 1000; sample size, 2 µl.

Table 4 summarizes the urinary excretion of the metabolites in the 24-h urine
of humans receiving a single oral dose of 2 g of sodium cyclamate in this study.

The distribution and excretion of ^{14}C-labeled sodium cyclamate in animals
was studied by Miller *et al.*[18]. Thin-layer chromatography on Silica Gel G was used
to establish the chemical and radiochemical purity of the labeled material as well as
for the identification of sodium cyclamate in excreta. The solvents of choice were
95% ethanol–1 N NH_4OH (95:5) and acetonitrile–conc. NH_4OH–water (60:10:3).

The R_F values for the sodium cyclamate found in these systems were 0.80 and 0.62, respectively. In both rats and dogs following either acute or semi-chronic administration, at least 98–99% of the excreted radioactivity, either fecal or urinary, was present as unchanged cyclamate.

Derse and Daun[32] described a GLC method for the determination of cyclamates in animal excreta. A Barber-Colman Model 5000 gas chromatograph was used equipped with a flame-ionization detector and a 6 ft. × 6 mm i.d. borosilicate U tube

TABLE 4

URINARY EXCRETION OF THE METABOLITES IN THE 24-HOUR URINE OF HUMAN RECEIVING ORALLY 2 g SODIUM CYCLAMATE

Subject[a]	μg excreted			
	Cyclohexylamine	Cyclohexanol	Cyclohexanone	Conjugated cyclohexanol
K.I. (27–52) ♂	82	150	216	1,710
K.S. (25–52) ♂	628	134	212	3,990
S.K. (38–51) ♂	10,800	45	82	2,660
A.S. (21–47) ♀	2,370	25	20	3,080
Y.O. (21–51) ♀	40	20	10	5,140

[a] Bracketed quantities are subject's age in years followed by body weight in kilograms.

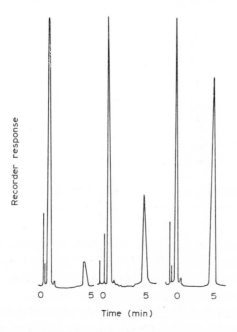

Fig. 10. Chromatograms represent 1.0, 3.0, and 5.0 mg calcium cyclamate dihydrate after hydrolysis and extraction into 1.0 ml methylene chloride.

containing a packing of 20% Dowfax 9N9 and 2.5% sodium hydroxide on 60 to 80 mesh Gas Chrom R. GLC operating conditions were: injector 230 °C, detector 240 °C, column 150 °C, nitrogen 100 ml/min, helium 50 ml/min, air 600 ml/min, attenuation 100, injection 2 µl.

Figure 10 illustrates chromatograms of calcium cyclamate dihydrate after hydrolysis and extraction into methylene chloride. (The retention time of the cyclo-hexylamine is *ca.* 5 min after the solvent.)

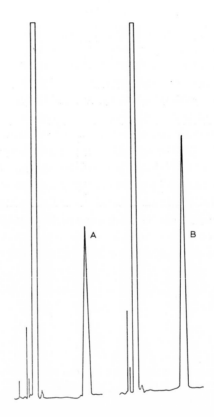

Fig. 11. A, cyclohexylamine peak obtained from composite of animal on 5% calcium cyclamate; B, cyclohexylamine peak obtained from fecal composite of same animal.

Figure 11 shows gas chromatograms of cyclohexylamine peaks obtained from composite of animal on 5% calcium cyclamate (A) and fecal composite of same animal (B). The limit of detection of cyclamate in tissues (determined as cyclohexyl-amine by GLC) was in the 1–5 p.p.m. range.

Sonders *et al.*[33] studied the excretion of ring-labeled sodium ^{14}C-cyclamate in the rat and reconfirmed both the utility of the GLC procedure of Derse and Daun[32] as well as the finding of nearly quantitative elimination of cyclamate in the rat independent of the dose of cyclamate or whether the administration was chronic (22 weeks) or acute.

The determination of cyclamates in soft drinks by GLC has been described by Rees[34] and Richardson and Luton[30]. The method of Rees is based on the reaction of cyclamate with nitrite, *viz*.

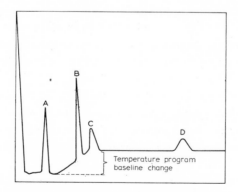

The cyclohexene formed is extracted and then determined on a silicone oil column. A Pye Argon chromatograph was used with a 4 ft. × 1/4 in. column containing 10% Apiezon on acid-washed siliconized Embacel 100–150 mesh at 50 °C. The argon inlet pressure was 10 p.s.i. at a flow rate of 40 ml/min; the detector voltage was 1000 V and the amplifier sensitivity was 10×. In addition to cyclohexene, gas chromatographic evidence indicated the presence of at least two other products of reaction, probably cyclohexanone and cyclohexanol, which had retention times of about 50 and 58 min, respectively. The GLC method permitted the analysis of cyclamates in the range 0–1 mg/ml of fruit drink. Richardson and Luton[30] modified the procedure of Rees[34] by ensuring that the pH of extractions was in accordance with Beck's[35] recommendations and by using sulfuric acid instead of hydrochloric acid in order to avoid the formation of monochlorocyclohexane. A Pye 104 Series Model 14 gas chromatograph was used with a flame-ionization detector and a 5-ft. column containing 10% polyethylene glycol 400 on 100–120 mesh Celite. The initial column

Fig. 12. Gas chromatography showing peaks of: A, cyclohexene; B, monochlorocyclohexane; C, cyclohexanone; D, cyclohexanol. For conditions see text.

temperature was held for 3 min at 50 °C then programmed at 10°/min from 50 to 120 °C. The carrier gas was argon at a flow rate of 45 ml/min, hydrogen, 45 ml/min, and air, 500 ml/min; attenuation was 1 × 10². Figure 12 illustrates a chromatogram of the reaction products of the acidification of cyclamate with sulfuric acid.

Kato et al.[36] described the GLC of cyclamate in food via its conversion to cyclohexyl nitrite on treatment with nitrous acid. The detection limits varied according to the type of food, but the range was generally in the order of 10–50 p.p.m. of cyclamate.

The preparation of uniformly labeled cyclohexylamine and cyclamate and the determination of its chemical and radiochemical purity by TLC and GLC was described by Alter and Forman[37]. Cyclohexylamine, uniformly labeled with ^{14}C, was prepared by the catalytic hydrogenation of uniformly labeled aniline using rhodium on alumina as catalyst. (The major by-products formed were cyclohexanol and dicyclohexylamine.) The cyclohexylamine was converted to uniformly labeled sodium cyclamate by treatment with a triethylamine–sulfur trioxide complex (prepared via the initial reaction of chlorosulfonic acid in dichloroethylene with triethylamine). GLC analyses were performed on a Barber-Colman Series 5000 instrument equipped with an 8-ft. column containing 10% Carbowax 20M–3% potassium hydroxide, a combustion furnace and a proportional counter for detection of ^{14}CO$_2$. TLC of cyclohexylamine sulfate was accomplished using silica gel with solvent systems ethanol–1 N ammonium hydroxide (95:5) and n-butanol–acetic acid–water (50:25:25) and detected by an iodine in chloroform spray. The former affords a wide separation between cyclohexylamine and dicyclohexylamine (R_F's 0.18 and 0.45, respectively). The purity of cyclamate was ascertained on silica gel using the following developers: (A) anhyd. ethanol, (B) 95% ethanol, (C) 95% isopropanol, and (D) acetonitrile–ammonium hydroxide–water (60:3:10). Detection was accomplished by spraying the plate first with a 1:20 Chlorox solution, allowing the plate to stand 15 min, and then spraying with ethanol to remove excess hypochlorite, drying and finally spraying with fresh starch–iodide solution producing a purple color for cyclamate.

Das et al.[38] described the detection and estimation of cyclamate using Silica Gel G with ethyl acetate–isopropanol–acetone–methanol–water (50:15:15:4:16). The plates are then exposed to bromine vapor, kept at room temperature for 5 min to remove excess bromine and then sprayed with an 0.05% alcoholic fluorescein solution, yielding pink spots on a yellow background (R_F cyclamate = 0.56). Cyclamate could be determined quantitatively by removing the thin-layer area of the pink spot, extracting with ethanol and measuring the fluorescence at 540 mμ. The solvent system employed has the advantage of separating cyclamate from other sweeteners, viz. sucrose (R_F = 0.25) and saccharin (R_F = 0.19).

Utility of paper chromatography for the separation and identification of sweeteners has been described by Mitchell[39]. Cyclamic acid, saccharin and dulcin in foodstuffs were extracted with ethyl acetate after acidification with sulfuric acid, the solvent evaporated off and the residue dissolved in ammoniacal 50% aq. ethanol and applied to paper. Separation was completed by ascending chromatography with a mobile phase of acetone–ethyl acetate–ammonia (8:1:1). Cyclamic acid and saccharin yield white spots when sprayed first with 0.005 M silver nitrate in ammoniacal ethanol and then with 0.0005 M pyrogallol in ethanol.

Ko et al.[40] separated the sweetening agents by paper chromatography using
n-butanol–10% ammonia (4:1) as the developer and a spray reagent containing
pyrogallol and silver nitrate for the detection of cyclamic acid. Komoda and Takeshita[41] used a similar procedure, but favored a mobile phase of isoamyl alcohol–
pyridine–water, which gave less interference from food preservatives such as salicylic, benzoic, and sorbic acids.

Das and Matthew[42] separated saccharin, cyclamate, and sucrose individually
and in mixture by descending chromatography (18 h) on Whatman No. 1 paper using
n-butanol–acetic acid–water (40:10:22). The detecting reagent consisted of phthalic
acid (0.9 g) and distilled aniline (1.6 g) in 100 ml of water-saturated butanol. The R_F
values found were sucrose 0.24, saccharin 0.17, and cyclamate, 0.64.

Ion exchangers have also been employed for the determination of cyclamates
in foods. Komoda and Takeshita[41] separated cyclamic acid, saccharin, and dulcin
from food by adsorption onto Amberlite CG400 Type 1, Cl form. The cyclamic acid
and saccharin were eluted with oxalic acid and alcohol, the eluate neutralized and
examined by paper chromatography as described above.

Asano et al.[43] used Amberlite CG46 (Cl form) for the sorption and eluted
cyclamic acid with 0.1 M hydrochloric acid, then titrated the cyclamic acid with
nitrous acid and determined the sulfate nephelometrically.

Yamaguchi[44] separated sulfamic and cyclamic acids on Amberlite IRA400
and IR46 with methanol and 0.05 M hydrochloric acid as eluting agent.

Other analytical procedures that have been employed include colorimetric[45–48]
(following conversion of cyclamate to cyclohexylamine), titrimetric[30, 34, 49], and
gravimetric[50, 51].

1.2 Cyclohexylamine

It has been estimated that about 30% of all people who consume cyclamate
convert it to cyclohexylamine. There is evidence (in the rat) that this conversion is
carried out by the gut flora, which acquire this ability when the diet contains cyclamate[52]. (This ability is largely lost on removing cyclamate from the diet.) Cyclohexylamine is considerably more toxic than the cyclamates (LD_{50} in rats is about
300 mg/kg, 20–40 times less than for cyclamates). Cyclohexylamine is weakly teratogenic in mice[53], induces chromosomal breaks in vitro in rodent cells in culture as
well as in vivo in rat spermatogonia[54, 55]. Cyclohexylamine (as well as cyclamate,
N-hydroxycyclohexylamine, and dicyclohexylamine) have also been reported to
induce chromosome damage in human leukocyte cultures[56].

Cyclohexylamine, which is produced on a commercial scale by the catalytic
hydrogenation of aniline at elevated temperatures and pressures, enjoys a broad
spectrum of utility in organic synthesis, the manufacture of insecticides, plasticizers,
corrosion inhibitors, rubber chemicals, dyestuffs, emulsifying agents, dry-cleaning
soaps, and in acid-gas absorbents. The toxicity[57–59], and pharmacology[60–62] of
cyclohexylamine have been described.

The metabolism of ^{14}C-cyclohexylamine in rabbits has been described by Elliott et al.[63]. Forty-five percent of the administered oral dose was shown to be excreted in the urine as unconjugated cyclohexylamine, 0.2% as N-hydroxycyclohexylamine in conjugated form and 2.5% as cyclohexanone oxime (which was probably an artifact arising from the glucuronide of N-hydroxycyclohexylamine in the hydrolysis procedure). TLC of all fractions of the radioactive urine were developed in n-butanol–acetic acid–water (4:1:5) and revealed the presence of one metabolite (R_F 0.53), giving a positive reaction with naphthoresorcinol, and two metabolites (R_F 0.37, major, and R_F 0.59 minor) both giving a positive reaction with ninhydrin (the minor one corresponding in position to cyclohexylamine).

Blumberg and Heaton[64] described the TLC analysis of cyclohexylamine in urines of psychiatric patients. The pH 9.3 extract from SA-2 ion-exchange paper soaked in urine and taken to dryness was dissolved in methanol, spotted on a Silica Gel G plate and developed in ethyl acetate–methanol–58% ammonia (85:10:5). In this system the cyclohexylamine spot has an average R_F value of 0.40, stains blue with bromocresol green, disappears with the application of potassium iodoplatinate and on drying, reappears chalky pink. R_F values in this area and similar color reactions have been observed with ephedrine, hydroxyamphetamine and phenylpropylamine. However, when the following solvent systems were used, the R_F values of each of these compounds were readily differentiated: ethanol–25% ammonia (4:1) and n-butanol–acetic acid–water (4:1:5). Of 83 known cyclamate-using psychiatric patients, 31 showed the presence of cyclohexylamine in their urine.

Oser et al.[28] described the utility of GLC for the elaboration of the conversion of cyclamate to cyclohexylamine in rats. A Varian Aerograph series 1520 gas chromatograph was used fitted with a hydrogen-flame-ionization detector and connected to an Infotronics digital computer. The column consisted of 20% Dowfax 9N9 + 2.5% NaOH on 60/80 mesh Gas Chrom Q.

The determination of cyclohexylamine by electron-capture gas chromatography has been reported by Weston and Wheals[65]. Cyclohexylamine is allowed to react with 1-fluoro-2,4-dinitrobenzene and the 2,4-dinitrophenyl derivative extracted into hexane and analyzed using an isothermal gas chromatograph fitted with an electron-capture detector[66] and a 140 cm × 1.5 mm glass column packed with silanized Chromosorb G, acid-washed 60–80 mesh coated with 1.0% GE-XE-60 silicone and 0.1% Epikote 1001. The column temperature was 215°C and the carrier gas was nitrogen at a flow rate of 180 ml/min.

The 2,4-dinitrophenyl derivative of cyclohexylamine had a retention time of 4.2 min under the conditions described. The retention time of the analogous aniline derivative was 3.6 min. The peak height response of a standard series was linear over the range 1–10 μg/ml of cyclohexylamine and the detection limit of the derivative was about 0.25 ng. Dicyclohexylamine, if present in the reaction mixture, did not interfere. The derivation technique described permitted the determination of cyclohexylamine in cyclamates down to 1 p.p.m. and in soft drinks down to 0.1 p.p.m.

Gunner and O'Brien[67] described the GLC of cyclohexylamine as its N-trinitro-phenyl derivative permitting detection in the sub-nanogram range (because of incor-poration of nitro groups) with an electron-capture detector. Quantitative analysis was achieved using 3,3,5-trimethylcyclohexylamine as an internal standard and was illustrated by the analysis of cyclohexylamine in cyclamate at a level of 1 p.p.m. Cyclohexylamine was reacted with picryl chloride (1-chloro-2,4,6-trinitrobenzene) in chloroform yielding the respective N-trinitrophenyl derivative which was analyzed on an Aerograph A-700 instrument equipped with a tritium-foil electron-capture detector and a 5 ft. × ⅛ in. stainless steel column packed with 3% SE-30 on 80 to 100 mesh Chromosorb G.

The column, injector port, and detector temperatures were maintained at 200, 220, and 200°C, respectively, and nitrogen was the carrier gas at a flow rate of 140 ml/min. Figure 13 illustrates the GLC separation of the N-trinitrophenyl deriva-tive of cyclohexylamine (A) and the N-trinitrophenyl derivative of the internal standard 3,3,5-trimethylcyclohexylamine (B).

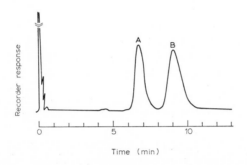

Fig. 13. GLC separation of A, TNP-CHA and B, TNP-TMCHA on 3% SE-30 on 80–100 mesh Chromosorb G column.

The gas chromatographic determination of cyclohexylamine in blood as the trifluoroacetyl derivative has been described by Litchfield and Green[68]. The method allowed the detection of 0.1 µg/ml of cyclohexylamine in blood when a sample volume of 10 ml was used. An Aerograph 1522 gas chromatograph fitted with a flame-ioniza-tion detector and a 5 ft. × ⅛ in. o.d. stainless steel column containing 10% Apiezon L on 80–120 mesh deactivated Celite was used. (The Celite support was deactivated by washing with 5% potassium hydroxide in methanol and removing the solvent by eva-poration.) The operating conditions were: column temperature 115°C, injector 155°C, detector 155°C, nitrogen flow rate 50 ml/min, and chart speed 20 in./h (Honeywell recorder 1 mV f.s.d.). Figure 14 illustrates chromatograms of 2 µl injec-tions of extracts from control blood (A), blood containing 0.1 µl/ml of cyclohexyl-amine (B), and blood containing 0.5 µl/ml of cyclohexylamine (C). The retention time of the trifluoroacetyl derivative (obtained *via* reaction with trifluoroacetic anhydride in dichloromethane) was 3 min.

A method for the determination of cyclohexylamine in cyclamates and artificially sweetened beverages has been reported[69, 70]. Cyclohexylamine was isolated by an extraction and/or distillation technique, determined by GLC and confirmed by trapping the eluted fraction, converting it into its dithiocarbamate salt and identifying the latter by infrared spectroscopy. Two instruments were employed, one for actual determinations and the second for trapping. (*1*) Perkin-Elmer Model 881

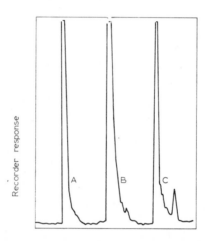

Fig. 14. Chromatograms of 2-μl injections of extracts from A, control blood; B, blood containing 0.1 μl ml^{-1} of cyclohexylamine; C, blood containing 0.5 μl ml^{-1} of cyclohexylamine.

with dual flame-ionization detection was used with a 12 ft. × ⅛ in. stainless steel column packed with 10% Carbowax 20M plus 2.5% NaOH on Anakrom SD, 90 to 100 mesh. The column was conditioned for 16 h at 200°C while maintaining a nitrogen carrier flow of 40 ml/min. The operating conditions for analysis were column temp. 100°C, injector 180°C, detector 175°C, carrier gas flow rate 72 ml/min, hydrogen flow rate 35 ml/min, air flow 550 ml/min, and attenuation range, ×5 ×50 at 1 × 10^{-11} A into a 1 mV full-scale recorder. (*2*) Hewlett-Packard Model 5750B research chromatograph with dual flame-ionization detectors and effluent splitters

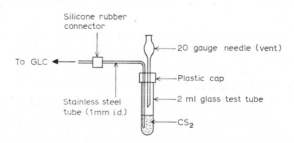

Fig. 15. GLC effluent collection trap for determination of cyclohexylamine.

(5:1 ratio) (Hewlett-Packard), or equivalent, with the same column and conditions as used with the Perkin-Elmer Model 881 above, with the following exceptions: carrier gas flow rate, 75 ml/min through the column and 62 ml/min through the splitter and attenuation range ×640 at 2×10^{-14} A into 1 mV full-scale recorder. Figure 15 illustrates the GLC effluent collection trap for the determination of cyclohexylamine.

Cyclohexylamine in cyclamates has also been determined spectrophotometrically at 418 nm as its picrylamide derivative[71] and absorptiometrically as its complex with 1-chloro-2,4,6-trinitrobenzene[48].

Crude cyclohexylamine produced commercially by the catalytic hydrogenation of aniline at elevated temperatures contains dicyclohexylamine, N-phenylcyclo-hexylamine (cyclohexylaniline) and unchanged aniline[72]. Since cyclamates are normally synthesized by the sulfonation of cyclohexylamine, the detection of the presence of the above impurities is of analytical concern. Howard et al.[73] reported a procedure for dicyclohexylamine in which extraction techniques were used to isolate the compound and GLC was utilized as the determinative method. Confirmation of the amine was achieved by mass spectrometry or by formation of the hydrochloride for IR spectrophotometry. A Hewlett-Packard Model 5750B with dual flame-ionization detectors and effluent splitters (5:1 ratio) was used with a 12 ft. × ⅛ in. stainless steel column packed with 90–100 mesh Anakrom SD support coated with 10% Carbowax 20M + 2.5% NaOH (from ethanol solutions of both coated materials). The operating temperatures were column 170°C, injector and detector, 240°C and flow rates helium carrier 60 ml/min, hydrogen 35 ml/min, air 550 ml/min; electrometer-recorder sensitivities were 4×10^{-11} A f.s.d. (range 10 at attenuation 4) and 8×10^{-11} A f.s.d. (range 10 at attenuation 8) into 1 mV recorder. New columns were conditioned 48 h at 200°C with a helium flow of 50 ml/min.

The GLC conditions for trapping were as specified above with the following exceptions: carrier gas flow rate 60 ml/min through the column, 50 ml/min through the splitter, and 20 lb. auxiliary gas (helium) to the detector as measured on the 2nd stage of a two-stage gauge; electrometer–recorder sensitivity was 64×10^{-11} A f.s.d. (range 10 at attenuation 64) into 1 mV recorder and effluent exit tube temperature of 200°C.

2. DULCIN

Dulcin (4-ethoxyphenylurea, sucrol, valzin) is prepared by treating p-phenetidine with phosgene and then with ammonia. It is a non-nutritive sweetening agent which until recently has been widely used in foods. The carcinogenicity of dulcin[74, 75] as well as its toxicity[76] has been reported.

The metabolism of dulcin in the rabbit has been described by Akagi et al.[77]. Following oral administration, 40% of the dose was excreted as phenolic substances, presumably as p-hydroxyphenylurea. The phenolic products were excreted as the unconjugated (7%), glucuronide (11%) and sulfate (23%). p-Hydroxyphenylurea was isolated as its diacetyl derivative. Paper chromatography was used for the separation and identification of the urinary metabolites. Toyo Roshi No. 51 paper was used with descending development in the following solvent systems: (*a*) isopropanol–ammonia–water (20:1:2), (*b*) amyl alcohol–water–methanol–benzene (2:1:4:2), (*c*) *n*-butanol–acetic acid–water (4:1:5), (*d*) *n*-butanol–propanol–water (3:2:5). For the detection of compounds on paper, the following reagents were used: (*1*) for detection of amino and ureido groups, *p*-dimethylaminobenzaldehyde (5 g in 100 ml of ethanol and 33 ml of conc. HCl), (*2*) for detection of glucuronic acid, naphthoresorcinol (2% in aq. 33% trichloroacetic acid, (*3*) for the same purpose, aniline hydrochloride (1.0% in methanol), (*4*) for detection of phenols and *p*-phenetidine, ferric chloride (2% aq.), (*5*) for detection of phenols, 2,6-dibromoquinone chloroimide (3% in methanol) followed by 10% sodium carbonate, (*6*) for the same purpose, phosphomolybdic acid (2% followed by ammonia gas), (*7*) for detection of aryl compounds, diazobenzenesulfonic acid (0.1 g in 20 ml of 10% sodium carbonate). Table 5 lists the R_F values of dulcin and its metabolites.

TABLE 5

R_F VALUES OF *p*-ETHOXYPHENYLUREA AND ITS METABOLITES[77]

Compound	R_F in solvent systems[a]			
	A	B	C	D
p-Ethoxyphenylurea (dulcin)	0.48	0.76	0.91	0.91
p-Ethoxyphenylurea sulfamate			0.65	0.50
p-Ethoxyphenylurea N-glucuronide		0.13	0.55	0.30
p-Hydroxyphenylurea	0.46	0.60	0.76	0.75
p-Hydroxyphenylurea O-sulfate		0.23		
p-Hydroxyphenylurea O-glucuronide		0.09	0.16	0.05
p-Phenetidine		0.81	0.96	0.94
p-Aminophenol		0.69	0.53	0.58
Urea		0.39	0.51	0.41

[a] Solvents: A, isopropanol–ammonia–water (20:1:2); B, amyl alcohol–water–methanol–benzene (2:1:4:2); C, *n*-butanol–acetic acid–water (4:1:5); D, *n*-butanol–*n*-propanol–water (3:2:5).

In a related area, Akagi *et al.*[78] extended earlier studies to determine whether arylureas other than *p*-ethoxyphenylurea could form N-glucosiduronate in animals. It was found that N-glucosiduronates of aryl ureas were excreted in the urine of rabbits dosed with arylureas and were identical with ammonium 1-[3-(aryl)-ureido]-1-deoxy-D-glucopyranosiduronates which were synthesized from glucuronic acid and arylureas in pyridine. It was thus suggested that N-glucosiduronate conjugation is a general pathway of arylurea metabolism in the rabbit.

In the above study, thin-layer chromatograms of silica gel with a solvent system of chloroform–methanol–half-saturated ammonium hydroxide (5:4:1), and detection with Ehrlich's reagent and naphthoresorcinol were used for the isolation and confirmation of the N-glucosiduronate of phenyl- and o-tolylurea from rabbit's urine.

The artificial sweeteners, dulcin, soluble saccharin and sodium cyclohexyl-sulfamate in soft drinks were identified by paper chromatography[40] on Whatman No. 4 paper with n-butanol–10% ammonium hydroxide (4:1). The color reagents used were 1% naphthylamine in acetic acid for soluble saccharin, 1% p-dimethyl-aminobenzaldehyde for dulcin and dilute solution of pyrogallol and silver nitrate for sodium cyclohexylsulfamate. The minimum detectable amounts were 2 μg for dulcin, 64 μg for soluble saccharin, and 12 μg for sodium cyclohexylsulfamate.

The determination of dulcin in soy sauce and similar seasonings by paper chromatography has been reported by Sasaki et al.[79]. The developing solvent was n-butanol–ammonium hydroxide (4:1) and p-dimethylaminobenzaldehyde in HCl was the spray reagent.

Kamp[80] separated and identified dulcin, sodium cyclamate, and saccharin sodium by thin-layer chromatography. The three compounds were separated on 0.25-mm layers of Silica Gel G by development with acetone–10% aq. ammonia (9:1) or acetone–10% aq. ammonia–ethyl acetate (8:1:1). Dulcin appeared after spraying with 65% nitric acid. The sodium cyclamate and saccharin sodium spots were detected by spraying successively with alkaline silver nitrate solution (340 mg of silver nitrate, 1 ml of water, 2.5 ml of 10% aq. ammonia, and 100 ml ethanol) and a pyrogallol solution (30 mg of pyrogallol and 100 ml of ethanol). The difference in the R_F values of dulcin found by Waldi[81] was attributed to the presence of the dulcin decomposition product N,N'-bis(p-ethoxyphenyl)urea. For separations of dulcin and saccharin sodium only, Silica Gel HF_{254} can be used; if sodium cyclamate is also present, Silica Gel G must be used.

Lee[82] separated dulcin, saccharin sodium and sodium cyclamate on Wako Gel B-5 chromatograms using fourteen n-butanol solvents. The spots were located by spraying successively with (1) 5% bromine in carbon tetrachloride, (2) 0.25% fluorescein in dimethylformamide–ethanol, and (3) 2% N-(1-naphthyl)ethylenediamine in ethanol. Dulcin and cyclamate were detected as violet and pink spots, respectively. The sensitivities of detection were dulcin 0.2 μg, saccharin sodium 5 μg and sodium cyclamate 5 μg.

Ludwig and Freimuth[83] elaborated the utility of TLC for the identification of a number of organic preservatives and sweetening agents using polyamide (Miramid FP, Leunawerke) layers containing a fluorescent pigment, development with (a) chloroform–petroleum ether (b.p. 30–90°C)–acetic acid (20:70:10) and (b) water–acetic acid (90:10). Fluorescence extinction by irradiation with UV light was used for detection. The minimum quantity of dulcin detected was 10 μg.

Salo and Salminen[84] separated dulcin, saccharin and cyclamate on mixtures of 10% acetylated cellulose and increasing amounts of polyamide using Shell solvent A–n–propanol–acetic acid–formic acid (45:6:3:6) as the mobile phase. Figure 16 depicts the change of R_F values of dulcin, saccharin, and cyclamate on a mixture of 10% acetylated cellulose with increasing percentages of polyamide.

Waldi[81] utilized Silica Gel G with chloroform–acetic acid (90:10) for development and Rhodamine B followed by silver nitrate for detection in the separation and identification of dulcin and saccharin from their acidic aqueous solutions.

Korbelak and Bartlett[85] reported the separation and identification of the sweeteners dulcin, cyclamate, saccharin and P-4000 (5-nitro-5-propoxyaniline) on Adsorbosil-1 (Applied Science) plates developed with n-butanol–95% ethanol–28%

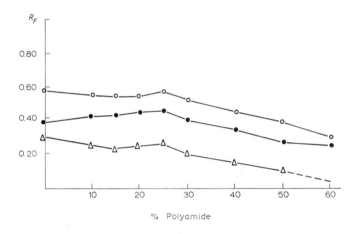

Fig. 16. R_F values of dulcin, saccharin, and cyclamate as a function of varying amounts of polyamide with 10% acetylated cellulose. ○, dulcin; ●, saccharin; △, cyclamate. Solvent: Shell Sol. A–n-propanol–acetic acid–formic acid (45:6:3:6).

ammonia–water (40:4:1:9). The plates were developed 10 cm and examined under short-wave (254 nm) ultraviolet light in which saccharin yielded a blue fluorescent spot and P-4000 a dark spot. The plate was then sprayed with 1% chloranil in benzene and heated in a 100°C oven for about 5 min (cyclamates and saccharin appeared as white spots on a lavendar background). Light spraying with 1% p-dimethylaminobenzaldehyde in 10% HCl then revealed dulcin as a bright yellow spot on a white background, while the P-4000 spot remained lavendar. The R_F values and sensitivity of the sweeteners were calcium cyclamate 0.30 and 5 μg, sodium saccharin 0.41 and 2 μg, dulcin 0.75 and 1 μg, and P-4000 0.8 and 1 μg.

Korbelak[86] utilized Silica Gel H or Adsorbosil-1 with the butanol–ethanol–ammonia–water developer[85] to separate dulcin, cyclamate, and saccharin in beverages. Successive spraying with solutions of bromine, fluorescein, and N-1-naphthylethylenediamine dihydrochloride revealed cyclamate as a bright pink spot, dulcin as a brownish or blue spot and P-4000 as a brown-pink spot. Concentrations as low as 0.004% saccharin, 0.08% cyclamate, 0.03% dulcin, and 0.002% P-4000 have been detected in extracts from commercial products spiked with sweeteners.

The TLC separation and detection of dulcin, saccharin, and cyclamate using 5 absorbents, Avicel SF (American Viscose Division of FMC), polyamide (Woelm), Silica Gel G (Merck), alumina (Woelm), and DEAE–cellulose (Serva) has been reported[87]. The developing solvents were: (A) ethyl acetate–conc. ammonia–acetone (1:1:8), (B) dimethylformamide–ethanol–water (5:4:1), (C) dioxan–pyridine–water (7:2:1), (D) pyridine–ethanol–water (6:3:1), and (E) tetrahydrofuran–pyridine–water (6:3:1). Tables 6 and 7 depict the R_F values of the synthetic sweeteners on Avicel (a microcrystalline cellulose) layers and on 5 different chromatographic media, respectively. Table 8 lists the limits of detection of the sweeteners on Avicel using 5 detection methods and Table 9 lists the detection limits of the sweeteners by Pinacryptol Yellow reagent on layers of different chromato-media.

TABLE 6

R_F VALUES OF SYNTHETIC SWEETENERS ON AVICEL LAYERS

Sample	Solvent system[a]				
	A	B	C	D	E
Cyclamate	0.15	0.74	0.29	0.47	0.31
Saccharin	0.31	0.83	0.39	0.64	0.41
Dulcin	0.82	0.91	0.80	0.79	0.89

[a] Solvents: A, ethyl acetate–conc. ammonia–acetone (1:1:8); B, dimethylformamide–ethanol–water (5:4:1); C, dioxan–pyridine–water (7:2:1); D, pyridine–ethanol–water (6:3:1); E, tetrahydrofuran–pyridine–water (6:3:1).

TABLE 7

R_F VALUES OF SYNTHETIC SWEETENERS ON LAYERS OF DIFFERENT CHROMATO-MEDIA

Chromato-media	Solvent system	R_F value		
		Cyclamate	Saccharin	Dulcin
Avicel SF	AcOEt–conc. ammonia–acetone (1:1:8)	0.04	0.23	0.88
Polyamide	Benzene–AcOEt–HCOOH (5:10:2)	0.41	0.69	0.81
Silica gel	Acetone–10% ammonia (9:1)	0.24	0.59	0.84
Alumina	iso-PrOH–conc. ammonia (4:1)	0.14	0.30	0.70
DEAE–cellulose	0.5 M HCOONH$_4$ (pH 6.5)	0.74	0.36	0.53

Dulcin, sodium cyclamate and saccharin sodium were separated by TLC[88] using the following solvent systems: (1) n-butanol–ammonium hydroxide (9:1), (2) isoamyl alcohol–ethanol–ammonium hydroxide (8:4:1), and (3) isoamyl alcohol–isopropanol–ammonium hydroxide (8:1:1). Both cyclamate and saccharin were detected as white spots on a pale black-brown background using a silver nitrate–

hydroquinone reagent. Dulcin was identified as a yellow spot with *p*-dimethylamino-benzaldehyde reagent. The limits of identification were 5 µg of sodium cyclamate, 1 µg of saccharin sodium, and 0.05 µg of dulcin.

The detection and determination of dulcin in foods by column and thin-layer chromatography have been reported by Takeshita *et al.*[89]. Hydrochloric acid–acetone extracts of foodstuffs were applied to an alumina column and eluted with ethyl acetate followed by acetone–methanol (9:1). The latter eluent was concentrated, then chromatographed on silica gel plates developed with ethyl acetate (R_F dulcin = 0.32).

TABLE 8

DETECTION LIMITS (µg) OF SYNTHETIC SWEETENERS ON AVICEL LAYERS

Sample	Method of detection[a]				
	I	II	III	IV	V
Cyclamate	1	2	4	8	–[b]
Saccharin	0.2	1	1	1	0.1
Dulcin	0.2	1	–	–	0.2

[a] Methods of detection: I, *Pinacryptol Yellow-UV*. The plate is sprayed with 0.1% (w/v) Pinacryptol Yellow soln. in 95% ethanol and allowed to dry in the dark for 10 min, and then examined in transmitted UV light (3650 Å). Cyclamate appears as an orange fluorescent spot on a light greenish-blue background. Saccharin is a non-fluorescent orange spot and dulcin a dark violet spot. II, *Bromine–fluorescein–naphthylethylenediamine*. After exposure to bromine vapor, the plate is sprayed with a 0.1% ethanolic soln. of fluorescein, air-dried, and then sprayed with a 2% ethanolic soln. of naphthylethylenediamine. Cyclamate appears as a yellow spot, saccharin is yellowish pink, and dulcin is yellowish orange on a dull orange background. III, *Methyl Red in phosphate buffer soln.* The well-dried plate is sprayed with a soln. consisting of 1 part of 0.1% (w/v) ethanolic soln. of Methyl Red and 2 parts of phosphate buffer soln. (pH 7.0). Cyclamate and saccharin give reddish orange spots on a yellow background. IV, *Silver nitrate–pyrogallol*. Silver nitrate (0.17 g) is dissolved in 1 ml of water and mixed with 5 ml of 10% ammonia; this soln. is diluted with ethanol to 200 ml. The plate is sprayed with the soln. prepared as above and then sprayed with a freshly prepared soln. of 0.01% (w/v) ethanolic pyrogallol. Cyclamate appears as a transient white spot on a brown background. V, *UV absorption or fluorescence*. The plate is examined in transmitted UV light (2537 Å), when saccharin appears as a blue fluorescent spot and dulcin as a dark spot.
[b] – denotes an undetected spot.

TABLE 9

DETECTION LIMITS (µg) OF SYNTHETIC SWEETENERS BY PINACRYPTOL YELLOW REAGENT ON LAYERS OF DIFFERENT CHROMATO-MEDIA

Sample	Chromato-media				
	Avicel SF	Polyamide	Silica gel	Alumina	DEAE–cellulose
Cyclamate	1	12	40	1	40
Saccharin	0.2	4	12	0.1	1
Dulcin	0.2	1	1	0.1	0.1

The chemical and gas chromatographic analysis of dulcin in human pathological tissues (blood, urine, internal organs) (as well as ingested rice-cakes) has been described by Nanikawa et al.[76]. A Shimadzu Model GC-1B gas chromatograph was used with a flame-ionization detector and a 225 cm × 0.4 cm column containing 1.5% SE-30. The column, injector port, and detector temperatures were 150, 180, and 200 °C, respectively. Figure 17 illustrates gas chromatograms of a reference control sample of dulcin and a sample of dulcin extracted from human brain.

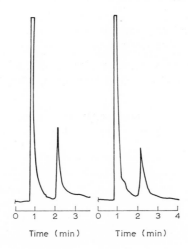

Time (min)   Time (min)

Fig. 17. Gas chromatograms of dulcin on 1.5% SE-30 at 150 °C. Left, dulcin as control; right, dulcin extracted from brain of a dead woman.

3. ALLYL ISOTHIOCYANATE

The natural source of allyl isothiocyanate (volatile oil of mustard) is sinigrin, one of the major glucosinolates in a variety of plants such as *Brassica oleraceae* (cabbages, kale, brussel sprouts, cauliflower, broccoli, and kohlrabi), *B. carinata* (Ethiopian rapeseed), *B. nigra* (black mustard), *Sinapsin alba* (white mustard), *S. arvensis* (Charlock) and *Amoracia rustica* (horseradish).

Allyl isothiocyanate is obtained from sinigrin by enzymic hydrolysis as shown.

$$CH_2=CH-CH_2-C \underset{N-O-SO_2O^-K^+}{\overset{S-C_6H_{11}O_5}{\Big<}} \xrightarrow{\overset{\text{thioglucosidase}}{+H_2O}}$$

$$\left[CH_2=CH-CH_2-C \underset{N^-}{\overset{S^-}{\Big<}} \right] + \text{glucose} + KHSO_4$$

$$CH_2=CH-CH_2-C\equiv N$$

$$CH_2=CH-CH_2N=C=S \qquad CH_2=CH=CH_2-S-C\equiv N$$

References pp. 421–429

Commercially, it is prepared by the reaction of allyl iodide with potassium thiocyanate and is used as a food additive (hot sauces, relish flavors, salad dressings, and synthetic mustard), medically as a counter-irritant, and has utility in polymer synthesis; it is also used as a meat preservative, plant regulator, insect attractant, and possesses antihelminthic and fungistatic properties.

The goitrogenic activity[90, 91], toxicity[92], cytotoxicity[93], and mutagenicity in *Drosophila*[94, 95] and *Ophiostoma*[96] of allyl isothiocyanate as well as its induction of chromosome aberrations in *Drosophila*[97] and *Allium*[98] and inhibition of protein synthesis by mouse pancreas *in vitro*[99] have been reported.

The decomposition of allyl isothiocyanate in aqueous solution has been elaborated by Kawakishi and Namiki[100]. The decomposition based on the addition reaction on –N=C=S and not on the hydrolysis of R–NCS (as in the case of *p*-hydroxybenzyl isothiocyanate) yielded 4 products (Fig. 18). Allyl isothiocyanate is decom-

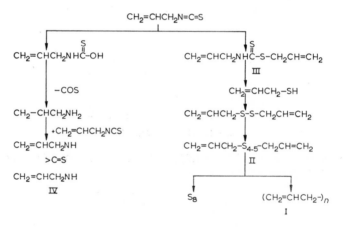

Fig. 18. Proposed degradation scheme of allyl isothiocyanate in aqueous solution.

posed to allyl-allyl dithiocarbamate (III) which was degraded to diallyl tetra- and pentasulfide (II). This polysulfide was further degraded to a paraffin like hydrocarbon (I) and sulfur. Moreover, N,N'-diallylthiourea (IV) was produced by the addition reaction of allylamine formed *via* the interaction of water with isothiocyanate.

The degradation products of the hydrolysis of allyl isothiocyanate were analyzed by column and TLC, UV, IR, and mass spectroscopy. Silica gel columns containing ether extracts of the degradation products were eluted with *n*-hexane, *n*-hexane–ether (9:1), *n*-hexane–ether (1:1), and finally ether and the above fractions concentrated and chromatographed on silica gel plates using *n*-hexane–acetone (9:1) for development and both Dragendorff's reagent and iodine–azide for detection.

The gas chromatographic studies on the acid components of Japanese horseradish *(Wasabia japonica)* were described by Kojima *et al.*[101, 102]. The acid components determined by GLC were allyl-, *sec.*-butyl-, 3-butyl-, and β-phenylethyl isothio-

TABLE 10

GAS CHROMATOGRAPHIC DATA OF STANDARD ISOTHIOCYANATES

Column: A and B, 10% SE-30/C-22; C, 2.5% DEGS/C-22.
Temperature: A, 124°C; B, 61°C; C, 102°C.
Carrier gas (N_2): A, 39.5 ml/min; B, 38.0 ml/min; C, 20.0 ml/min.

Isothiocyanates	B.p. (°C)	A		B		C	
		RT^a	RRT^b	RT	RRT	RT	RRT
Methyl	119	0.7	0.12	7.8	0.16	1.2	0.15
Ethyl	130–2	0.9	0.16	8.2	0.17	1.2	0.15
Isopropyl	136	1.0	0.18	8.7	0.18	1.4	0.19
Allyl	150	1.2	0.21	9.7	0.20	1.8	0.23
sec.-Butyl	156	1.4	0.25	12.5	0.26	1.6	0.21
3-Butenyl	171–2	1.8	0.32	15.6	0.33	3.0	0.38
Phenyl	221	5.7	1.00	47.6	1.00	7.8	1.00
Benzyl	243	11.7	2.05	99.5	2.10	35.7	4.58
β-Phenylethyl	247.5	19.8	3.50	171	3.60	72.0	9.22

^{a} RT = retention time (min).
^{b} RRT = relative retention time based on phenyl isothiocyanate (1.00).

cyanates. Table 10 lists the gas chromatographic data of a number of standard isothiocyanates chromatographed on three columns: (A) and (B) 10% SE-30/C-22 at
124° and 61°C, respectively, and (C) 2.5% DEGS/C-22 at 102°C. Figures 19 and
20 illustrate gas chromatograms of the acid components of *Wasabia japonica* chromatographed on 10% SE-30 at 124°C and on 2.5% DEGS at 102°C, respectively.

Twenty synthetic isothiocyanates including *n*-alkyl-, iso-alkyl-, *sec.*-alkyl-,
alkenyl-, aromatic, and other isothiocyanates were analyzed by GLC using 10% SE-30

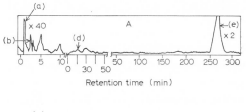

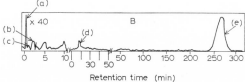

Fig. 19. Gas chromatogram of acid components in *Wasabia japonica* (Wasabi) I. A, Non-manured
Wasabi; B, manured Wasabi. Conditions: column: 10% SE-20/C-22 0.4 (i.d.) × 75 cm; temperature: 124°C; carrier gas: N_2, 39.5 ml/min; a, b, c, d, and e, peaks of allyl, *sec.*-butyl, 3-butenyl,
β-phenylethyl isothiocyanate, and unknown compound, respectively.

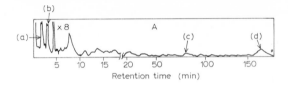

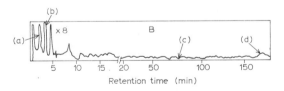

Fig. 20. Gas chromatogram of acid components in *Wasabia japonica* (Wasabi) II. A, Non-manured Wasabi; B, manured Wasabi. Conditions: column: 2.5% DEGS/C-22, 0.4 (i.d.) × 75 cm; temperature: 102°C; carrier gas: N_2, 20 ml/min; a, b, c, and d, peaks of allyl plus *sec.*-butyl, 3-butenyl, β-phenylethyl isothiocyanate and unknown compound respectively.

TABLE 11

GAS–LIQUID CHROMATOGRAPHIC DATA OF ISOTHIOCYANATES

Relative retention time based on phenyl isothiocyanate (1.00).

Sample no.	Isothiocyanates	Relative retention time			
		10% SE-30/C-22		2.5% DEGS/C-22	
		CB 102°C N_2 20 ml/min	CB 124°C N_2 39.5 ml/min	CB 57°C N_2 48 ml/min	CB 102°C N_2 20 ml/min
1	Methyl	0.08	0.12	0.06	0.15
2	Ethyl	0.11	0.16	0.07	0.15
3	*n*-Propyl	0.17	0.21	0.10	0.19
4	*n*-Butyl	0.30	0.35	0.19	0.28
5	*n*-Amyl	0.55	0.54	0.42	0.45
6	*n*-Hexyl	0.97	0.91	0.87	0.81
7	*n*-Octyl	3.33	2.87	4.36	2.42
8	Isopropyl	0.12	0.18	0.06	0.15
9	Isobutyl	0.23	0.28	0.14	0.24
10	Isoamyl	0.42	0.44	0.31	0.36
11	*sec.*-Butyl	0.21	0.25	0.13	0.21
12	*sec.*-Heptyl	1.27	1.03	0.96	0.81
13	*sec.*-Octyl	2.42	1.79	2.31	1.53
14	2-Ethyl-*n*-hexyl	2.20	1.47	2.52	1.56
15	Allyl	0.15	0.21	0.12	0.23
16	3-Butenyl	0.28	0.32	0.28	0.38
17	Phenyl	1.00	1.00	1.00	1.00
18	Benzyl	2.77	2.05		4.58
19	β-Phenylethyl	5.00	3.50		9.22

and 2.5% DEGS. Table 11 lists the GLC data for the isothiocyanates when chromatographed on 10% SE-30 at 102 and 124°C and on 2.5% DEGS at 57 and 102°C.

The GLC analysis of isothiocyanates in cruciferae oil seeds was reported[103]. A 2-m column packed with 10% DEGS on 80–100 mesh Gas Chrom P operated at 90°C with helium as carrier gas at 20 ml/min, air at 400 ml/min was used with flame-ionization detection for the determination of allyl-, butenyl-, and pentenyl isothiocyanates in rapeseed and black mustard seeds.

A GLC method for the determination of allyl isothiocyanate in mustard seed was reported by Andersen[104]. A gas chromatograph was used equipped with a flame-ionization detector and fitted with a 12-ft. column of 5% Carbowax 4000 on Fluoropak 80. The operating parameters were: column 145°C, injector 160°C, detector 200°C, and the carrier gas was nitrogen at a flow rate of 100 ml/min.

Binder[105] described a gas chromatographic method for the determination of a number of aliphatic and aromatic isothiocyanates. A Perkin-Elmer F 6/4 Frakto-

TABLE 12

RELATIVE RETENTION TIMES OF A NUMBER OF ISOTHIOCYANATES

R	Stationary phase[a]		
	AM	20 M	Tween 80
(a) At 100°C			
Methyl	0.16	0.42	0.29
Ethyl	0.24	0.40	0.42
Allyl	0.45	0.85	0.77
Isopropyl	0.26	0.29	0.38
n-Propyl	0.47	0.60	0.62
tert.-Butyl	0.27	0.23	0.29
sec.-Butyl	0.58	0.48	0.56
Isobutyl	0.70	0.68	0.72
n-Butyl	1.00	1.00	1.00
Cyclohexyl	6.06	4.50	4.14
n-Hexyl	4.40	3.04	2.85
Phenyl	6.06	6.59	5.16
(b) AM and 20 M at 170°C, Tween 80 at 150°C			
n-Butyl	0.26	0.25	0.27
Cyclohexyl	1.03	0.83	0.82
n-Hexyl	0.73	0.56	0.61
Phenyl	1.00	1.00	1.00
o-Tolyl	1.56	1.33	1.35
m-Tolyl	1.72	1.43	1.48
p-Tolyl	1.79	1.54	1.58
Benzyl	2.20	3.82	3.97
n-Decyl	5.52	2.59	3.04

[a] Stationary phases: AM = Apiezon M; 20 M = Carbowax 20 M.

meter gas chromatograph was used with flame-ionization and conductivity detectors and 200 cm × 0.3 cm stainless steel columns containing (*1*) 15% Apiezon M on 60 to 100 mesh Celite 545 and (*2*) 20% Carbowax on Chromosorb W. Columns (*1*) and (*2*) were used for separation and relative retention data (with temperature programmed from 75 to 200°C in 30 min). For purification of the isothiocyanates two columns were used, *viz*. 12% Carbowax 20M on Chromosorb G and 60–80 mesh Chromosorb G (a) acid-washed and (b) treated with dimethyl dichlorosilane. Table 12 lists the relative retention times of a number of alkyl and aryl isothiocyanates on Apiezon M, Carbowax 20M and Tween 80 columns and Fig. 21 illustrates a chromatogram of various isothiocyanates chromatographed on 20% Carbowax 20M with temperature programming from 75 to 200°C at 2.5°/min.

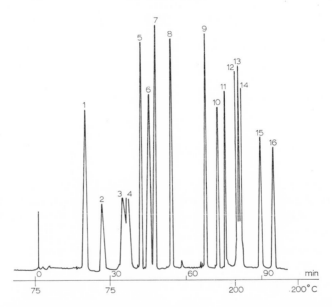

Fig. 21. Chromatograms of various isothiocyanates. Column: 200 cm × 0.3 cm stainless steel packed with 20% Carbowax 20M on Chromosorb W (HMDS treated). Initial oven temperature 75°C, then programmed to 200°C at 2.5°/min. Carrier gas nitrogen at 32 ml/min; flame-ionization detection sensitivity × 32. The peaks of the isothiocyanates were: 1, *tert*.-butyl; 2, isopropyl; 3, ethyl; 4, methyl; 5, *sec*.-butyl; 6, *n*-propyl; 7, isobutyl; 8, *n*-butyl; 9, *n*-hexyl; 10, cyclohexyl; 11, phenyl; 12, *o*-tolyl; 13, *m*-tolyl; 14, *p*-tolyl; 15, *n*-decyl; 16, benzyl. Volume injected: 0.6 µl.

Kjaer and Jart[106] separated a variety of isothiocyanates on 2-cm columns of Dixon helices coated with silicone grease at column temperatures of 30.5 and 68.5°C. Jart[107] described the GLC of 32 isothiocyanates (*e.g.* saturated and unsaturated, branched and unbranched, aliphatic, aromatic, ω-methylthio substituted and ester- or sulfone-containing). A Griffin and George Model II gas chromatograph was equipped with a katharometer detector and used with the following 6-mm diameter glass columns (*a*) 1.8 m containing 20% Apiezon L on Celite, (*b*) 1.8 m with

25% silicone elastomer (E301) on Celite, (c) 1.8 m with 25% silicone grease on fire-brick, (d) 1.8 m with 25% squalene on Celite, (e) 0.9 m with 30% dinonyl phthalate on firebrick, and (f) 1.8 m with 25% tritolyl phosphate on Celite.

The thin-layer chromatography of a variety of alkyl- and aryl isothiocyanates, their respective glucosinolate precursors and their respective thiourea derivatives was described by Wagner et al.[108]. Silica Gel G (Merck) was used with chloroform–ethyl acetate–water (3:3:4) and a spray reagent of 1% potassium ferricyanide–5% ferric chloride (1:1) for the separation and detection of the isothiocyanates as their thiourea derivatives. For the separation of glucosides, n-butanol–n-propanol–glacial acetic acid–water (3:1:1:1) was used with 25% trichloroacetic acid in chloroform and 1% potassium ferricyanide–1% ferric chloride (1:1) for detection.

The paper chromatography of isothiocyanates has been reported by Fisel et al.[109]. Treatment of the isothiocyanates (allyl-, phenyl-, o-, m-, and p-tolyl, and 1-naphthyl) on Whatman No. 1 paper with gaseous ammonia yielded the corresponding thiourea which when treated with bismuth nitrate yielded yellow complexes that could be satisfactorily separated by developing with water saturated butanol or with 40% aq. ethanol.

Other analytical procedures for the determination of allyl isothiocyanate include titrimetry[110] and colorimetry[111].

4. SAFROLE

Safrole (4-allyl-1,2-methylenedioxybenzene) is a major constituent of the essential oils of sassafras, star anise, and camphor and a minor constituent of oils of nutmeg, mace, cinnamon leaf, wild ginger, and California bay laurel. It has been widely used until recently as a flavoring agent in root beer, chewing gum, tooth-pastes, and in certain pharmaceutical preparations. It is also used to scent soaps and cosmetics and with isosafrole (4-propenyl-1,2-methylenedioxybenzene), is employed in the manufacture of piperonal (heliotropin) and the pesticidal synergist piperonyl butoxide. Both safrole and isosafrole are active insecticidal synergists for pyrethrum and carbaryl (1-naphthyl-N-methyl carbamates).

Safrole has been shown to be a rat hepatocarcinogen following long-term feeding at the 0.5 and 1% level[112-114], and to induce pulmonary adenomas and adenocarcinomas as well as hepatomas in mice[115]. The toxicity of safrole in dogs[116] has also been reported.

The metabolism of safrole in the rat following intravenous[117], oral[118, 119], and intraperitoneal[118, 119] administration has been reported. Biliary and urinary metabolites following i.v. administration of safrole were examined by TLC using Silica Gel DF-5 plates with the following solvent systems: (1) toluene–acetic acid–water (10:10:1), (2) ethyl acetate–acetic acid–methanol (70:10:20), (3) n-butanol–acetic acid–water (10:1:1), and (4) acetone–benzene (1:39). The detecting systems were (a) conc. sulfuric acid–n-butanol[120], (b) chromotropic reagent[121] and ultraviolet radiation sources of 366 and 253 nm. Ten, seven, and eleven biliary metabolites were found using solvent systems (1–3), respectively. Urinary metabolites following oral and i.p. administration of safrole[118, 119] to both rat and guinea pig were examined on Silica Gel GF plates using (a) benzene, (b) hexane–diethyl ether–acetic acid (90:10:1) and (c) hexane–diethyl ether–methanol–acetic acid (85:20:3:3). Basic ninhydrin-positive substances were isolated and identified as tertiary amino-methylenedioxypropiophenones, e.g. 3-N,N-dimethylamino-, 3-piperidyl- and 3-pyrrolidinyl-1-(3′,4′-methylenedioxyphenyl)-1-propanones. All three of these aminoketones decompose at alkaline pH to form 1-(3′,4′-methylenedioxyphenyl)-3-propen-1-one. Figure 22 illustrates the procedure utilized for the extraction of urine into acidic, basic and alkaloid fractions and Fig. 23 illustrates the structures of the urinary *tert.*-aminomethylenedioxypropiophenone metabolites obtained from rat and guinea pig

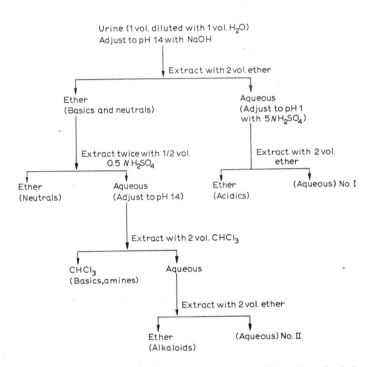

Fig. 22. Extraction of urine following oral or i.p. administration of safrole.

Kuwatsuka and Casida[122] utilized TLC for the elaboration of the metabolic fate of a number of labeled methylenedioxyphenyl (MDO-φ) derivatives including ^{14}C-safrole (prepared by the reaction of methylene ^{14}C-iodide with allylcatechol). Silica Gel G was used with a variety of solvent systems for the separation of MDO-φ derivatives (detected with the chromotropic reagent of Beroza[121]) and their corresponding catechols (detected with 2% ferric chloride in methanol) (Table 13).

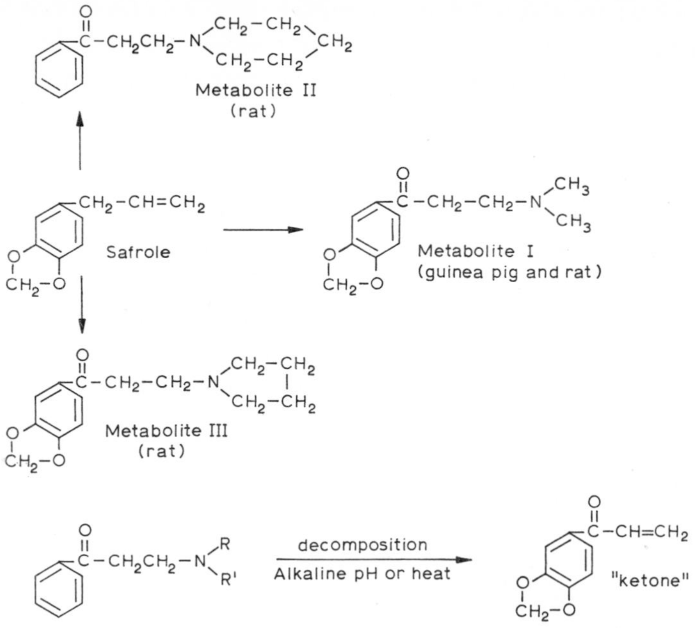

Fig. 23. Rat and guinea pig urinary *tert.*-aminomethylenedioxypropiophenone metabolites of safrole.

TABLE 13

CHROMATOGRAPHIC CHARACTERISTICS FOR SAFROLE, DIHYDROSAFROLE, MYRISTICIN, PIPERONYL BUT-
OXIDE, AND SULFOXIDE SYNERGIST

Thin-layer chromatography on Silica Gel G.

Solvent	Safrole	Dihydro-safrole	Myristi-cin	Piperonyl butoxide	Sulfoxide[a]	
					A	B
R_F values for catechols[b]						
Ether	0.93	0.93	0.90	0.82	0.27	0.27
Hexane–ether (1:1)	0.46	0.48	0.48	0.17	0.02	0.02
Benzene–ether (10:1)	0.28	0.28	0.28	0.06	0.03	0.03
R_F values for methylenedioxyphenyl compounds						
Hexane	0.25	0.29	0.08	0.00	0.00	0.00
Ether	0.98	0.98	0.98	0.98	0.53	0.45
Hexane–ether (10:1)	0.80	0.89	0.54	0.08	0.02	0.02
Hexane–ether (1:1)	0.98	0.98	0.93	0.65	0.10	0.10
Benzene	0.87	0.90	0.60	0.12	0.05	0.05
Benzene–ether (10:1)	0.98	0.98	0.89	0.45	0.10	0.10

[a] A and B are components obtained from the silicic acid column chromatography of technical
sulfoxide. The first ether-eluted component (that with the higher R_F value (0.53) on TLC) is designated
as sulfoxide A; the second with the lower R_F (0.45) as sulfoxide B.
[b] Catechol intermediates which on methylenation yielded appropriate methylenedioxyphenyl compounds.

Kamienski and Casida[123] described the importance of demethylenation in the *in vivo* and *in vitro* metabolism of MDO-φ synergists and related compounds in mammals (mouse, rat, and hamster). The techniques used for TLC[124] and detection of labeled and unlabeled products in co-chromatography included separation of metabolic products on Silica Gel H and Silica Gel F_{254} using for development the following solvent systems: (*A*) benzene (saturated with formic acid)–ether (1:3), (*B*) *n*-hexane–ether–formic acid (25:25:1), (*C*) *n*-butanol–acetic acid–water (3:1:1). Detection was achieved with (*a*) 0.3% ninhydrin in *n*-butanol–glacial acetic acid (100:3) and (*b*) chromotropic acid reagent[121]. Urinary metabolites of safrole were separated by TLC using systems (*a*) and (*b*) in two dimensions and revealed the presence of piperonylic acid in addition to 5 other metabolites. Safrole (as well as dihydrosafrole, myristicin, sulfoxide, and sulfoxide B) were found to be extensively demethylenated in living mice and are converted to more polar metabolites in the liver microsome–NADPH system. The metabolism of methylenedioxyphenyl compounds was suggested by Casida *et al.*[122, 124] to involve the mixed function oxidases and to proceed as illustrated in Fig. 24.

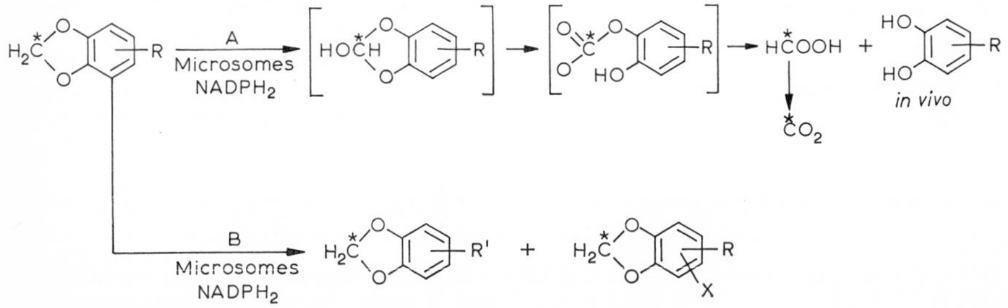

Fig. 24. Proposed pathway for the metabolism of methylene-^{14}C-dioxyphenyl compounds. Pathway A involves initial enzymatic hydroxylation of the methylene-^{14}C group, and this pathway accounts for (i) the release of formate-^{14}C on scission of the hydroxymethylene-^{14}C-dioxyphenyl group, and (ii) the metabolism *in vivo* of the formate-^{14}C intermediate to yield [14]CO_2. (The two intermediates in this pathway have not been isolated and are probably very unstable.) Pathway B results in modification of the ring substituent(s) (R′) or in introduction of an additional group (X) into the phenyl ring. In living mice and in the mouse liver microsomal system, pathway B represents a minor degradation route whereas, in houseflies, the significance of pathway B probably is greater because there is less extensive scission of the methylenedioxyphenyl group.

The methylene-^{14}C group is hydroxylated yielding formate-^{14}C in the microsome–reduced nicotinamide–adenine dinucleotide phosphate system *in vitro* and yielding expired [14]CO_2 in living mice and houseflies. Methylenedioxyphenyl compounds were thus said to serve apparently as alternate substrates (competitive inhibitors) for this enzymatic hydroxylation of microsomes and thus reduce the rate of metabolism and prolong the action of certain drugs and insecticide chemicals. It was pointed out in this classic study that it is not known whether any of the intermediates combine chemically with components of the microsomes such as the active sites of the enzymes. This mechanism of hydroxylation is believed to account, in part, for the action of methylenedioxyphenyl compounds as insecticide synergists and as inhibitors of drug detoxification.

Stahl[125] delineated the thin-layer chromatographic behavior of safrole, iso-safrole, and myristicin as well as a number of related methylenedioxyphenyl- and hydroxyphenylpropane derivatives. Separation was accomplished on $250\,\mu$ thick Silica Gel G layers prepared by the standard method and using benzene as the developing solvent. The chromogenic reagents were: (1) a mixture of antimony trichloride and antimony pentachloride in carbon tetrachloride and (2) anisaldehyde–sulfuric acid. Table 14 depicts the R_F values as well as the color reactions for a number of methylenedioxyphenyl- and hydroxyphenylpropane derivatives.

Nano and Martelli[126] separated a number of pairs of allylic–propenylic isomers (including safrole and isosafrole) by thin-layer chromatography on silica impregnated with silver nitrate. The capacity of olefins to complex selectively with silver cations has been previously indicated, e.g. allylic isomers can form π-complexes with silver nitrate. Table 15 illustrates the separation of allylic derivatives of benzene and cyclohexene from their propenylic isomers on both Silica Gel G (Merck) and silica impregnated with silver nitrate. The chromogenic reagents were, for the silica absorbent, phosphomolybdic acid at 100°C and for the silica + silver nitrate absorbent, vanillin reagent at 140°C.

The utility of the potent catalyst iron pentacarbonyl for the isomerization of safrole to isosafrole (which is an important step in the synthesis of piperonal) was recently investigated by Riezebos et al.[127]. Gas–liquid chromatographic analysis of the above reaction indicated an almost quantitative isomerism of safrole to isosafrole. In the original analytical procedure, a Pye Argon chromatograph was utilized with a 4 ft. × 4 mm column packed with 10% polypropylene sebacate on 100/120 B.S.S. mesh Celite, with a column temperature of 110°C. This resulted in rather long retention times (ca. 3.5h), with extremely good separation. Using an Apiezon L column (10% Apiezon L on 100/120 B.S.S. Celite) in the above apparatus at a temperature of 140°C, a chromatogram showing complete resolution was obtained in 40 min. In a recent modification[128] a Pye 104 series Model 24, dual flame unit was used for the analysis of the above isomerization. A coiled glass 5 ft. × 4–5 mm column packed with 10% Apiezon L on 100/120 mesh Celite was used at a column temperature of 120°C. The carrier gas was nitrogen 40 ml/min, hydrogen 40 ml/min and air 600ml/min. Complete resolution was achieved in 40 min.

A variety of MDO-φ derivatives including safrole and a number of synergists were analyzed by gas–liquid chromatography[129] utilizing an F & M Model 1609 flame-ionization instrument containing a modified flow system which permitted on-column injections into a glass column with the effluent passing directly to the hydrogen-flame detector (Applied Science). The columns employed were 6 ft. × $\frac{1}{4}$ in. glass coils of 3% Carbowax 20M on HMDS pretreated Chromosorb W (60–80 mesh) and of 4% QF-1 on HMDS-pretreated Chromosorb G (80–100 mesh). The analytical

TABLE 14

R_F VALUES AND COLOR REACTIONS OF METHYLENEDIOXYPHENYL DERIVATIVES AND RELATED COMPOUNDS

Layer: Silica Gel G[a]; standard methods; solvent: benzene. CS = Chamber saturation; NS = without chamber saturation. Running times for 10-cm migration distance: NS = 35 min, CS = 50 min, S chamber = 40 min.

Substances	R_F value × 100			Color reactions			
	Chamber		S cham-ber	Antimony (III) + (IV) chloride reagent 1 + 1			Anisaldehyde–H_2SO_4 after heating[c]
	CS	NS		Room temp.	After heating[b]	After 24 h under UV	
Safrole	57	90	89		Grey-violet	Red-brown, bright edge	Blue-green–grey
Isosafrole	57	90	89	Grey-violet	Blue-violet	Grey-violet, bright edge	Grey-violet
Anethole	56	87	83		Grey-violet	Pink-violet	Grey-red, blue edge
Methylchavicol	55	85	80		Olive-green–grey	Pink	Violet-grey (after 15 min)
Myristicin	46	70	61	Light-brown	Grey-brown	Dark, reddish edge	Brown-green–grey
Isomyristicin (trans)	46	70	61		Brown-violet	Brown-violet	Grey-brown
Resorcinol dimethyl ether	40	72	56	Grey-brown	Brown-green	Dark	Red
Apiole	35	58	45	Light-brown	Olive-green	Dark	Violet-grey
Isoapiole (trans)	35	58	45	Violet	Brown-violet	Dark	Blue-violet
Hydroquinone dimethyl ether	31	54	45		Yellow-green to brown	Dark	Weak violet-grey

[a] Batch No. 62,541.
[b] 10 min at 100–105 °C.
[c] No color reaction before heating.

operating conditions were 3% Carbowax 20M, column 160°C, injector 60 V, detector 220°C, nitrogen carrier 99.5 ml/min, hydrogen 110 ml/min, air 450 ml/min; 4% QF-1 fluorosilicone, column 100°C, injector 60 V, detector 220°C, nitrogen carrier 106 ml/min, hydrogen 110 ml/min, air 450 ml/min.

TABLE 15

SEPARATION OF ALLYLIC DERIVATIVES OF BENZENE AND CYCLOHEXENE
FROM THEIR PROPENYLIC ISOMERS

Compound	Silica Gel G[a]	Silica + AgNO_3[b]	Eluant
Safrole	0.57	0.29	Petroleum ether–benzene (1:1)
Isosafrole	0.57		
Pulegone	0.37	0.41	Benzene + 0.75% methanol
Isopulegone	0.43	0.18	
Estragole	0.66	0.51	Benzene
Anethole	0.68	0.67	
Eugenol	0.42	[c]	Benzene
Isoeugenol	0.42	[c]	
Eugenyl acetate	0.51	0.32	Benzene + 1% methanol
Isoeugenyl acetate	0.51	0.51	

[a] For Silica Gel G chromatoplates (300 μ thick) activation was 20 min at 120°C.
[b] SiO_2 + $AgNO_3$ chromatoplates were prepared by stirring 25 g of SiO_2 with 70 ml of 12.5% aq. $AgNO_3$ solution (for five 20 × 20 cm plates); activation was 30 min at 60°C.
[c] The compounds react with silver nitrate and reduce it on the plate.

Saiki et al.[130] described the GLC of a large variety of phenyl propene derivatives (components of natural volatile oils) using DEGS and SE-30 as stationary phases and hydrogen-flame-ionization detection. The operating parameters were: (1) 75 cm × 0.4 cm column of 10% SE-30 on C-22, column 178°C, detector 190°C, carrier gas nitrogen at 17 ml/min, hydrogen 50 ml/min; (2) 75 cm × 0.4 cm column of 20% DEGS on C-22, column 165°C, detector 210°C, carrier gas nitrogen at 50 ml/min, hydrogen 82 ml/min.

Gunner and Hand[131] described the TLC of a variety of MDO-φ derivatives (including safrole) using both Silica Gel G-HR plates and self-indicating plates of silica gel containing sodium chromotropate in aqueous sulfuric acid. The chromatographic solvents used were benzene–petroleum ether (1:1); benzene–1% methanol and benzene–2% acetone. Minimum detection limits (1–28 μg) for the derivatives and general plate appearance utilizing the silica gel–chromotrope plates were similar to those obtained using a chromotropic spray mode of detection.

5. EDTA

Ethylenediamine tetraacetic acid (EDTA, Versene) and its alkali metal salts are prepared by the reaction of ethylenediamine with hydrogen cyanide and base. The free acid is less stable than its salts and tends to decarboxylate when heated to temperatures of 150 °C.

EDTA and its alkali salts (*e.g.* disodium and calcium disodium salt) are widely used as sequestrants in food systems[132]. These derivatives are not considered as antioxidants in the classic sense in which hindered phenols such as BHT and BHA arrest oxidation by chain termination, or like ascorbates by scavenging oxygen. EDTA and its derivatives are of value in antioxidant systems due to their property of forming poorly dissociable chelate complexes with trace quantities (0.1–5 p.p.m.) of divalent and trivalent metals such as copper and iron in fats and oils. By chelating these metals, pro-oxidant catalytic effects are eliminated or minimized and higher efficiency can be derived from the antioxidant. In most cases the combined effect of sequestrants and antioxidant is synergistic and their combined utilization in foods containing fats and oils is reflected by superior initial products with extended shelf life.

EDTA has also been employed as a therapeutic agent in the treatment of lead[133], mercury[134], manganese[135] and radioactive metal poisoning[136], and is an anticoagulant in blood used for transfusion[137] and in other clinical procedures where control and investigations of metal ion concentration are desired[138]. Other areas of utility of EDTA and its derivatives include the stabilization of acrylonitrile for polymerization, the purification of antibiotics (dihydrostreptomycin), pesticidal compositions (*e.g.* hydrazine salt of EDTA), plant nutrients, plant-growth regulators, water treatment, stabilization of hydrazine fuel compositions, and in bactericidal and germicidal compositions.

The chemistry[139, 140], pharmacology[141], toxicity[142–144], metabolism in humans[145] and rats[146], its effect on chromosome radiation damages and induction of chromosome aberrations[147], breakage of chromosomes of *Tradescantia paludosa*[148], chromosomal changes in *Drosophila*[149, 150], *Vicia*[151] and *Hordeum*[151] and the synergistic effect in the production of chromosome aberrations in *Vicia*[152], and suppression of DNA synthesis in regenerating rat liver[153] have been described for EDTA.

The metabolism of ^{14}C-EDTA in the rat[146] and man[145] was elaborated by Foreman *et al.* The identification of the active material in the plasma and urine following i.v., i.p. and oral intubation of the labeled material (calcium salt of ethylenediamine tetra-2(^{14}C$_2$-acetic)acid) was carried out by means of descending paper chromatography using Whatman No. 1 paper with *n*-propanol–0.2 N HCl (1:1) in a 12–15-h development at 25 °C. Radioautograms were prepared by exposure of the developed chromatograms to Eastman X-ray film for 2–3 days. Radioautographs of chromatograms of both plasma and urine samples yielded a single band of activity (R_F 0.8, irrespective of the time of collection) which matched standard Ca-^{14}C-EDTA solutions in water or mixed with normal urine. The study indicated that essentially all of the chelate was eliminated unmetabolized (in man and rat) with a turn-over time of approximately 50 min. Less than 0.1 % of the material was oxidized and expired as CO_2 and after parental administration, 95–98 % of the compound was excreted in the urine within 6 h.

The analysis of EDTA in foods by gas–liquid chromatography has been described by Mihara et al.[154]. The sample containing EDTA or calcium salt was dehydrated under vacuum, 10 % HCl in methanol was added, the mixture heated for 2 h under reflux, the methanol removed by distillation. The sample was then diluted with water, extracted with ethyl acetate, neutralized with sodium bicarbonate and finally extracted with chloroform and concentrated. The chloroform extract was esterified with HCl–methanol and then chromatographed over a 1.08 m × 4 mm column containing 5 % QF-1 on Gas Chrom Q at 175°C or on a 2.5 m × 3 mm column of OV-1 on Gas Chrom Q at 185°C. The limit of detection was 0.0084 µg on QF-1 and 0.012 µg on OV-1.

The separation of a number of metal–EDTA complexes by TLC has been reported by Vanderdeelen[155] using Silica Gel H. The composition of the solvents together with the corresponding R_F values of the metal–EDTA complexes of Co, Cu, Ni, Mn, Cr, and Fe are shown in Table 16. By their specific and intense color, 5–10 µg of Co(III)–, Cr(III)–, Cu(II)–, Ni(II)–, and Fe(III)– EDTA could be identified as colored spots without auxiliary means. The position of Mn(II)–EDTA was located after spraying with a 20 % ammonium peroxydisulfate solution.

TABLE 16

R_F VALUES OF SOME METAL–EDTA COMPLEXES ON SILICA GEL H

Composition of solvent (%)						R_F values					
Water	Glycol mono-methyl ether	Ethyl methyl ketone	Butyl alcohol	Acetone	Ammonia (d = 0.91)	Co	Cu	Ni	Mn	Cr	Fe
45	20		25	10	0.25	0.83	0.79	0.66	0.59	0.45	0.27
40	20	20		20	0.15	0.86	0.79	0.69	0.71	0.44	0.17
40	30	15	10	5	0.10	0.78	0.69	0.65	0.58	0.47	0.26

Despite the variation in the composition of the solvent, the sequence of the R_F values is almost constant and arises because the metal ions are masked by the electron donors of the EDTA molecule. These complexes are very similar with regard to their chemical structure and molecular weight, hence suggesting the care in the selection of the composition and the pH of the developing solvent. This has already been mentioned by Sykora and Ebyl[156] who separated these complexes by paper chromatography.

Analytic applications of EDTA[157] as well as its analysis by titrimetric[158], potentiometric[159], polarographic[160], and complexometric and argentometric titrations[161] have also been reported.

Another chelating agent related to EDTA of recent environmental significance is NTA (nitrilotriacetic acid) which has been proposed as a substitute for phosphates in detergents. However, the recent finding of its teratogenicity in mice and rats[162] (as its cadmium and mercury complexes) has resulted in its ban in detergents in the U.S. Until its restriction, NTA had been going into detergents at the rate of 100 to 200 million pounds annually, or roughly 5–10% of the approximately 2.2 billion pounds of phosphates currently used each year.

6. CAFFEINE

The purine analog caffeine (1,3,7-trimethylxanthine) (I), theophylline (II) and theobromine (III) are three closely related alkaloids that occur in plants widely distributed throughout the world. Caffeine is found in the extensively used beverages, coffee, tea, cocoa, mate, as well as in some "soft drinks" (particularly the cola-flavored drinks made from the nuts of the tree *Cola acuminata* which contain about 2% caffeine). Caffeine also has broad medical utility, *e.g.* with analgesics or ergot alkaloids for relief of tension or migraine headache, with antihistamines to combat motion sickness, as a stimulant, and as a coronary artery dilator.

I   II   III

Caffeine has been reported mutagenic in *Drosophila*[163-165], bacteria *(E. coli)*[166], fungus *(Ophiostoma multiannulatum)*[167], human cells in tissue culture[168], onion tip roots[169], and bacteriophage T5[170], and produces chromosome aberrations in *Drosophila*[171], mice[172], and human cells in tissue culture[173]. However, the lack of mutagenicity of caffeine in *Drosophila*[171] and inconclusive results in mice[174, 175] have also been reported.

The teratogenic effect of caffeine in high doses in mice[176] as well as its ability to penetrate the pre-implantation blastocyst in the rabbit[177] have been described. Goldstein and Warren[178] have shown that caffeine has access to human adult and fetal gonads and that it achieves in these tissues concentrations substantially the same as in plasma. Bertoli *et al.*[179] reported a higher concentration of caffeine in testes of mouse than that recorded in other tissues following administration of ^{14}C-caffeine.

The pharmacology[180] and metabolism[181-184] of caffeine as well as its effect on lipid metabolism[185] have been described.

Caffeine is rapidly and essentially completely absorbed from the G.I. tr
man, then distributed in various tissues in approximate proportions to their
content; it passes rapidly into the central nervous system[181]. It is almost entirely
metabolized in man, about 1% being excreted in the urine. The rate of biotransformation is fairly uniform, with the average half-life being 3.5 h (15% metabolized/h).
Methylated uric acids as well as 1-methyl-, 7-methyl-, and 1,7-dimethylxanthine
(p-xanthine) have been identified as urinary excretion products following caffeine
ingestion[186]. Schmidt and Schoyer[187] found caffeine, theobromine, p-xanthine, and
1-methylxanthine in the urine of subjects given 300 mg caffeine daily.

The utility of thin-layer chromatography in the elaboration of caffeine and its
metabolites in human blood was described by Warren[184]. Plasma or red cells,
separated from heparinized blood, was extracted twice with chloroform (chloroform
extracts methylxanthines, but not uric acid). The chloroform extracts were clarified
by centrifugation and evaporated to dryness. The extracts were then chromatographed on plates of Silica Gel G (prepared with phosphate buffer, pH 6.8) and
developed first with petroleum ether (b.p. 60–80°C) (to remove the lipids) then re-
running the extracts in n-butanol–acetic acid–water (4:1:1). The R_F values found in
a 3-h development were caffeine 0.42, theobromine 0.38, and paraxanthine 0.47.
Detection was effected by spraying with iodine–potassium iodide in ethanol followed
by 95% ethanol containing 25% conc. HCl. By this procedure, caffeine was revealed
as a brown spot and the demethylxanthines as purple-blue spots against a yellow
background (sensitivity, approx. 1.0 µg). Since lipids are also stained by this procedure, the plates were also sprayed with 5% phosphomolybdic acid in ethanol and
heated to 120°C for 5 min revealing lipids as blue spots. The study of Warren revealed
that habitual consumers of beverages containing caffeine require about 7 days of
abstention from them before caffeine completely disappears from their blood extracts.
In such "decaffeinated" subjects, administration of 500 mg of caffeine orally leads to
the appearance of caffeine and paraxanthine in both red cells and plasma within 3h,
then disappears from the blood within 24 h. The first step in the metabolism of coffee
in man would appear to be the removal of the 3-methyl group and the formation of
p-xanthine (1,7-dimethylxanthine).

Grab and Reinstein[188] described the determination of caffeine in plasma by
GLC. The method involves the extraction of caffeine from plasma with chloroform
(after the aqueous phase was adjusted to pH 11.5–12.0), the chloroform extract evaporated to dryness, and the sample redissolved in carbon disulfide and analyzed
using a Varian Aerograph Series 1200 gas chromatograph with a flame-ionization
detector. The column employed was a 6 ft. × ⅛ in. o.d. stainless steel helix packed
with 3% OV-17 on Chromosorb W 100–120 mesh AW/DMGS H.P. (Applied
Science). The operating parameters were oven 200°C, injection port 260°C, detector
260°C, nitrogen carrier gas at 30 ml/min, and attenuations from 0.1 × 8 to 0.1 × 64.
Figure 25 depicts a chromatogram of both caffeine and hexobarbital.

Plasma levels of caffeine were determined after the ingestion of one cup of coffee and concentrations down to 0.25 mequiv./ml were accurately determined from 2 ml of plasma.

The column and paper chromatographic separation of purines in normal human urine has been reported[189]. The procedure involved the initial removal of uric acid, urea and salts by ion-exchange chromatography over Dowex-50 (H^+ form).

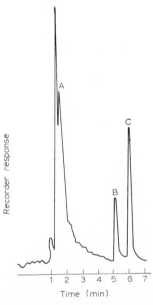

Fig. 25. Gas chromatogram of caffeine and hexobarbital on 3% OV-17. A, carbon disulfide; B, hexobarbital ($R_t = 4.3$); C, caffeine ($R_t = 5.2$). Column 3% OV-17.

The purines were separated from the remaining creatinine and amino acids by precipitation as their silver nitrate salts, and the concentrate then separated by two-dimensional paper chromatography on Whatman No. 3MM paper using n-butanol–0.6 M ammonium hydroxide (6:1) for the first development (36 h) then with n-butanol–88% formic acid–water (77:11:12) for the 90° development (8–14 h) (Fig. 26). Purine spots were visualized under UV light, cut out of the paper, eluted with water, and the eluates determined by ultraviolet spectroscopy. Table 17 lists the R_F values as well as the spectral constants of the purines isolated from urine and Fig. 26 illustrates a 2-dimensional chromatogram of urinary purine concentrates from a normal subject on unrestricted diet, or on purine-free diet with caffeine supplement.

The TLC separation and identification of caffeine, theobromine, and theophylline was effected[190] using Silica Gel G with chloroform–carbon tetrachloride–methanol (8:5:1) for development (53 min, 15 cm). The visualization of spots was best accomplished using a reagent consisting of 25 g iodine, 2 g potassium iodide dissolved in 100 ml of 95% ethanol–5% hydrochloric acid (1:1). The R_F values found were caffeine 0.54, theophylline 0.46, and theobromine 0.31.

A variety of chromatographic procedures have been used for the analysis of pharmaceutical formulations containing caffeine. Chromatography on alumina plates was used to separate a number of pyrazolone and purine derivatives[191] (antipyrine, amidopyrine, analgin, phenylbutazone, caffeine, theobromine, theophylline). The optimum solvent systems were benzene–ethanol (9:1) and benzene–methanol (9:1). Detection was accomplished with iodine–KI in 96% ethanol and 25% HCl–96% ethanol (1:1).

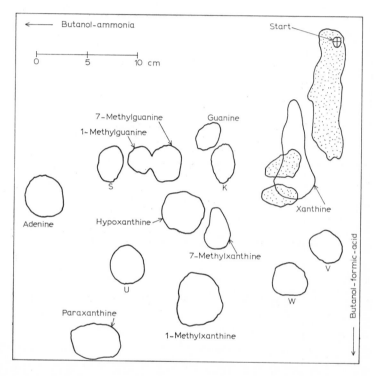

Fig. 26. Map of 2-dimensional chromatograms of urinary purine concentrates from normal subject on unrestricted diet, or on purine-free diet with caffeine supplement. Purine spots appear in ultraviolet light as dark spots on weakly fluorescent paper. Strongly fluorescent areas, also indicated, are shaded with dots. In the absence of caffeine, coffee, etc., spots for paraxanthine, 7-methylxanthine, and 1-methylxanthine are not observed.

Caffeine was separated from a variety of ergot alkaloids, barbiturates and miscellaneous drugs[192] on Silica Gel G plates developed with chloroform–methanol (9:1) and isopropanol–25% ammonia–chloroform (45:10:45). The spray reagents used were van Urk's reagent, iodoplatinate, and vanillin–sulfuric acid.

Purine derivatives and N-methylated cyclic ureides such as caffeine, theophylline, theobromine, 8-methylcaffeine, 8-chlorocaffeine, 8-methoxycaffeine, and 8-methoxytheobromine were separated on thin-layers of silica gel using solvent mixtures containing sodium-2-naphthol-6,8-disulfonate (as solubilizer)[193]. The spots were de-

TABLE 17

R_F VALUES AND SPECTRAL CONSTANTS OF PURINES ISOLATED FROM URINE BY TWO-DIMENSIONAL CHROMATOGRAPHY

Compound	R_F values[a]		pH	λ_{max} and relative extinction of peaks	
	From urine	Authentic		Substance from urine	Authentic compound
Hypoxanthine	0.13	0.12	2.0	250 (1.00)	250 (1.00)
	0.26	0.26	9.0	256 (0.97)	257 (1.00)
Xanthine	0.03	0.02	2.0	266 (1.00)	266 (1.00)
	0.19	0.20	9.0	240 (0.90); 276 (0.92)	240 (0.88); 277 (0.97)
7-Methylguanine	0.15		0–2	250 (1.00); ca. 275 (0.64)	250 (1.00); ca. 275 (0.60)
	0.12		6	248 (0.50); 283 (0.71)	245 (0.62); 281 (0.74)
			12–14	281 (0.71)	282 (0.73)
Adenine	0.33	0.32	2.0	262 (1.00)	264 (1.00)
	0.26	0.25	9.0	261 (0.97)	262 (1.00)
1-Methylguanine	0.17	0.17	2.0	251 (1.00); ca. 275 (0.62)	251 (1.00); ca. 275 (0.69)
	0.13	0.12	9.0	249 (0.80); 275 (0.72)	249 (0.90); 274 (0.75)
Guanine	0.12	0.11	2.0	248 (1.00); ca. 275 (0.58)	247 (1.00); ca. 275 (0.64)
	0.12	0.12	9.0	248 (0.71); 275 (0.64)	245 (0.83); 275 (0.72)
1-Methylxanthine	0.08	0.07	2.0	267 (1.00)	267 (1.00)
	0.44	0.43	9.0	243 (0.72); 276 (0.89)	243 (0.75); 277 (0.86)
7-Methylxanthine	0.06	0.06	6.0	268 (1.00)	268 (1.00)
	0.29	0.29	9.0	283 (0.70)	280 (0.70)
Paraxanthine	0.27	0.25	6.0	269 (1.00)	267 (1.00)
	0.53	0.53	9.0	285 (0.74)	282 (0.69)

[a] The R_F values given are mobilities with butanol–ammonia and butanol–formic acid, respectively, run on Whatman No. 1 filter paper by the descending technique, of materials eluted from 2-dimensional chromatograms, as compared with authentic specimens.

tected in UV light. Both two-dimensional paper chromatography and electrophoresis of a mixture containing caffeine, acetylsalicylic acid, aminopyrine, antipyrine, codeine phosphate and phenacetin has been reported[194]. A 0.01 ml aliquot of a mixture of the above was resolved when developed 8 h with *n*-butanol–acetic acid–water in the first dimension, dried, then developed 6 h with 50% *n*-butanol in the second dimension. The spots were detected by using 10% HCl and 1% iodine in alcohol for caffeine, 4% ferric chloride and Dragendorff's reagent for acetylsalicylic acid and bromine and ammonia for aminopyrine, antipyrine, phenacetin, codeine phosphate, and acetylsalicylic acid. Paper electrophoresis of the above drugs was accomplished using buffers of pH 2–8 and 250 V applied potential.

Caffeine was determined[195] in a variety of dosage forms, *e.g.* tablets, syrups, and ampuls, etc., *via* an initial extraction with chloroform either directly or from a mixture of 0.1 N NaOH or 0.5 N H_2SO_4, then chromatographed on silica gel with isobutanol–acetic acid–water (8:1:1). The spot corresponding to caffeine was scraped off and eluted with chloroform–methanol (9:1) then determined colorimetrically at 460 nm following successive extractions with acetic acid, pyridine, sodium hypochloride, and sodium hydroxide. Mixtures of caffeine, phenacetin, acetanilide, pyramidon, and exalgin were separated by programmed-temperature GLC[196] using columns of 0.8% and 1.6% butanediol and 1% XE-60 on silanized Chromosorb G.

The determination of caffeine in teas and coffees has been accomplished by GLC[197] and column-chromatographic nephelometric techniques[198], respectively.

The determination of caffeine has also been achieved by a variety of procedures including spectrophotometric[181,199,200], colorimetric[201], titrimetric[202], iodometric[203], and argentiometric[186].

7. NITROFURAZONE

Nitrofurazone (5-nitro-2-furaldehyde semicarbazone, Nitrofural, Furacin, Nifuson) is prepared from 2-formyl-5-nitrofuran and semicarbazide in the presence of sodium acetate. It has medical and veterinary potent utility as a local antibacterial agent (active against both Gram-positive and Gram-negative organisms) in nasal and ophthalmic solutions and ointments for skin, ear, eye, and G.U. applications with greatest utility in the treatment of swine enteritis and bovine mastitis. It is also used extensively in the U.S. as a feed additive in doses of 0.0055% to prevent avian coccidiosis and as a food preservative in Japan.

The chemistry and mode of action[204,205] of nitrofurazone, its effect on tumor growth[206], carcinogenic activity[207-209], effect on DNA synthesis[210] and phage induction[210], cross-resistance studies with *E. coli*[211], its reported mutagenicity[212] and non-mutagenicity in *E. coli*[213], as well as the testicular damage in rats by nitrofurazone and allied compounds[214,215] have been described.

Of the more than 40 million tons of feed produced annually in the United States, a very large percentage contains one or more drugs. The separation and iden-

TABLE 18

SOLVENT AND ADSORBENT SYSTEMS FOR THIN-LAYER CHROMATOGRAPHY OF DRUGS
USED IN MEDICATED FEEDS

System no.	Mobile solvent	Proportions	Min/10 cm	Adsorbent
1	n-Butanol–water–acetic acid[a]	80:20:2.5	90	Polyamide G or GF
2	Acetone–acetic acid	100:2	15	Alumina G or GF
3	Ethanol–ammonium hydroxide	80:20	80	Alumina G or GF
4	Diethyl ether–dimethylsulfoxide– n-butanol–acetic acid	98:1:1:0.2	15	Silica Gel G or GF
5a	Hexane–acetone–n-butanol	21:2:2	25	Silica Gel GF (flexible plastic sheet)
5b	Hexane–acetone–n-butanol	21:2:2	25	Silica Gel GF
6	Diethyl ether–ethanol–acetic acid	96:3:1	25	Polyamide GF
7	Acetone–n-butanol–acetic acid	78:20:2	20	Polyamide G or GF
8	Diethyl ether–n-butanol–acetic acid	73:25:2	20	Polyamide G or GF
9	n-Butanol–water–acetic acid	80:20:5	90	Silica Gel G or GF
10	Ethyl acetate–acetic acid	75:25	60	Silica Gel G or GF
11	Pyridine–benzene	50:50	45	Silica Gel G or GF
12	n-Butanol–acetic acid	75:25	60	Silica Gel G or GF

[a] The mixture was shaken in a separatory funnel and the lower layer was discarded.

TABLE 19

DETECTION SYSTEMS FOR THE TLC OF DRUGS IN MEDICATED FEEDS[a]

D1. A 5% w/v solution of potassium hydroxide in methanol was prepared 2 h before use. If the spots were not visible after spraying with this reagent, the plate was viewed under uv light.

D2. A saturated (ca. 0.2%) solution of barium diphenylamine-4-sulfonate[217] in methanol was prepared by first dissolving the salt in a few ml of N,N-dimethylformamide and then diluting to volume with the methanol. The plates were heated at 110 °C for 10 min after spraying, and then were viewed under uv light.

D3. The plates were subjected to uv light.

D4. A solution was prepared by dissolving 200 mg of 4-methylumbelliferone[217] in 35 ml of ethanol, and then diluting to 100 ml with water. After the plates were sprayed, spots were seen under uv light. Exposure of the plates to ammonia vapor may aid in visualization of the spots.

D5. A solution of 0.2% β-naphthoquinone-4-sulfonic acid (sodium salt)[217] in 5% sodium carbonate was prepared 10–18 min before use. The plates were viewed under uv light after spraying.

D6. A 1.0% solution of p-dimethylaminobenzaldehyde was prepared in ethanol which contained 1% hydrochloric acid.

D7. A saturated (ca. 22%) solution of antimony trichloride[217] in dried chloroform was prepared. This spray must be freshly prepared. The plates were sprayed, then heated at 110 °C for 15 min. If the spots were not visible, the plates were viewed under uv light.

D8. A 2% solution of vanillin[217] in isopropanol was prepared. The plates were heated at 110 °C for 10 min after spraying.

[a] uv light at 254 nm in all cases.

TABLE 20

R_F VALUES OF DRUGS COMMONLY USED IN MEDICATED FEEDS AT 1 μg SENSITIVITY

Compound	Trivial name	$R_F \times 100$	Solvent system no.
Phenylarsonic acids			
p-Aminobenzenearsonic acid	Arsanilic acid	69	1
		25	7
		35	8
		51	9
p-Ureidobenzenearsonic acid	Carbarsone	54	1
4-Nitrophenylarsonic acid		46	1
3-Nitro-4-hydroxyphenylarsonic acid		36	1
		20	10
		45	12
Nitrofuraldehydes			
3-(5-Nitrofurfurylideneamino)-2-oxazolidone	Furazolidone	84	2
		72	8
		33	9
		28	11
5-Nitro-2-furaldehyde acetylhydrazone	Nihydrazone	64	2
1-Ethyl-3-(5-nitro-2-thiazolyl)-urea	Nithiazide	50	2
5-Nitro-2-furaldehyde semicarbazone	Nitrofurazone	43	2
		38	11
Heterocyclic compounds			
Thiodiphenylamine	Phenothiazine	81	3
2-Sulfanilamidoquinoxaline	Sulfaquinoxaline	68	3
Hexahydropyrazine	Piperazine	58	3
1-(4-Amino-2-*n*-propyl-5-pyrimidinylmethyl)-	Amprolium	16	3
2-picolinium chloride hydrochloride		35	6
		87	7
Nitrophenyl compounds			
m,m'-Dinitrophenyl disulfide	Nitrophenide	95	4
3,5-Dinitrobenzamide	DNBA	73	4
		80	10
3,5-Dinitro-*o*-toluamide	Zoalene	73	4
4,4'-Dinitrocarbanilide	Nicarbazin	25[a]	4
and 2-hydroxy-4,6-dimethylpyrimidine		50[a]	
Aryl esters			
Diacetate of 3,4-bis(*p*-hydroxyphenyl)-2,4-hexadiene	Dienestrol diacetate	60	5a[b]
		57	5b[b]
		77	11
		73	12
Methyl 4-acetamido-2-ethoxybenzoate	Ethopabate	25	5a
		15	5b
		80	6
Miscellaneous			
Acetyl-(*p*-nitrophenyl)-sulfanilamide	APNPS	70	10

[a] R_F 25 is of a freshly prepared nicarbazin standard; R_F 50 is of a nicarbazin standard approximately 2 weeks old.
[b] 5a is Silica Gel GF on flexible plastic sheet; 5b is Silica Gel GF.

tification of these drugs (comprising a broad spectrum of structures including anti-biotics, arsenicals, steroids, heterocyclics, etc.) has largely been accomplished by thin-layer chromatographic techniques.

Antkowiak and Spatorico[216] reported a procedure for the identification of 18 ingredients and 7 mixtures commonly used in medicated feeds, using a variety of indicators and incorporation of lead–manganese–activated calcium silicate in the preparation of the absorbent to increase visibility of the spots and sensitivity of the procedure. Table 18 lists the solvent and adsorbent systems for the TLC of drugs used in medicated feeds; Table 19 the detection systems employed and Table 20 the R_F values of the medicated feeds at 1 μg sensitivity. Table 21 lists the detection systems of choice for 5 classes of compounds, phenylarsonic acids, nitrofuraldehydes, hetero-cyclics, nitrophenyl compounds, and aryl esters.

TABLE 21

DETECTION SYSTEMS FOR TLC OF DRUGS USED IN MEDICATED FEEDS

Class of compounds	Detection system[a]
Phenylarsonic acids	D4 or D1
Nitrofuraldehydes	D2 followed by D1
Heterocyclics	D3 followed by D8 followed by D1
Nitrophenyl compounds	D2
Aryl esters	D3

[a] Detection systems described in Table 19.

Hammond and Weston[218] described the column and TLC separations of 17 feed additives using two solvent systems with 4 chromogenic reagents and UV light for detection for the TLC separation and detection. Silica Gel G was used with chloro-form–methanol (9:1) and ethanol–1 N hydrochloric acid (1:1) as developing solvents and the following detecting reagents: (a) 1,2-diaminoethane, (b) modified Dragen-dorff's reagent, (c) Ehrlich's reagent, and (d) diazotization solutions (tin (II) chloride–sodium nitrite in 0.1 N HCl–N-1-naphthylenediamine dihydrochloride). Table 22 illustrates the identification of additives (R_F values and colors developed with the detecting reagents (a)–(d) and UV fluorescence) following an initial separation over alumina columns.

Knapstein[219] described the column and TLC separation and identification of nitrofurazone, zoalene, furoxone, furnicozone and amproleum in mixtures. Silica Gel G plates (0.75–1.0 mm) were developed with benzene–acetic acid–acetone–water (10:4:1:0.4) and 1 N HCl–96% ethanol (1:1) and detected with a modified Dragen-dorff reagent[220]. The R_F values found using the former solvent were nitrofurazone 0.84, zoalene 0.88, furoxon 0.57, and furnicozone 0.28; the R_F of amproleum using the latter solvent was 0.88.

TABLE 22

IDENTIFICATION OF ADDITIVES ON THIN-LAYER CHROMATOGRAMS

Additive	Fraction[a]	R_F value	Color developed with				Ultraviolet fluorescence
			Reagent A (diaminoethane)	Reagent B (Dragendorff's)	Reagent C (Ehrlich's)	Reagent D (diazotization solutions)	
Acinitrazole	1	0.52	Yellow	–	–	–	Dark spot
Aklomide	1	0.48	–	–	Yellow	Red-purple	Dark spot
Aminonitrothiazole	1	0.32	Red	Red	–	–	Dark spot
APNPS	4	0.40	Yellow	Yellow-brown	Yellow	–	Dark spot
Buquinolate	3	0.47	–	Dull orange	–	–	Violet fluor
Deccox	3	0.53	–	Yellow-orange	–	–	Blue fluor
Dimetridazole	1	0.61	–	Red	Orange	–	Dark spot
DNBA	1	0.40	Red-purple	–	–	Red-purple	Dark spot
Ethopabate	1	0.51	–	–	Dull red	–	Violet fluor
Furazolidone	1	0.45	Dark buff	Yellow	–	–	Brown
Meticlorpindol	4	0.32	–	Dull red	–	–	Dark spot
Nitrofurazone	4	0.25	Red buff	Dull yellow	–	–	Brown
Pyrimethamine	4	0.30	–	Orange	–	–	Violet fluor
Statyl	3	0.37	–	Dull orange	–	–	Blue fluor
Sulfaquinoxaline	4	0.39	–	Dull red	Yellow	–	Dark spot
Zoalene	1	0.37	Purple	–	–	Red-purple	Dark spot

– denotes no reaction.
[a] Fractions collected after passage of additive mixture in chloroform–acetonitrile over alumina column and eluted with chloroform then methanol.

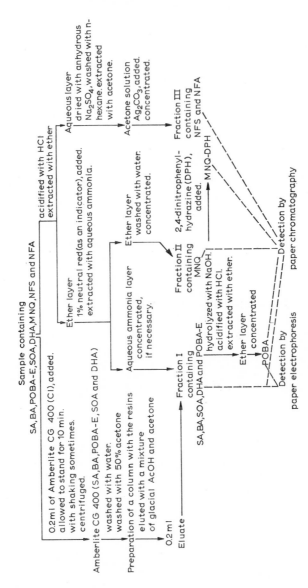

Fig.27. A systematic analysis of food preservatives in food. SA = salicylic acid; BA = benzoic acid; POBA-E = esters of *p*-hydroxybenzoic acid; SOA = sorbic acid; DHA = dehydroacetic acid; MNQ = methylnaphthoquinone; NFS = nitrofurazone; NFA = nitrofuracrylamide.

Komodo and Takeshita[221] utilized column and paper chromatography and electrophoresis to effect the separation of a variety of food preservatives. Figure 27 illustrates the systematic scheme of analysis of food preservatives in food samples. Nitrofurazone (NFS) (R_F 0.57) and nitrofurylacrylamide (NFA) (R_F 0.75) in fraction III were separated by ascending chromatography with n-butanol saturated with water. The spots were detected with UV light and a 10% sodium hydroxide spray. The food preservatives in fraction I were identified by paper electrophoresis with buffer solutions consisting of dimethylformamide–pyridine–acetic acid–water (pH 5.6) or n-butanol–aq. ammonia–acetic acid–water (pH 5.0) and also by paper chromatography using cyclohexanol–ammonium carbonate. The spots were detected with UV

TABLE 23

R_F VALUES OF NITROFURANS ON SILICA GEL AND ALUMINA

Compound		Constitutional formula
Nitrofurazone	(NFS)	O_2N—furan—$CH=N-NH-C(=O)NH_2$
Nitrofurylacrylamide	(NFA)	O_2N—furan—$CH=CH-C(=O)NH_2$
Furazolidone	(NFAO)	O_2N—furan—$CH=N-N-C(=O)O$, CH_2-CH_2
Nitrofurantoin	(NFAH)	O_2N—furan—$CH=N-N(C=O)(NH)$, $CH_2-C=O$
Dihydroxymethyl-furatrizine	(NFAT)	O_2N—furan—$CH=CH$—triazine—$N(CH_2OH)(CH_2OH)$

Compound	R_F		
	Silica gel		Alumina Solvent C[a]
	Solvent A[a]	Solvent B[a]	
NFS	0.19	0.10	0.09
NFA	0.33	0.19	0.40
NFAO	0.36	0.36	0.61
NFAH	0.57	0.36	0.01
NFAT	0.34	0.21	0.11

[a] Solvents: A, ether–acetone (3:2); B, benzene–acetone (3:2); C, acetone.

light and the chromogenic reagents, diazotized sulfanilic acid, diazotized nitroaniline, methyl red or ferric chloride. Methyl naphthoquinone (MNQ) in fraction II was separated by ascending paper chromatography using 95% methanol and detected with UV light and 10% sodium hydroxide.

The TLC separation of 5 nitrofuran derivatives, nitrofurazone (NFS), nitrofurylacrylamide (NFA), furazolidone (NFAO), nitrofurantoin (NFAH), and dihydroxymethylfuratrizine (NFAT), has been described by Komoda and Takeshita[222]. Table 23 lists the R_F values of nitrofurans obtained using silica gel with ether–acetone (3:2) and benzene–acetone (3:2), and aluminum oxide developed with acetone. Detection was accomplished using both UV light and 10% sodium hydroxide.

The gas chromatography of 60 nitrofuran derivatives has been reported by Nakamura et al.[223]. A Barber-Colman Model 10 gas chromatograph was used with 4 ft. × 5 mm columns containing 2% QF-1 on Chromosorb W (80–100 mesh) at 150, 180, and 210°C with an injector temperature of 300°C and the detector temperature the same as the column. The carrier gas was argon at an inlet pressure of 1.0 kg/cm². The detector voltage was 1250 V.

Nitrofurazone determination in feed has also been carried out by chromatography with an aluminum oxide column[224, 225] followed by reaction of the eluate with sodium hydroxide and subsequent spectrophotometric determination.

8. CYCASIN

Cycasin (methylazoxymethyl-β-D-glucopyranosyl) (I) and its aglycone methylazoxymethanol (MAM) (II) are naturally occurring alkylating agents which are found in the seeds, roots and leaves of cycad plants, primarily from the very widespread species *Cycas circinalis* and *Cycas revoluta*. The cycad nuts are a source of starch and are often used as

$$CH_3-N=N-CH_2O-\beta\text{-}D\text{-glucopyranosyl} \qquad CH_3-\overset{+}{N}=N-CH_2OH$$
$$\quad\quad\downarrow \qquad\qquad\qquad\qquad\qquad\qquad\qquad\qquad | $$
$$\quad\quad O \qquad\qquad\qquad\qquad\qquad\qquad\qquad\qquad O^-$$

$$\text{I} \qquad\qquad\qquad\qquad\qquad\qquad\qquad \text{II}$$

food after appropriate preparation. There is a high incidence of human and neurological diseases in areas of the world where cycads are utilized as food and medicines[226], principally in the tropics and subtropics and hence has largely implicated the principal constituents of cycads, cycasin, and MAM.

The hepatoxic and carcinogenic properties of cycasin in rats[227] has been shown to require the prior deglucosylation to MAM provided by the β-glucosidase activity of the intestinal flora[228, 229]. (MAM causes neoplasms after subcutaneous and intraperitoneal injection as well as feeding while cycasin causes tumors in mature animals only if fed[230].) MAM is also teratogenic in the golden hamster[228, 230, 231], and mutagenic in *Drosophila*[232] and *Salmonella typhimurium*[233, 234], induces chromosome aberrations in *Allium* seedlings[235] (the equivalent of 200 R of X-rays). The alkylation of liver RNA and DNA with cycasin and MAM has been described both *in vitro*[236] and *in vivo*[237]. The similarity of biological effects between cycasin and dimethylnitrosamine has been documented[237, 238]. The chemical structure of MAM suggests that it may have a mechanism of action similar to that of dimethylnitrosamine at the molecular level[238, 239]. Figure 28 illustrates the proposed degradation and metabolic pathway for dimethylnitrosamine and cycasin as suggested by Miller[238].

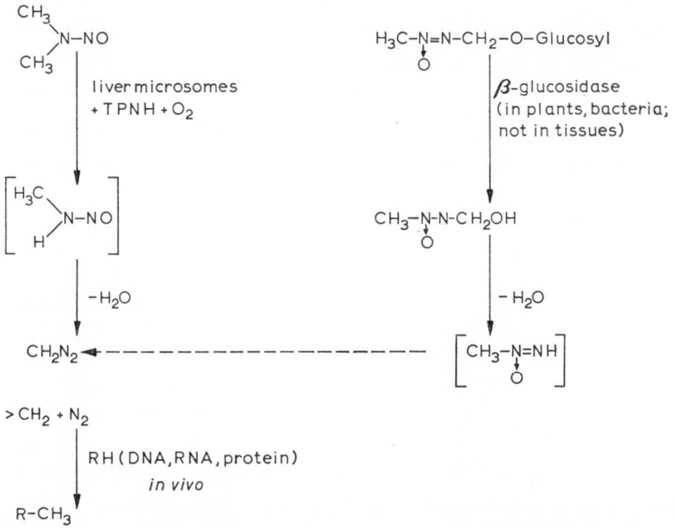

Fig. 28. Proposed degradation and metabolic pathways for dimethylnitrosamine and cycasin.

A variety of chromatographic procedures has been utilized for the separation and identification of cycasins and its aglycone MAM from natural sources as well as biological media.

Riggs[239] described the extraction of cycasin from *Cycas circinalis* L. The seeds were extracted with 80% ethanol, and the concentrate chromatographed on Whatman No. 4 paper developed with 90% aq. acetone (cycasin has an R_F value twice that of glucose). After elution with 50% aq. ethanol, cycasin was purified by partition chromatography on cellulose powder using 80% aqueous acetone. The cycasin thus obtained was found to be identical with that isolated from the seeds of *C. revoluta*[240].

The determination of cycasin in the kernels of *C. circinalis* was described by Dastur and Palekar[241]. The kernels were extracted with 80% methanol and an aliquot of the concentrate chromatographed on Whatman No. 1 paper, and developed with *n*-butanol–acetic acid–water (4:1:1). Detection was achieved with 0.2% naphthoresorcinol in acetone. The spots were eluted in warm 80% ethanol and determined colorimetrically. The values obtained were comparable to those found utilizing the chromotropic method of Matsumoto and Strong[242]. Freshly removed and processed kernels and dried nuts contained 0.96 and 0.55 g of cycasin/100 g of material, respectively.

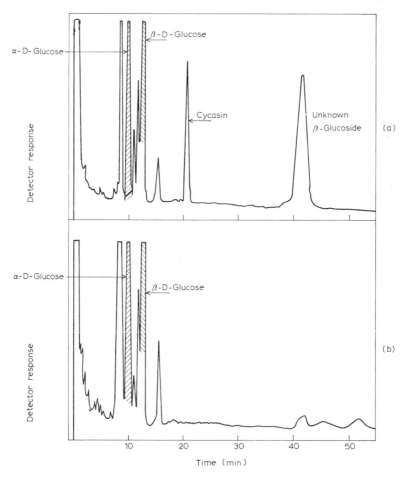

Fig. 29. a, Gas–liquid chromatographic record of trimethylsilylated residue of crude extract of cycad flour; b, the corresponding derivatives of the material from extracts of flour treated with β-glucosidase before gas chromatography. The column was 3 % OV-1 maintained at 230 °C in a model 402 Hewlett-Packard instrument. Other conditions are described in the text.

Matsumoto and Strong[242] described a systematic separation of the constituents of *C. circinalis* nuts coupled with a bioassay for toxicity. Ascending chromatography on Whatman No. 1 paper and development with *n*-butanol–acetic acid–water (4:1:1) or acetone–water (9:1) effected the separation of cycasin and sequoystol (a methyl ether of myoinositol and carbohydrates). Detection was achieved with

1 % ethanolic resorcinol–2 N HCl (1:9) or dipping the chromatogram through an acetone solution of silver nitrate followed by an alcoholic sodium hydroxide spray. The chromatogram was then fixed with 5 % sodium thiosulfate. Cycasin was detected as a yellow spot with resorcinol and both cycasin and sequoystol yielded black spots with the silver nitrate reagent.

The paper chromatographic separation of cycasin from sugars and its subsequent colorimetric and polarographic determination was described by Nishida et al.[243]. The R_F values (using n-butanol–acetic acid–water (4:1:1)) were cycasin 0.36, fructose 0.29, glucose 0.23, and sucrose 0.14.

Kobayashi and Matsumoto[244] described the utility of paper and column chromatography for the isolation and determination of cycasin in rat urine and excreta. Whatman No. 1 paper was used in the descending multiple developing technique of Jeanes et al.[245] using n-butanol–acetic acid–water (4:1:1) and n-butanol–pyridine–water (6:4:3). When cycasin was administered intraperitoneally to rats, it was almost qualitatively excreted unchanged in the urine with no apparent toxicity. Oral administration of cycasin, however, resulted in severe toxicity, and greatly reduced urinary excretion of cycasin.

The utility of TLC for the elaboration of the transplacental passage of cycasin and its aglycone in rats was described by Spatz and Laquer[246]. Rats exposed in utero to cycasin developed tumors in later life and malformations of the central nervous system and extremities were induced in fetuses when MAM was administered i.v. to pregnant hamsters on the eighth day of gestation[231]. Both cycasin and MAM can also pass the mammary gland and are excreted with the milk during lactation provided that they are administered within one day before littering or during lactation. The recovery of MAM from the fetuses of rats and hamsters thus supports a direct action of the carcinogen or teratogen on fetal tissues. The ascending TLC procedures utilized both silica gel Silicic ARTLC76F (Mallinckrodt) and Eastman Chromagram sheets (Type K301R silica gel with fluorescent indicator). The developing solvent in both cases was n-butanol–acetic acid–water (4:1:1). Benzene–carbon tetrachloride (2:1) and ethyl acetate–hexane (3:7) could also be employed[247]. Cycasin and MAM were identified using chromotropic acid reagent and UV fluorescence. Appropriate spots were removed from the plates, extracted with water, the silica gel removed by centrifugation at 20,000–30,000 rev./min in a refrigerated Spinco centrifuge, and a supernatant aliquot checked for azoxy compounds by measuring the changes in the UV absorption spectra at 211, 213, 215, 217, and 275 nm[242]. The maximum absorption occurs at 215 nm with an inflection at 275 nm.

Weiss[248] described the utility of GLC for the elaboration of cycasin samples using the procedure of Highet[249]. A Glowal Model 400 gas chromatograph was used equipped with an argon detector and a 6 ft. × $\frac{1}{8}$ in. column containing 1 % SE-30 on Gas-Chrom P. The column and injection port temperatures were 200 and 240 °C and cycasin was chromatographed as its trimethylsilyl derivative prepared by warming a pyridine solution of cycasin to 80 °C for 2 h in a sealed tube with hexamethyldisilazane and trimethylsilyl chloride. The GLC analysis of cycasin in cycad flour was reported by Wells et al.[250]. The flour was extracted with 70% ethanol and the residue from the dried extract directly trimethylsilylated and analyzed. The instruments employed were (1) a Hewlett-Packard Model 402 with hydrogen-flame detector and 6 ft. × $\frac{1}{8}$ in. glass column packed with 3% OV-1 on Gas-Chrom Z, 80–100 mesh at 230°C. The carrier gas was argon and the retention time of the internal standard androsterone relative to cycasin was 1.58; (2) Hewlett-Packard Model 810 gas chromatograph equipped with a hydrogen-flame detector and a 6 ft. × $\frac{1}{8}$ in. stainless steel column packed with HP-Chromosorb W, 100–120 mesh, coated with 3% OV-1. Helium was the carrier gas and the column temperature was 200°C. Figure 29 depicts gas chromatograms of trimethylsilylated residue of crude extract of cycad flour. The large unknown peak observed in Figure 29(a) which was susceptible to β-glucosidase activity may be a disaccharide.

9. MYCOTOXINS

Toxic mold metabolites (mycotoxins) represent a broad spectrum of biologically active substances that occur as a result of growth of saprophytic (spoilage) molds on various types of feed, food components, and products. Their discovery in agricultural commodities as well as the demonstration in animals of various biological ill effects resulting from the ingestion of contaminated feed has served to illustrate the potential health hazard which might arise from the contamination of the food supply with these metabolites.

9.1 Aflatoxins

Foremost in the above consideration are the aflatoxins which are produced by a limited number of strains of a few fungi, e.g. Aspergillus flavus, A. parasiticus, and Penicillium puberulum. Although the collective term "aflatoxin" was formerly widely used to describe the toxic products of these fungi, it is now well recognized that there are eight compounds of related molecular configuration, aflatoxins B_1, B_2, B_{2a}, G_1, G_2, G_{2a}, M_1, and M_2, as shown in Fig. 30. The biological and biochemical aspects of aflatoxins have been studied extensively[251–256] since they were shown to be the factors that caused "Turkey X disease" which killed 100,000 turkeys in Britain in 1960. The aflatoxins are potent carcinogens in many species[257, 258], e.g., in rats[259–261], ducks[262], rainbow trout[263]. The potency of aflatoxin as an acute toxin is illustrated by its LD_{50} value of 4.7 mg/kg when administered intraperitoneally to male weanling rats; hepatomas are induced in rats by diets containing less than 1 mg/kg (1 p.p.m.) and in rainbow trout by as little as 5 µg/kg[252].

The high order of carcinogenic potency of the aflatoxins has caused considerable speculation[257, 264–266] as to whether they might be involved in the etiology of human liver disease, including primary carcinoma. Wogan[267] has suggested that epidemiologic patterns of primary liver cancer incidence, together with what is known about the risks of aflatoxin contamination of foodstuffs and the potency of the compounds in animals, provide suggestive circumstantial evidence that the aflatoxins (or other mold toxins) may play a role in the etiology of the disease, particularly in certain areas of Asia and Africa.

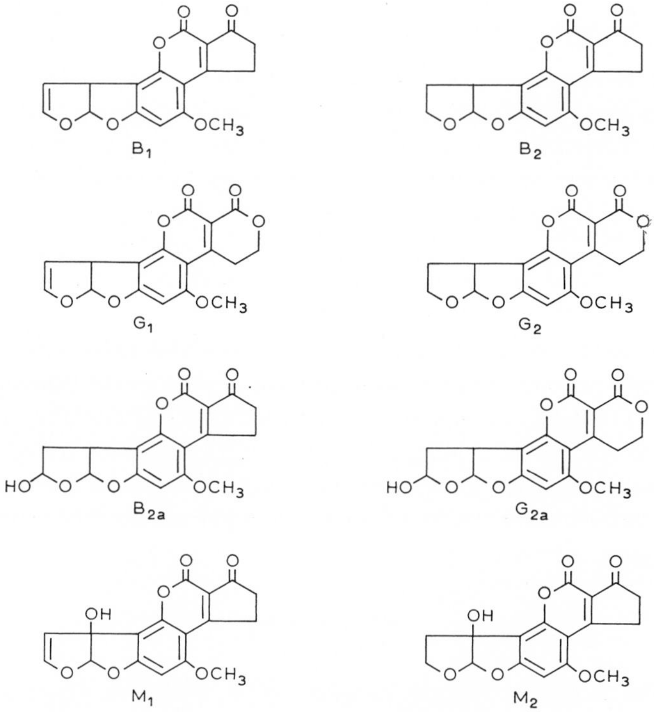

Fig. 30. Structure of the aflatoxins.

The teratogenic effect of aflatoxin B_1 has been demonstrated to be species specific[268] (e.g. teratogenic in hamsters[268, 269], but not in hamsters or mice[268]). Aflatoxin B_1 binds to DNA[270] and inhibits DNA[254, 271-274], RNA[275, 276] and protein synthesis[275, 277], and affects DNA polymerase of E. coli[271]. Aflatoxin B_1 has been shown to induce dominant lethal mutations in mice[278], Neurospora crassa[279], transforming DNA of B. subtilis[273], and induces chromosome aberrations in human leukocytes[280], cell line derived from the rat kangaroo[281], and in seedling roots of Vicia faba[282]. The ability of the aflatoxins to produce carcinogenic, teratogenic and mutagenic response would place the mycotoxin in a select group of compounds such as the alkylating agents which are known to inactivate resting DNA[256]. The inhibition of DNA synthesis and giant cell formation in tissue culture by aflatoxin in a manner similar to some of the alkylating agents has been noted[283].

The chemical, physical and biological properties and isolation of the afla-toxins[252, 284–286] as well as their occurrence in feeds and foods[286, 287] have been reviewed. The metabolism of aflatoxin B_1 has been studied in rats[258, 269, 288, 289], mice[252], sheep[290, 291], and cows[292].

TABLE 24

CONDITIONS PROPOSED FOR THE SEPARATION OF AFLATOXINS BY TLC ON SILICA GEL[286]

Silica gel		Developing solvent	Solvent path (cm)	Ref.
Type	Thickness (μm)			
G	508	Chloroform–methanol (95:5)	10	293
G	300	Chloroform–methanol (98:2)	10	294
G-HR	250	Chloroform–methanol (95:5)	15[a]	295
G-HR	250	Chloroform–methanol (93:7)	12–14[a]	296
G-HR	500	Chloroform–methanol (97:3)	14[a]	297
G-HR	400	Chloroform–methanol–acetic acid (95:4.5:0.5)	10[a]	298
G	250	Chloroform–acetone (9:1)	10–15[a]	299
G-HR	250	Benzene–ethanol–water (46:35:19)	12–14[a]	300
G-HR	250	Chloroform–acetone (9:1)	12–14[b]	301
Adsorbosil	500	Chloroform–acetone–isopropanol (850:125:25)	12–13[b]	302

[a] Lined and unequilibrated tank.
[b] Unlined and unequilibrated tank.

TABLE 25

EFFECT OF DEVELOPMENT CONDITIONS ON R_F VALUES OF AFLATOXINS ON SILICA GEL G-HR

Development solvent	Tank conditions	Silica Gel G-HR thickness (μm)	$R_F \times 100$ of aflatoxins[a]			
			B_1	B_2	G_1	G_2
Chloroform–methanol (97:3)	Lined[303]	500	50	45	40	35
Chloroform–methanol (93:7)	Lined[301]	250	41	37	30	28
Benzene–ethanol–water (46:35:19)	Lined[301]	250	58	53	47	43
Chloroform–acetone (9:1)	Lined[301]	250	36	30	24	20
Chloroform–acetone (9:1)	Unlined[301]	250	70	61	52	44
Chloroform–acetone (85:15)	Lined[304]	500	44	39	64	31
Chloroform–acetone (85:15)	Unlined[304]	500	61	53	46	69
Chloroform–acetone–isopropanol (825:150:25)	Lined[304]	500	54	48	42	37
Chloroform–acetone–isopropanol (825:150:25)	Unlined[304]	500	73	66	59	52

[a] 12–13-cm solvent path development.

The utility of chromatography procedures (primarily TLC) for the extract purification and separation of the aflatoxins as well as the elaboration of their metabolism
has been well documented. Table 24 summarizes the conditions proposed for the
separation of aflatoxins by TLC on silica gel. Table 25 depicts the effect of development conditions on R_F values of aflatoxins on silica gel. The apparent solvent systems
of choice for the isolation of the aflatoxins are chloroform–acetone or chloroform–
acetone–isopropanol mixtures. It is also of importance to note the dependence of
R_F values on variations in gel properties, particularly relative to humidity, and em-
phasizes the requirement of chromatographing authentic aflatoxin standards con-
comitantly with the unknowns.

The detection and semi-quantitative estimation of the aflatoxins has generally
been accomplished by visual estimation based on the fluorescence exhibited by the
aflatoxins under long-wave ultraviolet illumination. However, even under ideal conditions, there are recognized problems concerning the precision of visual measurements. For example, calculations of the estimation errors in four visual aflatoxin
procedures[294, 297, 305] would indicate a possible measurement error of ± 30–50%
when a given unknown is judged to match either one of two adjacent standards and
± 15–25% when the unknown is interpolated between two standards[306].

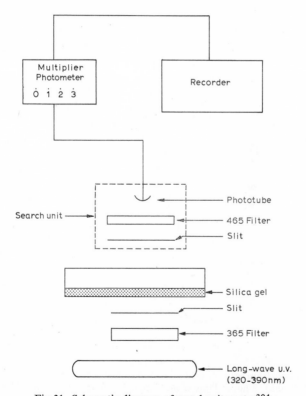

Fig. 31. Schematic diagram of TLC densitometer[304].

The accuracy and precision of aflatoxin measurements have been greatly improved by the use of more sensitive fluorodensitometric measurements of the aflatoxins directly on silica gel-coated plates[304, 307]. The basic system used for these measurements is illustrated in the schematic diagram of a TLC densitometer as shown in Fig. 31. The precision attainable with this technique was found to be markedly superior to that of visual estimation. With an aflatoxin standard containing B_1, B_2, G_1, and G_2 resolved on eight TLC plates, precision estimates ranged from ±4–10% for individual aflatoxins and $\pm6\%$ for total aflatoxins[304], and it was emphasized that the use of authentic reference standards of each aflatoxin is essential for fluorodensitometric measurements. Figure 32 illustrates recorder traces from scans of aflatoxins on TLC plates developed in two solutions (chloroform–acetone (85:15) and chloroform–acetone–isopropanol (825:150:25), respectively). Table 26 illus-

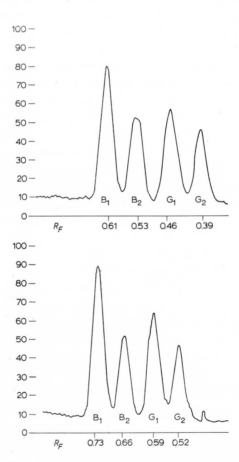

Fig. 32. Recorder traces from scans of aflatoxins on TLC plates developed in two solvents in unlined and unequilibrated tanks. Source: Pons *et al.* Top, chloroform–acetone (85:15); bottom, chloroform–acetone–2-propanol (825:150:25), 60 min.

TABLE 26

FLUORESCENCE PROPERTIES OF AFLATOXINS

Measurement conditions	Emission maxima ($m\mu m$)		Relative order	Relative intensity
	B_1-B_2	G_1-G_2		
Methanol[a]	425	450	$G_2 > B_2 > G_1 > B_1$	13.0:8.0:5.0:1
Methanol[b]	430	450	$G_2 > B_2 > G_1 > B_1$	14.5:8.8:1.7:1
Ethanol[b]	430	450	$G_2 > B_2 > G_1 > B_1$	4.7:2.7:1.4:1
Chloroform[b]	413	430	$G_2 > G_1 > B_2 > B_1$	34.0:31.0:1.3:1
Solid state[c, d]	427–432	450–455	$B_2 > G_2 > B_1 > G_1$	3.3:3.0:1.4:1

[a] Data from ref. 308, excitation max., 365 nm.
[b] Data from ref. 309, excitation max., 365 nm.
[c] Data from ref. 304, Silica Gel G-HR.
[d] Data from refs. 304, 310, excitation max., 368–375 nm.

trates the fluorescence properties of the aflatoxins and shows that both the order and the relative fluorescence intensities differ in solution and in the solid state on silica gel.

Van Duuren[311] described the use of the characteristic fluorescence and phosphorescence excitation and emission spectra of aflatoxins B_1 and G_1 in solution or adsorbed on potassium bromide as a sensitive test for the identification of the aflatoxins at concentrations as low as 10^{-2}–10^{-4} μg/ml. Table 27 illustrates the luminescence characteristics of aflatoxins B_1 and G_1 and Table 28 the limits of detection of aflatoxins B_1 and G_1 by both fluorometry and TLC with visible fluorescence.

A number of confirmatory tests have been devised to identify aflatoxin B_1 in unknown extracts[312, 313]. Portions of isolated toxin were treated with (a) formic acid–thionyl chloride; (b) acetic acid–thionyl chloride, and (c) trifluoroacetic acid, respectively[311]. Aliquots of the three fluorescent reaction products as well as the untreated standard were then chromatographed on Silica Gel G-HR plates developed with methanol–chloroform (5:95) and the spots then compared under UV light. The basis of the above confirmatory tests for aflatoxin B_1 depends on the capacity of the olefinic linkage of an enol ether moiety to react additively with a hydroxyl group under the catalytic influence of a strong acid[314]. This adduct formation is a general reaction that requires anhydrous conditions and is accompanied by few, if any, side reactions. The OR group which is introduced is invariably bonded to the ether end of the double bond and the hydrogen atom is attached to the other end as shown.

$$\text{—O—CH=CH—} + \text{ROH} \xrightarrow{\text{H}^+} \begin{array}{cc} \text{RO} & \text{H} \\ | & | \\ \text{—O—CH} & \text{—CH—} \end{array}$$

Aflatoxin contains such an enol ether unit in the furano ring and this unsaturation is isolated from the main chromophore of the compound. Hence, addition reactions at

TABLE 27

LUMINESCENCE CHARACTERISTICS OF AFLATOXINS B_1 AND G_1[311]

Measurement	Medium[a]	B_1		G_1	
		Excitation (nm)	Emission (nm)	Excitation (nm)	Emission (nm)
Φ Fluorescence, 23 $\pm$ 0.5°	Methanol (1)	363	426	363	450
Φ Fluorescence, 77°K	Ethanol–methanol (4:1) (2)	365	408	365	423
Φ Fluorescence, 23 $\pm$ 0.5°	Potassium bromide (3)	363[b]	495[b]	380[c]	520[c]
Phosphorescence, 77°K	Ethanol–methanol (4:1) (2)	365	498, 474	365	517 489 (sh), 481
Phosphorescence, decay time	Ethanol–methanol (4:1) (2)	$r_1 < 1$ Mc/sec, $r_2 = 1.2$ sec (excitation 365 nm; emission 498 nm)		$r_1 < 1$ Mc/sec, $r_2 = 2.0$ sec (excitation 365 nm; emission 516 nm)	
Φ Fluorescence, 23 $\pm$ 0.5°	Methanol (4)	$\Phi = 0.078$ (excitation 363 nm)		$\Phi = 0.24$ (excitation 363 nm)	

[a] Concentrations. B_1: (1), $3.1 \times 10^{-5}\,M$; (2), $2.5 \times 10^{-5}\,M$; (3), 65 µg/200 mg KBr; (4), $1.8 \times 10^{-6}\,M$. G_1: (1), $2.9 \times 10^{-5}\,M$; (2), $2.4 \times 10^{-5}\,M$; (3), 491 µg/200 mg KBr; (4), $1.7 \times 10^{-6}\,M$.
[b] Primary filter 7-39; secondary filter 3-75.
[c] Primary filter 7-51; secondary filter 3-75.

TABLE 28

LIMITS OF DETECTION OF AFLATOXINS B_1 AND G_1

	B_1	G_1
By fluorometry (in methanol)	$7.2 \times 10^{-8}\,M$ (2.2×10^{-2} µg/ml) with excitation at 264 nm and emission at 426 nm.	$1.7 \times 10^{-9}\,M$ (5.5×10^{-4} µg/ml) with excitation at 264 nm and emission at 450 nm.
By TLC[a] visible fluorescence	Blue fluorescence: 0.2 µg/spot. Noncharacteristic pale-white fluorescence: 0.01 µg/spot.	Green fluorescence: 0.1 µg/spot. Noncharacteristic pale-white fluorescence: 0.01 µg/spot.

[a] TLC on Silica Gel G (Merck) with chloroform–methanol (97:3) as developer. R_F values aflatoxin B_1 and G_1 were 0.52 and 0.46, respectively.

this isolated double bond were believed not to alter the fluorescence spectrum of the original metabolite significantly, but should alter its chromatographic behavior. The reaction product of B_1 with acetic acid–thionyl chloride exhibits two new intense fluorescent spots at the R_F of B_1–G_2. The reaction products of B_1 with both formic acid–thionyl chloride and trifluoroacetic acid exhibit a single intense fluorescent spot at approximately 10% of the R_F of B_1.

The acid-catalyzed addition of water to the vinyl ether double bond of aflatoxin B_1 was investigated by Pohland et al.[315]. The hemiacetal product was biologically inactive to the chick embryo and to tissue cultures at respective concentrations 80 and 50 times the minimum lethal dose of aflatoxin B_1. The structure of the hemiacetal was established and found to be identical to the hemiacetal reported as an intermediate in the total synthesis of racemic aflatoxin B_1.

Under mild acid conditions, aflatoxin B_1 is transformed into a hemiacetal (aflatoxin B_{2a}). The formation of this hemiacetal is regarded as a confirmatory test for aflatoxin B_1[312].

TLC techniques utilizing MN Silica Gel G-HR (Brinkmann) were used both for isolation and the elaboration of homogeneity of aflatoxin B_1 hemiacetal as well as its respective acetate derivative. For example, the acetylation of the hemiacetal with acetic anhydride in pyridine yielded the corresponding acetate which on TLC (silica gel, with ethanol–chloroform (19:93)) was found to be a mixture of the epimeric acetates (R_F 0.85 and 0.88). Aflatoxin B_1 hemiacetal showed only one spot (R_F 0.08) with methanol–chloroform (7:93) on a silica gel thin-layer chromatogram.

The use of oximes and 2,4-dinitrophenyl hydrazones of aflatoxins B_1 and B_2 were also suggested as confirmatory tests[316, 317]. Aflatoxins G_1 and G_2 do not form analogous derivatives.

The effects of prolonged contact of the aflatoxins with silica gel and exposure to UV light have been elaborated by Andrellos et al.[318]. It was found that irradiation of aflatoxins B_1 and G_1 with UV light (principal wavelength $\sim$365 nm) converted both compounds to new fluorescent photoproducts which have much lower R_F values than their respective precursors when chromatographed on Silica Gel G-HR and developed with methanol–chloroform (5:95) or methanol–chloroform (3:97). Exposure of aflatoxin B_1 spots on Silica Gel G-HR plates to irradiation yielded three additional blue-fluorescent spots. Aflatoxin G_1 underwent similar photochemical destruction on Silica Gel G-HR with the production of a similar series of new slow-moving spots of a blue-green fluorescence. Unlike the original aflatoxin, the principal photoproduct produced on silica gel plates were completely converted to non-fluorescent substance(s) when it was treated with strong acids, e.g. formic acid thionyl chloride, 3 N hydrochloric acid. Reaction with acetic anhydride in pyridine transformed the photoproduct into a new blue-fluorescent substance with an R_F of about 0.60 and an UV absorption maximum of 362 nm in neutral and in ammoniacal methanol solutions. These reactions were thus suggested to be diagnostic tests for the principal photoproduct from aflatoxin B_1.

Since fluorescence under UV irradiation is the basis for monitoring aflatoxins and making assay comparisons, there are a number of obvious implications. It was suggested that the presence of a significant amount of photoproduct in the aflatoxin B_1 recovered from preparatory thin-layer plates could lead to a negative bioassay (chick embryo)[319] because in this bioassay the photoproduct has proved to be non-toxic. The photoproduct could also lead to negative results in the test reactions for confirming identity[319] because strongly acidic media destroy the fluorescence of the photoproduct without giving rise to fluorescent reaction products. In the isolation, purification and crystallization of aflatoxins, photochemical destruction of the afla-toxins on silica gel could seriously impair the overall yield.

Although TLC on silica gel with chloroform-containing solvent mixtures has been the most widely used chromatographic technique for aflatoxin isolation, other absorbents and developers as well as paper and partition column chromatography have been suggested. The applicability of polyamide film (containing polyethylene terephthalate and polycaprolactam resin 1011) and polyamide-coated glass plates compared with Silica Gel H (Merck) and Silica Gel DO (Camag) was studied by Lee and Ling[320]. The developing solvents were 2% methanolic chloroform for the silica gel plates and benzene–cyclohexane–acetic acid–chloroform (5:2:2:1) for the poly-amide film and plates. Ascending development was carried out at 28 °C in the dark for periods ranging from 25–90 min and UV light (Black Ray-B-100A long-wave) used for detection. The limit of detection of aflatoxin by both polyamide-impregnated film and polyamide-coated plates was $6 \times 10^{-3} \, \mu g$ compared with $4 \times 10^{-4} \, \mu g$ detected on silica gel plates.

Kwon and Ayres[321] investigated the utility of anti-oxidants and chelating agents in the TLC of aflatoxin G_1. The purification of aflatoxin G_1 by standard TLC proce-dures resulted in the separation of 10 additional modified toxins which separated as fluorescent bands. The use of BHT (2,6-di-*tert.*-butylphenol) and EDTA in devel-oping and eluting solvents and EDTA in the preparation of the TLC plates plus the operation of the entire procedure in a dark room effectively prevented the above undesirable changes. MN Silica Gel G-HR (Brinkman) absorbent was used with 3% methanol in chloroform as developer and the toxins identified on the plates by com-parison with a standard toxin mixture, thence removed from the plate, extracted with chloroform by centrifugation, concentrated and the identity reconfirmed by UV absorption.

Partition-column chromatography of either crude or partially purified primary extracts provides a more efficient means of separation than liquid–liquid partition systems previously utilized in the extraction of aflatoxins from peanut[295, 322, 323] and cotton[303] samples. The partition-column chromatography of a methanol–water primary extract of peanut products on diatomaceous earth (Celite), then eluting inter-fering lipids and pigments with hexane and aflatoxins with a chloroform–hexane (1:1) has been described by Nesheim[296, 305]. The treatment of an acetone–hexane–water primary extract of peanut materials on an acid-treated Florisil adsorbent fol-lowed by elution of interfering materials with tetrahydrofuran and the aflatoxins with acetate has been reported[298]. The column chromatographic treatment of par-tially purified aqueous acetone extracts of cottonseed, peanuts and other agricultural products on (*1*) silica gel (Merck, 0.05–0.20 mm) followed by elution with diethyl ether to remove interfering fluorescent pigments, and with chloroform–methanol (97:3) for the elution of aflatoxins[297], and (2) a less active silica gel (Mallinckrodt CC-7, 100–120 mesh) with ether–hexane (3:1) and chloroform–acetone (8:2) as elution solvents[302] has also been described. Partition chromatogaphy on a cellulose column was suggested[300] for the further purification of partially purified aqueous acetone extracts of cottonseed products with the interfering pigments being eluted with hexane and the aflatoxins with hexane–chloroform (1:1).

The metabolic fate of the aflatoxins is of prime interest in regard to elaborating the biochemical mechanisms underlying their biological effects and pragmatically in determining the extent to which edible animal tissues or products become contaminated with aflatoxins or toxic aflatoxin derivatives when animals consume rations containing the compounds.

Early investigations by Carnaghan[308] and de Iongh[324] revealed that extracts of milk from cows fed aflatoxin-containing peanut meal could induce liver lesions in ducklings similar to those induced by aflatoxin itself. Based on evidence obtained by TLC and fluorescence analysis as well as bioassay, it was shown that the toxic product ("milk toxin") was a metabolic product of aflatoxin B_1[324].

Falk et al.[325] studied the metabolism of aflatoxin B_1 in the rat following intravenous (i.v.) administration in order to elaborate whether aflatoxin was eliminated unchanged in the bile and urine or whether its metabolism gave rise to products of harmless, toxic or even enhanced toxic properties. Fluorescence intensity in bile samples was found to rise rapidly to a maximum in 20–30 min followed by a rapid return to baselines in 120 min (about 60% of the fluorescence was found in the bile and the rest in the urine). In the bile, seven fluorescent spots were detected by PC utilizing n-butanol–acetic acid–water (4:1:5) for single development and in addition n-butanol–pyridine–water (3:2:2) for two-dimensional development. A strongly fluorescent spot, probably unchanged aflatoxin, appeared and waned during the first 20–30 min while several other spots, presumed to be conjugated metabolites of aflatoxin, increased in intensity at different rates with time. It was also found that the sensitivity of the fluorescent metabolites to UV light is an experimental concern in quantitative and qualitative studies and that stringent precautions are required to preclude artifactuous results.

The distribution and excretion of ^{14}C-labeled aflatoxin in the rat (following a single intraperitoneal dose of either methoxy-labeled or ring-labeled aflatoxin B_1) has been studied by Shank and Wogan[326] and Wogan et al.[288]. The results indicated certain significant differences by which ^{14}C derived from the two labeled forms are excreted and also provided an indication of one quantitatively important pathway by which aflatoxin is metabolized. Total excretion of radioactivity, through urine, feces, and CO_2 from both labeled forms of the compound amounted to 70–80% of the administered dose during the 24 h following dosing. Whereas 60% of administered radioactivity from the ring-labeled compound appeared in feces, only 22% of the methoxy-^{14}C was excreted by this route. The difference can be accounted for by the larger amount of $^{14}CO_2$ excreted by animals dosed with the methoxy-labeled compound (e.g. 27% compared to about 0.5% of the ring-labeled-^{14}C which appeared in CO_2). Thus O-demethylation appears to be a major pathway in the degradative metabolism of aflatoxin because of the significant amount of administered radioactivity found as $^{14}CO_2$. Most of the radioactivity of the ring-labeled compound was recovered in the excreta, with practically none appearing as exhaled CO_2, which suggests that the ring structures are metabolically stable. The relatively large amounts of radioactivity found in the feces and intestinal contents suggest the major role of biliary excretion. This view is supported by the studies of Falk et al.[325] as shown above. The chemical nature of the excreted metabolites and of the residual compounds has not as yet been elaborated. Minor amounts of aflatoxin appear to be excreted unchanged, but appear in the urine and the feces with greatly altered solubility and chromatographic properties.

Bassir and Osiyemi[289] investigated the biliary excretion in the rat after a single intraperitoneal injection of 50 µg of ^{14}C-aflatoxin B_1 with a specific activity of 30 mµC/mmole. It was found that aflatoxin was rapidly excreted through the bile (30% of dose excreted in the first 6 h) as was indicated earlier by the observations of Falk et al.[325]. Chloroform extracts of bile samples were assayed on Silica Gel G chromatoplates and developed with acetic anhydride–methanol–chloroform (1:2:97). The extracts contained a mixture of aflatoxin B_1 and aflatoxin M_1; the principle constituent was a conjugate compound which gave a positive ninhydrin reaction on PC and was alkali labile. On alkaline hydrolysis of this conjugate, it yielded two fractions, one of which was identifiable with taurocholate by TLC, while the other had a slightly lower R_F value than aflatoxin B_1 and gave a blue fluorescence in UV light. This last compound may have been a degradation product of aflatoxin B_1. Analysis of urine samples from the treated rat showed that approximately 20% of the radioactivity was excreted as a glucuronide conjugate in the first 6 h after dosing.

Studies of the metabolism in sheep[290] have led to the elaboration of the "milk toxin" described earlier. Following a single dose of mixed aflatoxins (B_1, B_2, G_1, and G_2) at a level of 1 mg/kg, chromatographic examination of extracts of liver, kidney, and urine revealed the presence of aflatoxins B_1 and G_1 and a fluorescent substance with chromatographic and spectroscopic properties identical with "milk toxin" isolated from toxic cow's milk. The liver contained a higher proportion of unaltered aflatoxins B_1 and G_1, whereas the metabolite was present in larger relative amounts in kidney and urine. Allcroft et al.[290] suggested the name aflatoxin M for this isolated urinary metabolite.

Holzapfel et al.[291] repeated the experiments of Allcroft et al.[290] and elaborated the metabolism of aflatoxin following i.p. administration to sheep. Aflatoxin M was obtained from urine following extraction with chloroform, column chromatography on silica gel, thence elution with 2% methanol in chloroform. TLC employing the systems of Allcroft et al.[290] confirmed the presence of aflatoxin M. A pure concentrate of aflatoxin M was obtained by chromatography on Silica Gel G (Merck) using chloroform–methanol (97:3) as developing solvent. Chromatography of the concentrate on Whatman No.1 paper impregnated with formamide–water (85:15) using ethyl acetate–benzene (9:1) as mobile phase gave two components, a blue-violet fluorescent substance R_F 0.34 designated aflatoxin M_1 and a violet fluorescent substance R_F 0.23 designated aflatoxin M_2 in the ratios of approximately 3:1.

In an ancillary study, Holzapfel et al.[291] elaborated the structure of aflatoxins isolated from moldy peanuts. In addition to aflatoxins B_1, B_2, G_1, and G_2, aflatoxins M_1 and M_2 were isolated and found to be identical (PC and TLC, UV, IR, and MS) with the analogous aflatoxins isolated from urine in their studies above. Structural studies of aflatoxins M_1 and M_2 including NMR and MS led to the conclusion that aflatoxin M_1 is hydroxyaflatoxin B_1 (I) and aflatoxin M_2 is identical with dihydroaflatoxin M_1 (II).

Aflatoxin M_1 in extracts of infected peanuts could be detected by TLC on Silica Gel G (Merck) employing chloroform–methanol (97:3) as the developing solvent. It could be assayed quantitatively by comparing the intensity of the fluorescent spot at R_F 0.4 with that of standard aflatoxin M_1. The fluorescence of aflatoxins M_1 and M_2 was approximately three times the intensity of aflatoxin B_1.

The isolation of crystalline aflatoxin M_1 from sheep urine and cow's milk after a dose of mixed aflatoxins has been recently reported by Masri et al.[292]. Silica gel chromatography of urinary and milk extracts yielded crystalline aflatoxin M_1 following prior solvent extraction and column development with chloroform and methanol–chloroform eluants. The identity of aflatoxin M_1 from urine and milk was established on the basis of melting point (310°C), UV spectra in methanol (λ_{max} 357, 265, and 226 nm), formation of acetyl derivatives which were separable by TLC from the unreacted material, IR spectra (KBr) similar to that of aflatoxin B_1 but showing a hydroxyl signal at 3430 cm^{-1} and proton magnetic resonance spectra (taken at 100 Mc/sec in dimethylformamide-d_7 (DMF-d_7)). In addition, the authors have tentatively identified aflatoxin M_1 (by TLC, UV spectrum, and acetyl derivative formation) in the inoculated mice used in their experiments suggesting the possibility that the appearance of aflatoxin M_1 in urine and milk was related to its presence preformed in the concentrates of aflatoxin B_1 rather than exclusively to its de novo synthesis from B_1. An estimate of aflatoxin M_1 in the dry milk appeared to be only 2–3% of the ingested dose of B_1, and the excretion of M_1 in the milk is completed within a few hours of ingestion of a B_1 concentrate. (The question of preferential excretion of aflatoxin M_1, if present in extracts of A. flavus into milk, had been raised

References pp. 421–429

by de Iongh et al.[324].) Aflatoxin B_1 itself was identified (TLC, UV spectrum, and formation of characteristic derivatives[327] in the milk) and is estimated at 190 p.p.b. in the dried milk representing about 0.3% of the ingested dose of aflatoxin B_1.

The analysis of aflatoxin M_1 in dried milk has been studied by Masri et al.[292]. The TLC procedures used two types of adsorbents, Silica Gel G-HR or a mixture of Silica Gel HR and anhydrous $CaSO_4$ (1:1) prepared to a thickness of 0.5 mm, air-dried 1 h and activated at 110°C for 1 h. The latter adsorbent greatly enhanced the separation of aflatoxin M_1 by substantially retarding an interfering substance with an R_F value only slightly lower than that of aflatoxin M_1 on Silica Gel G-HR plates. The developing solvent for aflatoxin M_1 was 5–6% methanol in chloroform. While this solvent was also acceptable for TLC of aflatoxin B_1 thus permitting simultaneous estimation of aflatoxins M_1 and B_1, superior separation of aflatoxin B_1 was obtained on Silica Gel G-HR plates developed with 3% methanol in chloroform or 10% acetone in chloroform. Aflatoxin M_1 was quantitatively measured spectrophotometrically on eluted thin-layer bands by measurement of the absorption at 357 nm.

Roberts and Allcroft[328] described a semi-quantitative estimation of aflatoxin M_1 in liquid milk by TLC. Acetone was used for precipitation of protein followed by partitioning in methanol extraction with petroleum ether to remove the fat, thence extraction of the aflatoxin from the methanol solution with chloroform. The chloroform concentrate and a standard aflatoxin B_1 solution were chromatographed on Silica Gel G (Merck) plates (250 µ), developed with methanol–chloroform (3:97) for about 30 min at room temperature and examined under UV light of wavelength 365 nm. Aflatoxin M_1 gives a blue-violet fluorescent spot at an R_F value of 0.25 and standard aflatoxin B_1 yields a similar blue-violet fluorescent spot at R_F 0.5. The fluorescence of aflatoxin M_1 is approximately three times as intense as that of aflatoxin B_1 (ref. 10), and this factor of three was used in calculating the amount of aflatoxin M_1 in milk samples. The amount of aflatoxin M_1 excreted in milk bears a linear relationship to the amount of aflatoxin B_1 ingested[329]. The lowest daily intake of aflatoxin B_1 which gave an identifiable amount of aflatoxin M_1 in the milk of a 500 kg cow was found to be in the order of 0.6–0.9 mg. At this intake, the aflatoxin content of the milk is about 0.0003 µg/ml which is considered by Roberts and Allcroft[328] to be about the lower limit of detection with the method described above. Results obtained for samples of liquid milk containing various levels of aflatoxin M_1 have shown good agreement with values obtained on dried samples of the same milk examined by the method of Purchase and Steyn[330] in which a Soxhlet extraction with an azeotropic mixture of acetone–chloroform–methanol (30:47:23) was used[331].

Aflatoxin M_1 has been recently detected by Campbell et al.[332] in human urine from subjects known to have consumed peanut butter contaminated with aflatoxin. All urine samples were extracted twice with one quarter volumes of chloroform and combined, concentrated, and chromatographed on Adsorbosil-1 (0.25 mm) plates with 10% acetone in chloroform at 65–70°F and a relative humidity of 20–40%. The identity of aflatoxin M_1 was confirmed by chromatography with an M_1 standard n four additional solvents, (*a*) 100% ethyl acetate, (*b*) the benzene-rich phase of benzene–ethanol–water (46:35:19), (*c*) 3% methanol in chloroform, and (*d*) diethyl ether–chloroform–acetic acid (40:40:20). The thionyl chloride test of Andrellos and Reid[312] with both the isolated urinary aflatoxin M_1 and a reference standard, yielded identical and single derivatives with each of the acid catalysts, trifluoracetic acid R_F 0.18, glacial acetic acid R_F 0.30, and formic acid R_F 0.30. The R_F of the unreacted M_1 was 0.33. The developing solvent was solvent (*b*), above.

The isolation, structure and biochemical properties of the aflatoxins B_{2a} and G_{2a} were described by Dutton and Heathcote[333]. Both of the hydroxy aflatoxins were isolated from a toxin-producing strain of *A. flavus* (C.M.I. 91019B) cultured on potato-dextrose agar and both column and TLC on silica gel were used for their isolation from concentrated culture fluids. Aflatoxins B_{2a} and G_{2a} were found to be much less toxic than B_1, B_2, G_1, and G_2 when tested on day-old Khaki Campbell ducklings. It has been shown that treatment of aflatoxin B_1 with cold dilute aqueous mineral acid produces aflatoxin B_{2a} and analogously aflatoxin G_{2a} arises from G_1[333].

It was suggested by Dutton and Heathcote[333] that an acid treatment of aflatoxin-contaminated feedstuffs could be of value in lowering their toxicity substantially since B_1 and G_1 are by far the major components produced by the mold.

The structures of aflatoxin B as well as dihydroaflatoxin B and tetrahydrodesoxyaflatoxin B have been elaborated by van Dorp *et al.*[334]. Column, thin-layer, and gas–liquid chromatography as well as IR and MS were utilized to elaborate the reduction and oxidation products of aflatoxin B. For example, the reaction mixture from the *prolonged* catalytic reduction of aflatoxin B in glacial acetic acid over palladium on charcoal (total uptake 3 moles of hydrogen in 12–16 h) was chromatographed on both thin layers of Silica Gel G (Merck) (developed with chloroform–2% methanol) and columns of Silica Gel G and Hyflo (eluted with chloroform–1% ethanol) and shown to contain tetrahydrodesoxyaflatoxin B (R_F 0.7 by TLC above).

In contrast with aflatoxin B_1, it was found possible to analyze the tetrahydrodesoxyaflatoxin by GLC[335]. A Pye argon chromatograph with an Ra D (^{210}Pb) β-ray ionization detector was used. The glass column of 115×0.4 cm i.d. was packed with acid washed and silanized 80–100 mesh Celite 545, coated with 2.5% silicone oil MS 550. Both column and detector temperatures were 220°C and the carrier gas was argon at a flow rate of 30 ml/min. Employing these parameters, the retention time of tetrahydrodesoxyaflatoxin expressed as carbon member for an *n*-fatty acid methyl ester was 26.8[335].

9.2 Ochratoxins

Although aflatoxin is probably the most common toxic metabolite, it is only one of a great number of toxins known to be produced by fungi. For example, in South Africa it was found that the substrates from 46 strains of 22 species of molds were lethal when fed to ducklings[336]. Strains of *A. ochraceus*, a ubiquitous storage mold, is often found in soil and on decaying vegetation, and has been shown to infect stored wheat with moisture content above 16%[337] and has been frequently recovered in small amounts from South African cereal and legume crops (*e.g.* wheat, rye, sorghum, rice, buckwheat, soybeans, and peanuts)[338]. The isolation and characterization of ochratoxin A (a toxic metabolite produced by *A. ochraceus* Wilh.) has been described by van der Merwe[338] and shown to be 7-carboxy-5-chloro-8-hydroxy-3,4-dihydro-3R-methylisocoumarin linked over its 7-carboxy group to L-β-phenylalanine (Fig. 33).

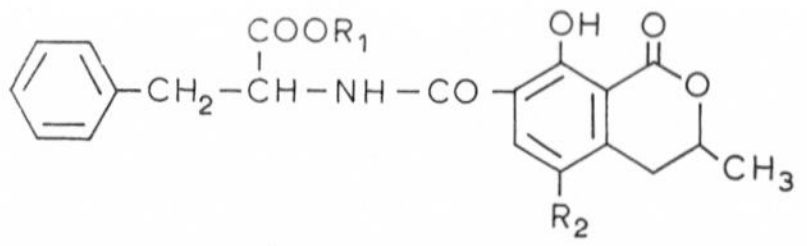

Ochratoxin A, R_1 = H R_2 = Cl
Ochratoxin B, R_1 = H R_2 = H
Ochratoxin C, R_1 = C_2H_5 R_2 = Cl

Fig. 33. Structures of the ochratoxins.

Toxicological aspects of ochratoxin A[339], its inhibition of mitochondrial respiration[340] and effect[341] on glycogen storage in the rat liver have been cited. The toxicity of ochratoxin A in ducklings is of the same order as that of aflatoxin B_1 (*e.g.* LD_{50}, 25 µg recorded 6 days after administration and calculated for a body weight of 50 g). The constitution of ochratoxins B and C metabolites of *A. ochraceus* Wilh. was also described by van der Merwe and Steyn[342]. Figure 33 illustrates the structures of the ochratoxins.

Ochratoxins B and C are respectively the dechloro derivative and the ethyl ester of the A compound, both are nontoxic, which suggests that the toxicity of ochratoxin A is dependent upon the combined presence of the chlorine atom and free carboxyl group.

A. ochraceus Wilh. (strain K-804) was grown in bulk on sterilized wet maize meal, the dried moldy meal extracted with chloroform–methanol (1:1) and the solvent removed *in vacuo*. The toxic extract in chloroform was extracted with 0.5 *M* aq. sodium bicarbonate and the acidified aqueous phase extracted with chloroform and the isolated acids chromatographed in acidic silica gel. Elution with benzene–chloroform (3:1) yielded a colorless band showing a bright green fluorescence under uv light and containing all the toxicity. Paper chromatography using 1-propanol–3 *N* aq. ammonium carbonate (3:1) and TLC on silica gel using benzene–methanol–acetic acid (12:2:1) of this band revealed the presence of several fluorescent components which were quantitatively separated by ion-exchange chromatography on Dowex formate columns. The ochratoxins which were thus isolated were further characterized on paper and thin-layer chromatograms. Ochratoxins A, B, and C appeared under uv light as bright blue fluorescent spots (R_F 0.65, 0.68, and 0.8, respectively) on Whatman No. 3M paper chromatograms using 1-propanol–3 N aq. ammonium carbonate (3:1) as developing solvent, and as bright green fluorescent spots (R_F 0.57, 0.55, and 0.73, respectively) on silica gel thin-layer plates using benzene–methanol–acetic acid (12:2:1) as developer.

The isolation and identification of the methyl and ethyl esters of ochratoxins A and B, respectively, from fungal cultures of $A.$ $ochraceus$ has been reported by Steyn and Holzapfel[343]. The esters possess a toxicity similar to that of ochratoxin A and function as cotoxins. Extraction of a maize culture of $A.$ $ochraceus$ with chloroform–methanol (1:2) or of liquid media with chloroform yielded a toxic concentrate which was washed with water and extracted with 0.5 M sodium bicarbonate and the aqueous phase acidified with 1 N HCl and extracted with chloroform to give ochratoxins A and B, penicillic acid, and hydroxyaspergillic acid.

Preparative silica gel chromatoplates developed with benzene–acetic acid (25:1) were used for the separation of components from the $A.$ $ochraceus.$ Two fluorescent bands which were separated, one light green (band 1) R_F 0.55 and the other blue (band 2) R_F 0.30 were both re-chromatographed on formamide-impregnated Whatman No. 1 paper using hexane as the mobile phase. Band 1 (from TLC above) was resolved into two main components by paper chromatography, $e.g.$ ochratoxin A ethyl ester (R_F 0.50, green) and ochratoxin A methyl ester (R_F 0.30, green). Band 2 (from TLC above) was also separated into two main components by paper chromatography, $e.g.$ ochratoxin B ethyl ester (R_F 0.18, blue) and ochratoxin B methyl ester (R_F 0.10, blue). The structures of the isolated compounds were proven by esterification of ochratoxin B, whereas mild alkaline hydrolysis of compounds IIA and B gave ochratoxin B (the ochratoxin esters were not considered to be artefacts since only chloroform was used in the extraction of liquid media). In general, the yield of ochratoxin A was 0.50–0.75 g/10 kg of moldy maize meal. Under similar conditions, the yield of the ochratoxin A esters was 0.25–0.40 g.

Scott and Hand[344] described a method for the detection and semi-quantitative estimation of ochratoxin A in flour and other cereal products which could be used in conjunction with analysis of the foodstuff for aflatoxins. The sample was extracted with aqueous methanol and n-hexane and the toxin partitioned on a Celite column. Ochratoxin A was separated by TLC on Silica Gel G-HR (activated at 80°C for 2 h) and developed with toluene–ethyl acetate–90% formic acid (5:4:1) in a 13–15-cm development. Detection was accomplished by simultaneous observation under both short- and long-wave UV light. In the above solvent system, the aflatoxins migrate less than ochratoxin A (R_F 0.7); the distances moved relative to ochratoxin A were: R_F values B_1 0.79, B_2 0.75, G_1 0.72, and G_2 0.62. The presence of ochratoxin A was confirmed by respotting the sample and standard on the Silica Gel G-HR plate and developing with freshly prepared benzene–methanol–acetic acid (24:2:1). Ochratoxin A has an R_F value of 0.6 in this system; R_F values of aflatoxins relative to ochratoxin A were B_1 0.74, B_2 0.63, G_1 0.52, and G_2 0.43.

The fate of ochratoxin A in rats has been studied by Nel and Purchase[345].

Following i.p. administration, ochratoxin A was found to be metabolized to an iso-coumarin (III) without a phenylalanine residue and a second unidentified green fluorescent metabolite.

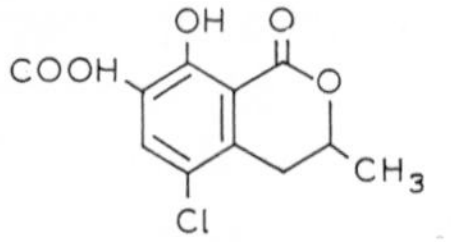

III

The fluorescent compounds found in the urine were compared with those from the feces on thin-layer chromatoplates using the solvent systems toluene–ethyl acetate–formic acid (5:4:1), benzene–methanol–acetic acid (12:2:1), and benzene–acetic acid (8:2) and were found to be the same. The total extracts from urine and feces were combined and the metabolite III isolated by TLC was shown to be the dihydroisocoumarin derivative by UV, MS and mixed melting point when compared with a reference sample obtained by the acid hydrolysis of ochratoxin. In the rat, ochratoxin produces enteritis, renal necrosis and an increase in the quantity of glycogen in the liver.

9.3 Sterigmatocystin

Sterigmatocystin is a mycotoxin that has gained increasing significance as a food or feed contaminant during recent years. Sterigmatocystin (IV) (R = H) originally characterized as a metabolite of *A. versicolor*[346] is of interest because of its structural affinities with the aflatoxins. It has been produced in high yields from *A. nidulans* and *Bipolaris*. Sterigmatocystin has been shown to be carcinogenic in the rat[347, 348] and inhibits RNA synthesis[349]. It is of interest to note that

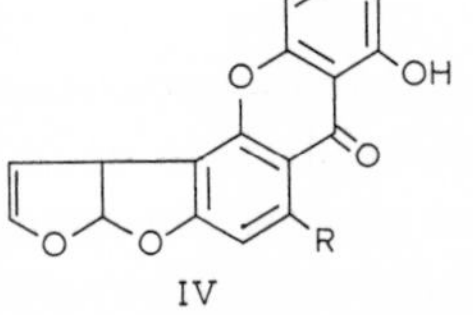

R = H = sterigmatocystin
R = OCH$_3$ = O-methyl–sterigmatocystin

IV

the male rats were more susceptible to hepatoma development than females and that histologically the tumors bore a marked resemblance to those occurring in man in South Africa[348].

Vorster and Purchase[350] described a sensitive method for the quantitative determination of sterigmatocystin in grain and oilseeds by TLC. Detection of as little as 0.0025 µg is achieved by conversion of sterigmatocystin into the monoacetate (IV) which possesses an intense light blue color on thin-layer chromatoplates when viewed under UV light. Thin-layer plates of Silica Gel D-5 (Camag) and activated for 2 h at 105°C were found to resolve the acetylated extract of maize and sorghum so clearly into its components that it suggested the possibility of densitometric evaluation of the chromatograms by the procedure of Pons *et al.*[304]. The thin-layer plates were developed with chloroform containing 0.75% ethanol as stabilizer and 2% methanol. When viewed under a UV source such as HPW 125W, Type 57202E/70, Philip's lamp (peak emission about 360 nm), the presence of sterigmatocystin was revealed by a light blue fluorescent spot at R_F 0.6. Aflatoxins do not interfere in the detection of sterigmatocystin acetate, since they are more polar and hence barely move on a thin-layer plate developed with chloroform containing 2% methanol.

A derivative of sterigmatocystin, O-methyl-sterigmatocystin (IV, R = OCH_3) has been isolated from the aflatoxin-containing fraction of *A. flavus* (cycad strain)[351], by column chromatography on aluminum oxide using heptane–chloroform (7:3) then chloroform–methanol (19:1). O-methyl-sterigmatocystin has structural similarity with aflatoxin B_1 and G_1 with regard to the dihydrodifurano ring system. A possible diketo configuration in the remainder of the molecule is blocked by the dimethyl ether configuration.

9.4 Aspertoxin

The mycotoxin aspertoxin (V) is a hydroxy derivative of O-methyl sterigmatocystin and has been obtained from aflatoxin-producing culture of the same mold *(A. flavus)*[352, 353]. Aspertoxin also bears a close resemblance to aflatoxin M_1 (I).

The previously cited structural resemblance of aflatoxin M_1 to sterigmatocystin and the fact that aflatoxins and sterigmatocystin are found in the same genus led to the early postulation that sterigmatocystin or a precursor of it may be a biosynthetic precursor of the aflatoxins[354]. The isolation of both O-methyl sterigmatocystin and aspertoxin from *A. flavus* cultures offers support for the above hypothesis.

Aspertoxin has been isolated following the initial column chromatography of a crude precipitate of *A. flavus* on acid alumina, elution with benzene–chloroform (1:1), yielding aflatoxins B_1, G_1 and B_2; aflatoxins G_2 and M_2 were removed with chloroform–acetone (9:1).

Aspertoxin was obtained after concentrating the eluate containing aflatoxins G_2 and M_2. Aspertoxin was originally believed to be aflatoxin M_1, because of an equivalence of R_F values (R_F 0.32) on silica gel chromatoplates when chloroform–acetone (9:1) was used as the developing solvent. Development of the plates with chloroform–pyridine (9:1) or chloroform–acetic acid (9:1), however, provided criteria for distinguishing the two compounds. In each case aspertoxin moved at an R_F of 0.55–0.60 whereas aflatoxin appeared at an R_F of 0.40. The crude precipitates of *A. flavus* extracts usually contain less than 0.5% aspertoxin by weight, whereas aflatoxin B_1 usually accounts for about 30%.

Aspertoxin has pronounced toxic effects on the developing chick embryo. Injection of 2.0 µg/egg into either the yolk or the air cell of fresh fertile eggs killed 100% of the embryos; a concentration of 0.7 µg/egg killed 50% of the embryos (aflatoxin B_1 has an LD_{50} air cell route of 0.25 µg/egg)[355].

9.5 Zearalenone

The toxin zearalenone, 6-(10-hydroxy-6-oxo-*trans*-1-undecenyl)-β-resorcylic acid lactone (VI), (estrogenic factor F-2) is produced by *Fusarium graminearum*[356]. (The *Fusaria* often attack wheat, barley and maize, many species being actively toxigenic.) Zearalenone is anabolic and uterotrophic in rats, mice, and guinea pigs, and is suspected of contributing to infertility in dairy cattle and swine.

VI

Mirocha *et al.*[357] described the isolation and identification of zearalenone produced by *F. graminearum* in stored corn. The extracts of the biologically active components (as well as authentic zearalenone samples) were examined by TLC on Silica Gel E (Merck) and developed with a solution of 5% ethyl alcohol in chloroform. The fluorescence was checked at the same R_F value (0.5) as the standard after illumination with an UV lamp. (Zearalenone fluoresces with maximum at 450 nm when excited by UV light of a wavelength of 310 nm.) GLC of zearalenone was accomplished *via* the respective tri- or dimethylsilyl ether derivatives. The trimethylsilyl (TMS) ether derivative was prepared by reacting 0.1 ml of pyridine, 0.1 ml of chlorotrimethyl silane and 0.2 ml of hexamethyl disilazane with 0.1 ml of a zearalenone extract and allowing 30 min for reaction before analysis. An Aerograph Hy-Fi gas chromatograph was used equipped with a hydrogen-flame detector and a 5 ft. × ⅛ in. stainless steel column packed with 5% SE-30 on 60–80 mesh Chromosorb W. The column and injection block temperatures were 238 and 280°C, respectively; the helium carrier gas and hydrogen flow rates were each 25 ml/min. The retention times of the TMS ether of zearalenone at column temperatures of 238 and 261°C were 12 and 2.5 min, respectively, and the limit of sensitivity was 0.4 µg.

TABLE 29

TLC CHARACTERIZATION OF CONSTITUENTS OF *F. graminearum* GROWN ON RICE SUBSTRATE

Constituent	R_F	Spectrum[a] (nm)	GLC[b]	DNPH[c]	H_2SO_4[d]
F-5-0, (red)	0.00	236,274	3	Negative	Blue
F-5-1A, (yellow)	0.06	236,274	3	Negative	Blue
F-5-1B, (fluoresces)	0.07	236,274	4	Negative	Pink
F-5-2, (fluoresces)	0.15	314,274,236	4	Negative	None
F-5-3, (fluoresces)	0.24	314,274,236	3	Negative	None
F-5-4, (fluoresces)	0.32	314,274,236	3	Negative	Yellow
F 5-5, (fluoresces)	0.50	314,274,236	2	Negative	None
F-5-6, (fluoresces)	0.60	314,274,236	5	Negative	None
F-5-7, (fluoresces)	0.67	314,274,236	3	Yellow	Yellow
F-2, (fluoresces)	0.75	314,274,236	1	Yellow	Green
F-3, (fluoresces)	0.80	274,236	1		

[a] Absorption maxima in ethanol.
[b] Number of constituents as determined by GLC.
[c] Reactivity with 2,4-dinitrophenylhydrazine.
[d] Color reaction with 50% H_2SO_4–methanol.

Mirocha *et al.*[358] described studies involving five other estrogenic compounds produced by *F. graminearum* growing on rice and corn. Table 29 depicts the characterization of the various constituents of a methylene chloride extract of *F. graminearum* after growth on a rice substrate and resolution by TLC. Only one of these constituents (F-2) (zearalenone) has been identified. Components F-5-4 and F-5-5 were found to react with N,O-bis(trimethylsilyl) acetamide to form TMS ethers, analyzed by GLC and were found to possess a retention time identical to that of F-3. An Aerograph Model 1520 gas chromatograph was used equipped with a hydrogen-flame detector and a 10 ft. $\times \frac{1}{8}$ in. stainless steel column containing 4% SE-30 on Chromosorb W. The column, injection port, and detector temperatures were 265, 280, and 265°C, respectively and the carrier gas was nitrogen at a flow rate of 30 ml/min.

Compounds similar in structure, but perhaps not in biological activity, to the estrogenic factor (F-2) have been reported before, *e.g.* curvularin, produced by the fungi *Curvularia* sp., *P. steckii* Zaleski, and *P. expansum* Link[359], and radicol produced by *Nectria radicicola*[360]. Radicol differs from F-2 (zearalenone) in conjugation and position of the ketonic function in the side chain as well as the presence of an epoxy group in the undecenyl side chain. It is identical with the antibiotic monorden (VII) produced by the fungus *Monosporium bonorden*[361].

VII

9.6 *Miscellaneous Fusarium toxins*

The mold *Fusarium nivale* is found on tall fescue, and important forage crop that has occasionally become toxic to ruminants. Yates *et al.*[362] isolated a butenolide D1-4-acetamido-4-hydroxy-2-butyric acid-γ-lactone (VIII) from *F. nivale* Fries cesati. Buckardt *et al.*[363] reported the synthesis of VIII *via* the reaction of 4-bromo-4-hydroxy-2-butenoic acid-γ-lactone and acetamide in chloroform, *viz.*

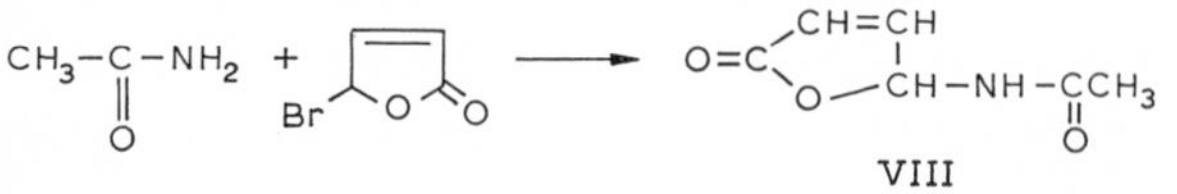

VIII

TLC on Silica Gel G (Merck) was used for the identification of reaction products. The components were detected by spraying with a mixture of 1 ml anisaldehyde, 10 ml acetic acid, 85 ml methanol, and 5 ml conc. sulfuric acid, followed by heating the chromatoplates for 10–20 min at 140°C. The fungus *F. nivale* has been extensively studied in Japan because of its damage to the wheat crop and the subsequent findings of human and animal poisonings due to "scabby grains" resulting from its infestation. A number of sesquiterpenoids have been isolated from wheat and rice, including the scirpenol derivatives nivalenol (3α, 4β, 7α, 15-tetrahydroxy-scirp-9-ene-8-one) (IX), fusarenon-X (3, 17, 15-trihydroxy-scirp-4-acetoxy-9-ene-8-one) (X), and T-2-toxin (4, 15-diacetoxy-8-(3-methyl-butyryl-oxy)-scirp-9-en-3-ol) (XI). Figure 34 illustrates the structures of the toxic scirpenols.

Tatsuno *et al.*[364, 365] described the isolation and detection of nivalenol from moldy rice infected with *F. nivale* using column, thin-layer, and gas–liquid chromatography. Silica Gel G chromatoplates were used with chloroform–methanol (5:1) for

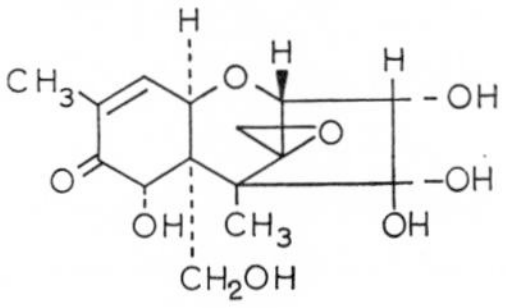

Nivalenol

IX

4 mg/kg i.p.

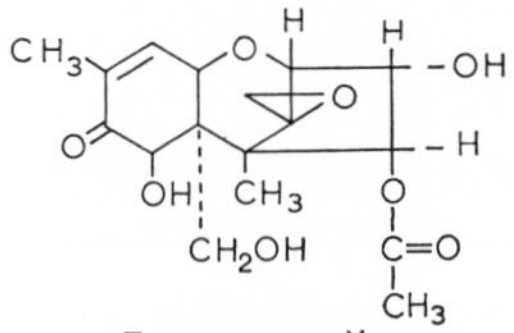

Fusarenon-X

X

3.3 mg/kg i.p.

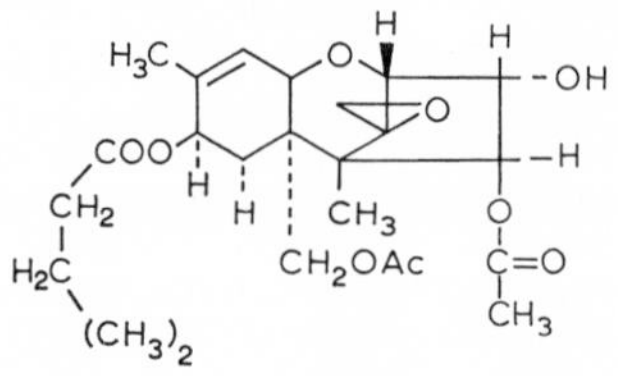

T-2 Toxin

XI

Fig. 34. Structures of the toxic scirpenols from *F. nivale*.

development following initial extraction and fractionation of the toxic principle on charcoal columns. The pure toxin on TLC had an R_F of 0.45 and gave a brown spot color with ammoniacal silver nitrate and sulfuric acid reagents and pink-violet with α-naphthol–sulfuric acid. The pure toxin exhibited a weak UV absorption spectrum at 220 nm (in methanol) and 260 nm (in acetonitrile). Nivalenol was chromatographed as its TMS derivative on an 1.5% OV-17 on Gas Chrom P column. Nivalenol has been found to inhibit completely DNA synthesis of HeLa cells at 5γ/ml[366] and ascites tumor[367] as well as inhibit the protein synthesis of rabbit reticulocytes[368] and cause cell degeneration of bone marrow, lymph nodes, intestines, testes, and thymus following i.p. administration to mice and induce radiomimetic damage in animal cells.

Fusarenon-X (X) has been isolated from a culture broth of *F. nivale*[369] and has been found to have marked cytotoxic properties in mice[369] and in a reticulocyte bioassay[370]. Following column chromatographic (charcoal) isolation in the methanol–chloroform fraction, the crude toxin was chromatographed on silica gel using chloroform–methanol (97:3 to 5:1) for development yielding a highly toxic fraction which eluted before the nivalenol fraction. Re-chromatography on silica gel yielded a pure product when developed with chloroform–acetone (5:1). Fusarenon-X gave a single spot at R_F 0.89 (chloroform–methanol) (5:1), 0.19 (chloroform–methanol) (97:3), or 0.37 (ethylacetate–*n*-hexane) on TLC with Silica Gel G. Fusarenon-X inhibits protein synthesis in rabbit reticulocytes and in Ehrlich ascites tumor[369] analogously as shown for nivalenol.

The structure of T-2 toxin (XI) has been established by Bamburg *et al.*[371]. The toxic effects on trout, rats and mice of T-2 toxin (produced by *F. tricinctum*) have been described by Marasas *et al.*[372].

9.7 Miscellaneous Penicillium toxins

Species of *Penicillium* have been identified as frequent contaminants of grain and feeds. *P. puberulum* Banier, for example, has yielded among its metabolic products aflatoxin[373], tropolone, puberulic and puberulonic acids[374], and penicillic acids[375].

P. cyclopium, which has a worldwide distribution, is frequently isolated from stored grain and cereals. Holzapfel[376] found that strains recovered from samples of a variety of South African products caused acute toxicoses in ducklings and rats and showed that the main toxic agent was the new compound cyclopiazonic acid (XII).

XII

Cyclopiazonic acid was isolated from cultures of a strain of *Penicillium cyclo-pium* Westling by chromatography on formamide impregnated cellulose and ion-exchange (Dowex-I) columns and preparative TLC on Silica Gel G.

Rice infected with the mold *Penicillium islandicum* Sopp. has been found to yield two toxic metabolites, *viz.* luteoskyrin (XIII), a yellow pigment bisdihydro-anthraquinone derivative, and islanditoxin (XIV), a chlorine-containing cyclopeptide.

Both luteoskyrin and islanditoxin are potent hepatotoxic agents in mice and rats[377, 378]. The chemical determination and isolation of islanditoxin has been de-scribed by Ishikawa *et al.*[379]. TLC of the peptide was accomplished on Silica Gel G

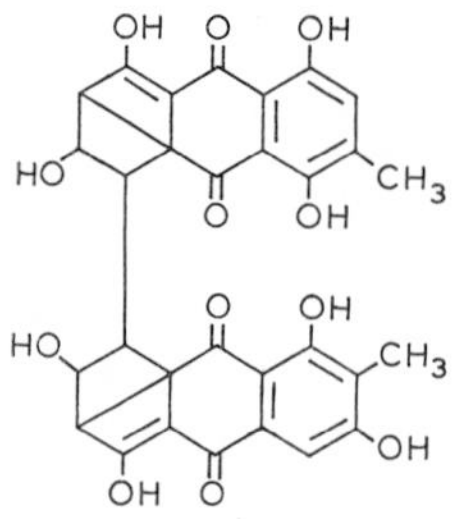

XIII

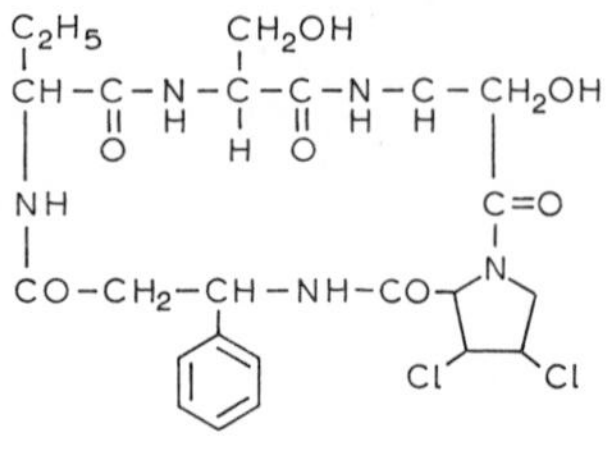

XIV

TABLE 30

THE SEPARATION OF MYCOTOXINS ON 0.25 mm LAYER OF SILICA GEL G IMPREGNATED
WITH OXALIC ACID

Mycotoxin	Fungus	R_F value ($\times 100$)	Fluorescence	Color reagents[a]	
				H_2SO_4	$FeCl_3$
Aspertoxin	*A. flavus*	12	Light yellow	Green-yellow	
Ochratoxin B	*A. ochraceus*	20	Blue		Red-brown
Secalonic acid D	*P. oxalicum*	23	Dark	Light brown	Light brown
8α-(3-methylbutyryl-oxy)-4β, 15-diacet-oxyscirp-9-en-3α-ol	*F. tricinctum*	28		Lead grey	
Aflatoxin G_1	*A. flavus*	30	Green	Green-grey	
Aflatoxin B_1	*A. flavus*	40	Blue	Green-grey	
6β-Hydroxyrosenono-lactone	*T. roseum*	44		Orange-red	
Ochratoxin A	*A. ochraceus*	48	Green		Red-brown
Cyclopiazonic acid	*P. cyclopium*	65	Dark	Red-brown	Red-brown
Zearalenone	*F. graminearum*	72	Faint blue	Light yellow	Red-brown
Sterigmatocystin	*A. nidulans* *A. versicolor* *Bipolaris* sp.	85	Orange	Green-grey	Green

[a] Color reagents: conc. sulfuric acid (after spraying, the plate was heated at 110°C for 10 min); 1% ethanolic ferric chloride.

using n-butanol–acetic acid–water (4:1:4) as developer. When the plate was sprayed with $2 N$ H_2SO_4 solution and heated at 100°C, the peptide gave a brown spot at R_F 0.7.

The evidence and the ability of various ubiquitous fungi, e.g. A. flavus and P. islandicum to elaborate potent carcinogens has, as stated earlier, prompted theories of a possible relationship between the diseases of unknown etiology (e.g. the high incidence of hepatocarcinogenicity in Africa[265]) and the consumption of mycotoxins. Rapid and sensitive analytical methods are thus essential for the detection of these hazardous agents both in agricultural commodities and consumer products. The separation and detection of a number of mycotoxins by TLC has been reported by Steyn[380]. The separations were performed on Silica Gel G (Merck) (prepared with 0.4 N oxalic acid in a 1:2 ratio), developed with chloroform–methanol (4:1). The spots were detected by exposure of the plates to longwave (366 nm) UV illumination and spraying with conc. sulfuric acid and 1% ethanolic ferric chloride. Table 30 depicts the separation of mycotoxins on 0.25 mm layers of Silica Gel G impregnated with oxalic acid and spot colors obtained.

Another suitable chromogenic reagent for the mycotoxins is 1% ceric sulfate in 6 N sulfuric acid. Some compounds screened in the above study gave a characteristic color with a specific reagent, e.g. cyclopiazonic acid gave a violet color on spraying with Ehrlich reagent, and was also found to turn violet-red on prolonged standing on silica gel plates impregnated with oxalic acid.

REFERENCES

1 J.M.Price, C.G.Biava, B.L.Oser, E.E.Vogin, J.Steinfeld and H.L.Ley, Science, 167 (1970) 1131.
2 R.O.Egeberg, J.L.Steinfeld, I.Frantz, G.C.Griffith, H.Knowles, Jr., E.Rosenow, H.Sebrell and T.van Itallie, J. Am. Med. Assoc., 211 (1970) 1358.
3 G.Rudali, E.Coezy and I.Muranyi-Kovacs, Compt. Rend., D269 (1969) 1911.
4 F.J.C.Roe, L.S.Levy and R.L.Carter, Food Cosmet. Toxicol., 8 (1970) 135.
5 R.Tanaka, J. Public Health Assoc. Japan., 11 (1964) 909.
6 D.Lorke, Arzneimittel-Forsch., 19 (1969) 920.
7 C.Klotzsche, Arzneimittel-Forsch., 19 (1969) 925.
8 K.Sax and H.J.Sax, Japan. J. Genet., 43 (1968) 89.
9 D.Stone, E.Lamson, Y.S.Chang and K.W.Pickering, Science, 164 (1969) 568.
10 E.Bajusz, Nature, 223 (1969) 407.
11 K.Hagmüller, H.Hellauer, R.Winkler and J.Zangger, Wien. Klin. Wochschr., 81 (1969) 927.
12 L.F.Dalderup and W.Visser, Nature, 221 (1969) 91.
13 O.G.Fitzhugh, A.A.Nelson and J.P.Frawley, J. Am. Pharm. Assoc. Sci. Ed., 11 (1951) 583.
14 P.O.Nees and P.H.Derse, Nature, 81 (1965) 81.
15 P.O.Nees and P.H.Derse, Calcium Cyclamate Feeding Study, Wisconsin Alumni Research Foundation, Madison, Wisconsin, 1965.
16 J.D.Taylor, R.K.Richards, R.G.Wiegano and M.S.Weinberg Food Cosmet. Toxicol., 6 (1968) 313.

17	J.D.Taylor, R.K.Richards and J.C.Davin, *Proc. Soc. Exptl. Biol. Med.*, 78 (1951) 530.
18	J.P.Miller, L.E.M.Crawford, R.C.Sonders and E.V.Cardinal, *Biochem. Biophys. Res. Commun.*, 25 (1966) 153.
19	R.C.Sonders and R.G.Wiegand, *Toxicol. Appl. Pharmacol.*, 12 (1968) 291.
20	S.Kojima and H.Ichibagase, *Chem. Pharm. Bull. (Tokyo)*, 16 (1968) 1851.
21	S.Kojima and H.Ichibagase, *Chem. Pharm. Bull. (Tokyo)*, 14 (1966) 971.
22	J.S.Leahy, M.Wakefield and T.Taylor, *Food Cosmet. Toxicol.*, 5 (1967) 447.
23	J.S.Leahy, T.Taylor and C.J.Rudd, *Food Cosmet. Toxicol.*, 5 (1967) 595.
24	R.C.Sonders and R.C.Wiegand, *Toxicol. Appl. Pharmacol.*, 11 (1968) 13.
25	*Food Additives and Contaminants Committee, Second Report on Cyclamates*, Ministry of Agriculture, Fisheries and Food, H.M.S.O., London, 1967.
26	S.Kojima and H.Ichibagase, *Chem. Pharm. Bull.*, 17 (1969) 2620.
27	L.Goldberg, C.Parekh, A.Patti and K.Soike, *Toxicol. Appl. Pharmacol.*, 14 (1969) 654.
28	B.L.Oser, S.Carson, E.E.Vogin and R.C.Sonders, *Nature*, 220 (1968) 178.
29	R.Higuchi, *J. Food Hyg. Soc. Japan*, 6 (1966) 448.
30	M.L.Richardson and P.E.Luton, *Analyst*, 91 (1966) 520.
31	K.Maruyama and K.Kawanabe, *J. Food Hyg. Soc. Japan*, 4 (1963) 265.
32	P.H.Derse and R.J.Daun, *J. Assoc. Offic. Anal. Chemists*, 49 (1966) 1090.
33	R.C.Sonders, R.G.Wiegand and J.C.Netnal, *J. Assoc. Offic. Anal. Chemists*, 51 (1968) 136.
34	D.I.Rees, *Analyst*, 90 (1965) 568.
35	K.M.Beck, *Food Technol.*, 11 (1957) 156.
36	S.Kato, T.Kaneko and A.Tanimura, *Shokuhin Eiseigaku Zasshi*, 11 (1970) 98.
37	A.Alter and J.C.Forman, *J. Labelled Compds.*, 4 (1969) 320.
38	D.K.Das, T.V.Matthew and S.N.Mitra, *J. Chromatog.*, 52 (1970) 354.
39	L.C.Mitchell, *J. Assoc. Offic. Agr. Chemists*, 38 (1955) 943.
40	I.S.Ko, I.S.Chung and Y.H.Park, *Rept. Natl. Chem. Lab. (Korea)*, 3 (1959) 72; *Chem. Abstr.*, 54 (1960) 10181L.
41	T.Komoda and R.Takeshita, *Shokuhin Eiseigaku Zasshi*, 3 (1961) 382; *Chem. Abstr.*, 60 (1964) 6130F.
42	D.K.Das and T.V.Matthew, *Inst. Chemists (India)*, 41 (1969) 192.
43	K.Asano, M.Taira, H.Nakanishi, E.Senda, Y.Shiraishi and R.Takeshita, *Nichidai Igaku Zasshi*, 22 (1963) 797; *Chem. Abstr.*, 61 (1964) 8813L.
44	H.Yamaguchi, *Nippon Kagaku Zasshi*, 82 (1961) 486; *Chem. Abstr.*, 56 (1962) 9927C.
45	M.Ichikawa, S.Kojima and H.Ichibagase, *Yakugaku Zasshi*, 84 (1964) 563; *Chem. Abstr.*, 61 (1964) 10045A.
46	S.Kojima and H.Ichibagase, *Yakugaku Zasshi*, 83 (1963) 1108; *Chem. Abstr.*, 60 (1964) 15060f.
47	D.E.Johnson and H.B.Nunn, *J. Assoc. Offic. Anal. Chemists*, 51 (1968) 1274.
48	E.Bradford and R.E.Weston, *Analyst*, 94 (1969) 68.
49	A.Vercillo and A.Manzone, *Rend. Ist. Super. Sanita*, 24 (1961) 644.
50	J.B.Wilson, *J. Assoc. Offic. Agr. Chemists*, 38 (1955) 559.
51	W.S.Cox, *J. Assoc. Offic. Agr. Chemists*, 35 (1952) 321.
52	A.G.Renwick and R.T.Williams, *Biochem. J.*, 114 (1969) 78P.
53	B.A.Becker and J.E.Gibson, *Toxicol. Appl. Pharmacol.*, 17 (1970) 551.
54	M.Legator, *Med. World News*, 9 (1968) 25.
55	M.Legator, K.A.Palmer, S.Green and K.W.Petersen, *Science*, 165 (1969) 1139.
56	D.R.Stoltz, K.S.Khera, R.Bendall and S.W.Gunner, *Science*, 167 (1970) 1501.
57	W.L.Sutton, *Ind. Hyg. Toxicol.*, 2 (1967) 2058.
58	G.V.Lomonova, *Federation Proc.*, 24 (1965) T96.
59	F.S.Mallette and E. von Haam, *Arch. Ind. Hyg. Occup. Med.*, 5 (1952) 311.
60	K.Kikuchi and T.Touchi, *Life Sci.*, 8 (1969) 843.
61	I.P.Lee and R.L.Dixon, *Toxicol. Appl. Pharmacol.*, 14 (1969) 654.
62	H.I.Yamamura, I.P.Lee and R.L.Dixon, *J. Pharm. Sci.*, 57 (1968) 1132.
63	T.H.Elliott, N.Y.Lee-Yoong and R.C.C.Tao, *Biochem. J.*, 109 (1968) 11P.
64	A.G.Blumberg and A.M.Heaton, *J. Chromatog.*, 48 (1970) 565.

65 R. E. WESTON AND B. B. WHEALS, *Analyst*, 95 (1970) 680.
66 M. J. DE FAUBERT MAUNDER, H. EGAN AND J. ROBURN, *Analyst*, 89 (1964) 157.
67 S. W. GUNNER AND R. C. O'BRIEN, *J. Assoc. Offic. Anal. Chemists*, 52 (1969) 1200.
68 M. H. LITCHFIELD AND T. GREEN, *Analyst*, 95 (1970) 168.
69 J. HOWARD, T. FAZIO AND R. WHITE, *J. Assoc. Offic. Anal. Chemists*, 52 (1969) 492.
70 T. FAZIO AND J. W. HOWARD, *J. Assoc. Offic. Anal. Chemists*, 53 (1970) 701.
71 T. MOVIS, private communication in M. L. RICHARDSON, *Talanta*, 14 (1967) 385.
72 T. S. CARSWELL AND H. L. MORRILL, *Ind. Eng. Chem.*, 29 (1937) 1247.
73 J. W. HOWARD, T. FAZIO, B. K. WILLIAMS AND R. H. WHITE, *J. Assoc. Offic. Anal. Chemists*, 52 (1969) 1197.
74 F. BAR AND F. GRIEPENTROG, *Naturwissenschaften*, 45 (1958) 390.
75 F. GRIEPENTROG, *Arzneimittel-Forsch.*, 9 (1959) 123.
76 R. NANIKAWA, S. KOTOKU AND T. YAMADA, *Japan. J. Legal Med.*, 21 (1967) 17.
77 M. AKAGI, I. AOKI AND T. UEMATSU, *Chem. Pharm. Bull. (Tokyo)*, 14 (1966) 1.
78 M. AKAGI, I. AOKI, T. UEMATSU AND T. IYANAKI, *Chem. Pharm. Bull. (Tokyo)*, 14 (1966)10.
79 T. SASAKI, Z. IIKURA AND T. YOKOTSUKA, *Chomi Kagaku*, 16 (1969) 6.
80 W. KAMP, *Pharm. Weekblad*, 101 (1966) 57.
81 D. WALDI, in E. STAHL (Ed.), *Thin-Layer Chromatography*, Academic Press, New York, 1965, p. 365.
82 S. C. LEE, *Hua Hsueh*, [3] (1966) 117; *Chem. Abstr.*, 67 (1967) 42602M.
83 E. LUDWIG AND U. FREIMUTH, *Nahrung*, 9 (1965) 569.
84 T. SALO AND U. SALMINEN, *Suomen Kemistilehti A*, 37 (1964) 161.
85 T. KORBELAK AND J. N. BARTLETT, *J. Chromatog.*, 41 (1969) 124.
86 T. KORBELAK, *J. Assoc. Offic. Anal. Chemists*, 52 (1969) 489.
87 K. NAGASAWA, H. YOSHIDOME AND K. ANRYU, *J. Chromatog.*, 52 (1970) 173.
88 S. KOJIMA AND H. ICHIBAGASE, *Yakuzaigaku*, 26 (1966) 115.
89 R. TAKESHITA, Y. SAKAGAMI AND T. YAMASHITA, *Eisei Kagaku*, 15 (1969) 66.
90 P. LANGER AND V. ŠTOLC, *Z. Physiol. Chem.*, 335 (1964) 216.
91 P. LANGER AND V. ŠTOLC, *Endocrinology*, 76 (1965) 151.
92 P. M. JENNER, E. C. HAGAN, J. M. TAYLOR, E. C. COOK AND O. G. FITZHUGH, *Food Cosmet. Toxicol.*, 2 (1964) 327.
93 K. HORAKOVA, L. DROBNICA, P. NEMEC, K. ANTOS AND P. KRAISTIAN, *Neoplasma*, 15 (1968) 160.
94 C. AUERBACH, *Biol. Rev. Cambridge Phil. Soc.*, 24 (1949) 355.
95 C. AUERBACH, *Publ. Staz. Zool. Napoli.*, 22 (Suppl. 1–2) (1950).
96 N. FRIES, *Physiol. Plantarum*, 1 (1948) 330.
97 C. AUERBACH AND J. M. ROBSON, *Nature*, 154 (1944) 81.
98 K. SHARMA AND A. SHARMA, *Nucleus (Calcutta)*, 5 (1962) 127.
99 S. LEBLOVA, *Naturwissenschaften*, 52 (1965) 429.
100 S. KAWAKISHI AND M. NAMIKI, *Agr. Biol. Chem.*, 33 (1969) 452.
101 M. KOJIMA AND I. ICHIKAWA, *J. Ferment. Technol.*, 47 (1969) 262.
102 M. KOJIMA, Y. AKAHORI AND I. ICHIKAWA, *J. Ferment. Technol.*, 46 (1968) 18.
103 K. MODZELEWSKA AND F. MORDRET, *Tluszcze Jadalne*, 14 (1970) 127; *Chem. Abstr.*, 73 (1970) 127, 119289Y.
104 D. L. ANDERSEN, *J. Assoc. Offic. Anal. Chemists*, 53 (1970) 1.
105 H. BINDER, *J. Chromatog.*, 41 (1969) 448.
106 A. KJAER AND A. JART, *Acta Chem. Scand.*, 11 (1957) 1423.
107 A. JART, *Acta Chem. Scand.*, 15 (1961) 1223.
108 H. WAGNER, L. HÖRHAMMER AND H. NUFER, *Arzneimittel-Forsch.*, 15 (1965) 453.
109 S. FISEL, F. MODREANU AND A. CARPOV, *Acad. Rep. Populare Romine, Filiala Iasi, Studii Cercetari Stiint. Chim.*, 8 (1957) 277; *Chem. Abstr.*, 54 (1960) 18155.
110 J. A. VINSON, *Anal. Chem.*, 41 (1969) 1661.
111 I. NISHIOKA, R. MATSUO, Y. OHKURA AND T. MOMOSE, *Yakugaku Zasshi*, 88 (1968) 1281.
112 E. I. LONG, A. A. NELSON, O. G. FITZHUGH AND W. H. HANSON, *Arch. Pathol.*, 73 (1963) 595.
113 E. I. LONG AND A. A. NELSON, *Federation Proc.*, 20 (1961) 287.
114 E. C. HAGAN, P. M. JENNER, W. I. JONES, O. G. FITZHUGH, J. G. BROWER AND W. K. WEBB, *Toxicol. Appl. Pharmacol.*, 7 (1965) 18.

424   FOOD AND FEED ADDITIVES AND CONTAMINANTS

115   S.S.Epstein, K.Fujii, J.Andrea and N.Mantel, *Toxicol. Appl. Pharmacol.*, 16 (1970) 321.
116   E.C.Hagan, W.H.Hansen, O.G.Fitzhugh, P.M.Jenner, W.I.Jones, J.M.Taylor, E.L.Long, A.A.Nelson and J.B.Browwer, *Food Cosmet. Toxicol.*, 5 (1967) 141.
117   L.Fishbein, J.Fawkes, H.L.Falk and S.Thompson, *J. Chromatog.*, 29 (1967) 267.
118   E.O.Oswald, L.Fishbein and B.J.Corbett, *J. Chromatog.*, 45 (1969) 437.
119   E.O.Oswald, L.Fishbein, B.J.Corbett and M.P.Walker, *Biochim. Biophys. Acta*, 230 (1971) 237.
120   W.L.Anthony and W.T.Beher, *J. Chromatog.*, 13 (1964) 567.
121   M.Beroza, *J. Agr. Food Chem.*, 11 (1963) 51.
122   S.Kuwatsuka and J.E.Casida, *J. Agr. Food Chem.*, 13 (1965) 528.
123   F.X.Kamienski and J.E.Casida, *Biochem. Pharmacol.*, 19 (1970) 91.
124   E.G.Essac and J.E.Casida, *J. Insect. Physiol.*, 14 (1968) 913.
125   E.Stahl, *Thin-Layer Chromatography*, Academic Press, New York, 1965, p.198.
126   G.M.Nano and A.Martelli, *J. Chromatog.*, 21 (1966) 349.
127   G.Riezebos, A.G.Peto and B.North, *Rec. Trav. Chim.*, 86 (1967) 31.
128   G.Riezebos, private communication.
129   W.L.Zielinski, Jr. and L.Fishbein, *Anal. Chem.*, 38 (1966) 41.
130   Y.Saiki, A.Ueno, H.Sasaki, T.Morita, M.Suzuki, T.Saito and S.Fukushima, *Yakugaku Zasshi*, 88 (1968) 185.
131   S.W.Gunner and T.B.Hand, *J. Chromatog.*, 37 (1968) 357.
132   T.E.Furia, *Food Technol.*, 18 (1964) 50.
133   M.Rubin, S.Gignac, S.P.Bessman and E.L.Belknap, *Science*, 117 (1953) 659.
134   A.Guarino and S.Biondi, *Folia Med. (Naples)*, 40 (1957) 111; *Chem. Abstr.*, 51 (1957) 9927.
135   J.F.Fried, A.Lindenbaum and J.Schubert, *Proc. Soc. Exptl. Biol. Med.*, 100 (1959) 570.
136   H.Foreman, *J. Am. Pharm. Assoc. Sci. Ed.*, 42 (1953) 629.
137   F.Proescher, *Proc. Soc. Exptl. Biol. Med.*, 76 (1951) 619.
138   H.Spencer, V.Vankinscott, I.Lewin and D.Lazlo, *J. Clin. Invest.*, 31 (1952) 1023.
139   J.Bailar, *The Chemistry of the Coordination Compounds*, Reinhold, New York, 1956.
140   S.Chaberek and A.E.Martell, *Organic Sequestering Agents*, Wiley, New York, 1959.
141   L.S.Goodman and A.Gilman (Eds.), *The Pharmacological Basis of Therapeutics*, 3rd edn., MacMillan, New York, 1967, p.934.
142   S.S.Yang, *Food Cosmet. Toxicol.*, 2 (1964) 763.
143   M.J.Seven, *Metal Binding in Medicine*, Lippincott, Philadelphia, 1960, Chap.10.
144   J.E.Wynn, B.Van'triett and J.F.Borzelleca, *Toxicol. Appl. Pharmacol.*, 16 (1970) 807.
145   H.Foreman and T.T.Trujillo, *J. Lab. Clin. Med.*, 43 (1954) 566.
146   H.Foreman, M.Vier and M.Magee, *J. Biol. Chem.*, 203 (1953) 1045.
147   L.S.Tsarapin, *Zashchitn. Vosstanov. Luchevykh Povrezhdeniyakh, Akad. Nauk SSSR*, 142 (1966); *Chem. Abstr.*, 67 (1967) 18400Y.
148   N.L.Delone, *Biofizika*, 3 (1958) 717.
149   N.B.Kristol Yubova, *Dokl. Akad. Nauk SSR*, 138 (1961) 681.
150   B.P.Kaukman and M.R.McDonald, *Proc. Natl. Acad. Sci. U.S.*, 43 (1957) 262 and 255.
151   R.Wakonig and T.J.Arnason, *Proc. Genet. Soc. Can.*, 3 (1958) 37.
152   R.Rieger, H.Nicoloff and A.Michaelis, *Biol. Zentr.*, 82 (1963) 393.
153   I.Lieberman, R.Abrams, N.Hunt and P.Ove, *J. Biol. Chem.*, 239 (1963) 3955.
154   M.Mihara, R.Amano, T.Kondo and H.Tanabe, *Shokuhin Eiseigaku Zasshi*, 11 (1970) 88; *Chem. Abstr.*, 73 (1970) 97446T.
155   J.Vanderdeelen, *J. Chromatog.*, 39 (1969) 521.
156   J.Sykora and V.Eybl, *Collection Czech. Chem. Commun.*, 32 (1967) 352.
157   T.S.West and A.S.Sykes, *Analytical Applications of Diaminoethane Tetraacetic Acid*, 2nd edn., British Drug Houses, Poole, Dorset, England, 1960.
158   B.G.Blijenberg and B.Leijnse, *Clin. Chim. Acta*, 26 (1969) 577.
159   B.Van'triet and J.E.Wynn, *Anal. Chem.*, 41 (1969) 158.
160   W.Hoyle, I.P.Sanderson and T.S.West, *J. Electroanal. Chem.*, 2 (1961) 166.
161   G.Schwarzenbach and H.Flaschka, *Die komplexometrische Titration*, Ferdinand Enke Verlag, Stuttgart, 1965.
162   Anon, *Chem. Eng. News*, 49 (1971) 15.

REFERENCES   425

163   L. E. ANDREW, *Am. Naturalist*, 93 (1959) 135.
164   W. OSTERTAG AND J. HAAKE, *Z. Vererbungslehre*, 98 (1966) 299.
165   T. ALDERSON AND A. H. KHAN, *Nature*, 215 (1967) 1080.
166   E. A. GLÄSS AND A. NOVICK, *J. Bacteriol.*, 77 (1959) 10.
167   G. ZETTERBERG, *Hereditas*, 46 (1960) 229.
168   W. OSTERTAG, *Mutation Res.*, 3 (1966) 249.
169   B. A. KIHLMAN AND A. LEVAN, *Hereditas*, 35 (1949) 109.
170   H. E. KUBITSCHEK AND H. E. BENDIGKEIT, *Mutation Res.*, 1 (1964) 113.
171   A. M. CLARK AND E. G. CLARK, *Mutation Res.*, 6 (1968) 227.
172   W. KUHLMAN, H. G. FROMME, E. M. HEEGE AND W. OSTERTAG, *Cancer Res.*, 28 (1968) 2375.
173   W. OSTERTAG, E. DUISBERG AND M. STÜRMANN, *Mutation Res.*, 2 (1965) 293.
174   M. F. LYON, J. S. R. PHILLIPS AND A. G. SEARLE, *Z. Vererbungslehre*, 93 (1962) 7.
175   B. M. CATTANACH, *Z. Vererbungslehre*, 93 (1962) 215.
176   H. NISHIMURA AND K. NAKAI, *Proc. Soc. Exptl. Biol. Med.*, 104 (1960) 140.
177   S. FABRO AND S. M. SIEBER, *Nature*, 223 (1969) 410.
178   A. GOLDSTEIN AND R. WARREN, *Biochem. Pharmacol.*, 11 (1962) 166.
179   M. A. BERTOLI, G. DRAGONI AND A. RODARI, *Med. Mucl. Radiobiol. Latina*, 11 (1968) 231.
180   L. S. GOODMAN AND A. GILMAN (Eds.), *The Pharmacological Basis of Therapeutics*, 3rd edn., MacMillan, New York, 1967, p. 354.
181   J. AXELROD AND J. REICHENTHAL, *J. Pharmacol. Exptl. Therap.*, 107 (1953) 519.
182   F. S. FISHER, E. S. ALGERI AND J. T. WALKER, *J. Biol. Chem.*, 179 (1949) 71.
183   B. SCHMIDT, *3rd Intern. Colloq. Chem. Beverages, Trieste*, 1967, p. 286.
184   R. N. WARREN, *J. Chromatog.*, 40 (1969) 468.
185   C. J. ESTLER AND H. P. T. AMMON, *Experientia*, 22 (1966) 589.
186   H. H. CORNISH AND A. A. CHRISTMAN, *J. Biol. Chem.*, 228 (1957) 315.
187   G. SCHMIDT AND R. SCHOYER, *Deut. Z. Ges. Gerichtl. Med.*, 57 (1966) 402.
188   F. L. GRAB AND J. A. REINSTEIN, *J. Pharm. Sci.*, 57 (1968) 1703.
189   B. WEISSMANN, P. A. BROMBERG AND A. B. GUTMAN, *Proc. Soc. Exptl. Biol. Med.*, 87 (1954) 257.
190   U. M. SENANAYAKE AND R. O. B. WITESEKERA, *J. Chromatog.*, 32 (1968) 75.
191   V. M. PECHENNIKOV AND A. Z. KNIZHNIK, *Nauchn. Tr. Aspir. Ordinatorov, 1-i Mosk. Med. Inst.*, (1967) 146; *Chem. Abstr.*, 70 (1969) 50490A.
192   W. N. FRENCH AND A. WEHRLI, *J. Pharm. Sci.*, 54 (1965) 1515.
193   M. STUCHLIK, I. CSIBA AND L. KRASNEK, *Cesk. Farm.*, 18 (1969) 91.
194   V. VUKCEVIC-KOVACEVK AND K. K. ANAM, *Bull. Sci. Conseil. Acad. RPF Yougoslavie, Sect. A*, 13 (1968) 77; *Chem. Abstr.*, 71 (1969) 42378A.
195   R. BONTEMPS, S. LECLERCQ AND P. TEIRLINCK, *J. Pharm. Belg.*, 23 (1968) 512.
196   A. MONARD, *J. Pharm. Belg.*, 23 (1968) 323.
197   J. M. NEWTON, *J. Assoc. Offic. Anal. Chemists*, 52 (1969) 653.
198   A. PEREIRA, JR. AND M. M. PEREIRA, *Garcia Orta*, 15 (1967) 41; *Chem. Abstr.*, 71 (1969) 122435G.
199   J. I. ROUTH, N. A. SHANE, E. G. ARREDONDO AND W. D. PAUL, *Clin. Chem.*, 15 (1969) 661.
200   R. SMITH AND D. REES, *Analyst*, 88 (1963) 310.
201   U. TANAKA AND Y. OHKUBO, *J. Coll. Agr. Tokyo Imp. Univ.*, 14 (1937) 153.
202   H. RABER, *Sci. Pharm.*, 34 (1966) 202.
203   K. A. CONNORS, in T. HIGUCHI AND E. BROCHMANN-HANSSEN (Eds.), *Pharmaceutical Analysis*, Wiley, New York, 1961, p. 240.
204   *The Nitrofurans*, Eaton Laboratories, Norwich, New York, 1958.
205   J. A. BUZARD, *Giorn. Ital. Chemioterap.*, 5 (1962) 1.
206   C. E. FREIDGOOD AND L. B. RIPSTEIN, *Cancer Res.*, 11 (1951) 248.
207   R. J. STEIN, D. YOST, F. PETROLIUNAS AND A. VON ESCH, *Federation Proc.*, 25 (1966) 291.
208   J. E. MORRIS, J. M. PRICE, J. J. LALICH AND R. J. STEIN, *Cancer Res.*, 29 (1969) 2145.
209   E. ERTÜRK, J. M. MORRIS, S. M. COHEN, J. M. PRICE AND G. T. BRYAN, *Cancer Res.*, 30 (1970) 1409.
210   D. R. MCCALLA, *Can. J. Biochem.*, 42 (1964) 1245.
211   D. R. MCCALLA, *Can. J. Microbiol.*, 11 (1965) 185.

212   A. ZAMPIERI AND J. GREENBERG, *Biochem. Biophys. Res. Commun.*, 14 (1964) 172.
213   W. SZYBALSKI, *Ann. N.Y. Acad. Sci.*, 76 (1958) 475.
214   T. MIYAJI, M. MIYAMOTO AND Y. UEDA, *Acta Pathol. Japan.*, 14 (1964) 261.
215   D. G. MONTEMURRO, *Brit. J. Cancer*, 14 (1960) 319.
216   J. L. ANTKOWIAK AND A. L. SPATORICO, *J. Chromatog.*, 29 (1967) 277.
217   D. WALDI, in E. STAHL (Ed.), *Thin-Layer Chromatography*, Academic Press, New York, 1965, p. 485.
218   D. W. HAMMOND AND R. E. WESTON, *Analyst*, 94 (1969) 921.
219   H. KNAPSTEIN, *Z. Anal. Chem.*, 217 (1966) 181.
220   E. STAHL, *Dünnschicht-Chromatographie*, Springer, Berlin, 1962, p. 504.
221   T. KOMODA AND R. TAKESHITA, *Shokuhin Eiseigaku Zasshi*, 3 [4] (1962) 374.
222   T. KOMODA AND R. TAKESHITA, *Shokuhin Eiseigaku Zasshi*, 13 (1967) 201.
223   K. NAKAMURA, Y. UTSUI AND Y. NINOMIYA, *Yakugaku Zasshi*, 86 (1966) 404.
224   H. F. BECKMAN, *J. Agr. Food Chem.*, 6 (1958) 130.
225   J. BRÜGGEMANN, K. BRONSCH, H. HEIGENER AND H. KNAPSTEIN, *Anal. Chem.*, 10 (1962) 108.
226   M. G. WHITING, *Econ. Botany*, 17 (1963) 270.
227   G. L. LAQUER, O. MICKELSEN, M. G. WHITING AND L. T. KURLAND, *J. Natl. Cancer Inst.*, 31 (1963) 919.
228   M. SPATZ, E. G. MCDANIEL AND G. L. LAQUER, *Proc. Soc. Exptl. Biol. Med.*, 121 (1966) 417.
229   M. SPATZ, D. W. E. SMITH, E. G. MCDANIEL AND G. L. LAQUER, *Proc. Soc. Exptl. Biol. Med.*, 124 (1967) 691.
230   G. L. LAQUER AND H. MATSUMOTO, *J. Natl. Cancer Inst.*, 37 (1966) 217.
231   M. SPATZ, W. J. DOUGHERTY AND D. W. E. SMITH, *Proc. Soc. Exptl. Biol. Med.*, 124 (1967) **476.**
232   H. J. TEAS AND J. G. DYSON, *Proc. Soc. Exptl. Biol. Med.*, 125 (1967) 988.
233   D. W. E. SMITH, *Science*, 152 (1966) 1273.
234   M. G. GABRIDGE, A. DENUNZIO AND M. S. LEGATOR, *Science*, 163 (1969) 689.
235   H. J. SAX AND K. SAX, *Science*, 149 (1965) 541.
236   H. MATSUMOTO AND H. H. HIGA, *Biochem. J.*, 98 (1966) 20.
237   R. C. SHANK AND P. N. MAGEE, *Biochem. J.*, 105 (1967) 521.
238   J. A. MILLER, *Federation Proc.*, 23 (1964) 1361.
239   W. V. RIGGS, *Chem. Ind. (London)*, (1956) 926.
240   K. NISHIDA, A. KOBAYASHI AND T. NAGAHAMA, *Bull. Agr. Chem. Soc. Japan*, 19 (1955) 77.
241   D. K. DASTUR AND R. S. PALEKAR, *Nature*, 210 (1966) 841.
242   H. MATSUMOTO AND F. M. STRONG, *Arch. Biochem. Biophys.*, 101 (1963) 299.
243   K. NISHIDA, A. KOBAYASHI AND T. NAGAHARA, *Kagoshima Daigaku Nogakubu Hokoku*, 15 (1956) 118.
244   A. KOBAYASHI AND H. MATSUMOTO, *Arch. Biochem. Biophys.*, 110 (1965) 373.
245   A. JEANES, C. S. WISE AND R. T. DIMLER, *Anal. Chem.*, 23 (1951) 415.
246   M. SPATZ AND G. L. LAQUER, *Proc. Soc. Exptl. Biol. Med.*, 127 (1968) 281.
247   M. SPATZ, personal communication.
248   U. WEISS, *Federation Proc.*, 23 (1964) 1357.
249   R. J. HIGHET, personal communication.
250   W. W. WELLS, M. G. YANG, W. BOLZER AND O. MICKELSEN, *Anal. Biochem.*, 25 (1968) 325.
251   R. SCHOENTAL, *Ann. Rev. Pharmacol.*, 7 (1967) 343.
252   G. N. WOGAN, *Bacteriol. Rev.*, 30 (1966) 460.
253   G. N. WOGAN, *Cancer Res.*, 28 (1968) 2282.
254   M. LEGATOR, *Bacteriol. Rev.*, 30 (1966) 471.
255   G. N. WOGAN, in L. A. GOLDBLATT (Ed.), *Aflatoxin*, Academic Press, New York, 1969, p. 151.
256   M. S. LEGATOR, in L. A. GOLDBLATT (Ed.), *Aflatoxin*, Academic Press, New York, 1969, p. 107.
257   H. F. KRAYBILL AND M. B. SHIMKIN, *Advan. Cancer Res.*, 8 (1964) 191.
258   P. M. NEWBERNE AND W. H. BUTLER, *Cancer Res.*, 29 (1969) 236.
259   W. H. BUTLER, M. GREENBLATT AND W. LIJINSKY, *Cancer Res.*, 29 (1969) 2206.
260   M. C. LANCASTER, F. P. JENKINS AND J. McL. PHILP, *Nature*, 192 (1961) 1095.
261   R. SCHOENTAL, *Brit. J. Cancer*, 15 (1961) 812.
262   R. B. A. CARNAGHAN, *Brit. J. Cancer*, 21 (1967) 811.

263 J.E.HALVER, in G.N.WOGAN (Ed.), *Mycotoxins in Foodstuffs*, M.I.T. Press, Cambridge, 1965, p.209.
264 R.SCHOENTAL, *Bull. World Health Organ.*, 29 (1963) 823.
265 A.G.OETTLE, *J. Natl. Cancer Inst.*, 33 (1964) 383.
266 H.F.KRAYBILL AND R.E.SHAPIRO, in L.A.GOLDBLATT (Ed.), *Aflatoxin*, Academic Press, New York, 1969, p.401.
267 G.N.WOGAN, *Progr. Exptl. Tumor Res.*, 11 (1969) 134.
268 J.ELLIS AND J.A.DIPAOLO, *Arch. Pathol.*, 83 (1967) 53.
269 J.A.DIPAOLO, J.ELIS AND H.ERWIN, *Nature*, 215 (1967) 638.
270 M.B.SPORN, C.W.DINGMAN, H.L.PHELPS AND G.N.WOGAN, *Science*, 151 (1966) 1539.
271 J.B.WRAGG, V.C.ROSS AND M.S.LEGATOR, *Proc. Soc. Exptl. Biol. Med.*, 125 (1967) 1052.
272 E.B.LILLEHOF AND A.CIEGLER, *J. Bacteriol.*, 93 (1967) 464.
273 V.M.MAHER AND W.C.SUMMERS, *Nature*, 225 (1970) 68.
274 A.K.ROY, *Biochim. Biophys. Acta*, 169 (1968) 206.
275 J.I.CLIFFORD AND K.R.REES, *Nature*, 209 (1966) 312.
276 H.V.GELBOIN, J.S.WORTAM, R.G.WILSON, M.FRIEDMAN AND G.N.WOGAN, *Science*, 154 (1966) 1205.
277 R.H.SMITH, *Biochem. J.*, 88 (1963) 50P.
278 S.S.EPSTEIN AND H.SHAFNER, *Nature*, 219 (1968) 385.
279 T.ONG, *Mutation Res.*, 9 (1970) 615.
280 D.A.DULIMPIO, A.C.JACOBSON AND M.LEGATOR, *Proc. Soc. Exptl. Biol. Med.*, 127 (1968) 559
281 S.GREEN, M.LEGATOR AND C.JACOBSON, *Mammalian Chromosomes Newsletter*, 8 (1967) 36.
282 L.J.LILLY, *Nature*, 207 (1965) 433.
283 J.GABLIKS, W.SCHAEFFER, L.FRIEDMAN AND G.WOGAN, *J. Bacteriol.*, 90 (1965) 720.
284 G.N.WOGAN, *Mycotoxins in Foodstuffs*, M.I.T. Press, Cambridge, 1968.
285 G.N.WOGAN, *Trout Hepatoma Res. Conf. Papers, 1967. Res. Rept. No.70*, U.S. Govt. Printing Office, Washington, D.C., 1968, p.121.
286 L.A.GOLDBLATT (Ed.), *Aflatoxin*, Academic Press, New York, 1969.
287 E.BORKER, N.F.INSALATA, C.P.LEVI AND J.S.WITZEMAN, *Advan. Appl. Microbiol.*, 8 (1966) 315.
288 G.N.WOGAN, G.S.EDWARDS AND R.C.SHANK, *Cancer Res.*, 27 (1967) 1729.
289 O.BASSIR AND F.OSIYEMI, *Nature*, 215 (1967) 882.
290 R.ALLCROFT, H.ROGER, G.LEWIS, J.NABNEY AND P.E.BEST, *Nature*, 209 (1966) 154.
291 C.W.HOLZAPFEL, P.S.STEYN AND I.F.H.PURCHASE, *Tetrahedron Letters*, 25 (1966) 2799.
292 M.S.MASRI, D.E.LUNDIN, J.R.PAGE AND V.C.GARCIA, *Nature*, 215 (1967) 753.
293 *Tropical Products Institute Rept. No.613*, May, 1965.
294 H. DE IONGH, J.G.VAN PELT, W.O.ORD AND C.B.BARRETT, *Vet. Record*, 76 (1964) 901.
295 W.T.TRAGER, L.STOLOFF AND A.D.CAMPBELL, *J. Assoc. Offic. Agr. Chemists*, 47 (1964) 993.
296 S.NESHEIM, *J. Assoc. Offic. Agr. Chemists*, 47 (1964) 586.
297 W.A.PONS, JR., A.F.CUCULLU, L.S.LEE, A.O.FRANZ AND L.A.GOLDBLATT, *J. Assoc. Offic. Anal. Chemists*, 49 (1966) 554.
298 M.R.HEUSINKVEID, C.C.SHERA AND F.J.BAUR, *J. Assoc. Offic. Agr. Chemists*, 43 (1965) 448.
299 R.H.ENGEBRECHT, J.L.AYRES AND R.O.SINNHUBER, *J. Assoc. Offic. Agr. Chemists*, 48 (1965) 815.
300 L.STOLOFF, A.GRAFF AND H.RICH, *J. Assoc. Offic. Anal. Chemists*, 49 (1966) 740.
301 R.M.EPPLEY, *J. Assoc. Offic. Anal. Chemists*, 49 (1966) 473.
302 W.A.PONS, JR., A.F.CUCULLU, A.O.FRANZ AND L.A.GOLDBLATT, *J. Am. Oil Chemists' Soc.*, 45 (1968) 694.
303 W.A.PONS, JR. AND L.A.GOLDBLATT, *J. Am. Oil Chemists' Soc.*, 42 (1965) 471.
304 W.A.PONS, JR., J.A.ROBERTSON AND L.A.GOLDBLATT, *J. Am. Oil Chemists' Soc.*, 43 (1966) 665.
305 S.NESHEIM, *J. Assoc. Offic. Agr. Chemists*, 47 (1964) 1010.
306 W.A.PONS, JR. AND L.A.GOLDBLATT, in L.A.GOLDBLATT (Ed.), *Aflatoxin*, Academic Press, New York, 1969, p.77.
307 J.L.AYRES AND R.O.SINNHUBER, *J. Am. Oil Chemists' Soc.*, 43 (1966) 423.

308 R.B.A.Carnaghan, R.D.Hartley and J.O'Kelly, *Nature*, 200 (1963) 1101.
309 J.A.Robertson, W.A.Pons, Jr. and L.A.Goldblatt, *J. Agr. Food Chemists*, 15 (1967) 798.
310 J.A.Robertson and W.A.Pons, *J. Assoc. Offic. Anal. Chemists*, 51 (1968) 1190.
311 B.L.van Duuren, T.Chan and F.M.Irani, *Anal. Chem.*, 40 (1968) 2024.
312 P.J.Andrellos and G.R.Reid, *J. Assoc. Offic. Agr. Chemists*, 47 (1964) 801.
313 L.Stoloff, *J. Assoc. Offic. Anal. Chemists*, 50 (1967) 354.
314 R.E.Borman and W.D.Fordham, *J. Chem. Soc.*, (1952) 3945.
315 A.E.Pohland, M.E.Cushmac and P.J.Andrellos, *J. Assoc. Offic. Anal. Chemists*, 51 (1968) 907.
316 E.V.Crisan and A.T.Grefig, *Contrib. Boyce Thompson Inst.*, 24 (1967) 3.
317 E.V.Crisan, *Contrib. Boyce Thompson Inst.*, 24 (1968) 37.
318 P.J.Andrellos, A.C.Beckwith and R.M.Eppley, *J. Assoc. Offic. Anal. Chemists*, 50 (1967) 346.
319 M.J.Verrett, J.P.Marliac and J.McLaughlin, *J. Assoc. Offic. Agr. Chemists*, 47 (1964) 1003.
320 H.T.Lee and K.H.Ling, *J. Formosan Med. Assoc.*, 66 (1967) 92.
321 T.W.Kwon and J.C.Ayres, *J. Chromatog.*, 31 (1967) 420.
322 J.A.Robertson, Jr., L.S.Lee, A.F.Cucullu and L.A.Goldblatt, *J. Am. Oil Chemists' Soc.*, 42 (1965) 467.
323 A.E.Watking, G. Bleffert and M.Kiernan, *J. Am. Oil Chemists' Soc.*, 45 (1968) 880.
324 H. de Iongh, R.O.Vles and J.G.van Pelt, *Nature*, 202 (1964) 466.
325 H.L.Falk, S.J.Thompson and P.Kotin, *Am. Assoc. Cancer Res. Proc.*, 6 (1 965) Abst. No.18.
326 R.C.Shank and G.N.Wogan, *Federation Proc.*, 24 (1965) 627.
327 H.Kurata, H.Tanabe, K.Kanota, S.Udagawa and M.Ichinoe, *J. Food Hyg. Soc. Japan*, 9 (1968) 29.
328 B.A.Roberts and R.Allcroft, *Food Cosmet. Toxicol.*, 6 (1968) 339.
329 R.B.A.Carnaghan, *Nature*, 208 (1965) 308.
330 I.F.H.Purchase and M.Steyn, *J. Assoc. Offic. Anal. Chemists*, 50 (1967) 363.
331 R.Allcroft and B.A.Roberts, *Vet. Record*, 82 (1968) 116.
332 T.C.Campbell, J.P.Caedo, Jr., J.Bulatao-Jayme, L.Salamat and R.W.Engel, *Nature*, 227 (1970) 404.
333 M.F.Dutton and J.G.Heathcote, *Chem. Ind. (London)*, (1968) 418.
334 D.A.van Dorp, A.S.M.van der Zijden, R.K.Beerthius, S.Sparreboom, W.O.Ord, K. de Jong and R.Keuning, *Rec. Trav. Chim.*, 82 (1963) 587.
335 D.A. van Dorp, personal communication.
336 De.B.Scott, *Mycopathol. Mycol. Appl.*, 25 (1965) 213.
337 C.M.Christensen, *Cereal Chem.*, 39 (1962) 100.
338 K.J. van der Merwe, P.S.Steyn, L.Fourie, De.B.Scott and J.J.Theoron, *Nature*, 205 (1965) 1112.
339 I.F.H.Purchase and J.J.Theron, *Food Cosmet. Toxicol.*, 6 (1968) 479.
340 J.H.Moore and B.Truelove, *Science*, 168 (1970) 1102.
341 M.J.Pitout, *Toxicol. Appl. Pharmacol.*, 13 (1968) 299.
342 K.J. van der Merwe and P.S.Steyn, *J. Chem. Soc.*, (1965) 7083.
343 P.S.Steyn and C.W.Holzapfel, *J. S. African Chem. Inst.*, 20 (1967) 186.
344 P.M.Scott and T.B.Hand, *J. Assoc. Offic. Anal. Chemists*, 50 (1967) 366.
345 W.Nel and I.F.H.Purchase, *J. S. African Chem. Inst.*, 21 (1968) 87.
346 E.Bullock, J.C.Roberts and J.G.Underwood, *J. Chem. Soc.*, (1962) 4179.
347 F.Dickens, H.E.H.Jones and H.B.Wyanforth, *Brit. J. Cancer*, 20 (1966) 134.
348 I.F.H.Purchase and J.J. van der Watt, *Food Cosmet. Toxicol.*, 6 (1968) 555.
349 W.Nel and H.E.Pretorius, *Biochem. Pharmacol.*, 19 (1970) 957.
350 L.J.Vorster and I.F.H.Purchase, *Analyst*, 93 (1968) 694.
351 H.J.Burkhardt and J.Furgacs, *Tetrahedron*, 24 (1968) 717.
352 J.V.Rodricks, K.R.Henery-Logan, A.D.Campbell, L.Stoloff and M.J.Verrett, *Nature*, 217 (1968) 668.
353 A.C.Waiss, M.Wiley, D.R.Black and R.E.Lundin, *Tetrahedron Letters*, 25 (1968) 2975.
354 J.S.E.Hocker and J.G.Underwood, *Chem. Ind. (London)*, (1964) 1865.

355 C. T. DWARAKANATH, E. T. RAYNER, G. E. MANN AND F. D. DOLLEAR, *J. Am. Oil Chemists' Soc.*, 45 (1968) 93.

356 M. STOB, R. S. BALDWIN, J. TUITE, F. N. ANDREWS AND K. G. GILLETTE, *Nature*, 196 (1962) 1318.

357 C. J. MIROCHA, C. M. CHRISTENSEN AND G. H. NELSON, *Appl. Microbiol.*, 15 (1967) 497.

358 C. J. MIROCHA, C. M. CHRISTENSEN AND G. H. NELSON, *Cancer Res.*, 28 (1968) 2319.

359 S. SHIBATA, S. NATORI AND S. UDAGANA, *List of Fungal Metabolites*, Univ. of Tokyo Press, 1964.

360 B. N. MIRRINGTON, E. RITCHIE, C. W. SHOPPEE, W. C. TAYLOR AND J. STERNHELL, *Tetrahedron Letters*, 7 (1964) 365.

361 F. MCCAPRA, A. I. SCOTT, P. DELMOTTE AND J. DELMOTTE-PLAQUEE, *Tetrahedron Letters*, 15 (1964) 869.

362 S. G. YATES, H. L. TOOKEY, J. J. ELLIS AND H. J. BURKHARDT, *Tetrahedron Letters*, 25 (1967) 621.

363 H. J. BURKHARDT, R. E. LUNDIN AND W. H. MCFADDEN, *Tetrahedron*, 24 (1968) 1225.

364 T. TATSUNO, *Cancer Res.*, 28 (1968) 2393.

365 T. TATSUNO, M. SAITO, M. ENOMOTO AND H. TSUNODA, *Chem. Pharm. Bull. (Tokyo)*, 16 (1968) 2519.

366 K. OHTSUBO, M. YAMADA AND M. SAITO, *Jap. J. Med. Sci. Biol.*, 21 (1968) 185.

367 Y. UENO AND K. FUKUSHIMA, *Experientia*, 24 (1968) 1032.

368 Y. UENO, M. HOSOYA AND T. TATSUNO, *J. Japan Biochem. Soc.*, 39 (1967) 718.

369 Y. UENO, I. UENO, T. TATSUNO, K. OHOKUBO AND H. TSUNODA, *Experientia*, 25 (1969) 1062.

370 Y. UENO, M. HOSOYA, Y. MORITA, I. UENO AND T. TATSUNO, *J. Biochem. (Tokyo)*, 64 (1968) 479.

371 J. R. BAMBURG, N. V. RIGGS AND F. M. STRONG, *Tetrahedron*, 24 (1968) 3329.

372 W. F. O. MARASAS, J R. BAMBURG, E. B. SMALLEY, F. M. STRONG, W. L. RAGLAND AND P. E. DE-GURSE, *Nature*, 214 (1967) 817.

373 F. A. HODGES, Z. R. ZUST, H. R. SMITH, A. A. NELSON, B. H. ARMBRECHT AND A. D. CAMPBELL, *Science*, 145 (1964) 1439.

374 A. W. JOHNSON, N. SHEPPARD AND A. R. TODD, *J. Chem. Soc..* (1951) 1139.

375 C. L. ALSBERG AND O. F. BLACK, *U.S. Dept. Agr. Bur. Plant. Ind. Bull.*, (1962) 270.

376 C. W. HOLZAPFEL, *Tetrahedron*, 24 (1968) 2101.

377 K. URAGUCHI, T. TATSUNO, F. SAKAI, M. T. SUKIOKO, M. MIYAKE AND T. ISHIKO, *Japan. J. Exptl. Med.*, 31 (1961) 19.

378 T. TATSOUNO, *Food Cosmet. Toxicol.*, 2 (1964) 678.

379 I. ISHIKAWA, Y. UENO AND H. TSUNODA, *J. Biochem. (Tokyo)*, 67 (1970) 753.

380 P. S. STEYN, *J. Chromatog.*, 45 (1969) 473.

Chapter 7

MISCELLANEOUS TOXICANTS

1. HYDRAZINE AND ITS DERIVATIVES

Hydrazine is prepared commercially by several procedures including (*a*) the reaction of ammonia and sodium hypochlorite, (*b*) the amination of the intermediate hydroxylamine-O-sulfonic acid (prepared from hydroxylamine sulfate and sulfur trioxide), and (*c*) the Bayer process *via* diazocyclopropane or ketazine. Hydrazine and its derivatives are widely used in the preparation of agricultural chemicals (plant-growth regulators, defoliants), medicinals (antitubercular and hypotensive agents, psychic energizers, topical antiseptics), textile agents, explosives, rocket fuels, plastics, preservatives, blowing agents, solder fluxes, and in photography and metal processing. Aqueous solutions of hydrazine are strong oxygen scavengers which are used as anti-corrosion agents.

The chemistry of hydrazine[1, 2], its physiology[3], acute and chronic toxicity[4-7], hepatotoxicity[8, 9], carcinogenicity[10-12], mutagenicity[13, 14], and induction of chromosomal alterations in mouse Ascites tumor[15] and *Vicia faba*[16] has been described. The metabolic fate of hydrazine and its derivatives has also been delineated[17-20].

Hydrazine is extensively metabolized by few animals and appears rapidly in the urine of dog and rabbit. In all animals studied, except the dog, which has limited capabilities to acetylate hydrazines, hydrazine is metabolized to acetylhydrazine and so rapidly converted to the diacetyl derivative that the monoacetyl hydrazine intermediate cannot be detected in the urine[18, 19]. (However, diacetylhydrazine is excreted unchanged.)

Unsymmetrical 1,1-dimethylhydrazine (UDMH) is prepared by the electrolytic reduction of dimethylnitrosamine and in addition to its primary use in rocket fuels has patented applications as a solvent for acetylene, and as a gasoline additive. Its toxicological and pharmacological properties[3, 21-24], carcinogenicity[11], mutagenicity[25], and metabolism[17, 26, 27] have been reported. The initial metabolic reaction involving methyl-substituted hydrazines is apparently demethylation followed by oxidation since large proportions of the activity from both UDMH and monomethylhydrazine (MMH) labeled with ^{14}C were expired in 10–24 h as one carbon fragment. Approximately 30% of a subacute dose of UDMH-^{14}C injected intraperitoneally into rats was recovered as respiratory $^{14}CO_2$ in the first 10 h. The possibility of UDMH being a possible metabolite of dimethylnitrosamine has been raised[28] in light of the finding by Süss[29] that N-nitrosomorpholine is metabolized to the corresponding hydrazine by liver homogenates.

A variety of GLC procedures has been utilized for the analysis of hydrazine and substituted hydrazines in admixture. The separation of hydrazine, monomethylhydrazine (MMH) and water on a column using 10% Dowfax 9N9 on Teflon 6 was described by Jones[30]. Dowfax 9N9 has a chemical composition of

$$C_9H_{19}\text{---}\langle\;\rangle\text{---}(OCH_2CH_2)_9\text{--}OH$$

A Perkin-Elmer Model 154-C vapor fractometer used a dual-chamber thermistor thermal conductivity cell and equipped with a 6 ft. × ¼ in. stainless steel column containing 10% Dowfax 9N9 on Teflon 6. The column was stabilized overnight at 150°C at a 20 ml/min helium flow rate and during analysis, column and injector port temperature of 110°C and a helium flow rate of 40 ml/min were maintained. The retention times were: water 2 min, MMH 3 min, and hydrazine 4.5 min. (The column was found inadequate for the analysis of hydrazine mixtures containing UDMH since its retention time was the same as that of water.)

The GLC separation of mixtures of hydrazine, MMH, and UDMH has been described by Bighi et al.[31-33]. The analyses[31] were carried out using a Fractovap-type B instrument (C. Erba) with a thermal conductivity detector and a 2 ml column containing 25% Carbowax 400 on Celite 22 (30–60 mesh) alkanized with 1% KOH and a helium flow rate of 109 ml/min.

Table 1 lists the retention volumes of hydrazine, methylhydrazine, and 1,1-dimethylhydrazine calculated for different temperatures. A plot of the experimental

TABLE 1

RETENTION VOLUMES (V_r) AS A FUNCTION OF $1/T$

$1/T^a$	$(CH_3)_2N\text{--}NH_2$	$CH_3\text{--}NH\text{--}NH_2$	$N_2H_4 \cdot H_2O$
2.790×10^{-3}	1035	2905	
2.755×10^{-3}	891	2450	
2.717×10^{-3}	747	2043.4	5346
2.681×10^{-3}	650.5	1730	4288
2.645×10^{-3}	540	1423.4	3440
2.610×10^{-3}	440.5	1230	2864

^{a} Temperature in degrees absolute.

values of the retention volumes on a semi-logarithmic scale against the reciprocal of the absolute temperature gave linear graphs as shown in Fig. 1 and illustrates the considerable differences in the retention volumes of the three hydrazines making their qualitative separation possible in the temperature range under consideration. The V_r values of hydrazine fall off more steeply with increasing temperature than do those of methyl- and dimethylhydrazine. This is suggested to result from the hydrogen-bond formation between hydrazine and the polyethylene glycol of the stationary phase, and that this interaction decreases with increasing temperature. Figure 2 illustrates a chromatogram of the separation of the three hydrazines on Carbowax 400

$+ 7\%$ KOH at 130°C. The limiting concentrations for quantitative determinations were 2.5×10^{-6} moles for hydrazine, 1.95×10^{-7} moles for 1,1-dimethylhydrazine, and 0.46×10^{-7} moles for methylhydrazine.

Bighi and Saglietto[34] used various alkanized adsorbents to investigate the GLC behavior of hydrazine and hydrazine hydrate, the resolution of the two isomeric dimethylhydrazines and the separation of 1,1-dimethylhydrazine from its oxidation

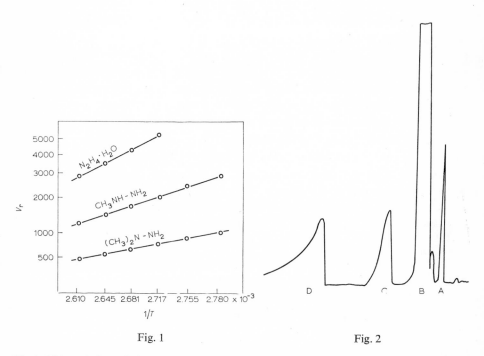

Fig. 1   Fig. 2

Fig. 1. The variation of the retention volumes with the reciprocal of the absolute temperature.

Fig. 2. Gas-chromatographic separation. A, $(CH_3)_2NNH_2$; B, CH_3CH_2OH; C, CH_3NHNH_2; D, $N_2H_4 \cdot H_2O$. Experimental conditions: Support: Celite C-22 (30–60 mesh); stationary phase: Carbowax 400 (25%) + 7% of KOH; carrier gas: helium; column temperature: $130° \pm 0.2$°C; inlet pressure: 0.7 kg/cm²; flow rate: 204 ml/min; current of bridge circuit: 24.5 mA; chart speed: 1.25 cm/min; length of column: 2 m; internal diameter of column: 0.6 cm. Retention times: dimethylhydrazine 1 min 59 sec, methylhydrazine 6 min 36 sec, hydrazine 1 min 42 sec.

products tetramethyl tetrazene and nitrosodimethylamine. The two alkali treatments used were: Column A, treatment of Celite C-22 with 25% Carbowax 400 dissolved in chloroform (10 g of stationary phase in 100 ml solvents). After 4 h at 120°C, a 7% solution of KOH in methanol (2.8 g of KOH in 100 ml solvent) was added to the above. Column B, treatment of Celite C-22 with KOH in methanol, drying at 120°C then permeated with Carbowax 400 in chloroform in quantities used in the treatment of Column A above. Both columns were 2 m × 0.6 cm internal diameter, the carrier gas was helium and the analyses were carried out on a Fractovap model B instrument (C. Erba) equipped with a thermal conductivity detector. The retention times of hydrazine hydrate (10% by volume in absolute ethanol) at a column temperature of $130 \pm 0.2°C$ and current in bridge circuit of 24.5 mA and a helium flow rate of 416 ml/min were 1 min 19 sec for column A and 4 min 52 sec for column B (Fig. 3). The more pronounced basicity of the absorbent system alkalized after the

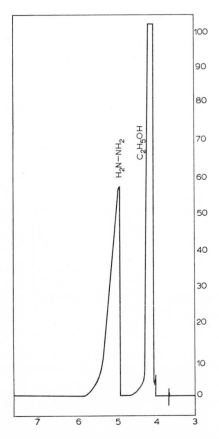

Fig. 3. GLC of hydrazine on 25% 400 Carbowax–Celite C-22 alkanized with 70% KOH (column A) at $130 \pm 0.2°C$.

support had been treated with the stationary phase (column A) diminishes the retention times of both compounds due to the increased repulsive forces between the more strongly alkalized adsorbent and the distinctly basic samples. The lowest concentrations to which the apparatus responded using column A at 130°C were hydrazine 4×10^{-6} moles (0.000128 g) and hydrazine hydrate 2.5×10^{-6} moles (0.000125 g).

The separation of 1,1-dimethyl- and 1,2-dimethylhydrazines was carried out preferentially on column A at $100 \pm 0.2°C$ (Fig. 4). The retention times of both

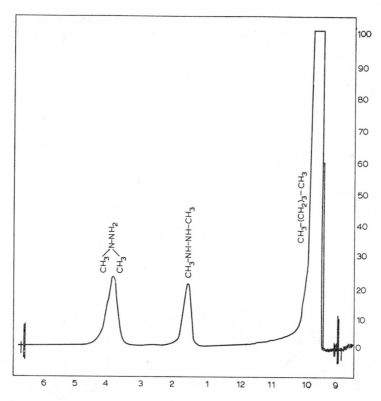

Fig. 4. GLC of 1,1-dimethylhydrazine and 1,2-dimethylhydrazine on column A at 100 ± 0.2°C.

dimethylhydrazines using both columns A and B were: 1,1-dimethylhydrazine column A 6 min 31 sec, column B 7 min 24 sec; 1,2-dimethylhydrazine column A 4 min 24 sec, column B 7 min 24 sec. Table 2 lists the retention times of 1,1-dimethylhydrazine and its oxidation products tetramethyl tetrazene and dimethylnitrosamine. Figure 5 illustrates their separation on column B at 90 ± 0.2°C.

TABLE 2

RETENTION TIMES OF 1,1-DIMETHYLHYDRAZINE AND ITS OXIDATION PRODUCTS
AT 90 ± 0.2°C[a]

	Column A	Column B
1,1-Dimethylhydrazine	43 sec	1 min 11 sec
Tetramethyl tetrazene	1 min 16 sec	2 min 48 sec
Dimethylnitrosamine	3 min 12 sec	5 min 36 sec

[a] Helium carrier gas flow 500 ml/min. Current in bridge circuit 18.5 mA; column conditions, see text.

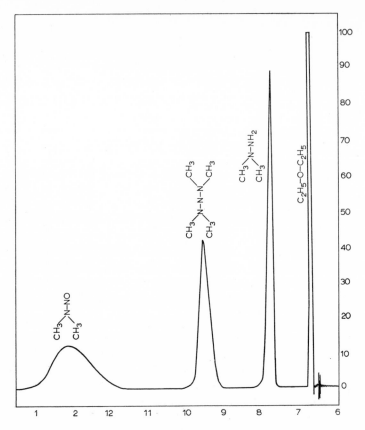

Fig. 5. GLC of 1,1-dimethylhydrazine and its oxidation products tetramethyl tetrazene and dimethyl-nitrosamine on column B at 90 ± 0.2 °C.

The optimum resolution of 1,1-dimethylhydrazine and tetramethyl tetrazene was achieved on column A at 90 ± 0.2 °C with helium flow rate of 300 ml/min and current in the bridge circuit of 18.5 mA (Fig. 6). The retention times were 1,1-dimethyl-hydrazine 1 min 38 sec and tetramethyl tetrazene 2 min 24 sec.

The composition of binary mixtures of hydrazine and methyl-substituted hydrazine was determined by GLC analysis of their thermal decomposition products[35]. Decomposition takes place in the presence of a copper oxide catalyst and the products are separated on a 5A molecular sieve. The methyl-substituted component is determined by comparison of the methane peak area to a calibration curve obtained from synthetic blends of the two components, and hydrazine or methyl-substituted hydrazine were also determined specifically in aqueous solutions of ammonia, amides, and amines. A Micro-Tek Model 2500-2 gas chromatograph was used with a thermal conductivity detector. The injector port was a $\frac{1}{4}$ in. o.d. stainless steel "tee"; the injector tube was 6 in. × $\frac{3}{8}$ in. o.d. copper tube packed with copper oxide (wire form,

Allied Chemical and Dye Corp.) and the column was a 6 ft. × ¼ in. o.d. stainless steel column packed with 30–60 mesh 5A molecular sieve (Chromline Labs). Figure 7 illustrates a chromatogram of a hydrazine–monomethylhydrazine (MMH) mixture. Thermal decomposition of hydrazine in the presence of a catalyst results in the formation of ammonia in amounts which vary with the temperature and catalyst[1]. The variation of nitrogen produced with sample size is linear at a temperature of 140°C

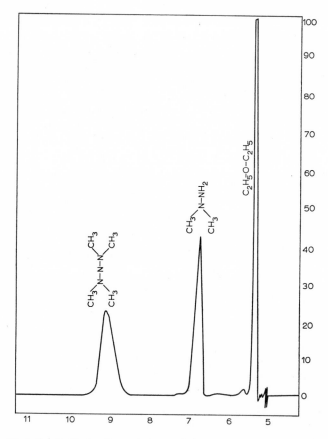

Fig. 6. GLC of 1,1-dimethylhydrazine and tetramethyl tetrazene on column A at 90 ± 0.2°C.

when hydrazine is passed through copper oxide. Nitrogen and methane are among the products formed upon the decomposition of MMH and UDMH. Linearity of methane production with sample size occurs at a copper oxide temperature of approximately 300°C.

The utility of hydrazine as a rocket fuel is enhanced by the suspension of powdered metals such as aluminum, following gelations of the hydrazine with organic compounds of high molecular weight.

Hydrazine decomposition (in the above process) is believed to proceed by one or a combination of the following mechanisms.

$$3 \, N_2H_4 \rightarrow N_2 + 4 \, NH_3 \tag{1}$$

$$2 \, N_2H_4 \rightarrow N_2 + H_2 + 2 \, NH_3 \tag{2}$$

An analytical technique employing gas chromatographic bubble-gas analysis was developed[36] to determine which of these reactions predominated and to clarify the physical and chemical conditions, *e.g.* pressure, temperature, and reagent purity, as

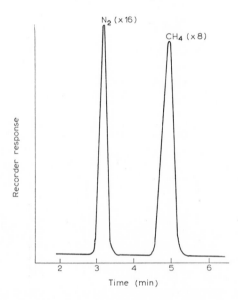

Fig. 7. Typical chromatogram of 1 µl injection of a hydrazine–MMH blend. Copper oxide temperature, 300 °C; molecular sieve temperature, 60 °C; helium flow rate, 45 ml/min; bridge current, 400 mA; detector block temperature, 130 °C.

well as the kinetics of the gel decomposition. Because it has proven difficult to separate the decomposition gases nitrogen, ammonia, and hydrogen with one column, a Varian Aerograph 1520-1 dual-column gas chromatograph was fitted with the two columns in tandem (Fig. 8). The gases were detected by the technique of Murakami[37] in which both sides of a dual thermal-conductivity cell are used alternatively as sensing elements. Ammonia was separated on a 10% Quadrol on Fluoropak 80 column (5 ft. × ¼ in. o.d. stainless steel), while nitrogen and hydrogen were separated on a 15 ft. × ¼ in. o.d. stainless steel column containing molecular sieve 5A. A carrier gas flow rate of 50 ml/min was maintained. The swelling of aluminized and non-aluminized hydrazine gels and the consequent trapping of gas bubbles were observed in sensitive 2.5 l borosilicate glass dilatometers (Fig. 9) in experiments conducted at

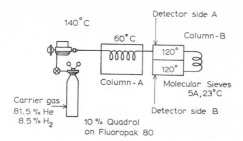

Fig. 8. Schematic diagram of gas chromatograph.

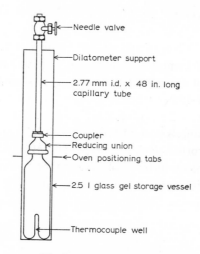

Fig. 9. Glass dilatometer.

temperatures varying from 90 to 110°F. Table 3 lists the relative sensitivities and retention times of the decomposition gases and Fig. 10 illustrates a chromatogram of a bubble gas that evolved within an aluminized gel. Reaction (1) is believed to be the principal decomposition reaction occurring within the gel structure as the primary decomposition in all of the analyses was nitrogen. Comparison of the products arising from neat and aluminized hydrazine gels showed that the presence of aluminum powder did not alter the decomposition reaction mechanism.

TABLE 3

RELATIVE SENSITIVITIES AND RETENTION TIMES ON DECOMPOSITION GASES

Decomposition gas	Relative sensitivity	Retention time (min)
N_2	100.00	19.0
NH_3	81.21	4.6
H_2	1.97	3.6

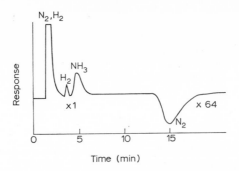

Fig. 10. Gel No. 127 bubble-gas chromatogram.

The utility of 5-nitro-2-hydroxybenzal derivatives of unsymmetrical hydrazine for the GLC separation and identification of hydrazine as well as N-nitroso compounds (the latter *via* their initial reduction to hydrazines) has been described by Neurath and Lüttich[38]. A Perkin-Elmer fractometer type F6 instrument was used with a flame-ionization detector, a 2.5 mV recorder (Siemens-Kompensograph 288 × 288) and a 2 m × 4 mm i.d. stainless steel tube packed with 2.5% silicone grease (E. Merck) on 60–80 mesh Chromosorb W (acid-washed and DMCS treated).

Table 4 lists the retention times of a variety of 5-nitro-2-hydroxy derivatives of unsymmetrical hydrazines (prepared by the procedure of Neurath *et al.*[39]). Figure 11

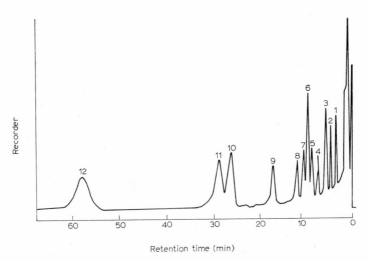

Fig. 11. Gas chromatographic separation of 5-nitro-2-hydroxybenzal derivatives of unsymmetric hydrazines. Sample size: 1–5 μg of each derivative in 5 μl acetic ester. Constant column temperature, 220 °C; injection temperature, 300 °C; carrier gas, helium; flow rate, 30 ml/min. 1, 1,1-dimethyl-hydrazine; 2, 1-methyl-1-ethylhydrazine; 3, 1,1-diethylhydrazine; 4, 1-methyl-1-isobutylhydrazine; 5, 1-methyl-1-butylhydrazine; 6, 1,1-dipropylhydrazine; 7, 1-methyl-1-pentylhydrazine; 8, 1,1-diiso-butylhydrazine; 9, 1,1-dibutylhydrazine; 10, 1-ethyl-1-heptylhydrazine; 11, 1-methyl-1-benzylhydra-zine; 12, 1,1-diphenylhydrazine.

illustrates a gas chromatographic separation of twelve 5-nitro-2-hydroxybenzal derivatives of unsym. hydrazine at a constant column temperature of 220°C, injection
temperature 300°C and carrier gas, helium of flow rate 30 ml/min. The retention
time of the reagent 5-nitro-2-hydroxybenzaldehyde was *ca.* 1 min and did not interfere with the analysis of the derivatives. The logarithm of the relative retention
values was found to be a linear function of the number of carbon atoms.

TABLE 4

RETENTION TIMES OF 5-NITRO-2-HYDROXYBENZAL DERIVATIVES OF UNSYMMETRIC HYDRAZINES

Parent hydrazine	Retention time (min)
1,1-Dimethylhydrazine	3.8
1-Methyl-1-ethylhydrazine	4.9
1,1-Diethylhydrazine	5.9
1-Methyl-1-butylhydrazine	8.5
1-Methyl-1-isobutylhydrazine	7.5
1,1-Dipropylhydrazine	9.3
1,1-Diisopropylhydrazine	8.5
1-Methyl-1-pentylhydrazine	10.2
1-Methyl-1-isopentylhydrazine	10.2
1-Ethyl-1-*sec.*-butylhydrazine	8.9
1-Ethyl-1-*tert.*-butylhydrazine	8.2
1,1-Dibutylhydrazine	16.0
1,1-Diisobutylhydrazine	11.5
1-Ethyl-1-heptylhydrazine	23.8
1,1-Dihexylhydrazine	51.8
1-Methyl-1-allylhydrazine	6.0
1-Ethyl-1-phenylhydrazine	23.4
1,1-Diphenylhydrazine	50.8
1-Methyl-1-benzylhydrazine	26.1
1-Amino-pyrrolidine	10.6
1-Amino-piperidine	12.0
1-Amino-2-methylpyrrolidine	11.1
1-Amino-2,5-dimethylpyrrolidine	11.0
1-Amino-4-methylpiperazine	15.0
4-Amino-morpholine	11.5

The paper chromatography (circular and ascending) of a number of alkyl-,
mono-, and diacylhydrazines has been reported by Hinman[40]. The alkylhydrazines
were separated using the upper layer from a mixture of *n*-butanol–acetic acid–water
(4:1:5); for the acylhydrazines the preferred solvent system was the upper layer from
isoamyl alcohol–acetic acid–water (10:1.5:10). Detection was best accomplished
using Ehrlich's reagent[41], although ammoniacal silver nitrate or ninhydrin could be
used in some cases. Table 5 lists the R_F values of some hydrazine derivatives obtained
by both circular and ascending chromatography.

A number of non-chromatographic procedures have been utilized for the ana-

TABLE 5

R_F VALUES OF SOME HYDRAZINE DERIVATIVES ON WHATMAN NO. 1 PAPER[a]

Compound	Isoamyl alc.–acetic acid–H_2O circular	1-Butanol–acetic acid–H_2O		Detecting agents	
		Circular	Ascending	$[Ag(NH_2)_2]^+$	Ehrlich's reagent
1,2-Diformylhydrazine	0.26			+	+
1,2-Diacetylhydrazine	0.48			+	+
1,2-Dipropionylhydrazine	0.81			+	+
1,2-Dibutyrylhydrazine	0.91			–	+
1,2-Dibenzoylhydrazine				–	–
1,2-Diformyl-1,2-dimethylhydrazine	0.60			–	+
1,2-Diacetyl-1,2-dimethylhydrazine	0.74			–	+
1-Formyl-2,2-dimethylhydrazine	0.72			+	+
1-Acetyl-2,2-dimethylhydrazine	0.76			+	+
1-Benzoyl-2,2-dimethylhydrazine	0.88			+	+
Hydrazine dihydrochloride	0.00	0.39	0.16	+	+
Methylhydrazine sulfate	0.13	0.47	0.17	+	+
1,1-Dimethylhydrazine hydrochloride	0.29	0.53	0.20	+	+
1,2-Dimethylhydrazine dihydrochloride	0.27	0.57	0.23	+	+
Trimethylhydrazine hydrochloride		0.42	0.23	+	+
1,2-Diethylhydrazine dihydrochloride	0.30	0.55	0.25	+	+
Ammonium chloride		0.33		+	+
Methylamine hydrochloride		0.38		+	+
Dimethylamine hydrochloride		0.41		+	+

[a] Circular paper developed at 24–28°C in 1–2 h. Ascending paper developed at 18–20°C to 30 cm.

lysis of hydrazine and its derivatives and in admixture. These include titrimetric[42-46], colorimetric[42,47,48], proton magnetic resonance[49] and differential reaction-rate acetylation[50].

2. NAPHTHYLAMINES AND THEIR METABOLITES

2-Naphthylamine is prepared by heating 2-naphthol with ammonium sulfite and ammonium hydroxide at 150°C and is used as an antioxidant for hydrocarbons and polyacetaldehyde polymers in the softening of textile materials, dyeing of vinylon yarn and as the derivative phenyl-2-naphthylamine as a corrosion inhibitor in anti-freeze and as an antiaging agent for rubber. Formaldehyde condensation products of 1- and 2-naphthylamines which are used in the rubber industry as antioxidants contain a small proportion (*ca.* 2.5%) of uncombined naphthylamines. 2-Naphthyl-amine has been detected in the air near the retorts of gas works[51]. Both 1- and 2-naphthylamines have been formed on pyrolysis of amino acids[52] and also isolated in cigarette smoke[53,54].

The carcinogenicity of 2-naphthylamine is well established[55-58]; it is the most commonly recognized cause of occupational bladder cancer[59]. The isolation of 2-naphthylamine above represents the first successful identification of a bladder carcinogen in a non-occupational respiratory environment and is of possible signifi-cance because of epidemiological data which suggest an association between cigarette smoking and urinary bladder cancer in man[60].

The metabolism of 2-naphthylamine has been described [61-64], and the carci-nogenic action on the bladder of men and dogs is believed to be exerted *via* its meta-bolites. Figure 12 shows the metabolites of 2-naphthylamine and 2-acetamidonaph-thalene in which the unsubstituted ring is involved. The metabolites shown are formed in reactions which occur in the ring remote from the amino group and are analogous to the metabolites of the non-carcinogenic naphthalene. All these metabolites are thought to be true detoxification products[57]. The metabolites of 2-naphthylamine in which the amino group is involved are shown in Fig. 13. Five of the metabolites shown in Fig. 13 have been found to be carcinogenic by mouse bladder implantation tests, *e.g.* 2-amino-1-naphthyl glucuronide, bis(2-amino)-1-naphthyl phosphate, 2-formamido-1-naphthylsulfate, 2-naphthyl hydroxylamine (N-hydroxy-2-naphthyl-amine), and 2-nitroso-naphthylamine. Both N-hydroxy-2-naphthylamine and 2-ni-trosonaphthalene are analogous to the N-hydroxy derivatives of 2-acetamido-fluorene and may react with nucleic acid in the same way or by a different route. It is also of interest to note that N-hydroxy-2-, and N-hydroxy-1-naphthylamines are mutagenic in *E. coli*[65-68] and that the *in vivo* binding of metabolites of 2-naphthylamine to mouse liver DNA, RNA and protein has been recently reported[69].

The nature of the metabolism of 2-naphthylamine and N-hydroxy-2-naphthyl-amine derivatives in the dog, guinea pig, hamster, and rabbit was elaborated with the aid of paper and thin-layer chromatography[61]. Whatman No. 1 paper was em-

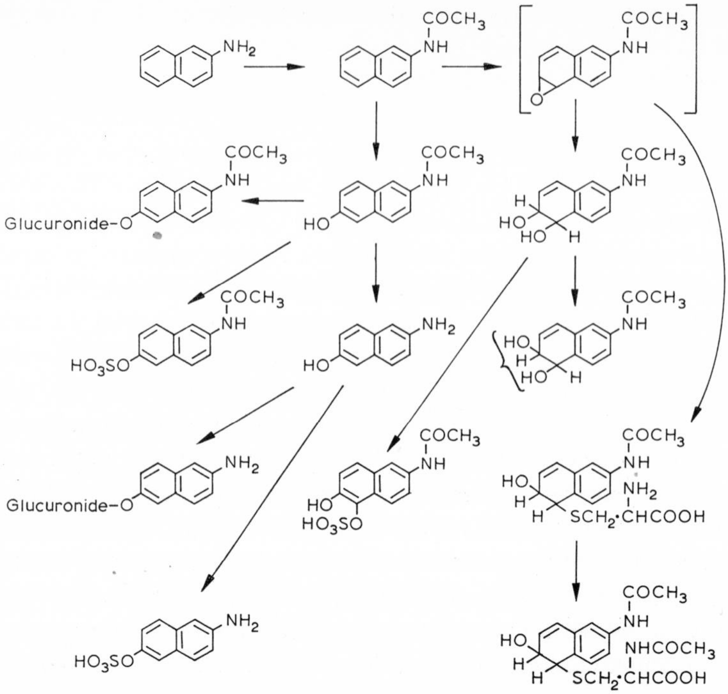

Fig.12. Metabolites of 2-naphthylamine and 2-acetamidonaphthalene in which the unsubstituted ring is involved[57].

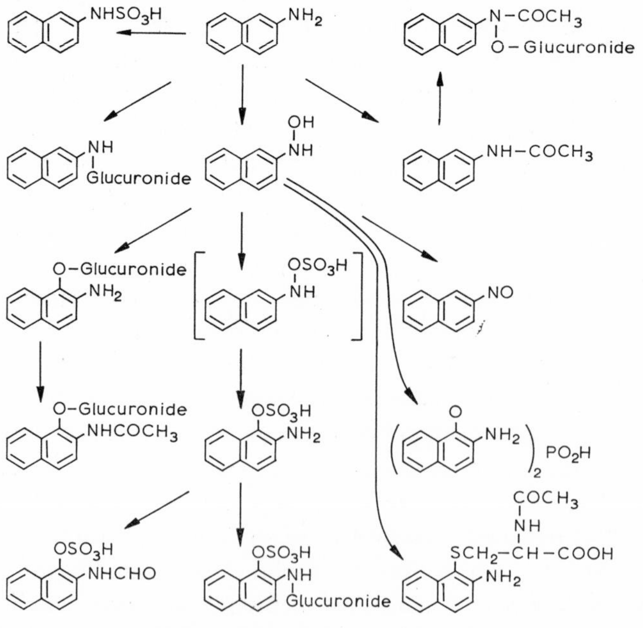

Fig.13. Metabolites of 2-naphthylamine in which the amino group is probably involved[57].

ployed for descending development in the following solvent systems, (a) n-butanol–n-propanol–aq. 0.1 N ammonia (2:1:1), (b) n-butanol–acetic acid–water (2:1:1), (c) n-butanol–acetic acid–water (12:3:5), (d) n-butanol saturated with 0.1 N ammonia, and (e) n-butanol–benzene (4:1) saturated with 0.1 N ammonia. Two-dimensional chromatography was carried out by successive use of solvents (a) and (b). TLC was carried out on silica gel (E. Merck) using the following solvent systems, (f) n-butanol–n-propanol–water (2:1:1), (g) light petroleum (b.p. 40–60°C)–acetone (7:3), (h) light petroleum (b.p. 40–60°C)–acetone (4:1), (i) chloroform–ethyl acetate–acetic acid (6:3:1), (j) benzene–ethanol (19:1), (k) chloroform–methanol (49:1), (l) chloroform–methanol (19:1), and (m) n-butanol–acetic acid–water (10:1:1). Solvents (f) and (m) were used for the TLC of water-soluble compounds, e.g. sulfuric esters.

The detecting reagents employed were as described by Boyland et al.[63] and also included (a) 5% aq. sodium aminoprusside, (b) 5% aq. ferric chloride, (c) ammoniacal silver nitrate, and (d) 15% titanium chloride. Urine was applied directly to chromatograms or concentrated by adsorption at pH 5.0 on charcoal, from which the metabolites were eluted with methanol containing about 5% of aq. NH_3 (sp. gr. 0.88). Bile was treated in the same way except that 5% phenol was used for elution. Tables 6 and 7 list the R_F values of 2-naphthylamine and some derivatives on paper and thin-layer chromatograms and their color reactions, respectively. Table 8 depicts the paper chromatography of some of the rat biliary metabolites of 2-naphthylamine. Table 9 summarizes the metabolites of 2-naphthylamine and some derivatives detected in urine of different species.

The metabolism of 1- and 2-naphthylamines in the dog was studied by Deichmann and Radomski[64]. The salient differences found following a TLC examination of the urinary metabolites were, (a) 2-naphthylamine is metabolized to a unique diester [di(2-amino-1-naphthyl)-phosphate], while 1-naphthylamine is metabolized to a monophosphate ester of 2-amino-1-naphthol, (b) 2-naphthylamine is metabolized exclusively to ortho-hydroxy conjugates and 1-naphthylamine is largely metabolized to para-hydroxy conjugates, (c) while both amines are N-hydroxylated, 2-naphthylamine is further oxidized to 2-nitrosonaphthalene (this further N-oxidative step did not appear to occur with the more stable N-1-hydroxynaphthylamine and the highly carcinogenic 4-aminobiphenyl is also metabolized to both the N-hydroxy and nitroso compounds). Table 10 depicts the procedure of the extractions and chromatography of dog urine following oral administration of ^{14}C-1- and 2-naphthylamines.

The unconjugated metabolites were separated using TLC with benzene–acetone (8:1) as developer and sulfate and glucuronide conjugates of the hydroxylated amine were separated using the more polar solvent system n-butanol–n-propanol–0.2 N ammonia (2:1:1).

The increased metabolic N-oxidation of 2-naphthylamine in dogs after phenobarbital pre-treatment has been demonstrated by Uehleke and Brill[73]. Both column and thin-layer chromatography and UV spectroscopy were used for the separation and identification of the products. In a typical experiment, the residue obtained from

TABLE 6

R_F VALUES OF 2-NAPHTHYLAMINE AND SOME DERIVATIVES ON PAPER AND THIN-LAYER CHROMATOGRAMS[a]

Compound	R_F on paper chromatograms in solvent						R_F on thin-layer chromatograms in solvent						
	(a)	(b)	(c)	(d)	(e)	(f)	(g)	(h)	(i)	(j)	(k)	(l)	(m)
2-Naphthylamine	0.84	0.86	0.93	0.80	0.92	0.95	0.80	0.60	0.50	0.63	0.80	0.78	
2-Acetamidonaphthalene	0.95	0.92	0.91	0.91	0.90	0.85	0.57	0.25	0.87	0.35	0.38	0.70	
2-Naphthylhydroxylamine						0.95	0.70	0.35		0.38	0.33	0.62	
N-Acetyl-2-naphthylhydroxylamine	0.90	0.95	0.92	0.83	0.85	0.95	0.51	0.20	0.08	0.34	0.29	0.45	
2-Amino-1-naphthol		0.85					0.65	0.30					
2-Acetamido-1-naphthol	0.87	0.88	0.95	0.77	0.95	0.95	0.65	0.30	0.75	0.32	0.30	0.60	
Oxidation product of 2-acetamide-1-naphthol[b]	0.87	0.88	0.95	0.77	0.95	0.75	0.30	0.00	0.50	0.18	0.07	0.45	
N-Benzyloxycarbonyl-2-naphthylhydroxylamine	0.94	0.96	0.92	0.92	0.92	0.98	0.75	0.45	0.92	0.50	0.63	0.90	
2-Nitrosonaphthalene[c]						0.90	0.95	0.75	0.95	0.98	0.98	0.98	
2-Amino-1-naphthyl hydrogen sulfate	0.44	0.71	0.65	0.24	0.24	0.57							0.60
2-Acetamido-1-naphthyl hydrogen sulfate	0.60	0.71	0.73	0.52	0.49	0.66							0.66
N-Acetyl-2-naphthylhydroxylamine-O-sulfonic acid[d]	0.60					0.66							
N-Benzyloxycarbonyl-2-naphthyl-hydroxylamine-O-sulfonic acid						0.77							
2-Naphthylhydroxylamine-N-sulfonic acid	0.50	0.56	0.52	0.22	0.24	0.51							0.55
2-Amino-1-naphthyl dihydrogen phosphate	0.07	0.70	0.65	0.06	0.05	0.40							0.70
2-Amino-1-naphthylmercapturic acid	0.43	0.92	0.91	0.18	0.16	0.36							0.68
Metabolite, probably S-(2-acetamido-5,6-dihydro-6-hydroxy-5-naphthyl)-cysteine:													
Before treatment with acid	0.20	0.70	0.48	0.09	0.05								
After treatment with acid	0.34	0.73	0.57	0.20	0.16								

(2-Amino-1-naphthylglucosid)uronic acid	0.19	0.65	0.56	0.02	0.02	0.31	0.00	0.36
(2-Acetamido-1-naphthylglucosid)-uronic acid	0.28	0.81	0.74	0.07	0.06	0.35	0.00	0.45
(N-Acetyl-2-naphthylhydroxylamine-O-glucosid)uronic acid	0.30	0.78	0.76	0.08	0.07	0.31	0.00	0.40
(2-Naphthylamine-N-glucosid)uronic acid	0.28			0.06	0.04	0.33		
(2-Naphthylhydroxylamine-N-glucosid)uronic acid[e]						0.33		

[a] Composition of solvents are given in text and other details in Boyland and Manson[70].

[b] This substance, blue in color, formed during chromatography of 2-acetamido-1-naphthol. On paper chromatograms it did not separate from 2-acet-amido-1-naphthol.

[c] On paper chromatograms it gave an unidentified substance that did not react with sodium amminoprusside.

[d] Converted into 2-acetamido-1-naphthyl hydrogen sulfate and 2-acetamido-1-naphthol on paper chromatograms during the usual length of run (16 h); the conversion was least with solvent (a). The compound also decomposed on thin-layer chromatography in solvents (i) and (m).

[e] On paper chromatograms it was converted into (2-naphthylamine N-glucosid)uronic acid.

TABLE 7

COLOR REACTIONS OF 2-NAPHTHYLAMINE AND SOME DERIVATIVES ON CHROMATOGRAMS[a]

Compound	HCl, NaNO$_2$ and hexylresorcinol	Red color with p-dimethylamino-cinnamaldehyde	Reduction of ammoniacal AgNO$_3$	Sodium amminoprusside	Alkaline hexyl-resorcinol	Other tests
2-Naphthylamine	Yellow	Immediate	Slow	Green	None	
2-Acetamidonaphthalene	None	None	None	None	None	Gives the reactions of 2-naphthylamine after acid hydrolysis
2-Naphthylhydroxylamine	Yellow	Immediate	Immediate	Mauve	Blue-green	Yellow with 2 N NaOH
N-Acetyl-2-naphthyl-hydroxylamine	Transient green, becoming grey	Slow	Slow	None	None	Mauve with FeCl$_3$
2-Amino-1-naphthol	Blue[b]	Immediate	Immediate	None	Blue[b]	Blue with 2 N NaOH or NH$_3$ vapor
2-Acetamido-1-naphthol	Blue[c]	None	Immediate	Blue	None	Red with diazotized sulphanilic acid; blue with FeCl$_3$ or HCl + NaNO$_2$
N-Benzyloxycarbonyl-2-naphthylhydroxylamine	Transient green, becoming grey	Slow	Slow	None	None	Mauve with FeCl$_3$
2-Nitrosonaphthalene	None	Slow	Slow	Mauve	Grey-green	Mauve with TiCl$_3$; brownish red with aniline
2-Amino-1-naphthyl hydrogen sulphate	Mauve	Immediate	Slow	Green	None	Spot turns reddish brown in sunlight
2-Acetamido-1-naphthyl hydrogen sulphate	None	None	None	None	None	Stable in 2 N NaOH; decomposes in 2 N HCl to give 2-acetamido-1-naphthol
N-Acetyl-2-naphthyl-hydroxylamine-0-sulphonic acid	Blue-green[d]	Slow	None	None	Blue-green	For breakdown in acid and alkali see the Methods and Materials section
N-Benzyloxycarbonyl-2-naphthylhydroxylamine-0-sulphonic acid	Blue-green[d]	Slow	Slow	None	Blue-green	After treatment with 2 N HCl for 1 h gives blue color with aq. NaNO$_2$
2-Naphthylhydroxylamine-N-sulphonic acid	Yellow	Immediate	Immediate	Mauve (Slow)	None	Yellow color a few min after spraying with 2 N HCl

Compound						
2-Amino-1-naphthyl dihydrogen phosphate	Reddish brown (mauve if diazotized with NaNO$_2$ and acetic acid)	Immediate	Slow	Green	None	Positive reaction to K$_2$Cr$_2$O$_7$–AgNO$_3$ and platinic iodide reagents
2-Amino-1-naphthylmercapturic acid	Yellow	Immediate	Slow	None	None	Mauve with ninhydrin; pink fluorescence after treatment with 2 N HCl at room temp. followed by NH$_3$ vapor; diazotizes and couples with hexylresorcinol (orange-red color) after treatment with hot 2 N HCl; positive reaction to K$_2$Cr$_2$O$_7$–AgNO$_3$ and platinic iodide reagents
Metabolite, probably S-(2-acetamido-5,6-dihydro-6-hydroxy-5-naphthyl)-L-cysteine	None	None	None	None	None	
(2-Amino-1-naphthyl-glucosid)uronic acid	Reddish brown (mauve if diazotized with NaNO$_2$ and acetic acid)	Immediate	Slow	Green	None	
(2-Acetamido-1-naphthyl-glucosid)uronic acid	None	None	None	None	None	Gives the reactions of the amino compound after treatment with hot 2 N HCl
(N-Acetyl-2-naphthyl-hydroxylamine O-glucosid)-uronic acid	Blue-green[d]	Slow	None	None	Blue-green	Yellow with 2 N NaOH; after treatment with 2 N NaOH or 2 N HCl the spot is diazotizable owing to the formation of 2-naphthylamine
(2-Naphthylamine N-glucosid)uronic acid	Yellow	Immediate	Slow	Green	None	
(2-Naphthylhydroxylamine-N-glucosid)uronic acid	Yellow	Immediate	Immediate	Mauve	Blue-green	

[a] Composition of detecting reagents given in Boyland et al.[71] and in text.
[b] This color is due to oxidation in the presence of alkali.
[c] This color is due to the action of nitrous acid and no further color is given with hexylresorcinol.
[d] This color is due to breakdown to 2-naphthylhydroxylamine, which reacts with hexylresorcinol in alkaline solution.

TABLE 8

PAPER CHROMATOGRAPHY OF SOME METABOLITES OF 2-NAPHTHYLAMINE IN BILE IN THE RAT[a]

Compound	R_F Solvent (a)	R_F Solvent (b)	Fluorescence	Reaction with ninhydrin	Reaction with $K_2Cr_2O_7$–$AgNO_3$	
N-Acetyl-S-(2-acetamido-5,6-dihydro-6-hydroxy-5-naphthyl)-L-cysteine:						These metabolites give 2-acetamido-6-naphthol on treatment with 2 N HCl at room temperature; after heating with 2 N HCl they diazotize and couple with hexylresorcinol to give orange-red colors.
Before acid treatment	0.26	0.80	None	None	+	
After acid treatment[b]	0.41	0.94	Orange-pink	None	+	
Metabolite, probably N-acetyl-S-(2-acetamido-5,6-dihydro-6-hydroxy-5-naphthyl)-glutathione:						
Before acid treatment	0.12	0.65	None	+	+	
After acid treatment[b]	0.24	0.66	Orange-pink	+	+	
Metabolite, probably N-acetyl-S-(2-acetamido-5,6-dihydro-6-hydroxy-5-naphthyl)cysteinylglycine:						
Before acid treatment	0.16	0.65	None	+	+	
After acid treatment[b]	0.30	0.70	Orange-pink	+	+	
Unidentified metabolites:						These metabolites do not give 2-acetamido-6-naphthol or show a change in R_F value or the development of an orange-pink fluorescence on treatment with 2 N HCl; after heating with 2 N HCl they diazotize and couple with hexylresorcinol to give orange-red colors.
Major spots	0.00	0.07	None	+	+	
	0.03	0.15	None	+	+	
	0.09	0.30	None	+	+	
Minor spots	0.03	0.14	None	+	+	
	0.06	0.30	None	+	+	
	0.06	0.40	None	+	+	

[a] Composition of solvents given in text.

[b] The spots were treated for a few minutes with 3 N HCl at room temperature and then neutralized with NH$_3$ vapor.

TABLE 9

METABOLITES OF 2-NAPHTHYLAMINE AND SOME DERIVATIVES DETECTED IN URINE OF DIFFERENT SPECIES[a]

Compound administered: Position of hydroxyl or thio ether group in the metabolite: Group in 2 position of metabolite

Species	Metabolite	2-Naphthylamine						2-Acetamidonaphthalene					
		N-		1-		6- and 5,6-		N-		1-		6- and 5,6-	
		Hydroxyl-amino	N-Acetyl-hydroxyl-amino	Amino	Acet-amido	Amino	Acet-amido	Hydroxyl-amino	N-Acetyl-hydroxyl-amino	Amino	Acet-amido	Amino	Acet-amido
Dog	Free hydroxyl	+	0	0	0	tr.	0	0	0	tr.	0	0	+
	Sulfuric ester			++	0	tr.	0	0	0	0	0	0	++
	Glucosiduronic acid			+	0		0		+				++
	Mercapturic acid						0						++
Guinea pig	Free hydroxyl	0	0	0	0	0	+	0	0	0	0	0	+
	Sulfuric ester			0	+	0	+			+	+	0	++
	Glucosiduronic acid						++						++
	Mercapturic acid						tr.						tr.
Hamster	Free hydroxyl	0	0	tr.	0	0	0	0	0	tr.	0	0	+
	Sulfuric ester			+	+	0	+		0	+	+	0	++
	Glucosiduronic acid			0			++						++
	Mercapturic acid						0						0
Rabbit	Free hydroxyl	0	0	tr.	0	0	+	0	0	tr.	0	0	+
	Sulfuric ester			+	0	0	+			+	0	0	+
	Glucosiduronic acid			0			++						++
	Mercapturic acid						++						++
Rat	Free hydroxyl	0	0	+	0	tr.	+	0	0	+	0	tr.	+
	Sulfuric ester			+	+	tr.	++		0	+	+	tr.	++
	Glucosiduronic acid			+			+++			+			++
	Mercapturic acid						++						++

[a] The table does not include 2-formamido-1-naphthylhydrogen sulfate[72], 2-nitrosonaphthalene[72], (2-naphthylamine-N-glucosid)uronic acid and naphthylsulfamic acid[72], and bis-(2-amino-1-naphthyl)hydrogen phosphate[72]. The N-acetyl compounds in the "6- and 5,6-" columns include conjugates of both 2-acetamido-6-naphthol and 2-acetamido-5,6-dihydro-5,6-dihydroxynaphthalene (found in rat, rabbit, hamster and guinea pig urine after dosing with 2-naphthylamine or 2-acetamidonaphthalene) and in dog urine after 2-acetamidonaphthalene only. 2-Acetamido-6-hydroxy-5-naphthyl hydrogen sulphate was found in rat and rabbit urine after dosing with 2-naphthylamine or 2-acetamidonaphthalene and in dog urine after 2-acetamidonaphthalene. + + indicates the principal metabolites or groups of metabolites when compared with those marked +; tr. indicates spots that gave only faint color reactions; 0 indicates metabolites that were looked for with the aid of reference compounds but were not detected. Vacant spaces have been left where reference compounds were not available for comparison but no evidence was obtained to indicate the presence of these metabolites. Dogs and hamsters were not dosed with N-acetyl-2-naphthylhydroxylamine.

TABLE 9 (cont.)

Compound administered: Position of hydroxyl or thio ether group in the metabolite: Group in 2 position of metabolite

Species	Metabolite	2-Naphthylhydroxylamine						N-Acetyl-2-naphthylhydroxylamine					
		N-		1-		6- and 5,6-		N-		1-		6- and 5,6-	
		Hydroxyl-amino	N-Acetyl-hydroxyl-amino	Amino	Acet-amido	Amino	Acet-amido	Hydroxyl-amino	N-Acetyl-hydroxyl-amino	Amino	Acet-amido	Amino	Acet-amido
Dog	Free hydroxyl	0			0		0						
	Sulfuric ester	0		++	0	tr.	0						
	Glucosiduronic acid			+		tr.	0						
	Mercapturic acid			+									
Guinea pig	Free hydroxyl			0	0		tr.		0				+
	Sulfuric ester			+		0	tr.		0	+	0	0	tr.
	Glucosiduronic acid	0		0	+	0	++		+	+	0	0	++
	Mercapturic acid						tr.						tr.
Hamster	Free hydroxyl			tr.	0		+						
	Sulfuric ester	0		++		0	+						
	Glucosiduronic acid			+	+	0	++						
	Mercapturic acid						0						
Rabbit	Free hydroxyl	0		++	0		tr.		0				0
	Sulfuric ester	0		++	0	0	tr.		0	tr.	0	0	tr.
	Glucosiduronic acid			+		0	tr.		++	tr.	0	0	tr.
	Mercapturic acid						tr.						0
Rat	Free hydroxyl			++	0		+		0				tr.
	Sulfuric ester			++		0	++		0	tr.	+	0	tr.
	Glucosiduronic acid	0		+	+	0	++		+	tr.	+	0	tr.
	Mercapturic acid						++						0

TABLE 10

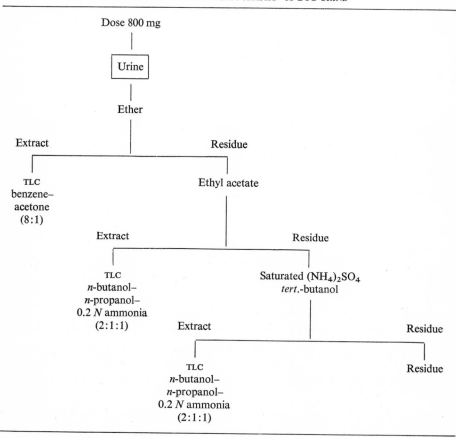

[a] Thin-layer chromatography on Silica Gel G.

the CCl_4 extracts of the urine (following pH adjustment to 4.5–5.0 with 1 N H_2SO_4 and made 6.5 mM in $K_3[Fe(CN)_6]$) of a phenobarbital pretreated dog receiving 730 mg of 2-naphthylamine was chromatographed on a column of silica gel (Woelm) and developed with n-hexane yielding 2-nitrosonaphthalene in a bright green front-running band (a total of 5.4 mg of product, m.p. 61–62°C, following removal of hexane *in vacuo* at 20–25°C). TLC of 2-nitrosonaphthalene on Silica Gel HF_{254} (Merck) developed with (*1*) petroleum ether (40–60°C)–acetone (4:1) and (*2*) n-hexane–benzene (5:1) gave R_F values of 0.55 and 0.30, respectively. The color development was performed with the TPF reagent[74] (5% aq. trisodium pentacyanoamine ferrate).

Extraction of the urines without added ferricyanide for the oxidation of the N-hydroxy-2-naphthylamine to the nitroso compound revealed the presence from 13.5 to 22.5% of 2-nitrosonaphthalene. The TLC separation of N-hydroxy-2-naphthylamine and 2-nitrosonaphthylamine from ether extracts of the urines with and without addition of ferricyanide confirmed a similar ratio of free N-hydroxy-2-naphthylamine and 2-nitrosonaphthylamine.

Radomski and Brill[75] further elaborated the role of N-hydroxy metabolites of aromatic amines in the induction of bladder cancer and also found by TLC and GLC that in the dog N-hydroxynaphthylamines are further oxidized to the respective 1- and 2-nitrosonaphthalenes. After the administration of 2-naphthylamine to dogs, the urine collected by catheter was adjusted to pH 4, then ferricyanide was added to convert the N-hydroxy-2-naphthylamine to 2-nitrosonaphthalene. The urine was then extracted with petroleum ether, and the extract determined by electron-capture gas chromatography. The peak obtained represented the sum of N-hydroxy and nitroso compounds present and is referred to as total N-oxidation products. Ferricyanide could not be used in the measurement of the N-oxidation products of 1-naphthylamine. Adjustment of the urine pH to 6.5, however, allowed the extraction of both N-hydroxy and nitroso metabolites with petroleum ether. N-hydroxy-1-naphthylamine was quantitatively converted to 1-nitrosonaphthalene on the GLC column at 125°C. The sensitivity of the electron-capture detector was 100 pg of these substances, thus enhancing the measurement of 40 p.p.b. in 5 ml of urine. This procedure was also found applicable to the determination of the N-hydroxy metabolites of 4-aminobiphenyl, a bladder carcinogen found to be distinctly more potent than 2-naphthylamine[76].

The quantitative determination of both 1- and 2-naphthylamines in cigarette smoke was described by Masuda and Hoffmann[54]. The basic non-volatiles of the smoke of 300 cigarettes were reacted with pentafluoropropionic anhydride and the resulting neutral components were chromatographed on Florisil. The concentrates of the N-pentafluoropropionamides of the naphthylamines were analyzed by electron-capture gas chromatography and the main stream smoke of an 85 mm U.S. non-filter cigarette was found to contain 27 and 22 ng of 1-naphthyl- and 2-naphthylamines, respectively.

A Wilkins gas chromatograph Model 1525 with flame-ionization detector was used to assess the purity of the free 1-naphthyl- and 2-naphthylamines and a Varian Aerograph Model 1200 with an electron-capture detector for the quantitative analysis, and a Perkin-Elmer gas chromatograph Model 800 with flame-ionization detector for the isolation of unknowns. The β-radiation of the ^{14}C-labeled internal standards (1-naphthylamine-1-^{14}C and 2-naphthylamine-8-^{14}C) was counted with a Nuclear Chicago Scintillation System 720. The naphthylamines were purified on basic alumina, recrystallized and checked for purity by GLC at 175°C on a 2.5 m × 3 mm column packed with 10% Siponate DS-10 on Chromosorb W. The pentafluoropropionamides of 1- and 2-naphthylamines were separated at 145°C on 1.7 m × 3 mm glass columns filled with 7.5% QF-1 and 5% DC-200 on Gas Chrom Q (the retention times were 7.1 and 9.5 min, respectively). For the gas chromatographic separation of the N-1- and N-2-naphthylpentafluoropropionamides from a concentrate from cigarette smoke, a 2 m × 6 mm glass column was used at 160 °C and the liquid phases and support were as described above. The column exit was connected with a glass splitter which led 10 % of the effluent to a flame-ionization detector and the remainder was collected.

Figure 14 illustrates the analytical separation procedure for 1- and 2-naphthylamines from cigarette smoke. Figure 15 shows gas chromatograms for concentrates of N-1-, and N-2-naphthylpentafluoropropionamide from cigarette smoke.

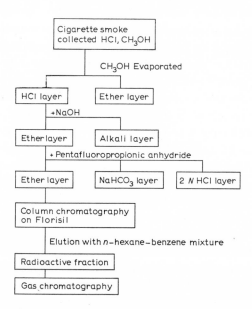

Fig. 14. Analytical separation precedure for 1-naphthylamine and 2-naphthylamine from cigarette smoke, internal standard added.

Masuda et al.[52] described the formation and detection of naphthylamines from the pyrolysis of amino acids using column, paper, thin-layer, and gas chromatography. From 100 g of L-glutamic acid pyrolyzed at 700 °C, 1.3 mg of 1-naphthylamine and 0.13 mg of 2-naphthylamine were isolated; from 100 g of L-leucine, 0.7 mg and 0.01 mg of the isomers, respectively, were isolated. However, when the amino acids were pyrolyzed at 500 °C, neither aromatic amine was found. Both 1- and 2-naphthylamines were purified by column chromatography on alumina and recrystallized until TLC showed no contaminants. TLC utilized Silica Gel G (Merck) with n-hexane–ether (2:3) and n-hexane–ether–acetic acid (25:25:1) for development, and paper chromatography was performed on partially acetylated paper prepared by acetylating Toyo Roshi No. 50 paper according to the method of Spotswood[77] and developed with ethanol–toluene–water (17:4:1). The gas chromatographic equipment was a Shimadzu Model GC-1B with a hydrogen-flame-ionization detector and equipped with a 1.5 m × 4 mm stainless steel column packed with 10 % sodium dodecyl benzenesulfonate on acid-washed Chromosorb W (60–80 mesh). Temperatures of the column and detector were maintained at 219.5 and 220°C, respectively, and the flow rate and inlet pressure of the nitrogen carrier gas were 32 ml/min and 2.0 kg/cm². Figure 16 illustrates gas chromatograms of naphthylamine extracts from

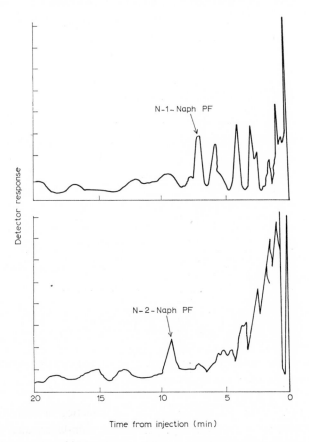

Fig. 15. Gas chromatograms for concentrates of N-1-, and N-2-naphthylpentafluoroproprionamide from cigarette smoke.

pyrolyzed L-glutamic acid and Table 11 lists the retention times of reference aromatic amines on 10% sodium dodecyl benzenesulfonate chromatographed at 219.5°C. Table 12 lists the R_F values of reference aromatic amines on silica gel plates developed with n-hexane–ether–acetic acid (25:25:1). The presence of naphthylamines in the pyrolyzates was confirmed by the identity of R_F values on TLC chromatograms and by the three positive color reactions, e.g. a purplish-red spot with N-(β-diethylaminoethyl)-1-naphthylamine reagent on partially acetylated paper, and both a brownish-yellow spot with p-diethylaminobenzaldehyde and a brownish-red spot with diazotized benzidine reagent on a TLC plate.

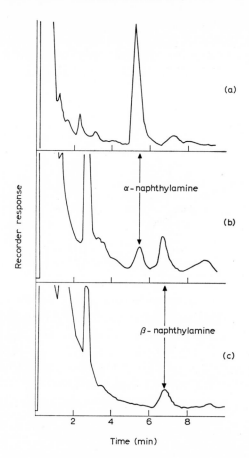

Fig. 16. Gas chromatography of naphthylamine extracts from pyrolyzed L-glutamic acid. (a), (b), Extracts from the first thin-layer plate corresponding to 1- and 2-naphthylamine, respectively; (c), Extract from the second thin-layer plate corresponding to 2-naphthylamine.

TABLE 11

RETENTION TIMES OF AROMATIC AMINES CHROMATOGRAPHED ON 10% SODIUM DODECYLBENZENESULFONATE AT 219.5 °C

Compound	Retention time (min)
Aniline	0.7
o-Toluidine	0.8
p-Toluidine	0.9
o-Aminodiphenyl	3.4
1-Naphthylamine	5.5
2-Naphthylamine	6.8
p-Aminodiphenyl	12.2
p-Nitroaniline	30.3

TABLE 12

TLC OF AROMATIC AMINES ON SILICA GEL DEVELOPED WITH *n*-HEXANE–ETHER–ACETIC ACID
(25:25:1)

Compound	R_F
o-Aminodiphenyl	0.67
1-Naphthylamine	0.38
1-Aminoanthracene	0.37
3-Aminopyrene	0.35
o-Toluidine	0.33
2-Naphthylamine	0.29
2-Aminoanthracene	0.29
p-Aminodiphenyl	0.27
Aniline	0.26
2-Aminofluorene	0.15
p-Toluidine	0.13
o-Toluidine	0.05
Benzidine	0.02

1-Naphthylamine is prepared by the reduction of 1-nitronaphthalene with iron and hydrochloric acid. The commercial-grade product contains up to 10% of 2-naphthylamine and is used in the preparation of azo dyes and growth regulators and is also used as a corrosion inhibitor, anti-oxidant for hydrocarbons and with epichlorohydrin in the synthesis of epoxy resins. Although 1-naphthylamine (or its hydroxylated metabolite, N-hydroxy-1-naphthylamine) is considered to be noncarcinogenic, Belman *et al.*[65] reported that N-hydroxy-1-naphthylamine injected intraperitoneally in rats is much more carcinogenic than N-hydroxy-2-naphthylamine.

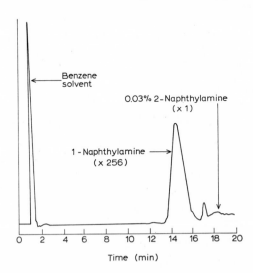

Fig.17. Gas chromatographic separation of 1- and 2-naphthylamines on 3% Siponate DS-10 at 240°C.

The determination of small concentrations of 2-naphthylamine in 1-naphthyl-amine has been achieved primarily by GLC procedures. Quick et al.[78] separated the naphthylamines on 3% dodecylbenzenesulfonate. An F & M Model 720 gas chromato-graph was used equipped with a thermal conductivity detector and a 15 ft. $\times \frac{1}{4}$ in. o.d. aluminum tubing packed with 3% Siponate-DS-10 (dodecylbenzene sodium sulfo-nate, Alcolac Chem. Corp.) on Haloport F. The operating conditions were column 240°C, injection port 310°C, detector 330°C, detector current 150 mA, carrier gas helium at 100 ml/min, and sample size 0.01–0.05 ml (sample of naphthylamine dis-solved in benzene in ratio of 2:1 of benzene to sample). The GLC separation of 1- and 2-naphthylamines is shown in Fig. 17.

The reactivity of acetone when used to dissolve the naphthylamines for analysis (and allowed to stand at room temperature) was demonstrated by the finding of a condensation product (mass 183) which was collected at the exit port of the gas chromatograph. The product is suggested to arise by a reaction similar to the Leuckart reaction[79] which produces Schiff bases (anils) from ketones and amines as shown.

The GLC determination of 2-naphthylamine in 1-naphthylamine by means of either N-naphthyl acetamide or the N,N-dimethylnaphthylamines has been described by Marmion et al.[80]. Acetylation of the primary amines was effected directly on the GLC column by consecutive injection of the amine and acetic anhydride. The N,N-dimethylnaphthylamines were synthesized by the external reaction with dimethyl-sulfate. For the analysis of the N-naphthylamides, an Aerograph Antoprep Model A-700 gas chromatograph was used with a standard thermal conductivity detector and connected to a 0–10 mV recorder. A Kintel 111 BF amplifier was placed in the circuit to enhance sensitivity. The chromatographic column was an 8 ft. $\times \frac{1}{4}$ in. o.d. stainless steel tube packed with 60–80 mesh acid-washed Chromosorb W coated with 15% SE-30. The instrument parameters were detector current and temperature, 200 mA and 232°C, column temperature 220°C, injection port temperature 287°C, carrier gas, helium at 50 ml/min, recorder speed 40 min/h, amplification 10 × 1. 0.025 ml of a 20% chloroform solution of sample was injected into the column fol-lowed immediately with 0.050 ml of acetic anhydride. The area of the N-2-naphthyl-acetamide peak was compared to that of a standard sample containing a known amount of 2-naphthylamine. A typical chromatogram of the GLC of the N-naphthyl-acetamide is shown in Fig. 18.

The GLC of the N,N-dimethylnaphthylamines was carried out using an F & M Model 500 gas chromatograph equipped with a hot-wire thermal conductivity detec-tor and a 10 ft. $\times \frac{1}{4}$ in. o.d. stainless steel tube packed with 15% Carbowax 20M coated on 70–80 mesh Anakrom AB (Analabs). The operating parameters were

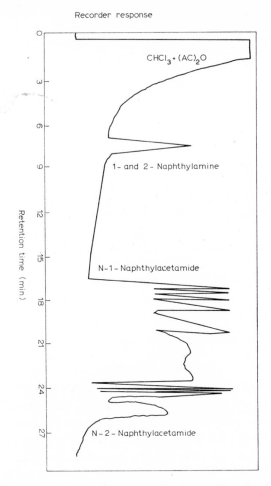

Fig. 18. Gas chromatography of naphthylamines as N-naphthylacetamides.

detector current and temperature 230 mA and 250°C, column temperature 225°C, injector port temperature 320°C, carrier gas helium 60 ml/min, recorder speed 30 in./h. Figure 19 shows a chromatogram of the N,N-dimethyl derivatives of the naphthylamines following methylation with dimethyl sulfate. This preferred analytical procedure was found repeatable to about 0.04% absolute 2-naphthylamine in the 0–2% range. The procedure utilizing the direct acetylation on column was found repeatable to about 0.02% absolute 2-naphthylamine in the 0–1% range, but although fast and simple, caused detector deteriorations.

   A microanalysis of 2-naphthylamine in commercial 1-naphthylamine using TLC and spectrofluorometry was described by Matsushita[81]. Silica gel (Woelm) plates were prepared with 1 N sodium sulfate and activated at 110°C for 60 min and developed with cyclohexane–ether–acetic acid (50:50:2) to effect the separation of

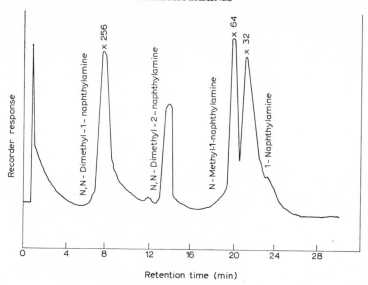

Fig. 19. Gas chromatography of the naphthylamines following methylation with dimethyl sulfate.

TABLE 13

TLC OF SOME AROMATIC AMINES[a]

Compound	Solvents						
	A		B		C		
	Water (1)	Zn (2)	Water (1)	Zn (2)	Water (1)	Zn (2)	Cd (3)
Aniline	0.63	0.33	0.53	0.28	0.47	0.25	0.21
o-Toluidine	0.71	0.57	0.60	0.45	0.51	0.41	0.29
m-Toluidine	0.67	0.35	0.55	0.27	0.47	0.24	0.15
p-Toluidine	0.64	0.26	0.50	0.17	0.43	0.15	0.10
2,4-Xylidine	0.71	0.42	0.58	0.31	0.48	0.16	0.25
2-Methyl-5-isopropylaniline	0.76	0.64	0.65	0.54	0.56	0.45	0.32
N-Methylaniline	0.78	0.58	0.68	0.56	0.62	0.59	0.57
N,N-Dimethylaniline	0.82	0.73	0.74	0.67	0.69	0.65	0.61
N,N-Diethylaniline	0.84	0.45	0.78	0.28	0.73	0.35	0.35
N,N-Diethyl-p-toluidine	0.84	0.48	0.79	0.24	0.71	0.13	
p-Phenetidine	0.59	0.23	0.49	0.11	0.31	0.06	0.06
Methyl-p-aminobenzoate	0.65	0.69	0.49	0.50	0.46	0.52	0.45
N-Benzylaniline	0.87	0.87	0.79	0.85	0.71	0.79	0.79
Diphenylamine	0.83	0.86	0.76	0.85	0.71	0.78	0.80
1-Naphthylamine	0.70	0.67	0.60	0.59	0.52	0.53	0.44
2-Naphthylamine	0.66	0.52	0.55	0.44	0.47	0.32	0.21
4-Amino-1-naphthol	0.37	0.09	0.34	0.12	0.27	0.07	0.07

Solvents: *A* chloroform–methanol (10:1), *B* benzene–methanol (5:1), and *C* benzene–methylethyl ketone (3:1).
[a] Absorbents: *(1)* silica gel prepared in water; *(2)* and *(3)* one part silica gel mixed with 2 parts by weight of 10% zinc nitrate or cadmium nitrate, respectively.

30 μg of commercial 1-naphthylamine and 0.5, 1.0, and 1.5 μg of standard 2-naphthyl-amine in ethanol–acetic acid (95:5). The plates were developed in 26 min (10 cm). The 2-naphthylamine spots were detected with long-wave ultraviolet light (365 nm) and extracted with 3 ml portions of ethanol–acetic acid (95:5), and the fluorescence measured at 400 nm. Commercial 1-naphthylamine for industrial use and practical grade reagent for laboratory use were found to contain about 4 % and 3 % 2-naphthyl-amine, respectively.

The TLC separation of aromatic amines has also been accomplished[82] using Silica Gel G (Warner-Chilcott) prepared in (1) water and (2) with 10 % zinc or cad-mium nitrate in 1–2 parts silica gel. The developing solvents were chloroform–methanol (10:1), benzene–methanol (5:1), and benzene–methylethyl ketone (3:1). Table 13 lists the R_F values obtained using silica gel plates prepared as described above.

TABLE 14

R_F VALUES × 100 AND SPOT COLORS OF VARIOUS NAPHTHALENE DERIVATIVES ON WHATMAN NO. 4 PAPER

Compound	Solvent systems[a]						Detection[b]	
	1	2	3	4	5	6	Fluorescence	Color
1-Naphthol	90	89	66	31	10	7	Blue-green	Yellow
2-Naphthol	87	87	62	21	6	4	Blue	Orange
1,1′-Dinaphthol	95	95		8	0		Blue	Red-violet
2,2′-Dinaphthol	96	96		59	21			
Unknown substance in 2-naphthol	96	96		39	11		Blue	Orange
1-Iodo-2-naphthol				83	63			Orange
1,3-Dihydroxynaphthalene	30	21		0	0		Blue	Yellow
1,5-Dihydroxynaphthalene	32	23	3	1	0	0	Light blue	Red-orange
1,6-Dihydroxynaphthalene	27	22					Yellow	Orange
1,7-Dihydroxynaphthalene	27	22					Blue	Red-orange
1,8-Dihydroxynaphthalene	80	81	48	2	0	2	Light blue	Orange
2,3-Dihydroxynaphthalene	40	37					Violet	Red-violet
2,6-Dihydroxynaphthalene	18	18					Violet	Red-violet
2,7-Dihydroxynaphthalene	20	19					Violet	Orange-red
1-Naphthylamine	92	93	93	73	45	8	Blue	Violet-blue
2-Naphthylamine	89	92	92	68	38	5	Violet-blue	Violet-blue
1,5-Diaminonaphthalene	46	51	40	5	2	0	Violet	Blue-green
1,8-Daminonaphthalene	87	88	85	58	23		Blue-violet	Violet-blue
1,6-Aminonaphthol	32	31	5	2	0	0	Blue	Violet-red
1,7-Aminonaphthol	33	33					Blue	Blue-violet
1-N-Phenylnaphthylamine	96	96	95	94	94	52	Blue	Blue-violet
2-N-Phenylnaphthylamine	95	95	94	93	93	39	Blue	Blue-violet

[a] Solvents: (1) formamide/benzene–ethyl acetate (8:2), (2) formamide/chloroform–ethyl acetate (8:2), (3) formamide–chloroform, (4) formamide–carbon tetrachloride, (5) formamide–cyclohexane, and (6) dimethylformamide–cyclohexane.
[b] Fluorescence colors obtained after exposure of the paper to ammonia. Spot colors obtained follow-ing spray with diazotized p-nitroaniline.

The paper chromatography of a variety of naphthalene compounds including amines, diamines, hydroxy, and dihydroxy derivatives was described by Latinak[83]. Impregnated Whatman No. 4 paper was used with the following solvents: (*1*) formamide/benzene–ethyl acetate (8:2), (*2*) formamide/chloroform–ethyl acetate (8:2), (*3*) formamide–chloroform, (*4*) formamide–carbon tetrachloride, (*5*) formamide–cyclohexane, and (*6*) dimethylformamide–cyclohexane. The spots were detected using ultraviolet light and spraying with diazotized *p*-nitroaniline or with a mixture of ferric chloride and ferricyanide[84]. Table 14 depicts the R_F values and spot colors of various naphthalene derivatives on Whatman No. 4 paper.

3. URETHAN AND N-HYDROXYURETHAN

Urethan (ethyl carbamate) is prepared by a variety of reactions. The predominant commercial process involves the reaction of ethyl chloroformate with ammonia. Other synthetic procedures involve (*a*) the reaction of ethanol with carbamoyl chloride, cyanic acid or urea at elevated pressure, (*b*) the reaction of urea nitrate, ethanol and sodium nitrite, and (*c*) the ammonolysis of diethyl carbamate.

Derivatives of carbamic acid (urethans) are widely used in the plastics industry as monomers, co-monomers, plasticizers, and fiber and molding resins, in textile finishing, in agricultural chemicals as herbicides, in insecticides and insect repellants, fungicides and molluskicides, in pharmaceutical chemicals as psychotropic drugs, hypnotics and sedatives, anticonvulsants, miotics, anesthetics and antiseptics. Urethans are also used as surface-active agents, selective solvents, dye intermediates and corrosion inhibitors. N-nitrocarbamates have been suggested as diesel fuel additives and urethan–aldehyde condensation products have been used as latent flavor-developing substances in packaged foods. Urethan itself has been used medically as a mild hypnotic, sedative and antispasmodic and in the treatment of chronic myeloid leukemia and related blood diseases as well as an antidote for strychnine, resorcinol and picrotoxin poisoning. Urethan is a "multipotential" carcinogen for several strains of mice[85] (inducing malignant lymphomas of the thymus, hepatomas, mammary carcinomas, hemangiomas, and lung adenomas in mice[86-88] and in rats[89]), induces carcinomata of the fore-stomach in hamsters and is teratogenic in mice[90], hamsters[91,92], fish[93], and amphibia[94]. Urethan is mutagenic in *Drosophila*[95,96], bacteria[97,98], and plants[99,100], non-mutagenic in mice[101] and *Neurospora*[102,103], and induces chromosome aberrations in *Oenothera*[104].

The mutagenicity of congeners of urethan (*e.g.* alkyl carbamates) in bacteria is conflicting with both positive[105] and negative [106]effects reported. The chemistry[107], carcinogenicity[108,109], and metabolism of urethan and N-hydroxyurethan[109-111] have been reviewed. Urethan is metabolized by mammals to N-hydroxyurethan and N-acetyl-N-hydroxyurethan[110-112], and according to Boyland *et al.*, the carcinogenic and antileukemic effects attributed to urethan are probably caused by the hydroxyurethan metabolites which act as alkylating agents toward mercaptoamino acids and react with cytosine residues of RNA. The mechanism is similar to, but distinct from, that of the action of alkylating agents which react mainly with guanine of nucleic acid; however, in both cases, the same base pairs, guanine–cytosine are modified. However, it has been suggested by Kaye and Trainin[113] that the carcinogenic actions of N-hydroxyurethan can be attributed to its rapid conversion back to urethan. N-hydroxyurethan and urethan are reported to be almost equal in their carcinogenic activity[88] and in their ability to produce chromosome abnormalities in cells of the Walker rat carcinoma[112]. The enhanced anti-tumor activity of N-hydroxyurethan (compared to urethan) in various mouse tumors[114], as well as its teratogenicity in the rat[115], the induction of chromosome aberrations in mammalian cells in culture[116, 117] and *Vicia faba*[112], and its inactivation of transforming DNA[118, 119] have also been reported.

It has been suggested[120] that even N-hydroxyurethan does not itself react with DNA, but reacts with oxygen, yielding peroxy radicals and other derivatives, some of which are believed to be the active reagents.

The metabolism of urethan and related compounds in the rat, rabbit, and man and the probable role of N-hydroxyurethan in carcinogenesis and chemotherapy was elaborated by Boyland and Nery[110]. Urinary metabolites were isolated and detected on Whatman No.1 chromatograms employed for descending development in the following solvent systems: (*a*) *n*-butanol–*n*-propanol–aq. 2 *N* ammonia (2:1:1), (*b*) *n*-butanol–acetic acid–water (12:3:5), and on Silica Gel G (Merck) plates developed with (*c*) acetone–light petroleum (b.p. 40–60°C) (3:7), (*d*) chloroform–acetone–light petroleum (b.p. 40–60°C) (2:1:7) or (*e*) ethanol–benzene–aq. 50% acetic acid (7:1:1). For detection of compounds on chromatograms, the following reagents were used: (*1*) ammoniacal aq. 2% silver nitrate, (*2*) 0.1 *M* potassium dichromate–acetic acid (1:1) followed by 0.1 *M* silver nitrate[121], (*3*) platinic iodide[122], (*4*) *p*-dimethyl-

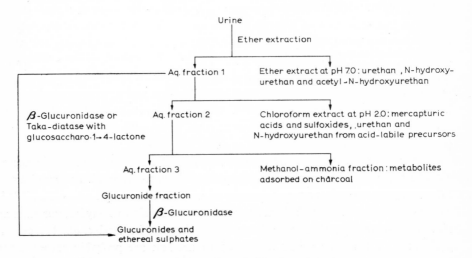

Fig. 20. Fractionation of metabolites from the urine.

amino cinnamaldehyde[123], (5) 0.05% ninhydrin in acetone, followed by heating the chromatograms at 85–90°C for 2–10 min, (6) NaI–HCl–starch[124], (7) 1% ferric chloride in aq. 50% ethanol, and (8) 1% sodium aminoprusside in water containing 0.1% of $MgCl_2 \cdot 6 H_2O$. Urethans and N-hydroxyurethans gave red spots with reagent (4) and N-hydroxyurethans gave purple colors with reagent (7) and reddish spots with reagent (8). N-hydroxyurethan, its N- and O-mono-substituted and N,O-disubstituted derivatives gave dark spots with reagent (1) after 5–10 min, 20–40 min, and 1–2 h, respectively, and substituted N-hydroxyurethans gave mauve colors with reagent (8). Figure 20 illustrates the scheme for the fractionations of metabolites from urine. Table 15 lists the R_F values of reference compounds and their occurrence as metabolites of urethan and N-hydroxyurethan in the urine of the rat and rabbit.

TABLE 15

R_F VALUES OF REFERENCE COMPOUNDS AND THEIR OCCURRENCE AS METABOLITES OF URETHAN AND N-HYDROXYURETHAN IN THE URINE OF THE RAT AND RABBIT

Compound	R_F in solvent system[a]				Occurrence as metabolite of			
	(a)	(b)	(c)	(d)	Urethan		N-Hydroxy-urethan	
					Rat	Rabbit	Rat	Rabbit
Urethan			0.35	0.11	+	+	+	+
N-Hydroxyurethan		0.79	0.23	0.04	+	+	+	+
O-Acetyl-N-hydroxyurethan			0.45	0.22	+	+	+	+
N-Acetyl-N-hydroxyurethan			0.48	0.24	+	+	+	+
NO-Diacetyl-N-hydroxyurethan			0.59	0.46	−	−	−	−
O-Carbethoxy-N-hydroxyurethan			0.50	0.26	−	−	−	−
NO-Dicarbethoxy-N-hydroxyurethan			0.61	0.48	−	−	−	−
S-Ethylcysteine	0.27	0.50			−	−	+	−
S-Ethylcysteine-S-oxide	0.10	0.26			−	−	−	−
N-Acetyl-S-ethylcysteine	0.33	0.82			+	+	+	+
N-Acetyl-S-ethylcysteine-S-oxide	0.29	0.74			−	−	+	+
S-Carbethoxycysteine	0.25	0.57			−	−	−	−
N-Carbethoxycysteine	0.30	0.53			−	−	−	−
NS-Dicarbethoxycysteine	0.37	0.83			−	−	−	−
N-Acetyl-S-carbethoxycysteine	0.35	0.75			+	+	+	+

[a] Solvents: (a) n-butanol–n-propanol–aq. 2 N ammonia (2:1:1), (b) n-butanol–acetic acid–water (12:3:5), (c) acetone–light petroleum (b.p. 40–60°C) (3:7), (d) chloroform–acetone–light petroleum (b.p. 40–60°C) (2:1:7). Solvents (a) and (b) used with Whatman No. 1 paper, and solvents (c) and (d) used for TLC with Silica Gel G chromatoplates.

This study revealed that urethan was metabolized by a process of N-hydroxylation in the rat, rabbit and man and that this occurred to a smaller extent when methyl, propyl, and n-butyl urethans are administered to the rat and rabbit. Other metabolites which were detected in urine of animals given urethan and N-hydroxyurethan were ethyl mercapturic acid, ethyl mercapturic acid sulfoxide and N-acetyl-S-carb-ethoxy cysteine. Substances which appeared to be S-ethyl glutathione and S-ethyl gluta-thione sulfoxide were detected in the bile of rats dosed with urethan or N-hydroxy-urethan. The metabolism of N-hydroxyurethan in relation to its carcinogenic action has been studied by Mirvish[125, 126]. [Carboxy-^{14}C]-N-hydroxyurethan was rapidly converted into ^{14}C-urethan following its i.p. administration in the mouse, with equal blood levels of the two compounds 80 min after injection, and a total estimated conversion of 70%. A second radioactive metabolite identified as N-hydroxyurethan-N-glucuronide (possibly mixed with the O-glucuronide) appeared in the blood, liver, and kidneys (where it constituted up to 34% of the radioactive material) and urine. Initially, blood extracts were concentrated under vacuum and chromatographed on Silica Gel G (Merck) using benzene–ethyl acetate (1:1) and the radioactivity ex-amined under a Vanguard glass-plate scanner[125]. Figure 21 shows a thin-layer

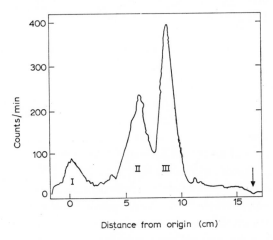

Fig.21. Thin-layer chromatogram of an alcoholic extract of blood withdrawn 2 h after injecting labeled N-hydroxyurethan. Forty μl were chromatographed, equivalent to 60 μl blood. Arrow indi-cates solvent front.

chromatogram of an alcoholic extract of blood withdrawn 2 h after injecting labeled N-hydroxyurethan. Peaks II and III were identified as N-hydroxyurethan and urethan, respectively, on the basis of (a) the R_F values (0.39 and 0.55, respectively), (b) reac-tions on the chromatograms to the chlorine test for N–H groups[127] (weak for N-hydroxyurethan, strong for urethan), and to ammoniacal silver nitrate as used to detect aryl hydroxylamines[128] (positive for N-hydroxyurethan), and (c) the isotope-dilution technique applied to urethan.

N-hydroxyurethan was also determined by a chemical method[129] (based on the colorimetric determination of liberated hydroxylamine by acid hydrolysis). TLC determinations of radioactive components in blood, tissue and urine extracts were also carried out using Silica Gel F (Merck) with development by the top layer of an ethyl acetate–heptane–water mixture (40:40:20). The isolation and purification of N-hydroxyurethan-N-glucuronide from urinary extracts was further achieved by column chromatography on silicic acid and elution with ethyl acetate (to remove urethan and N-hydroxyurethan), then with ethanol to remove the glucuronide. The ethanolic solution was concentrated and subjected to electrophoresis on Whatman 3MM paper using pH 3.3 buffer (0.5% pyridine–5% acetic acid) for 5 h at 250 V and 5°C. The moving band was eluted with 80% ethanol containing 0.025% pyridine and 0.25% acetic acid, then subjected to preparative TLC on Silica Gel G with 95% ethanol–water (94:6) as developing solvent which separated 3 distinct fractions. The middle fraction, which contained most of the radioactivity, was re-purified by a second electrophoresis to yield a gum consisting of about 50% of metabolite "X" (N-hydroxyurethan-N-glucuronide $(C_6H_9O_6 \cdot N(OH) \cdot COOC_2H_5)$), possibly mixed with the O-glucuronide $(C_6H_9ONH\ COOC_2H_5)$). The glucuronide was not hydrolyzed by β-glucuronidase and was hydrolyzed only slowly by hot acid, similar to known N-glucuronide metabolites of sulfonamides and the carbamates, meprobamate and ethionamate[130]. The above studies of Mirvish[125, 126] on the metabolism of [carboxy-^{14}C]-N-hydroxyurethan in the mouse showed about 70% conversion into urethan. Boyland and Nery[110] in their study on the metabolism of unlabeled N-hydroxyurethan in the rat found 2.0% of the injected material in the urine as urethan. This figure was fairly similar to that found in the studies of Mirvish (0.8%) and resembled the amount (1.7%) found by Boyland and Nery[110] on injecting urethan itself. The small amount of urethan excreted unchanged in the urine reflects the fact that over 95% of a dose of urethan is hydrolyzed to carbon dioxide, ammonia, and ethanol[131]. It was concluded by Mirvish that since 20% of N-hydroxyurethan catabolism proceeded by dehydroxylation to urethan, and N-hydroxyurethan is not more carcinogenic than urethan[88, 132], N-hydroxyurethan acts as a carcinogen by way of conversion into urethan, rather than the converse. A similar conclusion had also been reached by comparing the carcinogenicity and metabolism of N-hydroxyurethan in newborn mice[133].

The summary of the urinary metabolites of urethan and N-hydroxyurethan in the rat is shown in Table 16 and Fig. 22 illustrates the metabolic pathway and interconversions for urethan and N-hydroxyurethan in the rat.

A variety of additional chromatographic procedures have been reported for the separation and detection of urethan, N-hydroxyurethan and their related derivatives. The TLC of a number of alkyl N-hydroxyurethans and mono- and dihydroxyureas has been reported by Nery[134]. Silica Gel G was used with (a) acetone–light petroleum (b.p. 40–60°C) (3:7) and (b) ethanol–propanol (3:7) as developers. Detection was accomplished using (a) 2% aq. sodium aminoprusside solution containing 0.1% magnesium chloride hexahydrate which gave red-purple spots with N-hydroxyurethans, (b) 1% ferric chloride in aq. 50% ethanol, and (c) 0.01 N iodine in phosphate buffer, pH 8.0, followed by sulfanilamide and N-1-naphthyl ethylenediamine hydrochloride in the order given.

TABLE 16

URINARY METABOLITES OF URETHAN AND N-HYDROXYURETHAN IN THE RAT[a]

Metabolite	Formula no.[b]	Metabolites of urethan	Metabolites of N-hydroxyurethan	
		(1)	(2)	(3)
Urethan	I	2.2	2.5	0.8
N-Hydroxyurethan	II	0.10	35.3	1.3
N-Acetyl-N-hydroxyurethan	III	0.14	9.7	
O-Acetyl-N-hydroxyurethan	IV			
N-Acetyl-S-carbethoxycysteine	V	1.8	1.9	
N-Acetyl-S-ethyl cysteine	VI	0.13	0.6	
N-Hydroxyurethan-N-glucuronide	VII			7.0

[a] The results were obtained by Boyland and Nery[110], (columns (1) and (2) and Mirvish[126] (column (3)) and are expressed as percent of the injected doses which were 1.0, 0.4 and 0.53 mg/g for columns (1), (2) and (3), respectively.
[b] Formulas are shown in Fig.22.

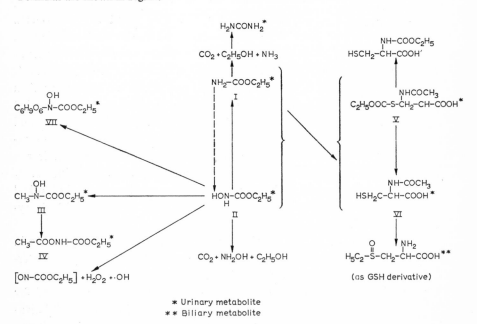

Fig.22. Possible metabolic pathway and interconversions for urethan and N-hydroxyurethan in the rat.

The R_F values of the following compounds in solvents (a) and (b), respectively, were: hydroxylamine 0, 0.21, methyl-N-hydroxyurethan 0.18, 0.65, N-hydroxyurethan 0.23, 0.79, n-propyl-N-hydroxyurethan 0.37, 0.81, n-butyl-N-hydroxyurethan 0.41, 0.86, benzohydroxamic acid 0.22, 0.86, hydroxyurea 0, 0.59, dihydroxyurea 0, 0.71, and N-phenyl-N'-hydroxyurea 0.28 and 0.87.

The gas chromatographic determination of acetyl and trimethylsilyl derivatives of alkyl urethans and their N-hydroxy derivatives was reported by Nery[135]. The analyses were performed on a Perkin-Elmer Model 800 dual-column gas chromatograph incorporating a dual flame-ionization detector and a Honeywell "Electronik" continuous balance recorder with a range from 0.25 to 2.5 mV. The dual columns consisted of 1 m × $\frac{1}{8}$ in. o.d. stainless steel coiled tubes packed with 1.5% SE-30 on a solid support of 80–100 mesh HMDS-treated Chromosorb W. The conditions used were hydrogen pressure 15 p.s.i., air pressure 30 p.s.i., nitrogen flow rate through both columns 30 ml/min and injector block temperature about 160°C (dial setting 4). The columns were used after equilibration for 24 h at an oven temperature of 200°C and the rate of temperature rise in both columns was programmed at 5°C/min. The TMS derivatives of the urethans and N-hydroxyurethans were dissolved in 0.5 ml of pyridine, and 0.2 ml of hexamethyl-disilazane, 0.1 ml of trimethyl chlorosilane, and 0.2 ml of triethylamine for four hours at 50°C, the mixtures centrifuged and 1 μl of the supernatant chromatographed. Figure 23 illustrates a gas chromatogram

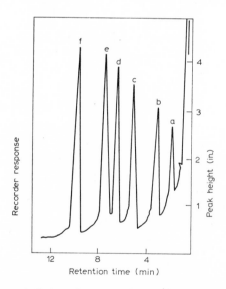

Fig. 23. Gas chromatogram of alkyl urethans as their TMS derivatives. Temperature programmed from 35 to 100°C at 5°/min. Attenuation ×200. Amount of each urethan analyzed, 5 × 10⁻⁸ moles. Alkyl groups: a, methyl; b, ethyl; c, propyl; d, isobutyl; e, butyl; f, pentyl.

of alkyl urethans as their TMS derivatives on SE-30 programmed 35°–100°C. Figure 24 depicts a programmed gas chromatogram of alkyl N-hydroxyurethans as their TMS derivatives. Figure 25 depicts a gas chromatogram of alkyl-N-hydroxy-urethans as their acetyl derivatives. The acetyl derivatives were prepared by reacting N-hydroxyurethans dissolved in 0.5 ml of pyridine with 0.2 ml of acetic anhydride and 0.3 ml triethylamine at 23°C for 4 h. Figure 26 shows the gas chromatographic

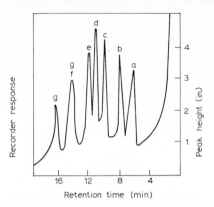

Fig.24. Gas chromatogram of alkyl-N-hydroxyurethans as their TMS derivatives. Temperature programmed from 35 to 130°C at 5°/min. Attenuation ×200. Amount of each hydroxyurethan analyzed, 4.25×10^{-8} moles. Alkyl groups: a–f, as for Fig.23; g, hexyl.

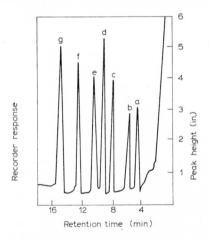

Fig.25. Gas chromatogram of alkyl-N-hydroxyurethans as their acetyl derivatives. Temperature programmed from 50 to 140°C at 5°/min. Attenuation ×20. Amount of each hydroxyurethan analyzed, 1.25×10^{-8} moles. Alkyl groups as in Fig.23.

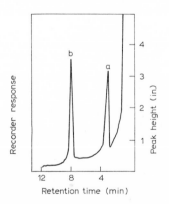

Fig.26. Gas chromatogram as a mixture of the TMS derivatives of urethan and N-hydroxyurethan. Temperature programmed from 35 to 90°C at 5°/min. Attenuation ×200. Amount of each substance analyzed, 5×10^{-8} moles. a, Urethan; b, N-hydroxyurethan.

separation of a mixture of urethan and N-hydroxyurethan as their TMS derivatives. Table 17 lists the relative retention times of the TMS and acetyl derivatives of a number of alkyl urethans and alkyl N-hydroxyurethans. With a programmed temperature use, the elution temperature varied linearly with the number of carbon atoms in the alkyl side chain of a homologous series and the isobutyl analogs were eluted at lower temperatures than the corresponding butyl analogs.

TABLE 17

GAS CHROMATOGRAPHY OF ALKYL CARBAMATES AND ALKYL N-HYDROXYCARBAMATES
AS THEIR TMS AND ACETYL DERIVATIVES

Compound	Boiling point (°C/mm Hg)	Relative elution[a]	
		Trimethylsilyl derivative	Acetyl derivative
A. ROCONH$_2$			
R = Methyl	177/760	0.60	N.D.[c]
Ethyl	184/760	1.0	N.D.
Propyl	195/760	1.73	N.D.
Isobutyl	206/760	2.20	N.D.
Butyl	203/760	2.53	N.D.
Pentyl	56[b]	3.33	N.D.
B. ROCONHOH			
R = Methyl	50–51[b]	0.79	0.77
Ethyl	86–88/0.6	1.0	1.0
Propyl	90–92/0.6	1.29	1.33
Isobutyl	41[b]	1.45	1.53
Butyl	100–102/0.8	1.58	1.70
Pentyl	115–118/0.04	1.84	2.07
Hexyl	42[b]	2.13	2.43

[a] Relative to the retention time of the corresponding derivative of the ethyl analogue. For the trimethylsilyl derivatives of urethan and N-hydroxyurethan, and for the acetyl derivative of N-hydroxyurethan, the retention times under the conditions described were 3.0, 7.6, and 6.0 min, respectively.
[b] Melting point (°C).
[c] N.D. = not determined.

The GLC of a number of alkyl and aryl urethans on two polar columns (Carbowax 20M and Versamid 900) and one non-polar column (SE-30) was described by Zielinski and Fishbein[136]. The analyses were carried out on (1) an F & M Model 720 gas chromatograph and equipped with a hot-wire detector and dual 8 ft. × ⅛ in. stainless steel columns containing (a) 10% Carbowax 20M on 60–80 mesh Chromosorb W and (b) 10% SE-30 on 60–80 mesh Chromosorb W HMDS-treated, (2) an Aerograph Hy-Fi Model 600-B gas chromatograph equipped with a flame-ionization detector and an 8 ft. × ⅛ in. stainless steel column containing 8% Versamid on 60 to 80 mesh Chromosorb W. Table 18 lists the operating parameters as well as the rela-

TABLE 18

GAS CHROMATOGRAPHY OF SIMPLE URETHANS

Compound	Mol. wt.	M.p. (°C)	Carbowax 20 M[a] R.E.[d]	Versamid 900[b] R.E.[d]	SE-30[c] R.E.[d]
Methyl carbamate	75	53–55	0.92	0.75	0.72
Ethyl carbamate (urethan)	89	51	1.0	1.0	1.0
Allyl carbamate	101	[e]	1.9	1.7	1.8
Isopropyl carbamate	103	95–96	1.0	1.1	1.3
n-Propyl carbamate	103	52–53	1.5	1.6	1.9
Methallyl carbamate	115	51–52	2.6	2.5	3.2
sec.-Butyl carbamate	117	93–94	1.5	1.7	2.7
Isobutyl carbamate	117	61–62	1.7	2.0	2.8
n-Butyl carbamate	117	53.5	2.1	2.5	3.4
β-Chloroethyl carbamate	124	71–73	0.45	0.55	0.32
Isopentyl carbamate	131	60–61	2.5	3.3	5.2
Amyl carbamate	131	56	3.0	3.9	7.0
Phenyl carbamate	138	149–150	3.7	6.2	2.3
Hexyl carbamate	145	67	4.2	6.0	12.4
Benzyl carbamate	152	86.5–87	3.0	3.4	3.6

[a] 10% w/w on 60–80 mesh Chromosorb W, 8 ft. × 0.125 in. o.d. stainless steel. Operating conditions: column 185 °C; injection port 265 °C; detector block 250 °C; filament current 150 mA; hot-wire detector; helium carrier 22 ml/min.
[b] 8% w/w on 60–80 mesh Chromosorb W (HMDS pretreated), 8 ft. × 0.125 in. o.d. stainless steel. Column conditions: column 185 °C; injection port 225 °C; detector 185 °C; carrier flow 10 ml/min of nitrogen, 300 ml/min of air, 24 ml/min of hydrogen; flame-ionization detector.
[c] 10% w/w on 60–80 mesh Chromosorb W, 8 ft. × 0.125 in. o.d. stainless steel. Column conditions: column 100 °C; injection port 265 °C; detector block 250 °C; helium carrier 24 ml/min; filament current 150 mA; hot-wire detector.
[d] Relative elution to ethyl carbamate as 1.0. Elution of urethan was 2.1 min on Carbowax 20 M; 2.5 min on Versamid 900, and 1.2 min on SE-30.
[e] Boiling point, 58–60 °C at 20 mm.

tive elution times of the urethans and Fig. 27 illustrates a gas chromatogram of a mixture of urethans.

The TLC separation of urethan in drug mixtures was described by Fresen[137]. Silica Gel G (Merck)–Kieselguhr G (Merck) (2:3) was used with the following solvents: (a) ether–petroleum ether (3:2), (b) ether–petroleum ether (1:4), and (c) carbon tetrachloride–chloroform–propionic acid (9:8:3). Detection was accomplished with iodine vapor and enhancement of the spots was attained by subsequent spraying with a 0.2% aqueous fluorescein reagent.

The separation of urethan from urea has been accomplished by paper chromatography[138] using Macherey-Nagel No. 214 paper with chloroform saturated with water as solvent and 1% p-dimethylaminobenzaldehyde in 10% hydrochloric acid for detection.

N-hydroxyurethans, monohydroxyureas, and dihydroxyureas in tissue extracts have been determined by diazotising sulfanilamide with the nitrite produced on oxi-

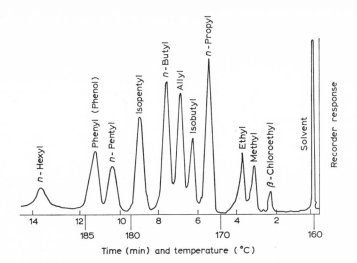

Fig. 27. Analysis of a urethan mixture. Column: 10% Carbowax 20M on 60–80 mesh Chromosorb W, $\frac{1}{8}$ in. × 8 ft. stainless steel. Conditions: 30 p.s.i.g. helium, programmed 160–185 °C at 2.1°/min; injection port temp. 265 °C; detector block temp. 250 °C; filament current 150 mA; hot-wire detector.

dation with coupling with N-1-naphthylethylenediamine[134]. Mixtures of hydroxylamine and hydroxamic acids were determined by (a) selective oxidative at pH 3.5 and pH 8.0 and (b) after separation of the components by TLC[134].

4. ORGANIC PEROXIDES

The utility of a number of organic peroxides in a wide variety of commercial and laboratory polymerizations is well established. Their formation and mechanisms of reaction have been reviewed[139, 140]. Organic peroxides find 90% of their market in the polymer industry (rubbers, elastomers, plastics, resins, etc.). The commercial organic peroxides are sources of free radicals (*e.g.* RO:OR → RO·+·OR). In the polymer industry, the uses of organic peroxides include (a) initiators for the free-radical polymerizations and/or copolymerization of vinyl and diene monomers, (b) curing agents for resins and elastomers, and (c) cross-linking agents for polyolefins. Organic peroxides have been used in the bleaching of various materials such as flour, gums, waxes, fats, and oils. They have also been used as cosmetic and pharmaceutical additives and intermediates and the miscellaneous uses of organic peroxides in the polymer industry include vulcanization of natural and butadiene rubbers, curing polyurethans and adhesives, preparing graft copolymers, flame retardant synergists for polystyrene, solidification of soils with calcium acrylate, cross-linking polyethylene and ethylene-containing copolymers.

A number of organic peroxides are of interest because of their mutagenicity and induction of neoplasia as well as their suggested formation in polluted air and oxidized fats.

References pp. 487–490

4.1 tert.-Butyl hydroperoxide

The commercial form of *tert.*-butyl hydroperoxide [(CH$_3$)$_3$COOH] consists of a 90% liquid and 60%–75% solutions containing di-*tert.*-butyl peroxide and is used as an initiator for vinyl monomer polymerizations and copolymerizations with styrene, vinyl acetate, acrylics, and as a curing agent for thermoset polyesters. The mutagenic effect of *tert.*-butyl hydroperoxide in *Drosophila*[141, 142], *E. coli*[143] and *Neurospora*[144] and its induction of chromosome aberrations in *Vicia faba*[99, 145, 146] and *Oenothera*[99] as well as its carcinogenicity in mice[147] has been described.

4.2 Di-tert.-butyl peroxide

Di-*tert.*-butyl peroxide [(CH$_3$)$_3$COOC(CH$_3$)$_3$] is commercially available as a 99% liquid and is widely used as an initiator for vinyl monomer polymerizations and copolymerizations of ethylene, styrene, vinyl acetate, and acrylics and as a curing agent for thermoset polyesters, styrenated alkyds and oils, and silicone rubbers. It is also used as an ignition accelerator for diesel fuels and is also used extensively in organic synthesis as a free-radical catalyst and as a source of reactive methyl radicals. Di-*tert.* butyl peroxide is mutagenic in *Neurospora*[102] but has been reported to be inactive toward transforming-DNA[148].

4.3 Cumene hydroperoxide

The commercial form of cumene hydroperoxide [C$_6$H$_5$C(CH$_3$)$_2$OOH] consists of a 70% liquid containing cumene, acetophenone, and cumyl alcohol. It is employed as an initiator for vinyl monomer polymerizations and copolymerizations with styrene, acrylics, butadiene–styrene, and as a curing agent for thermoset polyesters and styrenated alkyds and oils. Cumene hydroperoxide has been found mutagenic in *E. coli*[143] and *Neurospora*[102, 148]

4.4 Succinic acid peroxide and disuccinyl peroxide

$$\text{O}$$
$$\|$$
Succinic acid peroxide [HOOC–CH$_2$CH$_2$–C–O–OH] is commercially available as a 95% powder and is used as an initiator for vinyl monomer polymerizations and copolymerizations with ethylene and fluoroolefins. It has been found to inactivate T$_2$ phage[149] transforming-DNA of *H. influenzae*[150] and to be mutagenic in *E. coli*[143, 151]. Disuccinyl peroxide has analogous utility to that described for succinic acid peroxide, and has been found to inactivate transforming-DNA[119, 148–150].

4.5 Dihydroxydimethyl peroxide

Dihydroxydimethyl peroxide (HO–CH$_2$–O–O–CH$_2$–OH) is prepared *via* the reaction of formaldehyde and hydrogen peroxide in ether. Its formation in polluted air has been postulated by Kotin and Falk[152] as depicted in Fig. 28. Mutations in

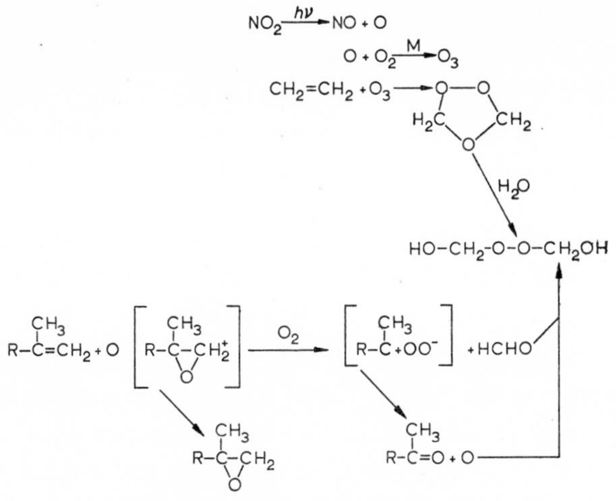

Fig. 28. Anticipated reactions leading to the formation of dihydroxydimethyl peroxide in polluted air[152].

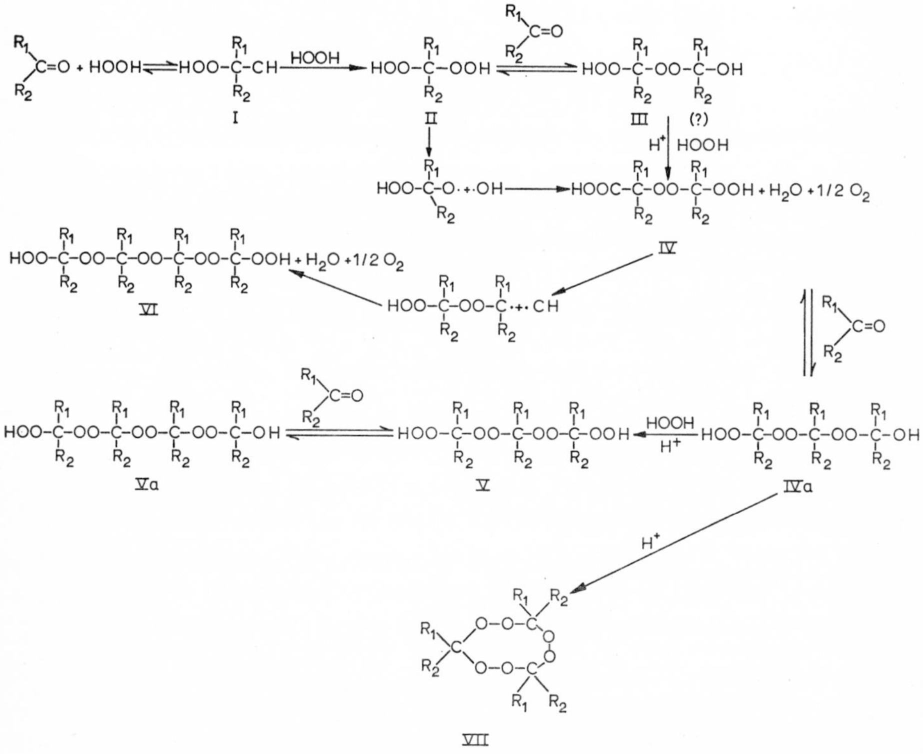

Fig. 29. Peroxides derived from simple ketones and hydrogen peroxide.

Drosophila have been produced by direct application of dihydroxydimethyl peroxide[153, 154]. Both dihydroxydimethyl peroxide and formaldehyde[155] produce mutations in earlier stages of spermatogenesis supporting the suggestion that the formation of a peroxide may also be involved in the mutagenic effects of formaldehyde on earlier stages of sperm development. A variety of chromatographic procedures have been developed for the separation and detection of organic peroxides.

The paper chromatographic separation of a variety of organic peroxides (with special emphasis on peroxides derived from simple ketones and hydrogen peroxide, Fig. 29) was achieved[156, 157] by descending development on Whatman No. 1 paper using the following solvents: (*1*) dimethylformamide–decalin[158], (*2*) N-methyl-N-formamide–decalin[158], and (*3*) *n*-butanol–ethanol–water (45:5:50). The detecting reagents were: (*a*) 0.1% *p*-aminodimethyl aniline hydrochloride, (*b*) glacial acetic acid–satd. aq. potassium iodide–starch (3:2:5), (*c*) 50% hydriodic acid–glacial acetic acid (10:90). Table 19 lists the R_F values of the peroxides in three solvent systems.

TABLE 19

PAPER CHROMATOGRAPHY[a] OF VARIOUS ORGANIC PEROXIDES ON WHATMAN NO. 1 PAPER

Peroxide	Solvents[b]		
	1	*2*	*3*
Hydrogen peroxide	0.00	0.00	0.49
p,p'-Dicyanobenzoyl peroxide	0.00	0.01	0.0–0.50
m-Monocyanobenzoyl peroxide	0.20	0.03	0.97
m-Monomethoxybenzoyl peroxide	0.29	0.30	0.94
Benzoyl peroxide	0.38	0.37	0.94
p,p'-Dichlorobenzoyl peroxide	0.56	0.40	0.94
2,4-Dichlorobenzoyl peroxide	0.69	0.44	0.94
2,5-Dimethylhexane-2,5-dihydroperoxide	0.00	0.18	0.89
2,5-Dimethyl-3-hexyne-2,5-dihydroperoxide	0.00	0.17	0.89
2,5-Dimethyl-3-hexyne-2,5-diperbenzoate	0.29	0.24	0.97
2,7-Dimethyl-3,5-octadyne-2,7-dihydroperoxide	0.00	0.10	0.92
Bis-(1-hydroxyheptyl) peroxide	0.00	0.40	*c*
tert.-Butyl hydroperoxide	0.00	0.00	*c*
Cumene hydroperoxide	0.09	0.30	0.91
tert.-Butylperoxymaleic acid	0.00	0.10	0.38
tert.-Butylperbenzoate	0.63	0.54	0.95
Lauroyl peroxide	0.98	0.88	0.97
tert.-Butyl peroxyisobutyrate	*c*		*c*
3,3-Dihydroperoxy pentane	0.02		
3,3'-Dihydroperoxy-3,3'-dipentyl peroxide	0.53		
1,1,4,4,7,7,10,10-Octaethyl-1,4,7-triperoxy 1,10-dihydroperoxide	0.86		
1,1,4,4,7,7-Hexaethyl-1,4,7-cyclononatriperoxane	0.91		

[a] Descending development (17–18 cm).
[b] Solvents: *1*, dimethylformamide–decalin[157]; *2*, N-methylformamide–decalin[157]; *3*, *n*-butanol–ethanol–water (45:5:50).
[c] 50 γ peroxide applied.

The peroxides were generally detected in quantities as low as 0.2–$0.5\,\gamma$. Of the systems used, dimethylformamide–decalin (solvent (*I*)) with detectors (*a*) or (*b*) gave the most useful results.

Cartlidge and Tipper[159] separated a number of peroxides on untreated and silicone- and ethyleneglycol-impregnated paper. Detection was accomplished with (*a*) aq. ferrous thiocyanate[160], (*b*) solution of *p*-phenylenediamine and acetaldehyde in aqueous acetic acid, (*c*) a dilute solution of *o*-toluidine and a small amount of fer-

TABLE 20

R_F VALUES OF PEROXIDES ON GLYCOL-TREATED PAPER[a]

Compound	R_F		
	I	*II*	*III*
Hydrogen peroxide	0.05	0	0
Methyl hydroperoxide	0.36	0	0
Ethyl hydroperoxide	0.38	0	0
tert.-Butyl hydroperoxide	Spreads across front	0.35	0.22
n-, *iso*-, *sec.*-Butyl hydroperoxide			0.31
tert.-Amyl hydroperoxide			0.43
n-Heptyl hydroperoxide, 3-methylhexyl-3-hydroperoxide			0.80
Cyclohexane hydroperoxide			0.30
Cyclohexene hydroperoxide			0.14
n-Heptane-3,3'-dihydroperoxide[a]			0.07
2,5-Dimethylhexane-2,5-dihydroperoxide			0.20
Cumene hydroperoxide			0.53
Peracetic acid	0.26	0.05	0
Perpropionic acid	0.54	0.16	0.11
Per-*n*-, per-isobutyric acid		0.37	0.30
Per-*n*-valeric acid		0.57	0.50
Per-*n*-hexoic acid		0.71	
Percrotonic acid[b]		0.23	
Diacetylperoxide	0.37	0.18	
Hydroxymethyl hydroperoxide	0.09	0	0
2-Hydroxyethyl hydroperoxide	0.22	0	0
Dihydroxymethyl peroxide	0.25		
tert.-Butyl, 2-hydroxyethyl peroxide			0.55
n-Heptyl, 2-hydroxyethyl peroxide			0.92
Cumyl, 2-hydroxy-2-phenylethyl peroxide			0.76
tert.-Butylperbenzoate			1.0
Di-*n*-heptyl peroxide			0.95
$(Me_3COO)_2CHMe$			0.95

[a] From the reaction of heptan-3-one and hydrogen peroxide.
[b] The main product of the gas-phase oxidation of crotonaldehyde at 150°C. Conditions: I, stationary phase (S) glycol from a 5% solution, moving phase (M) 10 vol.% *n*-butanol in 80–100°C petroleum ether; II (S) glycol from a 20% solution, (M) 50/50 chloroform in 80–100°C petroleum ether; III (S) as for II, (M) 5 vol.% ether in 80–100°C petroleum ether. Rate of movement of solvent front *ca.* 15 cm/h.

rous ammonium sulfate in aqueous acetic acid, or (d) a 10% solution of hydrogen iodide in glacial acetic acid.

The ferrous thiocyanate reagent detected hydrogen peroxide, hydroperoxides or peracids rapidly as red spots, whereas those due to other peroxides formed more slowly. With the diamine–aldehyde mixture the spots were pale red. The o-toluidine reagent gave a blue-green color only with compounds containing the –OOH group (or with substances which hydrolyze very readily in acid to yield this group). With the hydrogen iodide solution, nearly all peroxides gave a black or brown spot on a light-brown background. The sensitivity of developers decreased in the above order. Table 20 lists the R_F values of the peroxides obtained on 5% or 10% glycol-treated paper (the system of choice) developed in three solvent systems. The limits of detection of hydrogen peroxide and *tert*.-butyl hydroperoxide were 0.4 and 10 μg and the

TABLE 21

R_F VALUES OF PEROXIDES ON SILICONE-TREATED PAPER

Section (a). Stationary phase, silicone from a 5% solution. Moving phase, water–ethanol–chloroform (20:17:2 by vol.). Rate of movement of solvent front ≈ 1.8 cm/h.

Compound	R_F
Di-*n*-heptyl peroxide	0.018
(Me$_3$COO)$_2$CHMe	0.49
Hydrogen peroxide	0.68
tert.-Butyl, 2-hydroxyethyl peroxide	0.80
Cumene hydroperoxide	0.82
Di-2-hydroxyethyl peroxide	0.88
Alkyl hydroperoxides	1.0

Section (b). Stationary phase, silicone from a 20% solution. Moving phase, 80 vol.% methanol in water. Rate of movement of solvent front ≈ 5 cm/h.

Compound mixed with xanthydrol	R_F
tert.-Butyl hydroperoxide	0.27, 0.45
tert.-Amyl hydroperoxide	0.18, 0.43
n-Heptyl hydroperoxide	0.0, 0.09(a), 0.19, 0.30, 0.49(b), 0.67, 0.80 Spots a and b are the strongest

values for the other peroxides were of the order of 20 μg. Table 21 lists the R_F values of a number of peroxides on silicone-treated paper.

The separation of peroxides on partially acetylated Schleicher and Schüll 2043B paper using ethyl acetate–dioxan–water (2:4.5:4.6) (Table 22) was described by Rieche and Schultz[161].

TABLE 22

R_F VALUES OF ORGANIC PEROXIDES

Filter paper: Schleicher and Schüll 2043b partially acetylated.
Solvent: Ethyl acetate–dioxan–water (2:4.5:4.6).

Substance	R_F value
H_2O_2	0.83
Tetrahydrofuran hydroperoxide	0.76
Benzoyl peroxide	0.19
Cyclohexenyl hydroperoxide	0.46
Isochroman peroxide	0.22
Succinyl peroxide	0.82
Isochroman hydroperoxide	0.50
Cumol hydroperoxide	0.39
Lauroyl peroxide	0
Tetralin hydroperoxide	0.36
Stearoyl peroxide	0
Methylisochroman hydroperoxide	0.51
Isopropylisochroman hydroperoxide	0.36

Commercially important organic peroxides have been separated[162] on Silica Gel G using the following solvents: (*1*) toluene–carbon tetrachloride (2:1), (*2*) toluene–acetic acid (19:1), and (*3*) petroleum ether–ethyl acetate (49:1) and detected with N,N-dimethyl-*p*-phenylene diammonium dichloride (1.5 g in 128 ml methanol, 25 ml water and 1 ml acetic acid). Table 23 lists the R_F values of a number of peroxides using the procedure above.

TABLE 23

TLC OF PEROXIDES ON SILICA GEL G

Peroxide	Solvents[a]		
	1	*2*	*3*
Lauroyl peroxide	0.85	0.95	
2,4-Dichlorobenzoyl peroxide	0.81	0.88	
4-Chlorobenzoyl peroxide	0.74	0.94	
Benzoyl peroxide	0.55	0.70	
Methylisobutyl ketone peroxide	0.25	0.55	
tert.-Butyl perbenzoate	0.24	0.47	
tert.-Butyl peracetate	0.12	0.32	0.18
Cumol hydroperoxide	0.11	0.33	0.09
tert.-Butyl hydroperoxide	0.05	0.30	
Di-*tert.*-butyl peroxide	0.00	0.39	
2,2-Bis(*tert.*-butyl peroxy)-butane	0.10	0.35	
Hydrogen peroxide	0.00	0.00	0.00

[a] Solvents: 1, toluene–carbon tetrachloride (2:1); 2, toluene–acetic acid (19:1); 3, petroleum ether–ethyl acetate (49:1).

The separation of organic peroxides on Silica Gel KSK containing gypsum using toluene–methanol (20:3) for development was described by Bazlanova et al.[163]. For detection, both concentrated sulfuric acid and 2,4-dinitrophenylhydrazine (for cyclic ketones, the dissociation products of 1,1'-dihydroxyperoxide) were used. Table 24 lists the R_F values of organic peroxides separated as described above.

TABLE 24

TLC OF PEROXIDES ON SILICA GEL KSK DEVELOPED WITH TOLUENE–METHANOL (20:3)

Compound	R_F
Hydrogen peroxide	0.05
1,1'-Dihydroxydicyclohexyl peroxide	0.05
1,1'-Dihydroxy-2,2'-dimethyldicyclohexyl peroxide	0.05
1,1'-Dihydroxy-3,3'-dimethyldicyclohexyl peroxide	0.05
1,1'-Dihydroxy-4,4'-dimethyldicyclohexyl peroxide	0.05
1-Hydroxy-1'-hydroperoxy-dicyclohexyl peroxide	0.12
1-Hydroxy-1'-hydroperoxy-2,2'-dimethyldicyclohexyl peroxide	0.14
1-Hydroxy-1'-hydroperoxy-3,3'-dimethyldicyclohexyl peroxide	0.14
1-Hydroxy-1'-hydroperoxy-4,4'-dimethyldicyclohexyl peroxide	0.14
1,1'-Dihydroperoxy-dicyclohexyl peroxide	0.74
1,1'-Dihydroperoxy-2,2'-dimethyldicyclohexyl peroxide	0.75
1,1'-Dihydroperoxy-3,3'-dimethyldicyclohexyl peroxide	0.75
1,1'-Dihydroperoxy-4,4'-dimethyldicyclohexyl peroxide	0.75

A method was described[164] for the determination of dicumyl peroxide in polystyrene plastic materials that may contain other organic peroxides (e.g., tert.-butyl perbenzoate, benzoyl peroxide and cumene hydroperoxide, associated with the dicumyl peroxide). The dicumyl peroxide (DCP), which imparts a considerable degree of fire resistance to self-extinguishing grades of polystyrene and is often used in fire-resistant formulations, was extracted from the plastic with acetone and separated from the other additives by TLC on silica gel, thence transferred to a small reaction flask and the peroxide determined by a micro-titration procedure. Table 25 depicts

TABLE 25

DETECTION OF PEROXIDES AFTER SEPARATION BY TLC ON SILICA GEL GF$_{254}$ DEVELOPED WITH TOLUENE–CARBON TETRACHLORIDE (2:1)

Additive	Visual indication		R_F values
	N,N-dimethyl p-phenylenediamine	GF$_{254}$ plates and UV light	
Dicumyl peroxide	Positive	Positive	0.55
tert.-Butylperbenzoate	Positive	Positive	0.15
Benzoyl peroxide	Positive	Positive	0.30
Cumene hydroperoxide	Positive	Negative	0.05

the detection of peroxides after separation by TLC on Silica Gel GF_{254} developed with toluene–carbon tetrachloride (2:1).

The semiquantitative determination of *tert.*-butyl hydroperoxide in di-*tert.*-butyl diperphthalate and in 2,2-bis(*tert.*-butylperoxy)butane and diacyl peroxides by TLC was described by Glabik and Walczyk[165]. Silica gel plates were developed with benzene–chloroform–carbon tetrachloride–methanol (1:1:1:1) and detected with potassium iodide containing 1% acetic acid. The R_F values for the various peroxides were: *tert.*-butyl hydroperoxide 0.71, di-*tert.*-butyl peroxide 0.94, di-*tert.*-butyl diperphthalate 0.88, and 2,2-bis(*tert.*-butylperoxy)butane 0.895.

The determination of benzoyl peroxide degradation products in pharmaceuticals was effected using infrared spectroscopy and thin-layer chromatography[166]. Eastman Chromagram Silica Gel No. 6060 sheets with fluorescent indicator were used with toluene–glacial acetic acid–dichloromethane (50:1:2) for development. Significant amounts of benzoic acid and/or related acids were encountered as contaminants of all premixed commercially available pharmaceuticals containing benzoyl peroxide when stored for extended periods at room temperature. It is of interest to note that decomposition of a very dilute solution of benzoyl peroxide in benzene resulted in the formation of biphenyl, benzoic acid, carbon dioxide among other decomposition products[167].

Organic peroxides have been detected[168] by TLC (detection limit *ca.* 1 µg) using benzene–diethyl ether (peroxide-free) (20:1) as developing solvent and potassium iodide–starch or ferrous sulfate–ammonium thiocyanate for detection.

The GLC study of the composition of alkylbenzene oxidation products was described by Valendo and Norikov[169]. The separation and determination of the products of oxidation of cumene (acetophenone, 2-phenylpropan-2-ol, phenol and cumene hydroperoxide) and of ethylbenzene (acetophenone, α-methylbenzyl alcohol, phenol, and ethylbenzene hydroperoxide) was carried out at 70°C on a molybdenum-glass column (60 cm × 3 mm) of 20% Apiezon N on Celite 545 treated with 1% aq. sodium hydroxide or sodium carbonate with argon as carrier gas (150 ml/min), evaporator 70°C (the evaporator is a glass capillary of diameter 1.5 mm) and detection by flame ionization.

The GLC determination of the impurities in di-*tert.*-butyl peroxide, mainly hydroperoxide, was described by Adams[170]. A Perkin-Elmer Fractometer 451 gas chromatograph was used equipped with a flame-ionization detector and a 1 m × $\frac{1}{4}$ in. stainless steel column packed with 20% di-isodecyl phthalate on firebrick at 75°C.

Ewald *et al.*[171] described the GLC determination of alkyl hydroperoxides using Teflon and polyethylene columns containing 20% dinonylphthalate on 60–80 mesh Chromosorb W at 70–73°C and detection by flame ionization. Figures 30 and 31 illustrate the separation of mixtures of methyl and ethyl hydroperoxide and the chromatography of *n*-propyl, *n*-butyl and isobutyl hydroperoxides, respectively, on Teflon columns coated with 20% dinonylphthalate on Chromosorb and Fig. 32 shows the chromatography of *n*-propyl, *n*-butyl, and *tert.*-butyl hydroperoxide on poly-

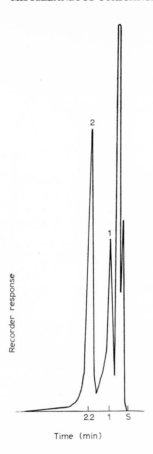

Fig. 30. GLC separation of methyl- and ethyl hydroperoxide. Column: 60 cm × 3.5 mm i.d. Teflon filled with 20% dinonylphthalate on 60–80 mesh Chromosorb W. Column temp.: 73.5°C. Detector: FID, 0.3 × 10⁻¹⁰ A. 1, Methyl hydroperoxide; 2, ethyl hydroperoxide.

ethylene containing the above liquid phase. Teflon tubing was found to be the best material for hydroperoxide analysis and copper and steel rated quite low because of tailing.

A flowing-liquid colorimetric detector was used for the gas chromatographic determination of hydroperoxides and nitrogen dioxide for possible application in air pollution studies[172]. Half the effluent from the column went to a flame-ionization detector, and half to a bubbler in which the hydroperoxides reacted with a flowing ferrous thiocyanate liquid reagent; the liquid passed through a colorimeter and the transmittance was recorded automatically. An Aerograph Model 204-2B gas chromatograph was operated isothermally at temperatures from 25–100°C using pre-purified nitrogen as carrier gas. The effluent from the column passed through a 1:1 stream splitter for the analysis as described above. The diagram of the flow system is shown in Fig. 33. The injection port was operated at room temperature. The influ-

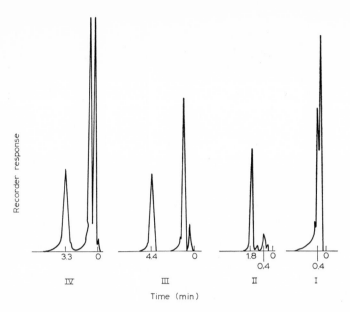

Fig. 31. Gas chromatography of alkyl hydroperoxides. Column: 60 cm × 1 mm i.d. Teflon filled with 20% dinonylphthalate on 60–80 mesh Chromosorb W. Column temp.: 70°C. Detector: FID. I, Propionaldehyde + *n*-propanol + propionic acid; II, *n*-propyl hydroperoxide; III, *n*-butyl hydroperoxide; IV, isobutyl hydroperoxide.

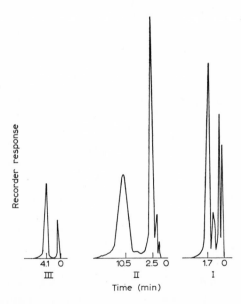

Fig. 32. Gas chromatography of alkyl hydroperoxides. Column: 60 cm × 1 mm i.d. polyethylene filled with 20% dinonylphthalate on 60–80 mesh Chromosorb W. Column temp.: 70°C. Detector: FID. Current: 10^{-9} A. I, II, III, as in Fig. 31.

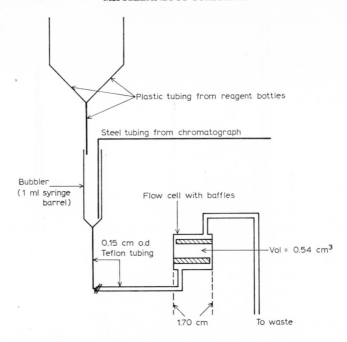

Fig. 33. Diagram of flow cell for colorimetric detector.

TABLE 26

LIQUID PHASES TESTED FOR ELUTION OF HYDROPEROXIDES AND NO_2

Liquid phase	Support	Comments
3.2% Squalene	CGD[a]	ROOH[b]: tailing; NO_2: 4.5% eluted
5.0% SE-30	CW[c]	Same as above
5.0% Kel-F 90	Silica gel	NO_2: no elution below 75 °C; 70% elution above 75 °C
2.8% PEG 1300	CGD[a]	ROOH: 12% eluted
5.0% Triton X-305	CGD[a]	ROOH: 8% eluted
3.0% Armeen SD	CGD[a]	ROOH: 10% eluted
1.0% PEG 400	Teflon	[g]
3.0% PEG 400	CGD[a]	ROOH: 4% eluted
2.9% PEG 400 + 0.3% TCEP[e]	CG[d]	ROOH: 10% eluted
1.95% PEG 400	CGH[f]	ROOH: 13% eluted; see Fig. 34

[a] CGD = acid-washed, DMCS-treated Chromosorb G.
[b] ROOH = hydroperoxides.
[c] CW = Chromosorb W.
[d] CG = acid-washed Chromosorb G.
[e] TCEP = 1,2,3-tris-(2-cyanoethoxy)propane.
[f] CGH = acid-washed, H_2O_2-treated Chromosorb G.
[g] Used for determination of heats of solution of hydroperoxides in PEG 400.

ence of column tubing indicated that both Teflon and stainless steel eluted about 60% of methyl hydroperoxide, while copper tubing eluted about 40%. Table 26 lists the liquid phases tested for elution of hydroperoxides and nitrogen dioxide. The polyethylene glycol 400 (PEG 400) columns were the best ones found. (The column containing 1% PEG 400 on Teflon was used to determine the heats of solution of methyl and ethyl hydroperoxides in PEG 400.) Figure 34 illustrates a gas chromato-

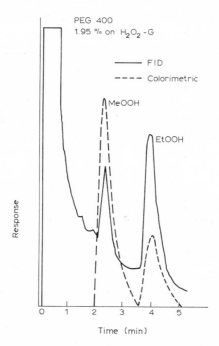

Fig. 34. Chromatogram of methyl and ethyl hydroperoxides in ethyl ether on 1.95% PEG 400 coated on acid-washed, H_2O_2-treated Chromosorb G.

gram of methyl and ethyl hydroperoxides obtained on 1.95% PEG 400 coated on acid-washed hydrogen peroxide-treated Chromosorb G comparing both FID with colorimetry detection. The apparent differences in sensitivity for the two hydroperoxides for the two detectors are principally due to the fact that the FID response for ethyl hydroperoxide is approximately four times that of methyl hydroperoxide because of the higher carbon response of the ethyl hydroperoxide molecule. The detection limit with the apparatus used was about 9 μg of hydroperoxide. Nitrogen dioxide eluted from the same stationary phases, but on most of them the colorimeter detector indicated that only about 1% of the eluted samples gave the NO_2 colorimetric reaction.

    Organic peroxides have also been determined by titrimetric[173, 174], spectrophotometric[175, 176], polarographic[177], and infrared spectroscopic[178] techniques.

5. HYDROGEN PEROXIDE

Hydrogen peroxide is prepared by the electrolysis of sulfuric acid solutions and subsequent hydrolysis of the formed peroxydisulfuric acid to yield hydrogen peroxide and sulfuric acid. The principal uses of hydrogen peroxide are in the bleaching of cotton textiles and wood and chemical (Kraft and sulfite) pulps. Hydrogen peroxide is also widely used in the conversion of soybean oil, linseed oil, and related unsaturated esters into their epoxides for use as plasticizers and stabilizers for polyvinyl chloride; the preparation of a variety of peroxy acids, hydroperoxides, diacyl peroxides and ketone peroxides which are used as oxidants and polymerization catalysts for cross-linking agents. Other uses of hydrogen peroxide include its use as blowing agent for the preparation of foam rubber, plastics and elastomers, bleaching, conditioning or sterilization of starch, flour, tobacco, paper, and fabric. Solutions of 3–6% hydrogen peroxide are employed for germicidal and cosmetic (bleaching use) although concentrations as high as 30% H_2O_2 have been used in dentistry. Concentrations of 35% and 50% H_2O_2 are used for most industrial applications.

Hydrogen peroxide is mutagenic in *Staphylococcus aureus*[179, 180], *E. coli*[181] and *Neurospora*[144, 182], inactivates transforming-DNA[183], and induces chromosomal aberrations in strains of ascites tumor in mice[184] and in *Vicia faba*[185].

The chemistry of hydrogen peroxide has been reviewed[186]. The determination of hydrogen peroxide has been achieved by a variety of techniques, including titration with potassium permanganate, polarography, or by procedures involving the oxidation of iodide ion to iodine, the reduction of ceric sulfate to cerous sulfate, and its spectrophotometric determination in biological systems[187].

The gas chromatographic analysis of hydrogen peroxide solutions has been described by Greiner[188]. The principle of the method is the quantitative catalytic decomposition of H_2O_2 into H_2O and O_2 on hot platinum gauze with quantitative analysis of the products. A Consolidated Electrodynamics Corporation Type X26-201 gas chromatograph equipped with thermal conductivity detectors and a 6 ft. $\times \frac{1}{4}$ in. column filled with 10% Carbowax 20M on Haloport F was used for analysis. A tube 5 in. long $\times \frac{1}{4}$ in. diameter between the injection port and the column was filled with a tightly rolled platinum gauze (52 mesh, 0.004 in. diameter). The column temperature was 150°C and the catalyst section was maintained somewhat above 150°C with a heating tape and the helium carrier gas flow rate was 100 ml/min. A variety of commercial solutions of H_2O_2 in water, normally 3, 30, and 97% weight, as well as commercial formaldehyde solution (37% CH_2O, 10–15% CH_3OH) admixed with H_2O_2–water were successfully analyzed by this procedure. The limit sample size was about 50 μg of H_2O_2.

REFERENCES

1   L. F. AUDRIETH AND B. A. OFF, *The Chemistry of Hydrazine*, Wiley, New York, 1951.
2   G. D. BYRKIT AND G. A. MICHALEK, *Ind. Eng. Chem.*, 42 (1950) 1862.
3   L. B. WITKIN, *Arch. Ind. Health*, 13 (1956) 34.
4   C. C. COMSTOCK AND F. W. OBERST, *Federation Proc.*, 11 (1952) 333.
5   P. MCGRATH, C. C. COMSTOCK AND F. W. OBERST, *Federation Proc.*, 11 (1952) 374.
6   S. KROP, *Arch. Ind. Hyg. Occup. Med.*, 9 (1954) 199.
7   F. BROIHAN, *Zentr. Arbeitsmed. Arbeitsschutz*, 7 (1957) 62.
8   A. M. DOMINGUEZ, J. S. AMENTA, C. S. HILL AND T. T. DOMANSKI, *Aerospace Med.*, 33 (1962) 1094.
9   C. BIANCIFIORI, E. BUCCIARELLI, D. B. CLAYSON AND F. E. SANTILLI, *Brit. J. Cancer*, 18 (1964) 543.
10 H. DRUCKREY, *Z. Krebsforsch.*, 67 (1965) 31.
11 F. J. C. ROE, G. A. GRANT AND S. M. MILLIKAN, *Nature*, 216 (1967) 375.
12 B. TOTH, *J. Natl. Cancer Inst.*, 42 (1969) 469.
13 R. RUDNER, *Biochem. Biophys. Res. Commun.*, 3 (1960) 275.
14 E. FREESE, E. BAUTZ AND E. B. FREESE, *Proc. Natl. Acad. Sci. U.S.*, 47 (1961) 844.
15 A. RUTISHAUSER AND W. BOLLAG, *Experientia*, 19 (1963) 131.
16 B. A. KIHLMAN, *J. Biophys. Biochem. Cytol.*, 2 (1956) 543.
17 L. B. COLVIN, *J. Pharm. Sci.*, 58 (1969) 1433.
18 H. MCKENNIS, JR., A. S. YARD, J. H. WEATHERBY AND J. A. HAGY, *J. Pharmacol. Exptl. Therap.*, 126 (1959) 109.
19 H. MCKENNIS, JR., J. H. WEATHERBY AND L. B. WITKIN, *J. Pharmacol. Exptl. Therap.*, 114 (1955) 385.
20 T. DAMBRAUSKUS AND H. C. CORNISH, *Toxicol. Appl. Pharmacol.*, 6 (1964) 653.
21 H. C. CORNISH AND R. HARTUNG, *Toxicol. Appl. Pharmacol.*, 15 (1969) 62.
22 E. H. JENNY AND C. C. PFEIFER, *J. Pharmacol. Exptl. Therap.*, 122 (1958) 110.
23 W. E. RINEHART, E. DONATI AND E. A. GREENE, *Am. Ind. Hyg. Assoc. J.*, 21 (1960) 207.
24 P. PETERSEN, E. BREDAHL, O. LAURITSEN AND T. LAURSEN, *Brit. J. Ind. Med.*, 27 (1970) 141.
25 F. LINGENS, *Z. Naturforsch.*, 19B (1964) 151.
26 K. C. BACK, M. K. PINKERTON, A. B. COOPER AND A. A. THOMAS, *Toxicol. Appl. Pharmacol.*, 5 (1963) 401.
27 F. N. DOST, D. J. REED AND C. W. WANG, *Biochem. Pharmacol.*, 15 (1966) 1325.
28 F. W. KRÜGER, M. WIESSLER AND U. RÜCKER, *Biochem. Pharmacol.*, 19 (1970) 1825.
29 R. SÜSS, *Z. Naturforsch.*, 20B (1965) 714.
30 R. M. JONES, *Anal. Chem.*, 38 (1966) 338.
31 C. BIGHI AND G. SAGLIETTO, *J. Chromatog.*, 18 (1965) 297.
32 C. BIGHI, G. SAGLIETTO AND A. BETTI, *Ann. Univ. Ferrara, Sez. 5*, 11 (1967) 163.
33 C. BIGHI, A. BETTI AND G. SAGLIETTO, *Ann. Chim. (Rome)*, 57 (1967) 1142.
34 C. BIGHI AND G. SAGLIETTO, *J. Gas Chromatog.*, 4 (1966) 303.
35 J. L. SPIGARELLI AND C. E. MELOAN, *J. Chromatog. Sci.*, 8 (1970) 420.
36 N. A. KIRSHEN AND G. H. OLSEN, *Anal. Chem.*, 40 (1968) 1341.
37 Y. MURAKAMI, *Bull. Chem. Soc. Japan*, 32 (1958) 316.
38 G. NEURATH AND W. LÜTTICH, *J. Chromatog.*, 34 (1968) 257.
39 G. NEURATH, B. PIRMANN AND M. DÜNGER, *Chem. Ber.*, 97 (1964) 1631.
40 R. L. HINMAN, *Anal. Chim. Acta*, 15 (1956) 125.
41 R. BLOCK, R. LESTRANGE AND G. ZWEIG, *Paper Chromatography*, Academic Press, New York, 1952, p. 72.
42 L. FEINSILVER, J. A. PERREGRINO AND C. J. SMITH, *Am. Ind. Hyg. Assoc. J.*, 20 (1959) 26.
43 J. D. CLARK AND J. R. SMITH, *Anal. Chem.*, 33 (1961) 1186.
44 H. E. MALONE AND R. BARRON, *Anal. Chem.*, 37 (1965) 548.
45 H. E. MALONE AND D. M. W. ANDERSON, *Anal. Chim. Acta*, 48 (1969) 87.
46 H. E. MALONE AND D. M. W. ANDERSON, *Anal. Chim. Acta*, 47 (1969) 363.
47 R. PREUSSMAN, H. HENGY AND A. VON HODENBERG, *Anal. Chim. Acta*, 42 (1968) 95.

48   M. PESEZ AND J. BARTOS, *Talanta*, 5 (1960) 216.
49   H. E. MALONE AND R. A. BIGGERS, *Anal. Chem.*, 36 (1964) 1037.
50   J. C. MACDONALD, *Anal. Chim. Acta*, 44 (1969) 391.
51   R. BATTYE, in *Transactions of the 15th International Congress on Industrial Medicine*, 1966, Vol. 3, p. 156.
52   Y. MASUDA, K. MORI AND M. KURATSUNE, *Int. J. Cancer*, 2 (1967) 489.
53   D. H. HOFFMANN, Y. MASUDA AND E. L. WYNDER, *Nature*, 211 (1969) 253.
54   Y. MASUDA AND D. HOFFMANN, *Anal. Chem.*, 41 (1969) 650.
55   E. BOYLAND, C. E. DUKES AND P. L. GROVER, *Brit. J. Cancer*, 17 (1963) 79.
56   E. BOYLAND, E. R. BUSBY, C. E. DUKES, P. L. GROVER AND D. MANSON, *Brit. J. Cancer*, 18 (1964) 575.
57   E. BOYLAND, *FEBS Symposium, 5th Meeting, Prague, July, 1968*, Academic Press, New York, 1969, Vol. 16, p. 183.
58   M. A. WALTERS, F. J. C. ROE, B. C. V. MITCHLEY AND A. WALSH, *Brit. J. Cancer*, 21 (1967) 367.
59   D. F. MCDONALD, *Ind. Hyg. Dig.*, (1969) 17.
60   *Rept. of the Advisory Comm. to the Surgeon General on Smoking and Health*, U.S. Publ. Health Serv. Publ. 1103, 1964; *The Health Consequences of Smoking*, U.S. Publ. Health Serv. Publ. 1696, 1967, Suppl., 1968.
61   E. BOYLAND AND D. MANSON, *Biochem. J.*, 101 (1966) 84.
62   E. BOYLAND, *The Biochemistry of Bladder Cancer*, Charles C. Thomas, Springfield, Ill., 1963.
63   E. BOYLAND, D. MANSON AND R. NERY, *Biochem. J.*, 86 (1963) 263.
64   W. B. DEICHMAN AND J. L. RADOMSKI, *J. Natl. Cancer Inst.*, 43 (1969) 263.
65   S. BELLMAN, W. TROLL, G. TEEBOR AND F. MUAKAKI, *Cancer Res.*, 28 (1968) 535.
66   F. MUKAI, S. BELMAN, W. TROLL AND I. HAWRYLUK, *Proc. Am. ASSOC. Cancer Res.*, 8 (1967) 49.
67   S. BELMAN, W. TROLL, G. TEEBOR, R. REINHOLD, B. FISHBEIN AND F. MUKAI, *Proc. Am. Assoc. Cancer Res.*, 7 (1966) 6.
68   G. PEREZ AND J. L. RADOMSKI, *Ind. Med. Surg.*, 34 (1965) 714.
69   P. E. HUGHES AND R. PILCZYK, *Chem. Biol. Interactions*, 1 [3] (1970) 307.
70   E. BOYLAND AND D. MANSON, *Biochem. J.*, 99 (1966) 189.
71   E. BOYLAND, D. MANSON AND S. F. D. ORR, *Biochem. J.*, 65 (1957) 417.
72   E. BOYLAND, C. H. KINDER AND D. MANSON, *Biochem. J.*, 78 (1961) 175.
73   H. UEHLEKE AND E. BRILL, *Biochem. Pharmacol.*, 17 (1968) 1459.
74   S. OTSUKA, *J. Biochem. Tokyo*, 50 (1961) 81.
75   J. L. RADOMSKI AND E. BRILL, *Science*, 167 (1970) 992.
76   W. B. DEICHMANN, J. L. RADOMSKI, A. D. ANDERSON, E. GLASS, M. COPLAN AND F. WOODS, *Ind. Med. Surg.*, 34 (1965) 640.
77   T. M. SPOTSWOOD, *J. Chromatog.*, 3 (1960) 101.
78   Q. QUICK, R. F. LAYTON, H. R. HARLESS AND O. R. HAYNES, *J. Gas Chromatog.*, 6 (1968) 46.
79   L. E. FIESER AND M. FIESER, *Advanced Organic Chemistry*, Reinhold, New York, 1961, p. 496.
80   D. M. MARMION, R. G. WHITE, L. H. BILLE AND K. H. FERBER, *J. Gas Chromatog.*, 4 (1966) 190.
81   H. MATSUSHITA, *Ind. Health (Kawasaki)*, 5 (1967) 260.
82   K. SHIMOMURA AND H. F. WALTON, *Separation Sci.*, 3 (1968) 497.
83   J. LATINAK, *Collection Czech. Chem. Commun.*, 24 (1959) 2939.
84   S. BARTLETT, *Chem. Ind. (London)*, (1951) 76.
85   A. TANNEWBAUM, *Natl. Cancer Inst. Monograph*, 14 (1964) 341.
86   J. W. ORR, *Brit. J. Cancer*, 1 (1947) 311.
87   N. TRAININ, A. PRECERUTTI AND L. W. LAW, *Nature*, 202 (1964) 305.
88   I. BERENBLUM, D. BEN-ISHAI, N. HARAN-GHERA, A. LAPIDOT, E. SIMON AND N. TRAININ, *Biochem. Pharmacol.*, 2 (1959) 168.
89   W. G. JAFFE, *Cancer Res.*, 7 (1947) 107.
90   J. G. SINCLAIR, *Texas Rept. Biol. Med.*, 8 (1950) 623.
91   J. A. DIPAOLO AND J. ELIS, *Cancer Res.*, 27 (1767) 1696.
92   V. H. FERM, *Lancet*, (1965-I) 1338.
93   H. I. BATTLE AND K. K. HISAOKA, *Cancer Res.*, 12 (1952) 334.
94   D. B. MCMILLAN AND H. I. BATTLE, *Cancer Res.*, 14 (1954) 319.

95 M. Vogt, *Experientia*, 4 (1948) 68.
96 M. Vogt, *Pubbl. Staz. Zool. Napoli*, 22 (1950) 114.
97 M. Demerec, G. Bertani and J. Flint, *Am. Naturalist*, 85 (1951) 119.
98 R. Latarjet, N. P. Buu-Hoi and C. A. Elias, *Pubbl. Staz. Zool. Napoli*, 22 (Suppl.) (1950) 76.
99 F. Oehlkers, *Heredity Suppl.*, 6 (1953) 95.
100 F. Oehlkers, *Z. Vererbungslehre*, 81 (1943) 313.
101 A. Bateman, *Mutation Res.*, 4 (1967) 710.
102 K. A. Jensen, I. Kirk, G. Kølmark and M. Westergaard, *Cold Spring Harbor Symp., Quant. Biol.*, 16 (1951) 245.
103 S. Rogers, *J. Natl. Cancer Inst.*, 15 (1955) 1675.
104 F. Oehlkers and G. Linnert, *Z. Vererbungslehre*, 83 (1951) 429.
105 V. Bryson, *Hereditas Suppl.*, (1945) 545.
106 R. DeGiovanni-Donnelly, S. M. Kolbye and J. A. DiPaolo, *Mutation Res.*, 4 (1967) 543.
107 P. Adams and F. A. Baron, *Chem. Rev.*, 65 (1965) 567.
108 A. Haddow, *Professor Khanolkar Felicitation Volume*, Indian Cancer Research Centre, Bombay Univ. Press, Bombay, 1963, pp. 158–181.
109 S. S. Mirvish, *Advan. Cancer Res.*, 11 (1968) 1.
110 E. Boyland and R. Nery, *Biochem. J.*, 94 (1965) 198.
111 R. Nery, *Biochem. J.*, 106 (1968) 1.
112 E. Boyland, R. Nery and K. S. Peggie, *Brit. J. Cancer*, 19 (1965) 878.
113 A. M. Kaye and N. Trainin, *Cancer Res.*, 26 (1966) 2206.
114 M. A. Hahn, C. C. Botkin and R. H. Adamson, *Nature*, 211 (1966) 984.
115 S. Chaube and M. L. Murphy, *Cancer Res.*, 26 (1966) 1448.
116 E. Borenfreund, M. Krim and A. Bendich, *J. Natl. Cancer Inst.*, 32 (1964) 667.
117 A. Bendich, E. Borenfreund, G. C. Korngold and M. Krim, *Federation Proc.*, 22 (1963) 582.
118 E. B. Freese, *Genetics*, 51 (1965) 953.
119 E. B. Freese, J. Gerson, H. Taber, J. H. Rhaese and E. Freese, *Mutation Res.*, 4 (1967) 517.
120 E. Freese, S. Sklarow and E. B. Freese, *Mutation Res.*, 5 (1968) 343.
121 R. H. Knight and L. Young, *Biochem., J.* 70 (1958) 111.
122 G. Toewwies and J. J. Kolb, *Anal. Chem.*, 23 (1951) 823.
123 J. Harley-Mason and A. A. P. G. Archer, *Biochem. J.*, 69 (1958) 60P.
124 J. F. Thompson, W. N. Arnold and L. Young, *Nature*, 197 (1963) 380.
125 S. Mirvish, *Biochim. Biophys. Acta*, 93 (1964) 673.
126 S. Mirvish, *Biochim. Biophys. Acta*, 117 (1966) 1.
127 H. N. Rydon and P. N. G. Smith, *Nature*, 169 (1952) 922.
128 J. Booth and E. Boyland, *Biochem. J.*, 91 (1964) 362.
129 S. S. Mirvish, *Analyst*, 90 (1965) 244.
130 H. Tsukamoto, H. Yoshimura and K. Tatsumi, *Life Sci.*, 2 (1963) 382.
131 H. E. Skipper, L. L. Bennett, C. E. Bryan, L. White, A. M. Newton and L. Simpson, *Cancer Res.*, 11 (1951) 46.
132 J. A. Miller, J. W. Cramer and E. C. Miller, *Cancer Res.*, 20 (1960) 950.
133 L. Boiato, S. S. Mirvish and I. Berenblum, *Intern. J. Cancer*, 1 (1966) 265.
134 R. Nery, *Analyst*, 91 (1966) 388.
135 R. Nery, *Analyst*, 94 (1969) 130.
136 W. L. Zielinski, Jr. and L. Fishbein, *J. Gas Chromatog.*, 3 (1965) 142.
137 J. A. Fresen, *Pharm. Weekblad*, 16 (1965) 532.
138 G. Szász, L. Khin and M. Szász, *Acta Pharm. Hung.*, 28 (1958) 55.
139 W. Machu, *Das Wasserstoffperoxyd und Die Perverbindungen*, Springer, Vienna, 1951.
140 C. A. Bunton, in J. O. Edwards (Ed.), *Peroxide Reaction Mechanisms*, Wiley, New York, 1962, p. 11.
141 L. S. Altenberg, *Proc. Natl. Acad. Sci. U.S.*, 40 (1940) 1037.
142 L. S. Altenberg, *Genetics*, 43 (1958) 662.
143 M. R. Chevallier and D. Luzatti, *Compt. Rend.*, 250 (1960) 1572.
144 F. H. Dickey, G. H. Cleland and C. Lotz, *Proc. Natl. Acad. Sci. U.S.*, 35 (1949) 581.
145 S. H. Revell, *Heredity Suppl.*, 6 (1953) 107.
146 A. Loveless, *Nature*, 167 (1951) 338.

147 H.Hoshino, G.Chihara and F.Fukuoka, *Gann*, 61 (1970) 121.
148 R.Latarjet, N.Rebeyrotte and P.Demerseman, in M.Haissinsky (Ed.), *Organic Peroxides in Radiobiology*, Pergamon, Oxford, 1958, p.61.
149 R.Latarjet, *CIBA Found. Symp. Ionizing Radiations Cell. Metab. 1957*, Little Brown, Boston, p.275.
150 D.Luzzati, H.Schweitz, M.L.Bach and M.R.Chevallier, *J. Chim. Phys.*, 58 (1961) 1021.
151 D.Luzzati and M.R.Chevallier, *Ann. Inst. Pasteur*, 93 (1957) 366.
152 P.Kotin and H.L.Falk, *Radiation Res. Suppl.*, 3 (1963) 193.
153 F.H.Sobels, *Drosophila Inform. Serv.*, 28 (1954) 150.
154 F.H.Sobels, *Nature*, 177 (1956) 979.
155 F.H.Sobels, *Am. Naturalist*, 88 (1954) 109.
156 N.A.Milas and I.Belič, *J. Am. Chem. Soc.*, 81 (1959) 3358.
157 N.A.Milas, R.S.Harris and A.Golubovic, *Rad. Res. Suppl.*, 3 (1963) 71.
158 R.Sundt and M.Winter, *Anal. Chem.*, 29 (1957) 851.
159 J.Cartlidge and C.F.H.Tipper, *Anal. Chim. Acta*, 22 (1960) 106.
160 M.J.Abraham, H.G.Davies, D.R.Llewellyn and E.M.Thain, *Anal. Chim. Acta*, 17 (1957) 499.
161 A.Rieche and M.Schucz, *Angew. Chem.*, 70 (1958) 694.
162 E.Knappe and D.Peteri, *Z. Anal. Chem.*, 190 (1962) 386.
163 M.M.Bazlanova, V.F.Stepanovskaya, A.F.Nesterov and V.L.Antonovsky, *Zh. Analit. Khim.*, 21 (1966) 507.
164 J.A.Brammer, S.Frost and V.W.Reid, *Analyst*, 92 (1967) 91.
165 B.Glabik and W.Walczyk, *Chem. Anal. (Warsaw)*, 12 (1967) 1299.
166 M.Gruber, R.Klein and M.Foxx, *J. Pharm. Sci.*, 58 (1969) 566.
167 D.DeTar and R.Long, *J. Am. Chem. Soc.*, 80 (1958) 4742.
168 D.Bernhardt, *Z. Chem.*, 8 (1968) 237.
169 A.Y.Valendo and Y.D.Norikov, *Izv. Akad. Nauk Beloruss. SSR., Ser. Khim. Nauk.*, 5 (1968) 120.
170 D.B.Adams, *Analyst*, 91 (1966) 397.
171 H.Ewald, G.Öhlmann and W.Schirmer, *Z. Physik. Chem. (Leipzig)*, 234 (1967) 104.
172 T.E.Healy and P.Urone, *Anal. Chem.*, 41 (1969) 1777.
173 R.D.Mair and A.J.Graupner, *Anal. Chem.*, 36 (1964) 194.
174 M.S.Kharash, in K.B.Wiberg (Ed.), *Oxidation in Organic Chemistry*, Academic Press, New York, 1965, p.265.
175 D.K.Banerjee and C.C.Budke, *Anal. Chem.*, 36 (1964) 2367.
176 C.D.Wagner, R.H.Smith and E.D.Peters, *Anal. Chem.*, 19 (1947) 976.
177 D.DuLog, *Z. Anal. Chem.*, 202 (1964) 258.
178 Y.N.Anisimov and S.S.Ivanchev, *Zh. Analit. Khim.*, 21 (1966) 113.
179 F.L.Haas, J.B.Clark, O.Wyss and W.S.Stone, *Am. Naturalist*, 74 (1950) 261.
180 O.Wyss, J.B.Clark, F.Haas and W.S.Stone, *J. Bacteriol.*, 56 (1948) 51.
181 V.N.Iyer and W.Szybalski, *Appl. Microbiol.*, 6 (1958) 23.
182 R.P.Wagner, C.R.Haddox, R.Fuerst and W.S.Stone, *Genetics*, 35 (1950) 237.
183 A.Zamenhof, H.E.Alexander and G.Leidy, *J. Exptl. Med.*, 98 (1953) 373.
184 J.Schöneich, *Mutation Res.*, 4 (1967) 385.
185 L.J.Lilly, *Nature*, 177 (1956) 338.
186 W.C.Schumb, C.N.Satterfield and R.L.Wentworth, *Hydrogen Peroxide*, Reinhold, New York, 1955.
187 M.Kminkova, M.Gottwaldova and J.Hanus, *Chem. Ind. (London)*, (1969) 519.
188 N.R.Greiner, *J. Chromatog.*, 31 (1967) 525.

SUBJECT INDEX

The numbers in parentheses refer to pages on which the formula, chemical or common name and literature citations for paper, thin-layer, gas-liquid and miscellaneous chromatography of the chemical carcinogens, mutagens and teratogens are given.